FRACTURE and *Life*

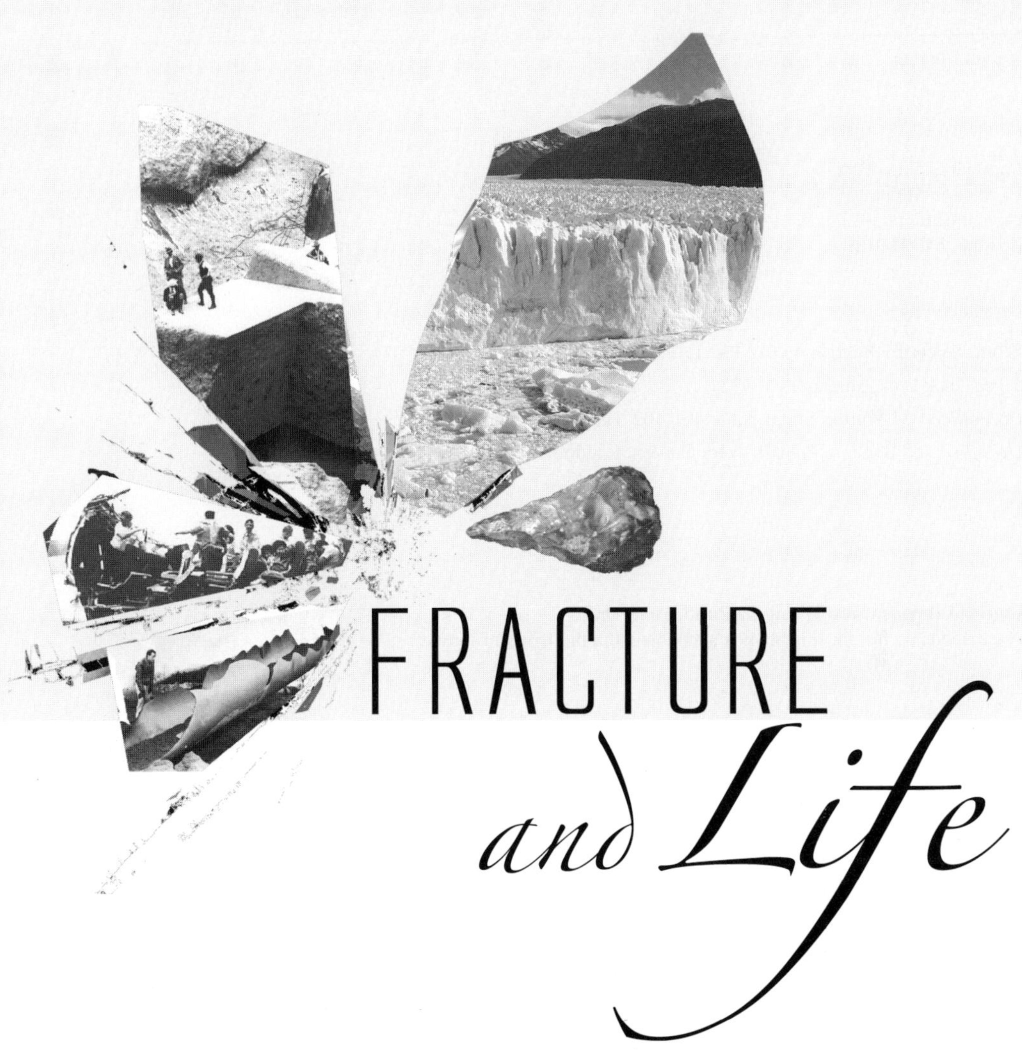

FRACTURE and *Life*

Brian Cotterell

University of Sydney, Australia

W **World Scientific**

NEW JERSEY · LONDON · SINGAPORE · BEIJING · SHANGHAI · HONG KONG · TAIPEI · CHENNAI

Published by

Imperial College Press
57 Shelton Street
Covent Garden
London WC2H 9HE

Distributed by

World Scientific Publishing Co. Pte. Ltd.

5 Toh Tuck Link, Singapore 596224

USA office: 27 Warren Street, Suite 401-402, Hackensack, NJ 07601

UK office: 57 Shelton Street, Covent Garden, London WC2H 9HE

British Library Cataloguing-in-Publication Data
A catalogue record for this book is available from the British Library.

FRACTURE AND LIFE

ISBN-13 978-1-84816-282-2
ISBN-10 1-84816-282-0

Printed in Singapore by B & JO Enterprise

To Maureen for her support and love

Foreword

Those of us who have worked on fracture for a long time often suspect that the subject has far-reaching implications in fields other than our own. Most come to the subject via various aspects of structural integrity or material development but we observe cracks in rocks and see mowers cutting grass, for example, and perceive that these could be described within a general framework of fracture mechanics. Putting this framework in place and explaining the arguments with supporting evidence is a huge task and it is this that Brian Cotterell has achieved.

We are given a historical review of the subject and intriguing explorations of the influence of fracture in making stone tools and designing classical buildings, for example. The whole area of the influence of fracture in biology is described via its effect on evolution. One is given a whole new perspective on the properties and design of teeth by this section. Biology is probably the next growth area in the subject and this book is a wonderful primer for anyone entering this new field. When this is followed by a review of the importance of fracture in the development of electronic materials one gains some perspective of the enormous range of the book.

I have, I hope, given some idea of the scope of *Fracture and Life*. It is an intellectual achievement of the highest order and required extraordinary diligence by the author to read, let alone review and summarise, the vast literature covered. The book is timely since the subject is changing and moving into new fields. The next generation now have a perfect starting point for this quest.

Gordon Williams,
Imperial College, London,
June 2009.

Preface

Fracture affects everything. On a grand scale fracture has played a part in the evolution of the world as we know it. The evolution of life has seen constant interplay between plants and animals avoiding being torn or eaten, and the need of other animals to eat. Human evolution has been greatly affected by the fact that stones were easily flaked to produce sharp tools. Without stone tools human evolution might have been radically different. Civilization has required the development of means to cut and fracture to fashion artefacts and structures as well as the development of the technology to avoid fracture. As civilizations became more sophisticated, so the need to control fracture grew. New technologies and materials brought new fracture problems. Fortunately, scientists and engineers are now largely very successful at controlling fracture so that most people do not even think about its possibility apart from breaking their own bones.

Man's understanding of fracture has developed with time. Even before we became human our hominid ancestors knew how to flake sharp stone tools. The very attribute that made stone tools easy to flake also made them easily broken and more durable metal tools finally replaced them. The ancient civilizations produced enduring stone buildings that required the development of the means to quarry and fashion stone. Building techniques had also to be developed to ensure that the buildings did not fracture and collapse. The control of fracture, until relatively recently, has been pragmatic. It was the Greeks who first began to try to understand fracture, but not until the Renaissance did the theory of fracture start to be developed. Practical problems caused the development of fracture theory. The Sun King, Louis XIV of France, wanted fountains of great height for Versailles and so Edme Marriotte developed an understanding of the mechanics of pressure piping so that he could avoid burst pipes. The Industrial Revolution saw an exponential growth in technology requiring professional engineers for the first time. From the Industrial Revolution to the mid-twentieth-century, fracture was to some extent out of control. Fortunately now fracture is well controlled and the new discipline of fracture mechanics, which began in the mid-twentieth century, has come to maturity. In my professional life I have witnessed and made a small contribution to the growth of this new discipline. It seems a good moment to record how fracture has affected our lives and how it has been understood.

The concept for this book first arose during a 1996 visit to Peter Rossmanith in Vienna, where the idea of jointly writing a history of fracture mechanics was conceived. Unfortunately for many reasons that book was not written then, but over the years it has been in the back of my mind. Since retiring I have had the time to revisit the concept. Although I have broadened the scope of the book, it still uses much of the framework that was worked out with Peter Rossmanith. What I have attempted to do is to show how fracture has affected our world and the efforts that have been made to understand, exploit, and control it. The book is written from a historical aspect but it is not a history as such. I have deliberately not given any mathematical derivations, concentrated on the physics and I have tried to keep the number of equations to a minimum. It is very much of a personal view. When Isaac Todhunter, the English nineteenth-century mathematician, wrote his classic history of the theory of elasticity, he could be exhaustive. That was not an option for this book. What I have tried to do is to cover what I see as the main developments in fracture. It has been very difficult to know what to exclude, not what to include. I know that in writing this book I will probably have made more enemies than friends. I have almost certainly unjustifiably excluded many whose work does form part of the main fracture developments and there are very many more researchers who have made a significant advance in fracture than I have been able to mention in this short book. It does not mean that because a particular researcher is not mentioned that I think their contribution was not important, in fact in many cases it just shows my own ignorance.

The book is written for a wide audience and I hope that it will be read by anybody whose interest or work touches on fracture. I am very much of the view that to really understand a piece of research it is necessary to know its background and what motivated the work. Also, genuine advances can be made by applying knowledge from one field to another. Because of the increasing complexity of knowledge, young researchers, while having expertise in their field, often do not have a wide knowledge. I would like to think that a researcher starting out to do research on any aspect of fracture would benefit from reading this book.

The book uses more general histories and reviews and I would like particularly to mention the exhaustive *History of the Theory of Elasticity and of the Strength of Materials, from Galilei to Lord Kelvin,* by Issac Todhunter, the reliable *History of Strength of Materials,* by Stephen Timoshenko, and the two wonderfully written books by Jim Gordon: *The New Science of Strong Materials, or Why You Don't Fall Through the Floor* and *Structures, or Why Things Don't Fall Down.* My writing style for this book has also been greatly influenced by the two books of Jim Gordon who had a marvellous personal style. To make the book more personal I have used people's preferred personal names where they are known to me. To maintain consistency I have given the personal names of Chinese people before their surname.

A book like this one relies on the work of others and I acknowledge my debt to a great number of people. My gratitude goes to Alan Wells who first introduced me to the wonderland of fracture when I joined the British Welding Research Association fifty years ago and taught me my first steps in fracture. Alan Wells was one of the greats of fracture and a true gentleman. Throughout my professional life I have gained much from formal and informal contacts in a wide field of fracture and I thank all colleagues and students with whom I have worked over the years. I thank Peter Rossmanith for the idea of writing a history of fracture and for his continued support. Gordon Williams is warmly thanked for writing the Foreword and making many valuable suggestions for the improvement of the book. My thanks also go to the following colleagues who have read various chapters of the book for me and made valuable suggestions: Tony Atkins, Yiu-Wing Mai, Jo Kamminga, Tony Kinloch, Peter Lucas, and Peter Rossmanith. They have all improved the book; the remaining errors are mine.

Brian Cotterell,
School of Aerospace, Mechanical and Mechatronic Engineering,
University of Sydney.

Contents

Plate 1. The Aloha Airways Boeing 737 which lost part of its top skin over Hawaii in 1988 showing convergent tears in the fuselage skin below floor level.

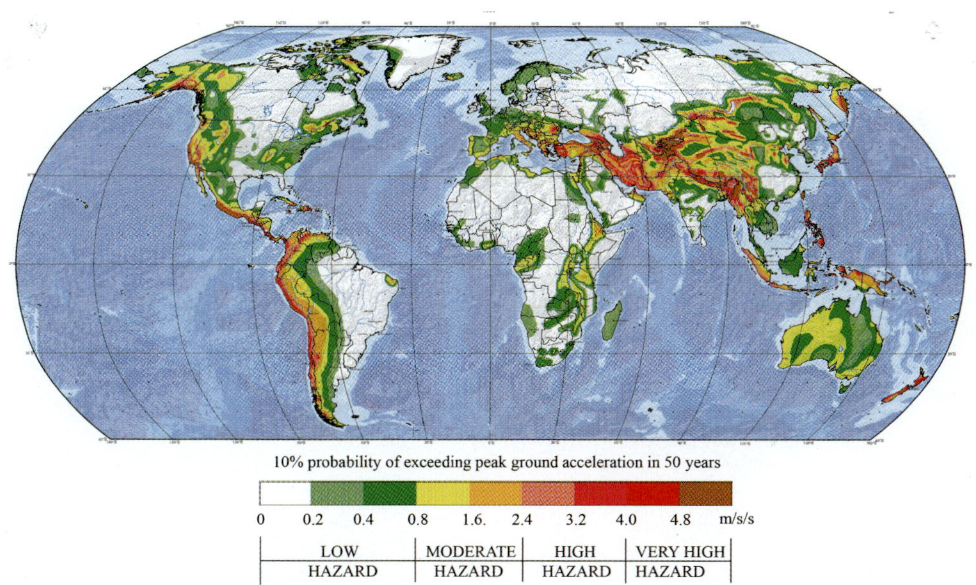

Plate 2. Global seismic hazard map (courtesy United States Geological Survey).

Plate 3. The Perito Moreno Glacier, March 2005 (courtesy Lucca Galuzzi, www.galuzzi.it).

Plate 4. Hatchet grinding grooves on sandstone rock in Kangaroo Valley, New South Wales (courtesy John Mulvaney).

Plate 5. Flint Acheulian hand-axe discovered by John Conyers in 1679 (with permission British Museum).

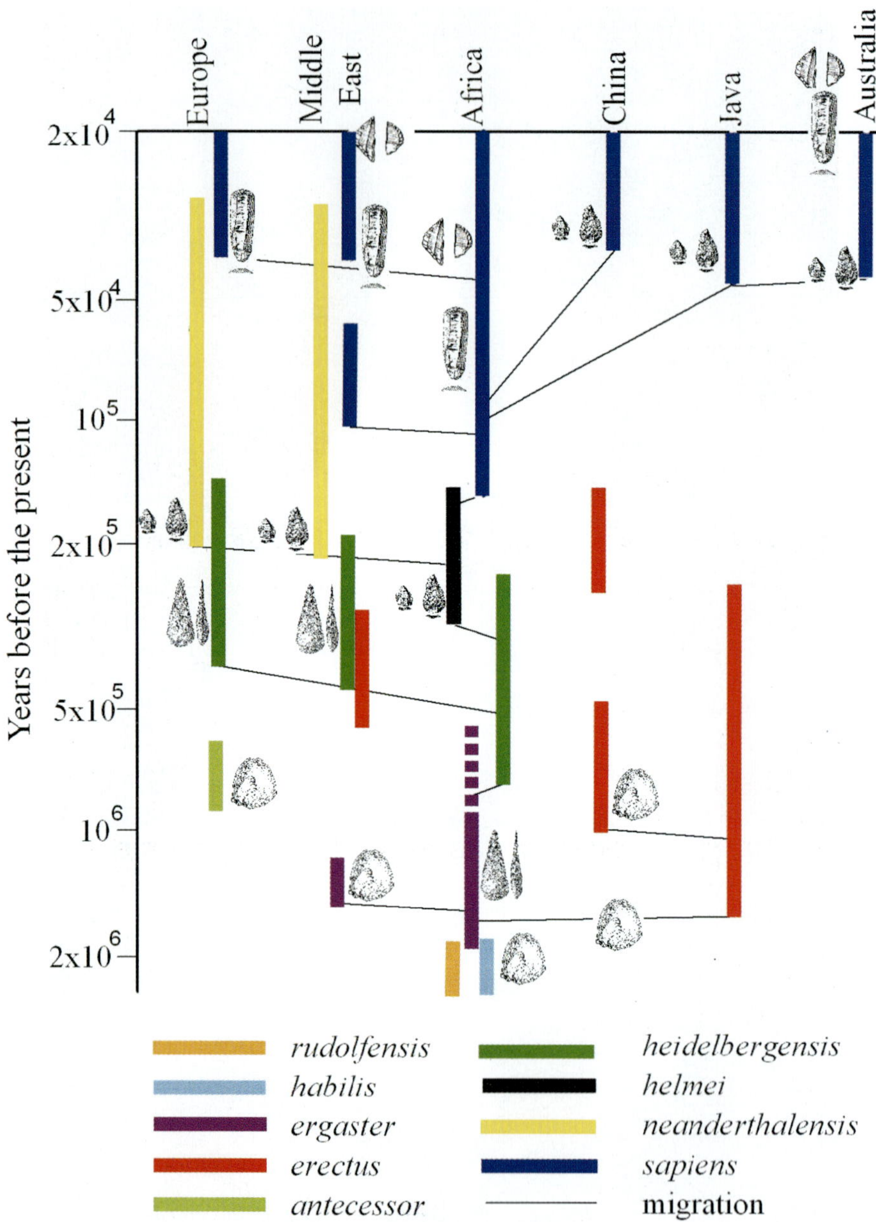

Plate 6. Chronological and geographic distribution of stone tools and the genus Homo (after Foley and Lahr 2003).

Plate 7. The Landscape Arch, Utah (courtesy Jay Wilbur).

Plate 8. The Postbridge, Dartmoor (courtesy Mark Robinson).

 Fracture and Life

Plate 9. The interior of the Pantheon, 18th century painting by Paolo Giovanni Pannini.

Plate 10. 4 Million pound (17MN) Universal Testing Machine originally purchased by Lehigh University in 1932 and subsequently moved to the University of California (1965 photograph courtesy University of California).

Plate 11. A late Corinthian bronze helmet ca 460 BC. The inscription around the rim records that the Argives won the helmet in battle from the Corinthians and dedicated it to Zeus in his sanctuary at Olympia (with permission British Museum).

Plate 12. Giant's Causeway, Co. Antrim, Northern Ireland (courtesy Code Poet).

Chapter 1

Introduction and Basic Solid Mechanics

The word 'failure' was first used in the sense of breakdown in an entity or process by John Smeaton (1724–1792), the first fully professional English engineer, in 1793 to describe the breaking of a bolt in the Eddystone Lighthouse, which he had built. Failure implies a breakdown in the function of any entity such as heart failure, corrosion of a boiler tube, or the collapse of the Tacoma Narrows Bridge in 1940, where the suspension bridge had insufficient torsional resistance and failed due to torsional vibrations induced by a 67 km/hr wind, which would not normally cause concern for the integrity of a bridge. Here the interest lies in one particular failure mode: fracture.

Fracture is associated by most people with the fracture of bones and that was indeed the way the word was first used in a translation by Robert Copland in 1541 of the *Therapeutic or Curative Method* by Claudé Galyen.[1] Usually fracture is unwanted and results in the failure of the object. Much of this book is about avoiding fracture. However, there are many cases where fracture is desired. The magnificent enduring stone edifices built in ancient times required knowledge of how to usefully fashion stone by controlled fracture. Fractures create new surfaces, which can be desired, as in cutting or machining. Traditionally, cutting and machining have been treated as separate subjects to fracture, but recently they have been seen to be just another aspect of fracture and will be discussed in Chapter 11.

Fractures have played a large part in shaping the world around us. The evolution of life has been controlled in part by the need either to avoid fractures and tears or to be able to exploit foodstuffs by tearing with tooth and claw. The ability to flake stone to make stone tools had a significant effect on the evolution of the human race. As civilisation grew, fracture was both avoided and exploited. With time, fracture needed to be understood. At first that understanding was empirical. From the time of the Greeks onwards attempts were made to understand how things fractured. Since the Industrial Revolution new technologies have brought fracture problems that needed to be solved. Corrosion and wear constitute today's largest failure cost, costing some $120 billion a year in the US alone, but the cost of fracture is not much less. Hence economically there is a great need to understand and control fracture. In this chapter the necessary basics to understand the subsequent chapters are presented.

1.1 What Holds a Solid Together?

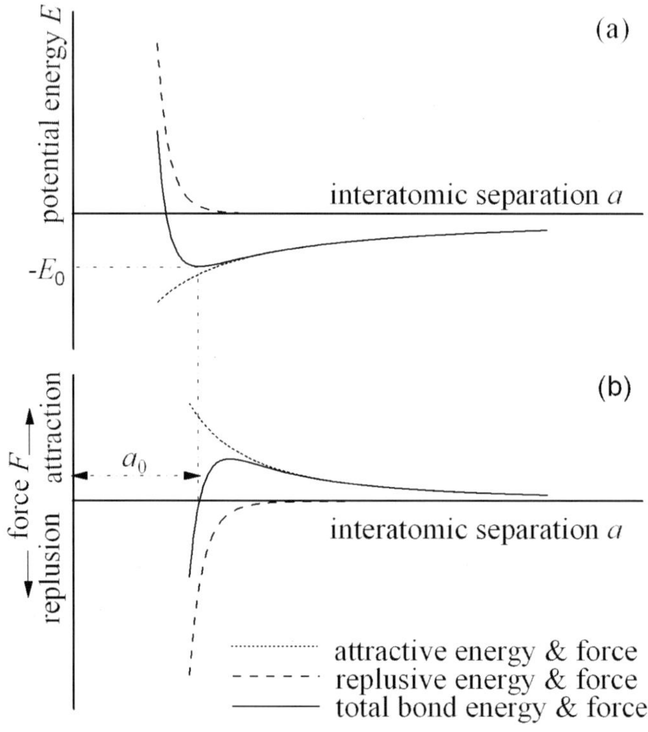

Fig. 1.1 (a) Energy of the ionic bond between two atoms as a function of interatomic distance. (b) Force between two atoms as a function of interatomic distance.

Before discussing how the fracture of a solid occurs we must first ask what holds a solid together. All matter is made up of atoms which are themselves composed of subatomic particles. The atoms are attracted to one another by a variety of forces, referred to as bonds, divided into strong primary bonds: ionic, covalent and metallic, and much weaker secondary or van der Waals bonds produced by asymmetries in atoms or molecules. Here we will discuss only the monovalent ionic bond, which is the simplest to understand. Atoms that have lost or gained an electron become positively or negatively charged ions attracting or repulsing each other. The energy change, E_c, as two unlike monovalent ions are brought together is given by

$$E_c = -\frac{q^2}{a},$$ (1.1)

where a is the interatomic distance, q is the electronic charge. If the ions are in close proximity to each other they interact and a repulsive energy, E_r, is generated given by

$$E_r = \frac{b}{a^n},$$ (1.2)

where $6<n<12$. The repulsive energy, E_r, has a much shorter range than E_c and the total bond energy between two unlike monovalent ions is thus given by

$$E = \frac{-q^2}{a} + \frac{b}{a^n}.$$ (1.3)

The bond energy between two atoms in a solid, shown in Fig. 1.1 (a), is similar to the expression given by Eq. (1.3) regardless of the bond type. The bond energy is a minimum, $-E_0$, at equilibrium when the interatomic distance is a_0.

1.1.1 Surface energy

On a surface an atom has fewer adjacent partners and hence it has a higher energy of approximately $-E_0/2$. The surface energy, γ, of a solid surface is this higher energy multiplied by the number of atoms in a unit area. Most materials have a surface energy in the range of 0.5–10 J/m^2. A fracture creates new surfaces and therefore the minimum or intrinsic energy necessary to create a fracture of unit area is the surface energy, 2γ. The factor 2 arises because there are two surfaces. The fracture energy of glass, which behaves as an elastic-brittle material, is only about four times its surface energy of 1 J/m^2. We all know how brittle glass is and we would live in a very brittle world if the energy necessary to create a fracture in materials generally was similar to that of glass. Fortunately the actual fracture energy of most materials is many orders of magnitude greater than the surface energy because it is usually not possible to produce a fracture without performing extrinsic work.

Since surface energy does not depend upon how the surface is produced it is based on the total surface area. When a solid is fractured two surfaces are produced of nominally the same area. It has become the practice to base the fracture energy, R, not on the total surface area created, but on the area of the pupative fracture that is half the nominal area of the fracture surfaces.[2] Hence

$$R = 2(\gamma + \gamma_p), \tag{1.4}$$

where γ_p is the extrinsic work performed in creating a fracture surface.

1.1.2 Interatomic force

The bond force, F, between atoms is the differential of the bond energy given in Eq. (1.3). Hence

$$F = \frac{dE}{da} = \frac{q^2}{a^2} - \frac{nb}{a^{(n+1)}}. \tag{1.5}$$

The first term in Eq. (1.5) is a special case of the electrostatic law that Charles-Augustin de Coulomb (1736–1806), the French physicist, published in 1785.[3] The variation of the bond force with interatomic distance is schematically illustrated in Fig. 1.1 (b). The bond force is zero at equilibrium when the bond energy is a minimum. It is ultimately the bond force that holds a solid together. In continuum mechanics the atomic structure is smeared to produce a continuum and the individual forces acting between atoms are replaced by the force per unit area acting across a plane, which is called stress. The cohesive stress, σ_{coh}, or force per unit area acting between two planes of atoms at a distance, a, apart has the same form as the force curve given in Fig. 1.1 (b) and the surface energy, γ, is given by

$$\gamma = \frac{1}{2} \int_{a_0}^{\infty} \sigma_{coh} da. \tag{1.6}$$

1.2 Stress and Strain

When a solid body has a force applied to it, the force is resisted by deformation. Except in very simple situations, quite complex mechanics is necessary to determine the details of the deformation. As well as developing resisting internal forces at the point of application of the force on the surface of a body, the solid also develops internal forces. These internal forces are described in terms of stress, which is a force per unit area and the deformation of the body in terms of the deformation per unit length or strain. The lack of a clear concept of stress until Augustin-Louis Cauchy (1789–1857) presented his memoir to the French

Academy of Sciences in 1822, created great difficulties for the early mechanics of solids.[4] If we imagine a small internal planar surface of area ΔA at a point P within a body the force, ΔF, acting on one side of the plane is communicated to the body on the other side (see Fig. 1.2). The stress vector, T, is defined as the limit

$$T = \mathrm{Lim}_{\Delta S \to 0} \left(\frac{\Delta F}{\Delta A} \right).$$

(1.7)

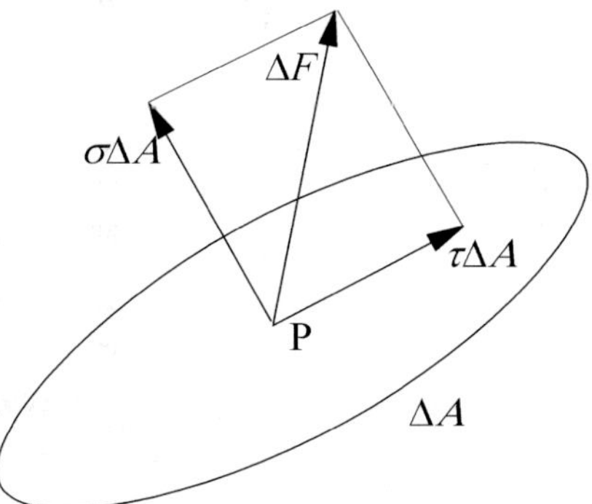

Fig. 1.2 Stress vector and stress at a point.

A force only depends upon the point of application, but a stress depends upon the orientation of the surface. If the surface has a different orientation then the stress will be different. The stress vector, T, can be resolved into a normal stress, σ, perpendicular to the surface, ΔS, and a shear stress, τ, parallel to the surface (see Fig. 1.2). The normal stress can either be tensile or compressive. The usual convention is that tensile normal stresses are positive and compressive normal stresses are negative. The direction of shear stresses does not have a physical significance, but for equilibrium there must be a complimentary shear stress, τ, as shown in Fig. 1.3 (b).

In an isotropic material, where the properties are the same in all directions, a normal stress, σ, acting in the x-direction causes a normal strain, ε, defined as

$$\varepsilon = \mathrm{Lim}_{\Delta x \to 0} \frac{\Delta u}{\Delta x},$$

(1.8)

where u is the displacement in the x-direction and a shear stress, τ, causes a shear strain, γ, defined as

$$\gamma = \text{Lim}_{\Delta y \to 0} \frac{\Delta u}{\Delta y}.$$

(1.9)

Normal and shear strains are shown in Fig. 1.3.

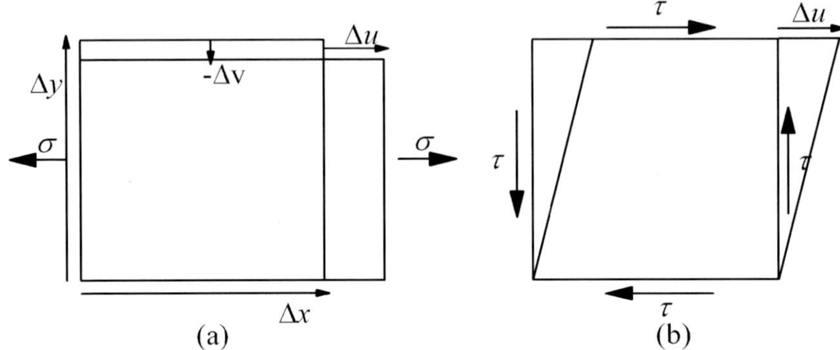

Fig. 1.3 (a) Normal strain. (b) Shear strain.

In most engineering applications the strains involved are less than 1% and in the definition of stress it does not make much difference whether stress is defined as the force per unit of undeformed or deformed area. However, if the strains are large then the difference does matter. If the stress is defined as the force per undeformed area it is called the *engineering* or *nominal* stress though frequently the qualifying nominal is omitted. *True* stress is the force per unit of the deformed area. The definition of strain, given above, based on the extension in the undeformed length is the *engineering* or *nominal* strain and is also not appropriate for large strains when the *true* strain increment, *de*, of a bar of undeformed length, l_0, and deformed length, l, is defined as

$$de = \frac{dl}{l},$$

(1.10)

and the *true* strain is thus

$$e = \int_{l_0}^{l} \frac{dl}{l} = \ln\left(\frac{l}{l_0}\right) = \ln\left(1+\varepsilon\right).$$

(1.11)

For small strains there is no difference between true and nominal strain. An alternative way of describing large strains, used for rubber and other materials that experience large strains, is the stretch ratio, λ, defined as

$$\lambda = \frac{l}{l_0}$$

(1.12)

1.2.1 *Principal stresses and Mohr's stress circles*

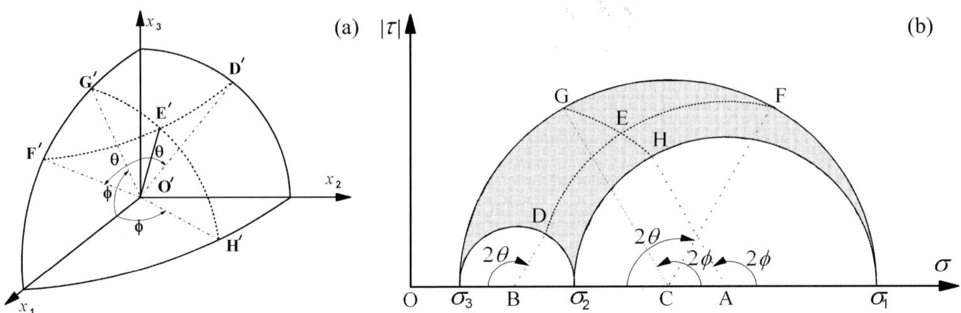

Fig. 1.4 (a) Physical plane. (b) Mohr's stress circles. (after Jaeger and Cook, 1979).

There are three mutually perpendicular planes, or *principal stress planes*, at every point in a body where the *principal stresses* are purely normal (tensile or compressive). The direction of these planes is defined in terms of the *principal axes*, which are the normals to the principal planes. There are equations which describe the resolution of the stresses at a point into any direction. However, Otto Mohr (1835–1918), who was especially interested in graphical methods, gave a geometric construction in 1882, where shear stress is plotted against the normal stress on a plane, which enables this resolution to be visualised.[5] The state of stress at a point where the three principal stresses are $\sigma_1 > \sigma_2 > \sigma_3$ and whose principal axes are x_1, x_2, and x_3, see Fig. 1.4 (a), can be represented by three families of circles with centres at A where $\sigma = (\sigma_1 + \sigma_2)/2$, B where $\sigma = (\sigma_2 + \sigma_3)/2$, and C where $\sigma = (\sigma_3 + \sigma_1)/2$, as is shown in Fig. 1.4 (b). Any plane whose normal is in the direction $O'E'$ can be defined by the angles θ and ϕ as shown in Fig. 1.4 (a), where the surface $F'G'D'E'H'$ is part of a unit sphere. The state of stress on a plane whose normal lies in a principal plane is given by the circles centred at A, B, and C with radii $(\sigma_1 - \sigma_2)/2$, $(\sigma_2 - \sigma_3)/2$, and $(\sigma_3 - \sigma_1)/2$ respectively. In the Mohr's stress circles the point representing the stress state, (σ, τ) rotates twice as fast as the normal in the physical plane and in the opposite sense, thus the stresses on a plane whose normal $O'D'$ makes an angle, θ, with the third principal stress direction, x_3, are given by the point D in the Mohr's stress circles which makes an angle of 2θ with the σ axes. Similarly the stresses on a plane whose normal $O'H'$ makes an angle 2ϕ with the first principal stress direction, x_1, are given by the point H in the Mohr's stress plane. The stress state for the plane whose normal is $O'E'$ is given by the point E which lies on the two part circles DEF and HEG centred on B and A respectively. Hence the stress state for any plane at a point where the principal stresses are σ_1, σ_2, and σ_3 lies in the segment that is shaded in Fig. 1.4 (b). More details can be found in the classic book on rock mechanics by John Jaeger (1907–1979) and Neville Cook (1938–1998) or other solid mechanics book.[6]

A stressed body is a strained body. Cauchy showed that strain could be resolved into components of normal and shear strain, which are mathematically similar to the components of stress. Hence principal strain planes can be found where the strain is purely normal and there are Mohr's strain circles analogous to his stress circles.

1.3 Elastic Deformation

When the atoms are in equilibrium in a solid that is unstressed, the force between them is zero and the interatomic distance is a_0. The force required to displace the atom a very small distance is proportional to the displacement and it is this fundamental behaviour that causes as a spring to increase in length in proportion to the force. This deformation behaviour is called linear elasticity.

Robert Hooke (1635–1703), a contemporary of Isaac Newton (1642–1727), discovered linear elastic behaviour in 1660 when he invented a spring escapement for clocks, but did not publish the result immediately because he wanted to obtain a patent. When Hooke finally committed himself to print in 1676 he did so in the form of a cryptic anagram. It was not until after the death of Henry Oldenburg (ca. 1619–1677), the Secretary of the Royal Society whom Hooke mistrusted and hated,[7] that he felt free to publish the solution, *ut tensio sic vis*, which translated reads: as the extension so the force.[8] Hooke's law was concerned with the overall behaviour of a body, but it also applies to internal stress and strain and the constant of proportionality between stress and strain is a material constant. This generalisation of Hooke's law was made by Thomas Young[9] (1773–1829) in a course on popular mechanics at the Royal Institution in 1802 though because the concept of stress had not been introduced his definition was different to what is used now.[10] The constant of elastic proportionality between normal stress and strain is named *Young's modulus, E*, in his honour. Thus:

$$E = \frac{\sigma}{\varepsilon}. \tag{1.13}$$

The units of Young's modulus are those of stress. Diamond has the highest Young's modulus, 1050 GPa, of any material and the Young's modulus of other ceramics is also very high, in the range 200–500 GPa. The Young's modulus of metals is generally lower than that of ceramics, but is still high and for most metals it is in the range 50–210 GPa. Polymers have a much lower Young's modulus in the range 1–3 GPa and rubbers both natural and synthetic have a very low Young' modulus in the range 2–100 MPa. In this chapter discussion is limited to isotropic elasticity where the principal axes of stress and strain coincide.

Shear stress, τ, is also proportional to shear strain, γ, and the constant of proportionality, μ, is called the shear modulus. Thus:

$$\mu = \frac{\tau}{\gamma}. \tag{1.14}$$

Cauchy showed in 1829 that only two elastic constants are required to describe the complete relationship between stress and strain for an isotropic elastic material.[11] However, a third elastic constant, named Poisson's ratio, ν, after the French mathematician Siméon Poisson (1781–1840), is frequently used. An elastic rod that is stretched by a tensile stress contracts laterally in proportion to the axial strain and the constant of proportionality between the lateral and axial strain is called the Poisson's ratio. Thus in the limit in Fig. 1.3 (a),

$$\nu = \mathrm{Lim}_{\Delta x, \Delta y \to 0} -\left(\frac{\Delta v}{\Delta y}\right)\bigg/\left(\frac{\Delta u}{\Delta x}\right). \tag{1.15}$$

Poisson derived the relationship between the shear modulus, Young's modulus and Poisson' ratio which is given by

$$\mu = \frac{E}{2(1+\nu)}. \tag{1.16}$$

Stress distributions are often planar in engineering. Except near any sharp re-entrant corners, the only appreciable stresses in thin plates, loaded at their edges, are those in the plane of the plate; such states of stress are called plane stress. The other archetypal two-dimensional stress state is plane strain where the strain perpendicular to the plane is zero and the normal stress, σ_z, perpendicular to the plane is given in terms of the normal stresses within the plane, σ_x and σ_y by

$$\sigma_z = \nu\left(\sigma_x + \sigma_y\right). \tag{1.17}$$

Under force boundary conditions the stresses σ_x and σ_y under plane stress and strain are identical. The strains under plane stress and strain have the same expression if the modified Young's modulus, $\overline{E}$, and the modified Poisson's ratio, $\overline{v}$, are used, where

$$\overline{E} = E \text{ and } \overline{v} = v \text{ for plane stress,}$$
$$\overline{E} = E/(1-v^2) \text{ and } \overline{v} = v/(1-v) \text{ for plane strain.} \tag{1.18}$$

The expressions for the elastic strains as a function of a general stress state can be found in any textbook on solid mechanics.

1.3.1 Elastic strain energy

During elastic deformation the work done in deforming a solid is stored as strain energy and can be recovered on unloading. In a brittle fracture all the energy required for fracture can come from the energy stored. When we drop an ice cube straight from the freezer into our gin and tonic, strain energy is stored as the outer layers expand when they warm, putting them into compression and the centre into tension as it is expanded by the outer layers. The ice cracks from trapped bubbles of air with an audible pop as the stored energy is released. Strain energy is the elastic energy stored by virtue of the deformation or strain in a body. Under a simple tension, σ, the strain energy density, U, the energy stored per unit volume is given by

$$U = \frac{1}{2}\sigma\varepsilon = \frac{\sigma^2}{2E} = \frac{E\varepsilon^2}{2}. \tag{1.19}$$

The strain energy density, U, stored per unit volume under a shear stress, τ, is given by

$$U = \frac{1}{2}\tau\gamma = \frac{\tau^2}{2\mu} = \frac{\mu\gamma^2}{2}. \tag{1.20}$$

The expression for the strain energy density function for a general stress state can be found in any textbook on solid mechanics.

1.4 Plastic Deformation and Hardness

Although all materials are elastic for small strains and recover deformation on the release of small loads, most deform plastically to some extent before fracture. By plastic deformation, we mean deformation that is permanent and not recovered on unloading. Plastic deformation dissipates energy and is one of the mechanisms that make fracture more ductile. Plastic deformation occurs by different mechanisms in different classes of materials. Metals are good general engineering materials because usually they deform plastically rather than fracture, but we will see in Chapters 7–9 that sometimes they do fracture in a brittle fashion instead of yielding plastically and can cause catastrophic failures. The stress at which plastic deformation first occurs in a tensile test is called the yield strength, σ_Y, in metals. Linear polymers can also deform plastically by large amounts. The polymer community call the maximum stress at which the polymer starts to draw the yield stress rather than the stress at which non-elastic deformation first occurs because, owing to the viscoelastic behaviour of most polymers, the initial yield is difficult to detect.

Plastic deformation limits the stress and the tensile true stress-strain curve for most metals can be represented by a power law known as the Ramberg–Osgood relationship after yielding.[12] The usual representation of the stress-strain curve for metals is slightly different to the Ramberg–Osgood relationship and is

$$\frac{\sigma}{\sigma_Y} = \frac{e}{e_Y}, \quad \text{for } e < e_Y,$$

$$\sigma = \left(\frac{e}{e_Y}\right)^n, \quad \text{for } e > e_Y, \tag{1.21}$$

where $e_y \approx \varepsilon_Y$ is the strain at yield and n is the strain hardening coefficient. For small strains Eq. (1.21) holds for both true and nominal stress and strain, but the cross-sectional area decreases with strain so that the nominal stress is less than the true stress. Plastic deformation in metals takes place with no change in volume so that during uniform deformation of a tension specimen

$$\frac{A}{A_0} = \frac{l_0}{l} = \exp(-e), \tag{1.22}$$

where A_0, l_0 and A, l are the original cross-sectional area and length, and current values, respectively.

Tensile stress-strain curves with low to high strain hardening coefficients and the idealisation of perfect plasticity where $n = 0$ are shown in Fig. 1.5. For ductile metals the nominal stress reaches a maximum at a true strain $e = n$ and a slightly higher nominal strain $\varepsilon = \exp(n) - 1$. At large strains greater than the strain for the maximum nominal stress the deformation becomes localised and a neck starts to form, which leads to ductile fracture. After a neck forms, Eq. (1.21) no longer applies. Less ductile metals fracture before the maximum load is reached. The ultimate tensile strength (UTS) of a material is defined as the maximum nominal tensile stress that can be sustained either due to plastic flow or fracture. Before the development of fracture mechanics, structures were designed so that the maximum stress was less than the UTS divided by a relatively arbitrary safety factor (SF). On unloading after yielding, only the elastic deformation is recovered.

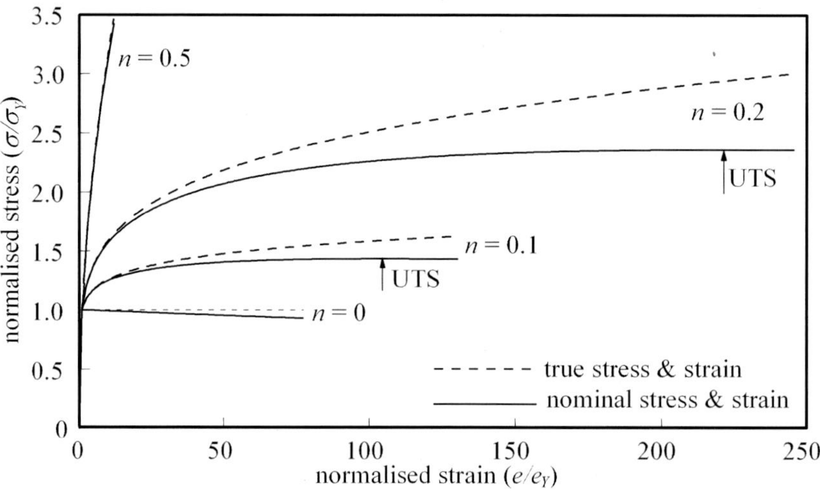

Fig. 1.5 Typical tensile stress-strain curves for metals for various strain hardening coefficients.

Crystalline metals and ceramics yield by the propagation of line defects in the crystal structure called dislocations by a shearing action at stresses orders of magnitude smaller than the theoretical yield strength of a perfect crystal. The *hydrostatic stress* or mean stress, σ_m, given by

$$\sigma_m = \left(\sigma_1 + \sigma_2 + \sigma_3\right)/3, \tag{1.23}$$

does not affect this type of yielding and under a complex stress system it is only the difference between the principal stress and the hydrostatic stress, called the *deviatoric stress*, that has an effect on yield. Various yield criteria were suggested for polycrystalline materials in the nineteenth-century, but it is the criterion advanced by Richard von Mises (1883–1953) in 1913 that is the most satisfactory. For an isotropic material, von Mises' criterion can be written in terms of an equivalent stress, σ_e, defined in terms of the three principal stresses $(\sigma_1, \sigma_2, \sigma_3)$ by

$$\sigma_e = \left\{ \left[\left(\sigma_1 - \sigma_2\right)^2 + \left(\sigma_2 - \sigma_3\right)^2 + \left(\sigma_3 - \sigma_1\right)^2 \right] / 2 \right\}^{\frac{1}{2}}, \tag{1.24}$$

with this definition yielding occurs if $\sigma_e \geq \sigma_Y$. Yield in polymers has a partly viscoelastic nature and is dependent on the rate of testing and the temperature; also it is to some extent dependent on the hydrostatic stress.

Hardness is the ability to resist indentation and in metals is measured by pressing a standard indenter in the form of a spherical ball, or pyramid into a flat surface. The hardness is a stress, but for historical reasons not measured on projected area but on the surface area of the indentation. The state of stress under an indenter has a high hydrostatic component, so that the yielding under the indenter is highly constrained. Because of the constraint, the hardness of a metal or ceramic is roughly three times its yield strength.

1.5 Strength Resilience and Fracture

Strength is not a fundamental property of a material. Fracture will occur in structures made of the same material but of different geometry and size, at different maximum stresses. A simple measure of the fracture performance of a material under tension that is better, but still not accurate, is its resilience. The specific resilience[13] of a material is defined as the work done in stretching a material to its breaking strength per unit volume. Although resilience is a better indicator of the tensile performance of a material, it is still size dependent.

1.5.1 Theoretical ideal strength

The variation in bond force with interatomic separation shown in Fig. 1.1 (b) can be transformed into a similar theoretical stress-strain relationship with the maximum bond force becoming equivalent to the theoretical ideal strength. The work of fracture is then the area under the stress-displacement curve and supplies the extra energy of the fracture surfaces. Using the fact that the initial slope of this curve is the Young's modulus, E, a simple approximate relationship for the theoretical ideal strength, σ_t, as a function of the surface energy, γ, E, and the equilibrium atomic spacing, a_0, can be obtained, which is given by

$$\sigma_t \approx \sqrt{\frac{\gamma E}{a_0}} \approx \frac{E}{10}. \tag{1.25}$$

Since the Young's modulus of ceramics and metals is of the order of 100 GPa the theoretical strength of these materials is very high. Only materials in the form of very fine fibres, or whiskers with diameters of less than 1 μm, approach the theoretical strength of a solid. Defects induce high local stresses, especially in materials like glass that are difficult to deform plastically, and enable failure to occur sequentially rather than over an area of macroscopic dimensions simultaneously and enable materials to fail at global stresses less than one hundredth of the theoretical ideal strength. Real materials are imperfect and are much weaker than the ideal.

1.5.2 Fracture of real materials

There are two basic conditions for fracture to occur: enough energy available for fracture and high enough stresses to cause fracture. For the more brittle materials the first condition is by far the most important, the second almost invariably being satisfied for most engineering materials once the first condition is met.[14] Fracture mechanics has been developed to understand these two basic conditions. The basics of fracture theory are given in this section, the details and history will be dealt with in later chapters.

1.5.2.1 Elastic fracture

Brittle materials fracture while a component is elastic except in the region of crack initiation. Such fractures can be predicted using linear elastic fracture mechanics (LEFM). Energy is required to create the fracture surfaces. The energy released per unit area of fracture is called the *energy release rate*,[15] G, and can be written in a general form as

$$G = \frac{\sigma^2 \pi a}{\bar{E}} Y, \tag{1.26}$$

where σ is a characteristic stress, $2a$ is the crack length, and Y is a non-dimensional geometric factor. For a classic Griffith's crack in an infinite plate under a uniform stress, σ, normal to the crack, $Y = 1$ and generally is of the order of one for other geometries. Only in the most brittle materials, such as glass, is this energy intrinsic to the creation of the new surfaces. Usually the energy extrinsic to the actual fracture process, required to deform the material non-elastically near the tip of a fracture, is very much larger than the intrinsic energy. In fracture mechanics the extrinsic energy is dissipated within a region around a crack tip, which is termed the *fracture process zone* (FPZ). The definition of what constitutes the FPZ varies with the type of material. For example, for materials like high-strength aluminium alloys that are elastic except for a small zone at the tip of the fracture the plastic zone is the FPZ, whereas for ductile materials such as low- and medium-strength steels, where there can be considerable plastic deformation away from the fracture tip and voids nucleate and grow before linking up in a crack, the region of plastic void growth is the FPZ. In fact, the definition of fracture process zone is used to suit the occasion and is reminiscent of Humpty Dumpty's words in Lewis Carroll's *Through the Looking Glass*: 'When I use a word…it means just what I choose it to mean — neither more nor less.'[16] The energy consumed in the FPZ to create a unit area fracture is the fracture energy, R. The most important of the necessary criteria for elastic fracture is

$$G \geq R. \tag{1.27}$$

Fractures initiated at inhomogeneities are either intrinsic ones like small micron-sized surface flaws in glass, or extrinsic ones like notches in the object that locally elevate the stresses. The second criterion for fracture is that the local stress is at least equal to the cohesive strength of the material. By modelling the deformation within the FPZ, both criteria of fracture can be included in fracture mechanics.

In LEFM the stresses, outside of a small FPZ at the tip of a crack, decay as the inverse of the square root of the distance, r, from the crack tip and the stresses acting normal to the prolongation of the crack can be written as

$$\sigma = \frac{K}{\sqrt{2\pi r}}, \tag{1.28}$$

where K is called the *stress intensity factor*. The general form of K is

$$K = \sigma\sqrt{\pi a}Y', \tag{1.29}$$

where Y' is a geometric factor. The stress intensity factor is related to energy release rate, G, by

$$G = \frac{K^2}{\overline{E}}. \tag{1.30}$$

Since fracture can occur if the energy release rate, G, reaches the critical value the fracture energy, R, it also occurs if the stress intensity factor reaches a critical value, K_c, which is called the *fracture toughness*; the plane strain fracture toughness is written as K_{Ic}.[17] The units of the fracture toughness are MPa√m. In the opinion of many fracture mechanists, it is the fracture energy, R, that is the fundamental measure of the fracture resistance of a material and the fracture toughness is a subsidiary unit. A friend of mine, Tony Atkins, states that God would not use so daft a unit as MPa√m. Under plane strain conditions and essentially elastic-brittle behaviour, the usual symbol for the plane strain fracture energy, or critical energy release rate, is G_{Ic}. Provided the FPZ is small compared to the other dimensions of a structure, especially any pre-crack or notch, then the fracture will occur if the energy criterion is satisfied.

In the more ductile materials the fracture toughness increases with crack growth, giving rise to what is known as *crack growth resistance*.

1.5.2.2 Plastic fracture

The more ductile materials deform plastically over a large region before fracture. LEFM cannot be used to model plastic fracture. *Elasto-plastic fracture mechanics* (EPFM) is more complex than LEFM and was developed after LEFM. The FPZ is frequently large in EPFM and the second criterion of fracture that the stress at a crack tip must exceed the cohesive strength of the material becomes more important. If the plastic deformation is very large before fracture so that the elastic strains can be neglected, then the mechanics of fracture at plastic collapse becomes simpler than LEFM, as is the case in a few examples in this chapter. A more general discussion of EPFM will be left to later chapters where it will be introduced from a historical perspective. These more ductile materials invariably exhibit crack growth resistance.

1.5.2.3 Size effect

Before the development of fracture mechanics, design against fracture was based on the UTS. Such designs had no intrinsic size effect. If the dimensions of a unit design were all scaled by a factor of two, then the safe loads predicted by such a concept would scale by a factor of four because stress has the units of force per unit area. If the unit design is safe and the defects that initiate the fracture are inherent, material defects that do not scale up with the design, the doubled-up design would be safe, providing the dimensions of the unit design were not comparable to the size of the FPZ. However, if fracture initiated from a design detail such as a sharp re-entrant corner which scales up with the design, then the scaled-up design on the basis of UTS is not safe. Fracture is size dependent. If the material of the design conforms to LEFM in the unit design and the design defect scales with the design then the safe loads for the doubled-up design are not four times the unit load but, from Eq. (1.26), only $4/\sqrt{2} = 2.83$ times the unit load.

Size effect is an important aspect of fracture mechanics and will be discussed further in later chapters.

1.5.2.4 *Toughness and the characteristic length of a material*

Is a material A, whose fracture energy is twice that of material B, twice as tough as material B? The answer is that it depends upon its yield strength and Young's modulus. A more useful parameter for judging the toughness of a material is its characteristic length, l_{ch}, defined by

$$l_{ch} = \frac{ER}{\sigma_Y^2}.$$ (1.31)

For materials that do not behave plastically, such as rocks, ceramics and concrete, the tensile strength is substituted for the yield strength, in Eq. (1.31). Classic LEFM only applies if a component or structure is large compared to the characteristic length. The characteristic length is comparable to the length of the FPZ and the larger it is, the better the material is at resisting fracture. Unfortunately, for most materials the fracture energy decreases with increase in yield strength. Without going into any design calculations, we know that if a material's characteristic length is comparable or larger than a characteristic dimension of the design, then it is likely to be safe if designed on UTS. A large concrete structure behaves in a brittle fashion and can be modelled using LEFM, but if a laboratory-sized scale model is tested, it behaves in a more ductile fashion because the characteristic length of concrete is of the order of a metre.

1.6 Simple Fracture Experiments

Although this chapter is necessary for those with little knowledge of solid mechanics so that the subsequent chapters can be better understood, it is a little dry. Thus a few simple fracture experiments are introduced at this point to make the chapter more interesting and enable some more fracture concepts to be presented in a painless fashion.

1.6.1 Paper tearing

We have all experienced trying to tear an article out of a newspaper that we wish to save only to be frustrated by the tear not going where we want it to go, but right through the article. Here we show that it is possible to predict the direction of tearing in many, but not all cases. The path of a tear or fracture is not always predictable. One of the things that fracture teaches us is that nothing is ever perfect. A tear in a piece of paper does not follow exactly the path we predict because of small imperfections in the paper which are unpredictable. However, the tear path can be predicted with high accuracy if, after a small deviation from the predicted path, the next most probable direction is back towards the original prediction. Such tear or fracture paths are termed path stable. However, in some cases after a small deviation from the predicted path the next most probable direction of a tear or fracture is away from the predicted path. Such tear or fracture paths are unstable and usually cannot be predicted over large distances with accuracy. This section is based on an article not by an expert fracture mechanist, but by Robert O'Keefe a science teacher at the United Nations International School in New York who obviously has a very good grasp of mechanics and produced an excellent article without the need to refer to a single fracture reference.[18]

The experiment here is to predict the path along which paper tears when it is torn by gripping it between the thumb and first finger of each hand along one edge at points A and B and pulling the hands apart as illustrated schematically in Fig. 1.6 (a). During tearing the paper bends so that the line of force passes through the tip of the tear at O′. It is always very much more difficult to predict where a tear or fracture will start in the absence of any gross imperfection or introduced cut or notch, than to predict its propagation. Thus in this case the experiment is made simpler by introducing a cut in the edge of the paper at O before it is torn. In this example of tearing, the paper deforms into a developable surface with negligible stretching. The energy stored due to the bending or the tension of the paper is negligible compared with the work done in tearing and can be neglected. Tearing takes place so that the work of fracture is minimised and it is this condition that enables the idea tear path to be calculated. The path of a tear that started at O and has reached point O′ is shown in Fig. 1.6 (b). The angle of the path to the edge of the paper at this moment is θ. When the tear reaches O′, AO′B lies on a straight line. Extending the tear increases by a small amount, $ds = $ O′O″, the distance between the two load points A and B increases by $dl = $ O$_a$O″ +O$_b$O″. If F is the force when the tear has reached O, the work done on extending the tear by *ds* is given by *Fdl*. If the paper is isotropic in its plane, then the work done is independent of the direction of the tear and the work is minimised if *ds/dl* is minimised. It is easily shown that this condition is reached when the angle AO′B is bisected by CO′.

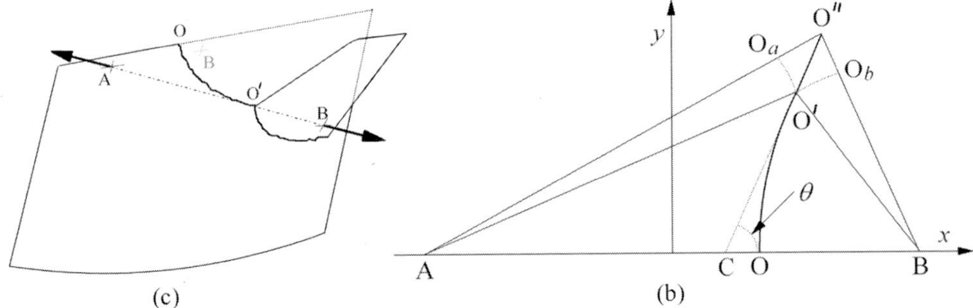

Fig. 1.6 Tearing a piece of paper held between the forefingers and thumbs at A and B, the pre-cut is at O and the current tip of the tear is at O′: (a) A 3-D view where AO′B are on a straight line. (b) The torn paper laid flat (after O'Keefe 1994).

There are complications: machine-made paper is not isotropic and also many papers delaminate when torn. The anisotropy in paper but not delamination can be accommodated. The one paper still manufactured that O'Keefe recommends that tears well contains 25% cotton,[19] but you could experiment to find a suitable alternative. Machine-made paper is more difficult to tear across the machine direction than along it. The difference in tear toughness means that the ideal path is not as predicted above. The tear toughness is anisotropic because the wood fibres are partially aligned in the machine direction and it is the pull-out of these fibres that contributes most to the tear toughness. If the toughness for tearing in the x and y directions are R_x and R_y respectively, the toughness for a tear at an angle θ to the x direction, R_θ, will be given reasonably accurately by

$$R_\theta = \left[\left(R_x \cos\theta \right)^2 + \left(R_y \sin\theta \right)^2 \right]^{1/2}. \tag{1.32}$$

To minimise the work done in tearing anisotropic paper $R_\theta(ds/dl)$ must be minimised. A Fortran programme has been written that performs this minimisation and predicts the tear path as a function of the position of the initial cut and the anisotropy parameter, $\lambda = R_x/R_y$.[20] The anisotropy parameter could be found by measuring the tear toughness in the two directions, but is easier to simply tear the paper from an asymmetrical initial cut and find the value of λ that most accurately predicts the tear path. The path for a tear from any other starting crack can then be predicted. The tearing paths from an initial cut located at a quarter of the distance between the two tearing positions is shown in Fig. 1.7 for two paper orientations and a range of values of the anisotropy parameter λ. If the paper is anisotropic the tearing path turns away from the isotropic path towards the machine direction.

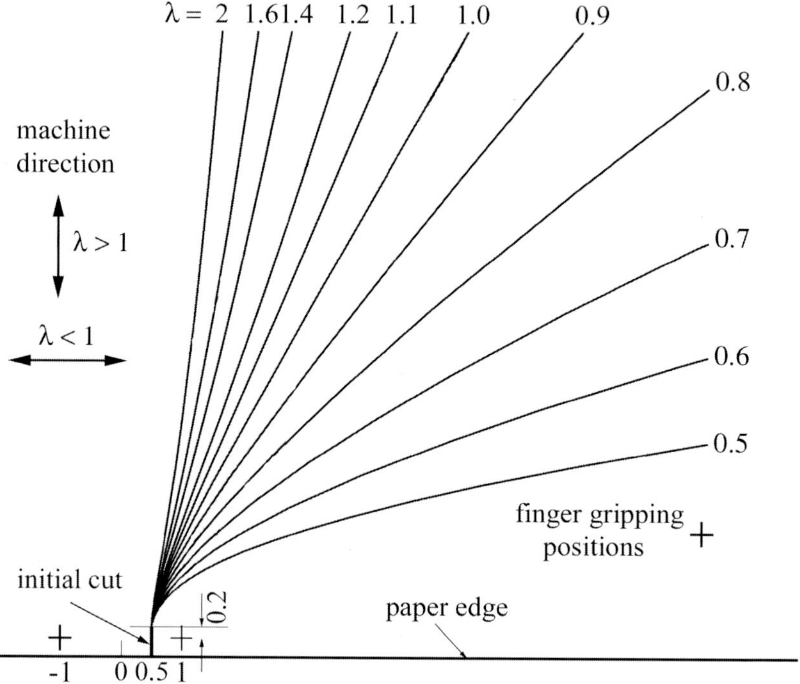

Fig. 1.7 The tearing paths for two different paper orientations for a range of the parameter λ.

1.6.2 The sardine can problem

Sardine cans opened by keys on which the top is rolled up, or by a pulling on a ring are scored along the edges to encourage tearing along these lines, but sometimes the scoring is not sufficient and instead of the top tearing along the edges an annoying convergent tear occurs. Tony Atkins has addressed this and other similar problems, which do have more serious applications.[21] Similar convergent tears occur when stripping wallpaper from a wall and can most easily be demonstrated on a piece of aluminium kitchen foil. First make two parallel cuts perpendicular to an edge and then holding the foil flat with one hand pull upwards on the parallel strip. Inevitably the tear will converge to a point in a short distance (see Fig. 1.8). Converging tears can also be seen in the fuselage of the Aloha Airways Boeing 737 which lost part of its top skin over Hawaii in 1988. Amazingly, the aeroplane did not crash and managed to land safely (see Plate 1). The sardine can phenomenon may have prevented more of the skin of the fuselage tearing off. The reason for the convergent tear path is that it is, again, the path that requires the least work.

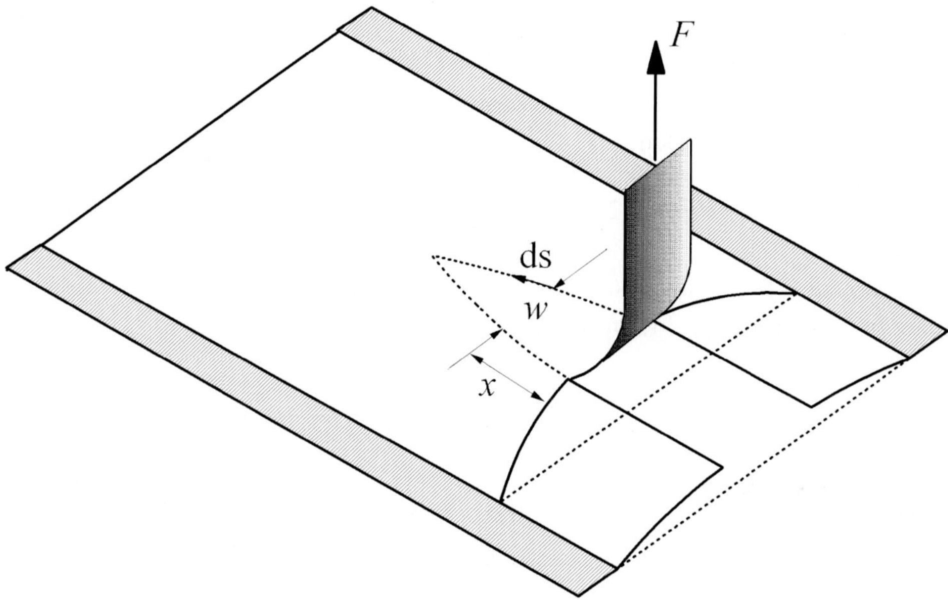

Fig. 1.8 Schematic tearing of a strip from aluminium foil.

In the previous example, any deformation in the paper was neglected and it was assumed that all the work went into tearing. If that were the case here then a parallel strip would result from the experiment. However, if you examine the strip you have torn from the aluminium kitchen foil you will find it is tightly curled because in tearing the strip the foil has been plastically bent.[22] Rather surprisingly, not only is the foil bent during tearing but also it is unbent. If you observe the foil during tearing, the torn strip will appear straight, the curvature only occurs elastically after the strip has been completely torn from the foil as the imposed moment is relaxed. To analyse this problem accurately[23] is difficult and the simpler analysis of the key-opened sardine can without guiding scores is offered here.[21]

It is assumed that there is a pre-cut parallel strip of width $2w_0$ and thickness h already wound up to the edge of the cut on a key of radius ρ. To turn the key by a small angle $d\theta$ and propagate the tear by ds requires a moment, M, which does work on the can of $Md\theta$. This work goes into the work of tearing, $2Rhds$ and the plastic work, dW_p. Without going into details this plastic work can be expressed approximately by

$$dW_p = \frac{\sigma_y h^2 w}{4\rho}dx, \tag{1.33}$$

where w is the current width of the strip being torn, and $dx = \rho d\theta$ is the small length of can rolled up. Hence

$$Md\theta = \left[2Rh\frac{ds}{dx} + \frac{\sigma_y h^2 w}{4\rho}\right]\rho d\theta . \tag{1.34}$$

The tear path which minimises the work done, minimises the terms within the square bracket in Eq. (1.34) and gives

$$w = w_0 - \lambda\left[\cosh\left(\frac{x}{\lambda}\right) - 1\right], \tag{1.35}$$

where

$$\lambda = \frac{8R\rho}{\sigma_y h}. \tag{1.36}$$

The shape of the tears, shown in Fig. 1.8, is a catenary.[24] The torn strip converges to a point after a tear distance, L, given by

$$L = \lambda \cosh^{-1}\left[1 + \frac{w_0}{\lambda}\right]. \tag{1.37}$$

The analysis for the shape of a strip torn by hand from a sheet of aluminium foil is similar, but the problem is that the radius of curvature to which the foil is bent is not known explicitly. Also, you may have noticed that the strip you have torn from the aluminium foil is coiled like a clock spring with the radius increasing with the width of the strip, which is an added complication. The net result is that the strips you tear from aluminium foil have straighter sides than a catenary-shaped strip.

1.6.3 Divergent concertinas tears

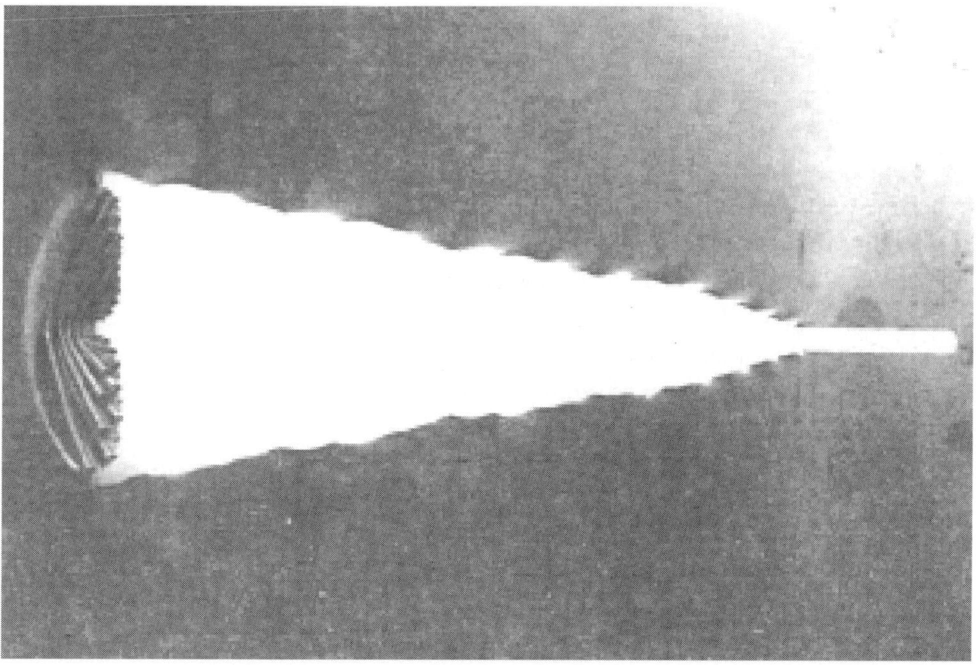

Fig. 1.9 A divergent concertina tear originating from a pre-cut slot in a thin low carbon steel sheet (Wiezbicki *et al.* 1998, with permission ASME).

If you try tearing aluminium foil by first puncturing it and then dragging a blunt penetrator such as a ball point pen or fingernail across it a divergent tear occurs with the foil folding up, similar to a concertina, in front of the penetrator.[21,25] A divergent concertina tear originating at a pre-cut slot in a thin low carbon sheet is shown in Fig. 1.9. Divergent concertina tears can arise in a number of trivial ways. Examine your newspaper and you will probably find tiny divergent concertina tears along the bottom edge of the pages caused by the pins which draw newspapers through the printing press. Divergent concertina tears can also occur in peeling soft fruit such as plums. However, the importance of divergent tears is that they can be produced on a large scale when a ship's hull is penetrated. In the notorious 1989 *Exxon Valdez* accident in Alaska, a divergent concertina tear was ripped from the ship's hull when it hit Prince William Sound's Bligh Reef and spilled an estimated 50 to 150 million litres of crude oil.

1.6.4 Wiggly cuts or the Kit Kat® problem

There is a difference in cutting with a sharp and blunt tool. The best material to demonstrate this difference is the thin polymer films used for food packaging.[26] The easiest packaging material use for this experiment is the packaging of a bar of Kit Kat® since the film needs to be free of contact with a hard surface and the packaging film between the long chocolate sections of Kit Kat® is ideal for this experiment. If the packaging film between the sections of chocolate is cut with a sharp knife the crack precedes the knife and a straight cut results. However, if the tip of a ball point pen is used instead of a knife the crack path oscillates from side to side (see Fig. 1.10). The reason why the film has to be unsupported is that the film deflects out of its plane during the formation of the oscillating path. This 'wiggly path' phenomenon is related to the much larger-scale wiggly fracture paths in failed gas transmission pipes observed during 1948–51, which ran at high speed for 50–1000 m. An example of this type of wiggly path is shown in Fig. 8.7. The wiggly path in the gas transmission line fractures was also due to out of plane deformation caused in this case by the force of the gas discharge. No direct engineering application is known for this quaint phenomenon which is introduced just for your amusement.

Fig. 1.10 The packaging of a Kit Kat® bar cut with a sharp knife (bottom cut) and torn with the tip of a ball point pen (top two cuts).

Tony Atkins was introduced to wiggly paths in polymer film at a 1996 Royal Society Soiree, showing that even scientists have their lighter moments. Once again, the fracture path is the one that minimises the work done. The size of the oscillating paths increases with the diameter of the cutting tool and, since packaging film is anisotropic, the details of the oscillations depend upon the degree of anisotropy.

1.7 Concluding Remarks

The development of fracture theory has often relied on simple concepts such as the energy necessary to produce a fracture and the minimum work concept. The present chapter will have helped the non-specialist have enough knowledge to appreciate the following chapters, especially Chapters two to four, where fracture concepts are used to explain how fracture influenced the evolution of the earth's features, the evolution of life, and human evolution.

In most cases fracture is something to be avoided and can lead to catastrophes, but in many other cases, such as the fashioning of stone tools, cutting and machining the knowledge of how to produce controlled fractures has been of great benefit.

1.8 Notes

[1] The original spelling was 'fractour'.

[2] Here the notation introduced by Charles Gurney (1913–1997) for fracture energy is used.

[3] Coulomb (1785).

[4] Cauchy (1823).

[5] Mohr (1882). With today's electronic calculators and computers there is little need for graphical methods for calculations, but it would be undesirable for them to lapse into disuse because they provide an insight that cannot be given by the equations alone.

[6] Jaeger and Cook (1979).

[7] Christiaan Huygens (1629–1685) perfected a hair spring watch in 1675 and recruited Oldenburg to help him get a patent in England much to the disgust of Hooke (Burgan 2007). However, Hooke himself was a difficult man.

[8] Hooke (1678).

[9] Young was a polymath of genius, apart from his mechanical studies, he studied ophthalmology and identified astigmatism. He was also a distinguished linguist and produced an almost correct translation of the Rosetta stone several years before Champollion's grammar was published, but he had to abandon Egyptology through lack of funds.

[10] Young (1845).

[11] Cauchy (1829).

[12] Ramberg and Osgood (1943).

[13] Sometimes called the modulus of resilience.

[14] Very extensible materials like rubber and skin are exceptions as is discussed in §3.3.1.4.

[15] Irwin named this energy release rate, G, in honour of Griffith. He also used the term *crack extension force* for G since the units are force per unit length.

[16] The use of the term fracture process zone is explored in Chapter 9.

[17] Some authors call R the fracture toughness and just call K_{Ic} the critical stress intensity factor.

[18] O'Keefe (1994).

[19] Watermark Antique Laid Electronic paper, heavy weight (O'Keefe states 20 lb, but 24 lb is the weight that is still available) manufactured by Southworth Company, www.southworth.com.

[20] The Fortran programme can be found at the website for this book http://www.icpress.co.uk/physics/p593.html.

[21] Atkins (1995).

[22] Even most paper does not tear without some plastic deformation and a strip torn from paper will converge and will be curled.

[23] Muscat-Fenech and Atkins (1994a).

[24] The shape of a chain or heavy cable hanging between two supports.

[25] Wierzbicki *et al.* (1998).

[26] Atkins (2007).

Chapter 2

Evolution of the Earth

In 1862 the famous nineteenth-century physicist William Thomson (1824–1907), later Lord Kelvin, gave the first scientific estimate of the age of the earth as between 20 and 400 million years. He made this estimate from calculating the time it would take a molten ball of rock to cool to its present surface temperature. What he did not know was that the earth was being heated by radioactive decay which made his estimate far too short. The discovery of radioactivity in the late nineteenth-century gave a method for accurately estimating the age of the earth. While still an undergraduate, Arthur Holmes (1890–1965) used radiometric methods to date the age of a rock from Norway to 370 million years. In 1913 he published a small book *The Age of the Earth* in which he gave the age as 1.6 billion years. The current accepted age of the earth is 4.6 billion years, using the decay of the isotopes of uranium (U^{238} and U^{235})[1] to the lead isotopes (Pb^{206} and Pb^{207}). Clair Cameron Patterson (1922–1995) determined this age of the earth from fragments of the Canyon Diablo meteorite in 1956.[2] Once the earth's crust had solidified, fracture played its part in shaping the world that we know today.

Sedimentary rocks are laid down in layers over time. It was realised quite early that the sequence of these layers give a time record of the history of the world. Stratigraphy is the study of rock layers and their sequence. The global stratigraphic record is apportioned into eons, eras, periods, epochs, and ages. The four major divisions[3] according to the International Commission on Stratigraphy are given in Table 2.1. The fossils embedded in sedimentary rock were the first markers used in stratigraphy. Even in the fifth-century BC Herodotus wrote that the presence of seashells on the hills in Egypt showed that in earlier times, ten thousand years in his estimate, Egypt was under the sea.[4] In the *Shen Xian Juan*, written some time before the Tang dynasty (618–906), it is stated that shells of oysters and clams could be seen in mountain rocks, which some thought came from the sea bed.[5] Leonardo da Vinci (1452–1519) clearly understood the process by which shells, fish, and leaves were fossilised and that the fossils of shells far from the sea and at high altitude were evidence of great changes in the earth. In his day, fossils were popularly thought not to be formed from shells and other animals or plants, but imitative artefacts of nature caused by some "plastic force". This idea dates back to Theophrastus, an Aristotelian philosopher of the forty-century BC. Leonardo refuted this idea by observing that many different existing species were represented in minute detail. The same popular notions were current in the seventeenth-century when Robert Hooke again refuted the idea that fossils were 'sports of nature'. In the seventeenth-century the Bible was relied on for the date of the creation of the earth and the biblical calculation by the Irish Archbishop, James Ussher (1581–1656), that the earth was created on Sunday October 23, 4004 BC was widely believed.

Table 2.1 Eons, eras, periods, and epochs (after Gradstein and Ogg 2004).

Eon	Era	Period	Epoch	Million of years before the present
Phanerozic	Cenzoic	Neogene	Holocene	0.011
			Pleistocene	1.816
			Pliocene	5.33
			Miocene	23
		Paleogene	Oligocene	34
			Eocene	56
			Paleocene	65
	Mesozoic	Cretaceous		146
		Jurassic		200
		Triassic		251
	Paleozoic	Permian		299
		Carboniferous		358
		Devonian		416
		Silurian		444
		Ordovician		488
		Cambrian		542
Proterozoic	Neoproterozoic			1000
	Mesoproterozoic			1600
	Paleoproterozoic			2500
Archean	Neoarchean			2800
	Mesoarchean			3200
	Paleoarchean			3600
	Eoarchean			3800

It was the Dane, Nicolas Steno (1638–1686), who laid foundations of stratigraphy. Not only did he recognise that rock layers containing fossils were formed from gradually accumulated sediment over long periods of time, but also that layers with different fossils formed a sequence with the oldest at the lowest level. Thus Steno formulated what is now known as the *Principle of Superposition*, which states that depending upon the fossil, rock layers record the succession of their formation. The first fossil traces of life are now known to date from some 3.5 billion years ago. However, uncontroversial fossils of invertebrate animals only date to about 600 million years ago during the Ediacaran period, named after the Ediacara Hills in South Australia, where they have been found in abundance.

At the end of the eighteenth-century there was a great controversy over the origin of rocks. Abraham Werner (1750–1817) in Freiberg was the founder of the Neptunists' School who maintained that nearly all rocks were formed as precipitates from water, even rocks such as granite and basalt. This school saw the molten lava from volcanoes as being not part of the normal geological process and probably due to the burning of subterranean beds of coal. The main opponent of Werner was James Hutton (1726–1797) in Edinburgh. Hutton led the Plutonists' School who correctly regarded granite, basalt and other similar rocks as having an igneous origin. Hutton also formulated the uniformitarian principle that natural processes have been uniform through long periods of time. At the time catastrophism, whereby every major feature such as a mountain range was formed by catastrophic events over a short period of time, was a commonly-held view. Jean-Louis Giraud (1752–1813) proposed that the fossils in the strata of southern France could be used as mileposts in time.[6] The first stratigraphic table of Britain was compiled in 1799 by William Smith (1769–1839), a canal surveyor who earned the nickname of Strata Smith. In 1830, Charles Lyell (1797–1875) published the first volume of *The Principles of Geology*.[7] From studies of fossils in the Italian Cenozoic (Tertiary in Lyell's day) strata, Lyell partitioned the Tertiary into three groups according to the proportions of living to extinct fossil shells in the strata. These epochs he called: Eocene (dawn of recent), Miocene (less recent), and Pliocene (more recent). These terms are still used together with later additions for Cenozoic epochs in the International Stratigraphic Chart.

2.1 Plate Tectonics

It is now known that the surface of the earth is in continuous, albeit very slow, motion. It is this motion that generates the strain energy that is released in earthquakes. The earth is composed of three main layers: the crust, the mantle and the core. The continental crust has a thickness of 30–50 km and is composed of granite and other igneous rocks, together with sedimentary rocks. The oceanic crust is much thinner at around 7 km and is composed of igneous rock. There is a discontinuity called the *Moho* between the crust and the mantle, which can be detected by seismological waves. The mantle extends to a depth of some 2,800 km and has a number of distinct layers. The upper layer consists of ultramafic rocks, which are basaltic rocks containing more than 70% iron and magnesium-rich minerals. The crust and the upper layer of the mantle form the lithosphere, which is fragmented into tectonic plates which are relatively rigid. The continental lithosphere has a thickness of about 150 km. The oceanic lithosphere is thinner at about 70 km except over mid-oceanic ridges, where it is only a few kilometres thick. The next layer in the mantle below the lithosphere is the asthenosphere which goes down to some 200–300 km. The rocks in the asthenosphere are closer to melting than in the mantle layers above and below. It is the convective movement in this layer at rates of the order of centimetres per year that causes the motion of the tectonic plates. Under the asthenosphere is the mesosphere where the rocks are denser and not so near their melting temperatures. Finally the core, composed of mainly iron plus silicon, sulphur, and nickel, has two parts: an outer liquid layer and the inner core where, though the temperatures are very high, the pressure is large enough to prevent melting. It is the plate tectonics moving on the asthenosphere that creates a changing world.

The present world contains the jigsaw puzzle pieces that made up the earlier world. The fact that parts of the present world could fit together was first recognized by the Dutch mapmaker Abraham Ortelius (1527–1598) who studied maps of the coasts of the three continents and wrote, in 1596, that the Americas were 'torn away from Europe and Africa…by earthquakes and floods'. In 1658 the French monk, François Placet, published a book: *The breaking up of the greater and lesser worlds: or, it is shown that before the deluge, America was not separated from the other parts of the world, suggesting that the old and the new worlds broke apart after Noah's Flood.* During an expedition to South America (1799–1805) the explorer, Alexander von Humboldt (1769–1859), recognised that as well as the apparent fit of the coast lines of South America and

Africa, the mountain ranges which ended on the east coast of South America were apparently continued on the west coast of Africa; there were also similarities in the geological strata. However, Humboldt again referred to Noah's Flood as the cause. These early speculations were catastrophic in nature and the first real enunciation of the theory of continental drift was made by the German Alfred Wegener (1880–1930) in his book *The Origin of Continents and Oceans* published in 1915. Wegener rejected the theory that the continents were once connected by land bridges which had since sunk into the sea by pointing out that the continents are made of a less dense rock than the ocean floor. Mountain ranges were formed, according to Wegener, not by the earth's crust wrinkling by contraction during cooling, but by folding as continents collided. Wegener's book was little known outside of Germany, because of the First World War, until a translation of third 1922 edition was published. In this edition he advanced geological evidence that 300 million years ago the continents were joined together forming a supercontinent, Pangaea (all lands), which then began to break up 200 million years ago and the continents slowly drifted to their present positions. At the time Wegener's theory was rejected by most geologists. In 1929, Arthur Holmes did elaborate on one of Wegener's ideas that convective forces could move the continents, but this work did not attract attention. Wegener froze to death during a mission to deliver supplies to a camp far inland after winter had set in while on an expedition to Greenland in 1930 to research his theory. It was more than 30 years before Wegener's continental drift theory got the recognition it deserved. It was the paleomagnetic discoveries in the 1960s that showed that continents had changed their positions relative to the magnetic pole that led to a final wide acceptance of Wegener's ideas.

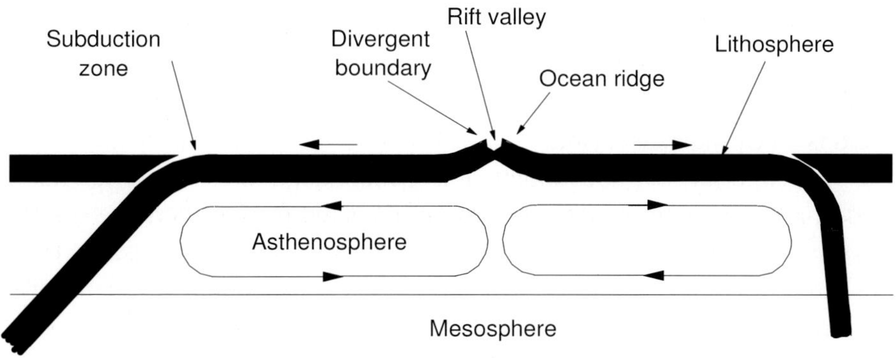

Fig. 2.1 Schematic concept of sea-floor spreading.

The key to continental drift lies in the concept of sea-floor spreading which is largely due to Harry Hess (1906–1969) who was interested in the geology of ocean basins.[8] Hess served in the US Navy during the Second World War and fought in the Pacific, but found time in between battles to conduct echo soundings of the ocean floor. The concept of sea-floor spreading as conceived by Hess is illustrated schematically in Fig. 2.1. In the middle of the oceans there are ridges which rise thousands of metres from the ocean floor, which are formed from the convection of the asthenosphere to form new oceanic lithosphere by extrusion of basaltic magmas. The tectonic plates meeting at the ocean ridge move away from the divergent boundary under forces exerted by the convective current in the asthenosphere; rift valleys commonly form near the crest of ocean ridges when the spread in the lithosphere is not fed completely by the athenosphere. Along the edges of the continents and island chains the lithosphere is forced down into the athenosphere at subduction zones where it melts. Oceanic subduction zones form deep trenches such as the Marianas Trench, which is more than 10 km deep. Subduction zones are also regions of earthquake activity. The schematic illustration in Fig. 2.1 is two-dimensional and the details at the edges of tectonic plates are not as simple as implied in this figure. There are seven major tectonic plates: the Eurasian, Antarctic, North American, South American, Pacific, African, and Indo-Australian whose size is from 10^7 to 10^8 km^2, eight intermediate sized plates (10^6 to 10^7 km^2), and more than twenty plates in the 10^5 to 10^6 km^2 size range. A tectonic map showing part of the Indo-Australian plate is shown in Fig. 2.2. At the ends of divergent boundaries or subduction zones there are transform faults where the plates slide relative to each other. For movement to take place along a transform fault the plates have to shear. Movement along the transform faults is intermittent. First strain builds up with time until the energy stored is enough to produce shearing. The sudden shearing of the plates gives rise to the release of very large amounts of energy in the form of earthquakes. So in Fig. 2.2 the transform fault in the south island of New Zealand is an earthquake region.

At the edge of oceanic subduction zone island arcs, such as the New Hebrides, are thrown up on the edge of the impacted plate. Where the subduction zone occurs at the edge of a continental plate, mountain ranges, such as the Andes, are thrown up. At continental subduction zones mountain ranges are also pushed up, such as the Himalayas: the greatest visible and the youngest creation of tectonic forces. However, the mechanism is different. The continental lithosphere is less dense than the oceanic lithosphere and its buoyancy limits its descent into the asthenosphere and the Himalayas are a collisional mountain range formed by thrusting and folding.

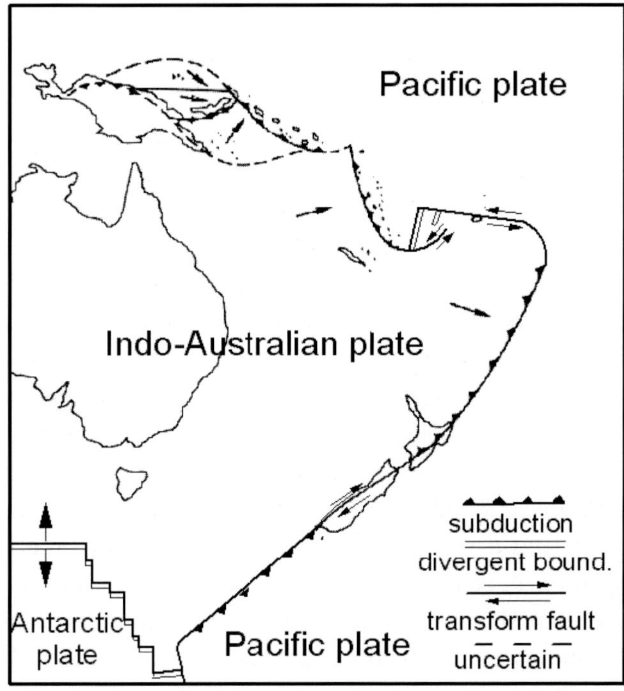

Fig. 2.2 Boundaries between parts of the Indo-Australian, Pacific and Antarctic tectonic plates (adapted from the US Geological Survey).

2.2 Folds and Faults

The differential movement in the earth's crust is accommodated by either ductile folding or brittle faulting (fracture) of the rock strata. Near the earth's surface the differential movements are horizontal and cause normal and shear strains. Very large deformations occur over time that cannot be accommodated simply by accumulating strain in the plane of the earth's surface. The surface has to either be thrust into folds or fracture to form faults, and there is competition as to which occurs. Faulting fractures require a critical stress to occur. Most stresses in the earth's crust are compressive and the compressive strength of rocks is far greater than their tensile strength. A comparison of the uniaxial tensile and compressive strengths of rocks is made in Table 2.2.

                                     Fracture and Life

Table 2.2 Tensile and compressive strength of rocks (after Jaeger and Cook 1979).

Rock	Tensile strength (MPa)	Compressive strength (MPa)	Ratio of compressive to tensile strength
Granite	21	229	11
Marble	7	90	13
Dolerite	40	488	12
Sandstone	4	50	14

In 1900 the Canadians, Frank Adams (1859–1942) and John Nicolson, performed compression tests on marble cylinders constrained by wrought iron jackets at both room temperature, and at 300°C. The cylinders could be deformed into barrels even at room temperature.[9] Rock can also flow viscously even at low temperature over long periods of time. The creep of rock at ambient temperature is surprisingly similar to the tensile creep of metals at high temperature, but at vastly different rates. In a creep test a specimen is subjected to a constant stress and the strain measured. A schematic illustration of the resulting creep curves is given in Fig. 2.3. Initially, on application of the stress there is an immediate elastic response, giving a strain $\varepsilon_0 = \sigma/E$. During primary creep the strain rate decreases, creep then occurs at a constant strain rate during secondary creep. Tertiary creep, the final stage, occurs when the creep rate increases prior to fracture, but has only been observed in very soft rock alabaster. Most creep in the earth's crust at low temperature occurs during the secondary stage when the strain rate, $\dot{\varepsilon}$ under uniaxial compression, σ, is given by

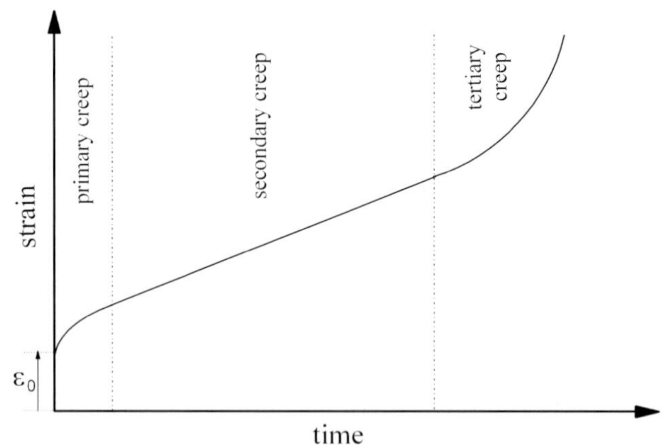

Fig. 2.3 Schematic illustration of creep in compression for rocks.

$$\dot{\varepsilon} = \frac{\sigma}{3\eta},$$ (2.1)

where η is the viscosity.[10] Because the creep rates are so small there are hardly any direct data on creep at ambient temperatures and most values are obtained by inference.[11] However Hidebumi Ito, in a heroic twenty-year creep bend test on granite from Akasaka in Japan, found that the viscosity, at ambient temperature, was $3\sim6\times10^{19}$ Pa s.[12] Primary creep, if it occurred at all, lasted for less than half a year. In the competition between folding and faulting it is the relaxation in stress with time that is important. Based on the creep rates measured by Ito, provided the strain rate in granite even at ambient temperature is less than about 2% per 1,000 years, the stress cannot exceed about 100 MPa.

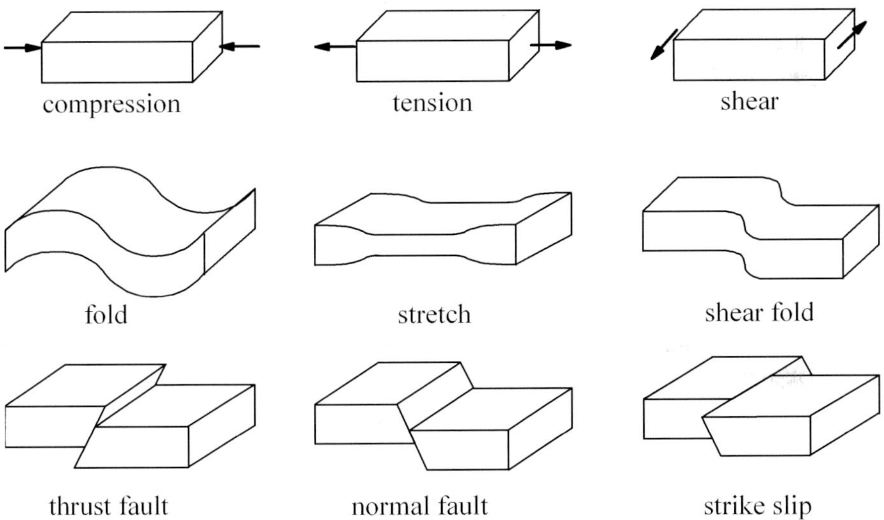

Fig. 2.4 Archetypal folds and faults.

The stresses perpendicular to the earth's surface are small near the surface and the stress system is predominantly two-dimensional. The stresses acting across any section perpendicular to the surface are a normal stress, either compressive or tensile, and a shear stress. In Fig. 2.4 the archetypal fold and fault systems are illustrated schematically for compressive stress, tensile stress and shear stress, which are highly simplified versions of usually complex systems.[13] Under compressive strain both folding and thrust faulting are common. Since the strength of rock is much less in tension than in compression, a normal fault is the far the more common way of accommodating tensile strain than stretching and thinning. Under tensile strain two normal fault systems can operate in the upper crust to form a rift valley as illustrated in Fig. 2.5. The Great Rift Valley, with a length of some 5,500 km and 600 m deep in parts, is the longest continental rift valley. The mechanism by which this valley formed was first recognised by the Scottish geologist John Walter Gregory (1864–1932), who gave it its name during an 1893 geological expedition from Mombassa to Lake Baringo in what is now Kenya. Gregory wrote that the valley was formed 'by the rocks sinking in mass while the adjacent land remained stationary'.

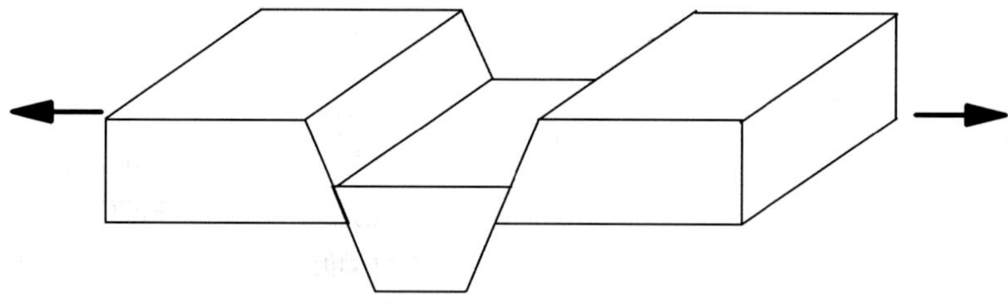

Fig. 2.5 Rift valley formation.

2.3 Earthquakes

Early Greek speculations on the cause of earthquakes were rooted in their four elements: air, earth, fire, and water. After initial linking of earthquakes to water attributed by Seneca to Thales of Miletus (ca. 624–545 BC), Aristotle (384–322 BC) linked earthquakes to the wind. According to Aristotle there were two types of air: humid vapours and dry air called *pneuma*. When *pneuma* comes out of the earth it produces winds, but when it is trapped in the earth it causes earthquakes. Up until the twelfth-century it was only parts of Aristotle's works on logic, known collectively as the *Organon*, that were known in the West. As with the other works of Aristotle, the *Metrologica*, which contained his ideas on earthquakes, came to the West by way of the Arabs. Aristotle's *Metrologia* became known in the West through the translation and commentary by Ibn Sina Avicenna (980–1037).Until the Renaissance, western science was more interested in teleological questions, or *why* things happen rather than *how* things happen. Teleological questions by their nature cannot be fully answered by science and are the realm of metaphysics and religion. Thus it is not surprising that for a very long time religious as well as naturalistic explanations of earthquakes coexisted. St. Thomas Aquinas (1221–1274) reconciled Aristotelian thought with religion. He saw God as the prime or teleological cause alongside Aristotelian secondary causes. Thus, Aquinas thought that there were two kinds of earthquakes: the natural explained by Aristotelian thought and the prodigious caused by God. This duality in cause has long persisted in the thoughts of many.

The introduction of gunpowder in the West and its use in mines and bombs in the sixteenth-century saw the classical theory of earthquakes undermined. By the seventeenth-century, René Descartes (1596–1650) was explaining in his *Principia Philosophica*, published in 1644, that earthquakes were caused by exhalations becoming trapped in cracks and cavities in the earth where they ignited to cause an explosion. The persistence of a duality of causes for earthquakes is seen in a book published in 1693 by the puritan preacher Thomas Doolittle (1630–1707).[14] In 1692 there was a minor earthquake, which was felt over a wide area in south-east England, and in Europe from Normandy to Holland. The earthquake caused little damage, but caused consternation in London, where the streets were filled with panic-stricken crowds.[15] Doolittle discussed this earthquake and the causes of earthquakes in general. He accepted the theory of Descartes for natural earthquakes felt over a small region, but considered earthquakes felt over large areas to be supernatural because it was impossible that exhalations could be trapped in caves over such a wide area and so he turned to the prime cause of Aquinas, God. He argued that only God could have caused such a widespread earthquake. Doolittle considered the earthquake to be 'great' because 'nothing is destroyed thereby, yet it is and might be truly called a great earthquake'. He prayed that all 'might search their hearts and ways, see what is amiss and mend it … that God may no more be provoked'.[14,16]

The modern study of earthquakes begins with John Mitchell (1724–1793) who, though he still accepted the Descartes' cause for earthquakes, recognised that it was wave-like deformations that transmitted the energy of the earthquake. Mitchell's work was still considered definitive in 1818 when it was reviewed in the *Edinburg Review*. Charles Lyell was aware that faulting occurred in earthquakes, but he thought that they were caused by volcanic activity or thermal expansion and contraction, not that the energy released by dynamic faulting was the direct cause. The American geologist Grove Karl Gilbert (1843–1918) in 1884 was the first to claim clearly that dynamic faulting was the cause of earthquakes, but limited his claim to the Great Basin of Nevada. Bunjiro Koto (1856–1935), unaware of the paper by Gilbert, also argued in 1893 that the Nobi earthquake in Japan in 1891 was caused by a dynamic normal fault. The disastrous 1906 San Francisco earthquake was caused by the dynamic shift of 6.8 m in the San Andreas strike fault that is the boundary between the Pacific and the North American tectonic plates. Most major earthquakes are caused by dynamic faulting, but explosive volcanic eruptions, large rock falls, landslides, and explosions can also cause earthquakes.

2.3.1 Seismology

The energy released in an earthquake is radiated by a number of waves that travel at different velocities. The fastest wave is the P-wave or longitudinal wave, where the vibration is in the direction of propagation. The slowest wave is the S-wave or shear wave where the vibration is transverse to the direction of propagation. The propagation velocities of P and S waves are independent of the wave length. In addition there are two other waves that propagate near the surface. These are Rayleigh waves, which have their largest magnitude at the surface and Love waves, which occur when the S-wave velocity of the surface layer is less than that of the underlying layer and waves are reflected at the underlying layer so that the surface layer acts as a wave guide. Love waves are dispersive, that is, their velocity of propagation depends upon their wavelength.

Seismographs record the waves propagating through the earth. Heng Zhang (78–139), a renowned Chinese mathematician, astronomer and geographer, is credited with inventing the first seismograph or *zao* (earthquake weathercock).[17] The details of this first seismograph are unknown but, from the description, it is clear that it was a pendulum device which released balls indicating the direction of the earthquake even when no perceptible shock could be felt. Modern seismographs enable the distance from the epicentre of the earthquake to be calculated from the difference in arrival time of P and S waves. The location of the earthquake is found from seismographs at different locations

Charles Richter (1900–1985) and Beno Gutenberg (1889–1960) developed what is known as the Richter Scale in 1935 to quantify the strength of an earthquake. The original scale was for just the Woods–Anderson seismograph and the magnitude, M, of the earthquake was given by the equation

$$M = \log_{10} A + 3\log_{10}\left(8\Delta t\right) - 2.92, \tag{2.2}$$

where A (mm) is the amplitude of the S-wave and Δt (s) is the time difference in the arrival of the P and S waves. The Richter Scale has now been extended to all seismographs. A major problem with the Richter Scale is that there is a saturation effect around 8.3–8.5 and the Richter Scale is being replaced by a moment magnitude scale, but at present it is the Richter Scale that is commonly quoted. The elastic energy, E_s (J), radiated by an earthquake is given by the Gutenberg-Richter empirical equation

$$\log_{10} E_s = 1.5M + 4.8. \tag{2.3}$$

The effects and frequency of earthquakes are given in Table 2.3.

Table 2.3 The Richter Scale (after the US Geological Survey).

Richter Magn.	Description	Effect of earthquake	Approximate frequency
<2	Micro	Not felt	8,000 per day
2–2.9	Very minor	Generally not felt, but recorded	1,000 per day
3–3.9	Minor	Often felt, but damage rare	150 per day
4–4.9	Light	Noticeable shaking, significant damage rare	20 per day
5–5.9	Moderate	Significant damage to poorly constructed buildings	800 per year
6–6.9	Strong	Damage to populated areas up to 150 km across	120 per year
7–7.9	Major	Serious damage over large areas	18 per year
8–8.9	Great	Serious damage over areas several hundred km across	1 per year
>9	Rare	Devastation over areas thousands of km across	1 per 20 years

2.3.2 *Earthquake hazards and prediction*

Earthquakes are one of the major natural disasters; nineteen earthquakes of magnitude seven or greater on the Richter Scale can be expected in any one year. A global seismic hazard map is shown in Plate 2 that gives the 1 in 10 probability of exceeding specified ground accelerations within 50 years. Maps of this kind enable building construction codes to be set for different regions. This response to earthquakes is probably the most effective. The other responses, through intermediate- or short-term earthquake prediction are not yet possible and may never be feasible.

The prediction of earthquakes falls into four categories. Earthquakes can be considered a random process in time so that if the earthquake frequency as a function of magnitude is known at any location, then the probability of an earthquake in a particular magnitude range within a certain time is given by Poisson's distribution. This category is the simplest and can be used to give long-term hazard estimates that can be used as a basis for land planning, building codes, and building insurance. More precise methods are needed for other than long-term prediction. If a degree of predictability is accepted then the earthquake hazard is time-dependent. The oldest prediction method is this category is the elastic rebound theory proposed in 1910 by Harry Fielding Reid (1859–1944) as a result of a study of the 1906 San Francisco earthquake. In this theory an earthquake reoccurs when the stress released in a previous earthquake is recovered. Stress cannot be measured directly, but geodetic measurements can be used to determine both the strain released during an earthquake and the subsequent build up of strain. There have been at least five historic magnitude 6 earthquakes in the Parkfield segment of the San Andreas Fault prior to the latest in 2004. Calculations using the rebound theory published in 2002 after the then previous magnitude 6 earthquake of 1966, showed that there was a 95% chance that an earthquake should have reoccurred between 1973 and 1987.[18] Jessica Murray and Paul Segall considered that Parkfield was an ideal location for the application of the rebound model and commented that 'The model's poor performance in a relatively simple tectonic setting does not bode well for its successful application to the many areas of the world characterised by complex fault interactions.' Earthquake forecasting relies on precursors to earthquakes. Historically, this method is the oldest, but in earlier times the precursors, such as peculiar weather or behaviour of animals, only fortuitously predicted earthquakes. Nevertheless, Charles Lyell thought that the universality of some of the myths indicated that there was some truth in them. The precursors used today have more grounding in fact, though some, such as electromagnetic precursors still have no mechanism to explain their link to earthquakes.[19] The problem even with precursors that are well linked to earthquakes is that they do not occur in every earthquake and so are unreliable. Significant crustal uplift can occur prior to an earthquake, for example, four hours prior to the Sado earthquake in Japan in 1802, there was uplift of a metre in the land. Seismic activity occurs between earthquakes and it is generally observed that after the main earthquake there are a number of aftershocks which decay in time to a period of quiescence. Some years prior to the next earthquake there is an increase in seismic activity, which is often followed by a second period of quiescence, at least over the central portion of the fault. Foreshocks can occur weeks or days before the earthquake. The fourth prediction category is deterministic prediction. Success in this category is a long way off and may not ever be achieved since so many variables affect earthquakes. Charles Richter wrote in 1964 that 'only fools and charlatans predict earthquakes'. There has been some progress since then, but there is still a debate as to the achievements that are possible in earthquake prediction.

2.4 Rock Fracture

Table 2.4 Fracture toughness of rock (compiled from Whittaker *et al.* 1992 and other sources).

Material		Fracture energy (J/m^2)	Fracture toughness $(MPa\sqrt{m})$
Igneous rocks	Granite	70–140	1.7–2.6
	Basalt	40–110	1.8–3.0
	Dolomite	35–75	1.7–2.5
	Syenite	20–50	1.2–1.9
Sedimentary rocks	Limestone	15–65	0.9–2.0
	Marble	13–16	1.0–1.1
	Sandstone	3–35	0.4–1.5
Ceramics	Glass	8	0.8
	Porcelain	14	1
	Alumina	38	4

In tension, rock is brittle. The fracture energy of the sedimentary rocks is similar to that of ceramics whereas igneous rocks are generally tougher (see Table 2.4).[20] However, most geological fractures occur when the principal stresses are compressive. The different types of compression fractures observed in the laboratory are schematically illustrated in Fig. 2.6. Under pure compression, fine-grained rock such as Solenhofen limestone and glass split vertically (Fig. 2.6 (a)). Splitting fractures are caused by secondary stresses induced because of the difficulty in obtaining pure compression in a test. The typical compression fracture in rock, shown in Fig. 2.6 (b), is by shear. On a geological scale this type of fracture produces a thrust fault. Under moderate confining pressure the single shear fracture, shown in Fig. 2.6 (c), also occurs – often at a slightly gentler slope. If the confining stress is large, multi-shear fractures, as shown in Fig. 2.6 (d), occur with considerably ductility. The ductility in rock comes from a completely different mechanism to the ductility in metals caused by plastic deformation. Prior to final failure microcracks orientated with their long axes in the direction of compression open, causing an increase in strain and confer ductility on an inherently brittle material. Microcracking causes dilatancy, that is, instead of the volume decreasing with compressive strain before the maximum stress is reached, it increases. The development of microcracks has been studied from compression tests on a quartzite specimens jacketed in thin copper tubes in stiff testing machine, which enables stable behaviour past the maximum load.[21] At up to about 90% of the ultimate strength, microcracks occur randomly. By 95% of the ultimate strength, the microcracks are clustered about the pupative fracture plane. After the ultimate strength has been reached, microcracks continue to form on a band containing the final fracture plane under the extra stability provided by the jacket. The specimen does not fracture completely until the stress has dropped very significantly from the ultimate strength value.

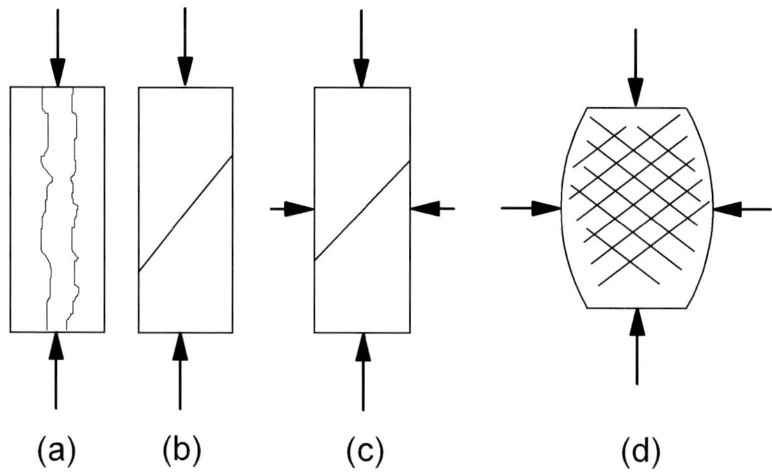

Fig. 2.6 Types of compression fracture: (a) splitting fracture of fine-grained rock. (b) shear fracture of coarse grained rock. (c) single shear fracture under moderate confining pressure. (d) multiple shearing and pronounced ductility under high confining pressure (after Jaeger and Cook 1979).

2.4.1 *The effect of confining pressure on the compressive strength*

The simplest and first description of the effect of confining pressure on the compressive strength of rocks was given by Coulomb in his famous 1773 paper presented to the French Academy of Science.[22] In this paper Coulomb proposed that under compression, fracture occurred by shear that was resisted not only by the inherent shear strength of the rock, S_0, but also by friction on the fracture plane so that fracture occurred when the shear stress on any plane, τ_n, reaches a critical value given by

$$|\tau_n| = S_0 - \mu\sigma_n,\tag{2.4}$$

where σ_n is the normal stress[23] on the fracture plane and μ is the coefficient of friction. Fracture only depends on the numerically largest and smallest principal stresses and occurs on a plane whose normal is in the direction of the intermediate principal stress, σ_2. The fracture criterion given by Eq. (2.4) is most easily visualised on the Mohr's stress circle as shown in Fig. 2.7 (a). The fracture plane and the shear stress acting on it are defined by the Mohr's stress circle that just touches the line defined by Eq. (2.4). There are two possible planes inclined at angles of $\pm(\pi/4+\phi/2)$ to the direction of the numerically largest principal stress, where $\phi = \tan^{-1}\mu$, as is indicated in Fig. 2.7 (b). It is observed that under triaxial compression the fracture plane is inclined at an angle of less than 45° to the maximum principal stress. In 1900 Mohr suggested a more general version of Coulomb's criterion where

$$\tau_n = f(\sigma_n). \tag{2.5}$$

Instead of the envelope of the Mohr's stress circles that represent the critical stress condition being a straight line it is a curve decreasing slope in slope as the compressive stress across a plane increases.[24]

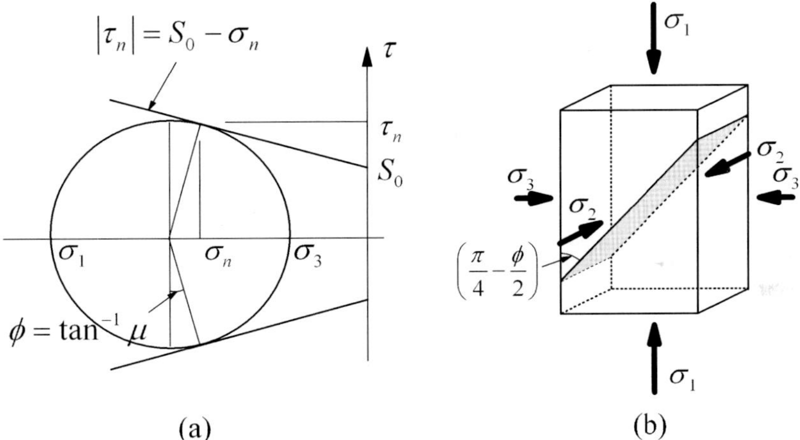

Fig. 2.7 Coulomb's criterion for compressive fracture: (a) Mohr's stress circle. (b) fracture plane.

The easiest triaxial compression test is to apply a compressive stress to a cylindrical specimen which is also subjected to fluid pressure to its surfaces so that the stress system is $\sigma_1 > \sigma_2 = \sigma_3$. Theodore von Kármán (1881–1963) was a student of Ludwig Prandtl (1875–1953) at Göttingen University when he made some of the first such triaxial tests on Carrara marble in 1911.[25,26] Another of Prandtl's students, Böker, repeated the experiments by first applying the compressive force and then increasing the fluid pressure until fracture so that the stress system was $\sigma_1 = \sigma_2 > \sigma_3$.[27] The two sets of results were significantly different, showing that the intermediate principal stress does affect fracture. Although these results were widely known, they were not seen as a challenge to the criteria of Coulomb and Mohr which do not depend on the intermediate principal stress. Not until the late 1960s, when Hidebumi Mogi at Tokyo University designed a special testing machine, was the effect of the intermediate principal stress thoroughly studied.[28] Mogi's experiments on Dunham dolomite showed that the intermediate principal stress had a significant effect on the compressive strength with an increase in strength when the intermediate stress was greater than the minimum principal stress. Mogi found that the empirical equation

$$\sigma_e = F\left(\sigma_{m,2}\right), \tag{2.6}$$

where the equivalent von Mises' stress, σ_e, is defined in Eq. (1.22) and $\sigma_{m,2}$ is defined as

$$\sigma_{m,2} = \left(\sigma_1 + \sigma_3\right)/2. \tag{2.7}$$

However, it is not clear why the equivalent stress should depend upon $\sigma_{m,2}$ rather than the hydrostatic stress. Bezalel Haimson, who has written a comprehensive review of the effect of the intermediate principal stress, has shown that strength of Westerly granite and amphibolite also follow Eq. (2.6).[29] However, for the very fine-grained brittle rocks, hornfels and metapelite, and probably other fine-grained rocks, the compressive strength is independent of the intermediate principal stress.[29]

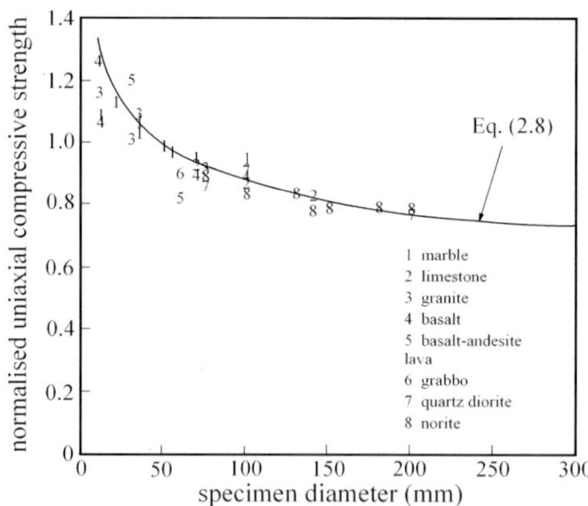

Fig. 2.8 Uniaxial compressive strength of a range of rocks normalized by the strength of a 50 mm diameter specimen (after Hoek and Brown 1997).

In intact rock there is a significant size effect and it has been suggested that the uniaxial compressive strength, σ_{cd}, for a rock specimen of diameter d, is related to the uniaxial strength, σ_{c50}, of a rock specimen 50 mm in diameter by[30]

$$\sigma_{cd} = \sigma_{c50} \left(\frac{50}{d} \right)^{0.18} . \tag{2.8}$$

Eq. (2.8) is compared with data from a range of rocks in Fig. 2.8.

On a geological scale, rock is not intact and contains joints or dividing planes. Unlike faults, joints are fissures where little movement has taken place across the blocks. Joints much reduce the compressive strength of rocks and there are empirical schemes, designed for rock engineers, for estimating the strength of rock masses that contain joints.[31]

2.4.2 Modelling the compression fracture of rocks

The mechanics of compression fracture in rocks are more complex than the mechanics of tension fracture because even in brittle rocks the first crack does not lead immediately to unstable fracture. Hence classic linear elastic fracture mechanics (LEFM) cannot accurately model compression fractures. Nevertheless the application of LEFM has enabled the qualitative behaviour of compression fractures to be understood to some extent. Alan Arnold Griffith (1893–1963), who laid down the foundations of fracture mechanics in 1920 and whose work will be discussed fully in §8.2.3, used Inglis' solution for the stresses around an elliptical hole[32] to compare the fracture strength of brittle specimens in tension and compression containing an infinitesimally thin elliptical hole orientated so that the stress concentration is a maximum.[33] Griffith assumed that the slit remained open in both cases. Under a biaxial stress system, $\sigma_1 > \sigma_2$ with the plane of the slit in the third direction he showed that, assuming a critical stress criterion, if $3\sigma_1 + \sigma_2 > 0$ a slit orientated normal to the maximum principal stress σ_1 gave the lowest strength, σ_c, and that fracture occurred when $\sigma_1 = \sigma_c$, and if $3\sigma_1 + \sigma_2 < 0$ the lowest strength was obtained when the slit was at an angle to the maximum and fracture occurred when

$$\left(\sigma_1 - \sigma_2 \right)^2 + 8\sigma_c \left(\sigma_1 + \sigma_2 \right) = 0. \tag{2.9}$$

Hence Griffith's theory predicted that the strength of a brittle rock in compression should be eight times that in tension. As can be seen from Table 2.2 the compressive strength of rock is 11–14 times its tensile strength. There are four reasons why Griffith's theory failed to give an accurate estimate of the compressive strength:

(a) Though a maximum stress criterion and an energy criterion for fracture in tension are equivalent, under compression the two criteria are not equivalent.[34]

(b) The early stages of crack growth in compression are stable.

(c) Under compression, cracks close and transmit stress across their faces.

(d) Many microcracks occur in compression before unstable fracture.

However, despite these problems Griffith's approach highlights the difference between tension and compression and gives a result that is not too inaccurate. In compression, fracture of brittle rock is due to the secondary tensile stresses whereas in tension it is the primary stresses that cause fracture. Frank McClintock and Joseph Walsh overcame the second of the problems with Griffith's approach by allowing the crack to close and transmit normal compressive stress and assuming that Coulomb friction occurs across the faces.[35] The predictions of this model are equivalent to Coulomb's prediction given by Eq. (2.4).

Chan'an Tang has developed a successful finite element model of rock fracture, Rock Failure Process Analysis (RFPA^{2D}).[36] In this model Tang uses an ideal elastic-brittle constitutive law for the local material where reduction in the material parameters after element failure which simulates the strain-softening behaviour of rock. Heterogeneity is introduced in the model by the use of Weibull's distribution for the strength of the elements. The details of this model are beyond the scope of this book, but John Harrison and his student Shih-Che Yang have reviewed this model and other similar ones.[37]

2.5 Ice

Ice sheets in Antarctica and Greenland still cover vast areas of the earth, even though they are shrinking due to global warming, and during the ice ages they covered very much more. The concept of an ice age was formed in the nineteenth-century on the basis of evidence that the Swiss glaciers were once much larger. In 1840, Louis Agassiz (1807–1873) a Swiss-American geologist and zoologist, published a book *Étude sur les glaciers* in which he concluded that Switzerland had in the past been another Greenland covered in ice. We now know that there have been at least four major ice ages in the earth's past. The earliest ice age is believed to have existed in the early Proterozoic age between about 2.7 and 2.3 billion years ago. Another ice age occurred 800–600 million years ago during the Cryogenian period. During the Carboniferous and early Permian periods 350–260 million years ago there was another major ice age. The last ice age began about 40 million years ago during the Eocene and intensified during the Pliocene. Since then, there have been cycles of glaciation with ice sheets advancing and retreating. The last glacial period ended about 10,000 years ago, but still 10% of the earth's land surface is covered by glaciers. In between the ice ages there have been periods when the earth was much warmer. Many factors have caused the cyclic variations in the temperature of the Earth including the composition of the atmosphere, changes in the Earth's orbit about the Sun, motion of the tectonic plates, impact of large meteorites, and eruption of super volcanoes. It is only during the last sixty years or so that the activities of man have had a significant effect on the climate. During the twentieth-century the near surface temperature rose by about 0.6°C, largely due to man-made carbon dioxide emissions. Between 1990 and 2100 it is estimated that the surface temperature will increase by 1.4–5.8°C.

2.5.1 Glaciers

Fracture plays a part in the dynamics of glaciers. On glaciers old snow is compressed by new snow and turns into ice granules which have connecting pores. As the ice granules become buried deeper the porosity decreases and the ice granules grow into large crystals. This process can take several years and even decades in Antarctica and high altitude Greenland glaciers where there is no surface melting. Ice creeps like rock but at a very much faster rate, so that the centre of a glacier flows faster than its edges. The surface of a glacier also flows faster than at the bedrock interface. The result of the different flow rates is to cause the tensile stress to build up in the direction of flow and cause transverse crevasses to form which can be very deep because the fracture toughness of ice is very small at about 120 kPa$\sqrt{m}$, which is only one-sixth the toughness of glass.[38] Longitudinal crevasses can also form where a glacier spreads out into a valley causing tensile stress perpendicular to the flow. At first the crevasse opening is quite small but can open to several metres by viscous flow. Crevasses can be very deep but not bottomless because the hydrostatic compressive stress, σ_h, induced by the weight of ice above increases with depth and is given by

$$\sigma_h = \rho g h, \tag{2.10}$$

where ρ is the density of the ice, g is the acceleration due to gravity, and h is the depth. This hydrostatic compressive stress reduces the effect of the tensile stress, σ_t, caused by the viscous flow. An approximate analysis can be made of the possible depth, h of a crevasse by assuming that the tensile stress is independent of depth. The stress intensity factor K at the tip of a crevasse is given by [39]

$$K = \left[1.12\sigma_t - 0.683\rho gh\right]\sqrt{\pi h}. \tag{2.11}$$

To give a qualitative feel to the possible depth of crevasse we equate this stress intensity factor to the fracture toughness of ice to obtain the tensile stress necessary to create a crevasse of depth h shown in Fig. 2.9. Small tensile stresses are capable of creating deep crevasses. What is interesting is that for a crevasse about 6 m deep the necessary tensile stress is a minimum. For a deeper crevasse the stabilising action of the hydrostatic stress becomes larger and limits the depth of a crevasse; shallower crevasses are not possible because once initiated they must propagate unstably to at least the depth at which the necessary driving tensile stress is a minimum. It would be interesting to study the distribution in the depth of crevasses on a particular glacier since this simple analysis suggests that the minimum depth should be about 6 m.

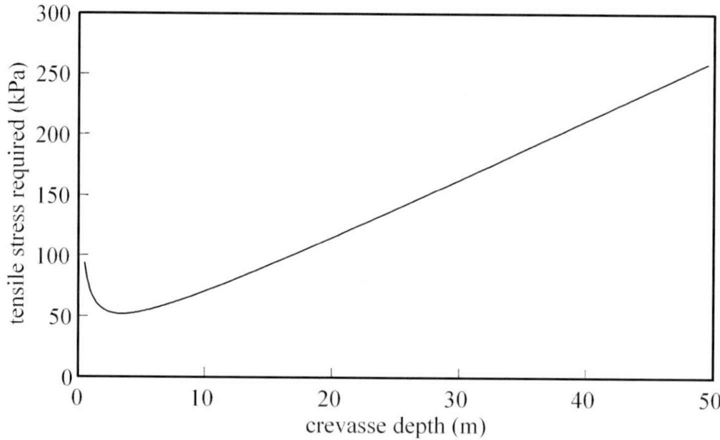

Fig. 2.9 Tensile stress required to form a crevasse of a given depth.

Lakes can be formed when a stream from a side valley meets a glacier. Usually these ice-dammed lakes fill up in summer until the head of water is sufficient to allow the water to escape under the ice. The Pategonian Perito Moreno Glacier flows over the L-shaped Lago Argentino and periodically forms a dam that separates the two arms of the lake. At its terminus the Perito Moreno Glacier is 5 km wide and 60 m above the surface of the lake. The water level on the Brazo Rico side of the lake can rise 30 m above the level on the other side. The pressure created causes a very spectacular fracture in the ice-dam on average every four to five years. The last rupture occurred in July 2008. A view of the glacier in March 2005 during the build up of the ice-dam is shown in Plate 3.

2.5.2 Icebergs

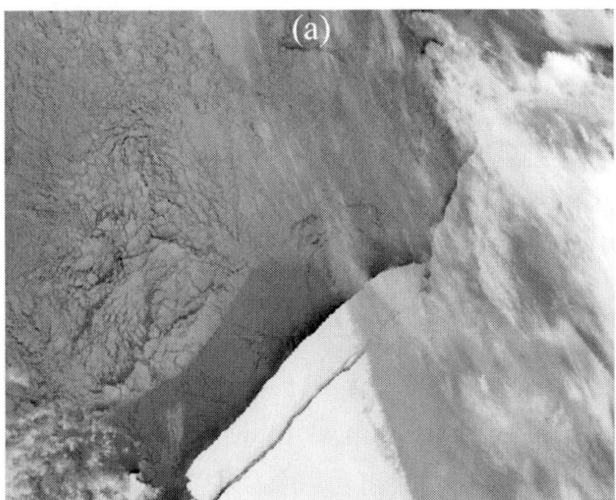
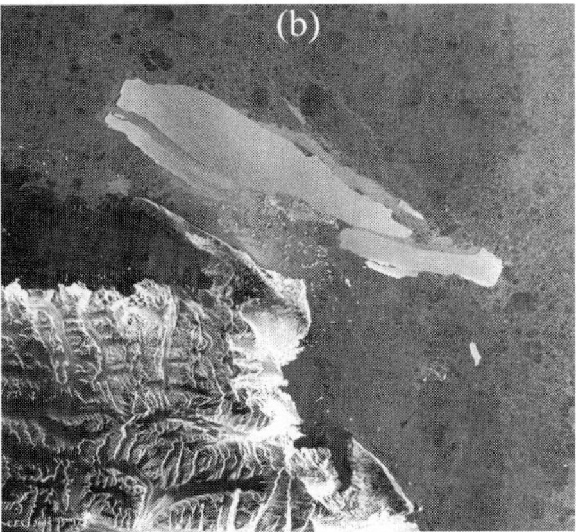

Fig. 2.10 (a) The calving of the B-15 iceberg from the Ross Ice Shelf in March 2000; on calving the iceberg was about 300 km long by 40 km wide. (b) The break-up of B-15A the largest section in October 2005 (courtesy European Space Agency).

Icebergs break off ice sheets as they are pushed out into the ocean in a process called calving. Although Arctic icebergs are the most dangerous to shipping, those formed in the Antarctic are the largest. The Ross Ice Shelf is the largest in Antarctica and Fig. 2.10 (a) shows the calving of one of its largest icebergs, the B-15 in March 2000, which was nearly 300 km long and 40 km wide. The mechanism of the calving and break-up of large icebergs is not fully understood. However, it has been suggested that sea swell induced vibrations cause fatigue and fracture at weak spots. The final break-up of B-15A the largest remaining section of the B-15 in October 2005, shown in Fig. 2.10 (b), has been traced to a storm in the Gulf of Alaska and the arrival of swell six days later in the Ross Sea.[40] Immediately prior to its break-up the B-15A grounded on shoals adjacent to the Possession Islands and suffered longitudinal fractures at the point of grounding. Earlier, in February 2005, the second most powerful vibrations in

B-15A were traced to Typhoon Olaf a category five hurricane that passed near to the Cook Islands.

Ice is a quasi-brittle material with a large fracture process zone, though its fracture toughness is comparable to glass. LEFM is only valid if the fracture process zone is small compared with the specimen size. The characteristic length,[41] l_{ch}, of sea ice is about 0.25 m and the apparent fracture toughness is not an accurate measure of the true fracture toughness until the specimen size is about $12l_{ch}$.[42] Large-scale fracture toughness tests in ice are difficult; fatigue tests are even more difficult. The main problem is that to predict crack growth rates in sea ice on the scale of Antarctic icebergs, very large specimens are needed. When the crack size is of the order of the characteristic length or smaller, one is not in the LEFM regime and instead of fatigue cracks growing at ever increasing rates, the crack growth rate decreases and the crack finally arrests.[43] Thus there are no current data on fatigue that can be used to judge the suggestion that fatigue plays a large part in the calving of icebergs and their break-up.

2.6 Concluding Remarks

We live in a dynamic earth that is ever-changing though at a slow pace so that even over many generations the change is not readily apparent, except in isolated catastrophic events. The changes cause the earth's crust to be stressed. If the stresses are small they can be accommodated by elastic deformation. Larger stresses either cause the crust to deform permanently with time or build up until the crust fractures. It is the balance of deformation and fracture that has fashioned the Earth as we know it today.

2.7 Notes

[1]   The half life of U238 is 4.5 billion years and that of U235 is 713 million years.
[2]   Because the earth was originally molten, its age cannot be accurately determined from terrestrial rocks.
[3]   The original geological eras proposed by Arduino in about 1759 were: Primary, Secondary, and Tertiary. In 1829 Desnoyers added the term Quaternary. The Tertiary and Quaternary eras correspond to the Cenozoic in The International Stratigraphic Chart. The division between the Tertiary and Quaternary Eras is less consistent. Dates from 3.6 to 1.8 million years ago have been used as the division between Tertiary and Quaternary Eras.
[4]   Herodotus, *The Histories*, II,13.
[5]   Needham and Wang (1959).

[6] Giraud (1780).

[7] Lyell (1830–33), twelve editions of the work were published between 1839 and 1876.

[8] Hess (1962).

[9] Adams and Nicolson (1900).

[10] The factor 3 occurs because the fundamental equation of creep is for the shear strain rate.

[11] Talbot (1999). Note the conversion of the viscosity measurement of Ito (1979) from poise to Pa s by Talbot is incorrect. In Talbot (1999) the estimates are actually an order of magnitude larger than the value given by Ito.

[12] Ito (1979).

[13] Obviously there can be a combination of normal and shear stress acting across a plane, but here only the archetypal systems are considered.

[14] Doolittle (1693).

[15] A chimney fell and a crack was opened in St Peter's Church in Colchester. Doolittle rather incongruously links this earthquake with the two great disasters in England during the seventeenth century: the Great Plague of 1665 and the Fire of London 1666.

[16] Doolittle imaged a benign God using the earthquake to bring his people back to the true path without retribution. Some modern day pastors unfortunately believe in a more vengeful God. The widespread bush fires in Victoria, Australia on the 7th February 2009, which led to many deaths and much destruction, were seen by a leader of the Pentecostal Church as just retribution for the decriminalisation of abortion in Victoria.

[17] Needham and Wang (1959).

[18] Murray and Segal (2002).

[19] Geller *et al.* (1997).

[20] Notice that there is much less difference between the fracture toughness of the rocks.

[21] Hallbauer *et al.* (1973).

[22] Coulomb (1776).

[23] Since tension is taken as positive, a compressive stress increases the critical shear stress.

[24] Mohr (1900).

[25] Kármán (1911).

[26] Prandtl and von Kármán are better known for their work on fluid dynamics.

[27] Böker (1915).

[28] Mogi (1967, 1971).

[29] Haimson (2006).

[30] Hoek and Brown (1980).

[31] Hoek and Brown (1997).

[32] See §8.2.2.

[33] Griffith (1924).

[34] Cotterell (1969, 1972).

[35] McClintock and Walsh (1962).

[36] Tang (1997).

[37] Yuan and Harrison (2006).

[38] Petrovic (2003).

[39] Tada *et al.* (1985).
[40] MacAyeal *et al.* (2006).
[41] See §1.5.2.4.
[42] Mulmule and Dempsey (2000).
[43] Bond and Langhorne (1997).

Chapter 3

Evolution of Life

The great geneticist, Ukrainian-born Theodosius Dobzhansky (1900–1975) asserted the importance of organismic biology, writing that: 'Nothing in biology makes sense except in the light of evolution.'[1] Certainly the strength and toughness of biological materials can only be explained by evolution. The first traces of life date to about 3.5 billion years ago. At first life was confined to the sea where plants remained until around 500 million years ago when a sufficient protective ozone layer had formed in the upper atmosphere and life could expand to the land. There was a substantial diversification of land plants during the Devonian period, 408–362 million years ago, and the first real trees appeared by the late Devonian period. During the Carboniferous period, 362–290 million years ago, conditions were warm and plants multiplied, but they mostly reproduced by spores. The first conifers appeared at the end of the Carboniferous period. Life on the land suffered a set back during the Permian period, 299–251 million years ago, when the climate became much colder and drier with widespread glaciation in the southern hemisphere. Flowering plants, angiosperms, first appeared in the Cretaceous period, 146–66 million years ago.

The mechanical properties of plant tissues depend upon the volume fraction of the cell walls. Pliable leaves have a low cell wall volume fraction, while on the other hand dense wood has a high cell wall volume fraction. There are many important mechanical properties of tissues apart from fracture, but tissues have had to evolve sufficient fracture resistance for their purpose. Leaves and grasses have needed to evolve defences against vertebrate and invertebrate animals, very frequently these defences are chemical, but the toughness of leaves and grasses is an important defence mechanism. For example, the toughness of maize leaves is an important defence against the European corn borer.

Animals first evolved, like plants, in the sea. The largest phylum, the arthropods, which includes insects and crustacea, has a segmented body with an exoskeleton and constitute 80% of extant animals. This phylum had their roots in the extinct trilobites, which evolved in the early Cambrian period, 542 to 488 million years ago. The first vertebrates, fish, also evolved during the Cambrian period about 530 million years ago. Animals probably first ventured on land around 375 million years ago. The move to land necessitated the development of strong exo- or endoskeletons. In the Carboniferous period the early reptiles moved away from water, helped by the evolution of the amniotic egg. Dinosaurs first appeared in the Triassic period 251–200 million years ago, and died out catastrophically at the end of the Cretaceous period. Placental mammals probably evolved in the late Cretaceous period and there are mammalian fossils from nearly 200 million years ago.

Animal tissues can be soft, such as skin, or hard like the cuticle in an insect's mandible, or bone in a vertebra, and as far as fracture behaviour is concerned, can be divided between those tissues having little or no minerals, and those tissues having a high volume fraction of minerals. As with plant tissue, fracture properties are only one of the mechanical properties of animal tissue. However, sufficient fracture toughness is always needed. Skin needs to be pliant, but it also must not tear easily. Exo- and endoskeletons must have sufficient stiffness to support the body, but they must be sufficiently tough not to fracture too readily.

Nature has evolved complex composite tissues to satisfy the necessary physiological and mechanical functions. Man has had to develop the materials he uses. At first man and his hominid ancestors made use of the materials he found around him: wood, bone, and stone, but since some 16,000 years ago, when the Jomon pottery was first made in Japan, he started to develop his own. However, it was not until the second half of the twentieth-century that the mechanical properties of materials were tailored by combining particles and fibres in various matrices to form high-performance composites using part of nature's philosophy. Nature's materials have hierarchical structures built from the atomic scale through the nano- and micro- to the macroscale, with the structure at each level being a composite in its own right. In recognition of the power of evolution, material scientists are now studying biological materials to produce biomimetic materials that mimic nature's design for engineering applications.

3.1 Biocomposites

Homogeneous materials have to rely on their innate toughness to resist fracture. In the absence of extraneous non-elastic work associated with a fracture, the theoretical energy necessary to separate two surfaces of unit area is very small. Glass is almost an ideally elastic-brittle whose toughness not much larger than the theoretical for an ideal solid. Metals are usually tough, but the surface energy of iron is only about twice that of glass. Metals owe their toughness to the plastic deformation that is necessary to produce a fracture. The thermoplastic polymers such as polyethylene are reasonably tough and biological tissues contain natural polymers: proteins are polymers of amino acids and polysaccharides are polymers of sugars, but they are too pliant to form a useful tissue on their own.

It is a general rule that strength and toughness are inversely related. What nature discovered by evolution was that this inverse relationship between strength and toughness can be beaten if strong fibres or plates are embedded in a soft matrix. Marc Meyers and his colleagues have written an excellent review of biocomposites which gives more details than is possible in this brief summary.[2]

3.1.1 Stiffness

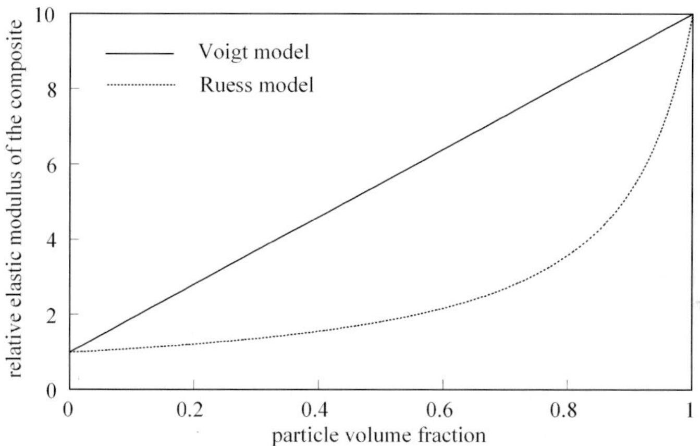

Fig. 3.1 The ideal Voigt and Reuss limiting moduli for a composite for $E_p/E_m = 10$.

The stiffness of a composite lies between that of the reinforcing particles and the matrix, and is governed by the rule of mixtures. There are two limiting ideals to the rule of mixtures. If the composite consists of aligned continuous fibres or plates with an elastic modulus E_p, embedded in a matrix with an elastic modulus, E_m, where the volume fraction of the fibres or plates is v_p. The elastic modulus of the composite, E_c, in the direction of the fibres is given by the Voigt model

$$E_c = E_p v_p + E_m \left(1 - v_p\right).$$

(3.1)

The elastic modulus the composite normal to the fibres or plates is given by the Ruess model

$$\frac{1}{E_c} = \frac{v_p}{E_p} + \frac{\left(1 - v_p\right)}{E_m}.$$
(3.2)

The two limiting values of the composite elastic modulus as a function of the reinforcing volume fraction are shown in Fig. 3.1 for a hypothetical composite where the elastic modulus of the particle is ten times that of the matrix. The actual modulus of a composite normally is between these two ideal values. However, the toughness of a composite is not generally governed by the rule of mixtures.

3.1.2 Toughness

High toughness can be achieved in composites by a number of synergistic mechanisms. For reinforcing particles in the form of macro-sized fibres and plates the main toughening mechanism is the work of pull-out of the particles – it is essential that the bond between the particles and the matrix is not too strong so that the particles pull out rather than fracture. Thus fibreglass (glass fibre reinforced epoxy composites) has a typical toughness of about 10,000 J/m^2, but the toughness of the epoxy is only of the order of 100 J/m^2 and the toughness of glass is only about 10 J/m^2. The toughness of biocomposites is usually achieved through much more complex and hierarchical structures than man-made composites.

The toughness requirements for biological materials to resist tearing or fracture vary depending upon the application. For resistance to impact, as in the case of the horns of bovids or the antlers of deer, high fracture energy, R, is required. However, for pliant materials such as skin and leaves to resist large deformations without tearing or for seed cases and bones etc. to resist large forces without fracturing, the criteria are different. Not all biological materials are linearly elastic, but to compare a range of different materials we can assign an equivalent elastic modulus. The energy release rate, G, for any component whether biological or not can be written either in terms of the force, F, being applied or the deformation, δ, by the equations

$$G = \lambda_P \frac{F^2}{l^4} \frac{a}{E} = \lambda_\delta \left(\frac{\delta}{l}\right)^2 Ea,$$
(3.3)

where l is a characteristic length of the component, a is the tear or crack length, and λ_p and λ_δ are constants. To prevent tearing or fracture the energy release rate, G, must be equal or greater than the fracture energy, R, of the material. Hence

$$F < \frac{l^2}{\sqrt{\lambda_p a}} \sqrt{RE}, \text{ or}$$

$$\delta < \frac{l}{\sqrt{\lambda_\delta a}} \sqrt{\frac{R}{E}},$$

$$(3.4)$$

if tearing or fracture is to be avoided. Therefore the parameter that is important for impact controlled fracture is R, for force controlled fracture it is $\sqrt{RE}$ and for displacement controlled fracture it is $\sqrt{R/E}$.

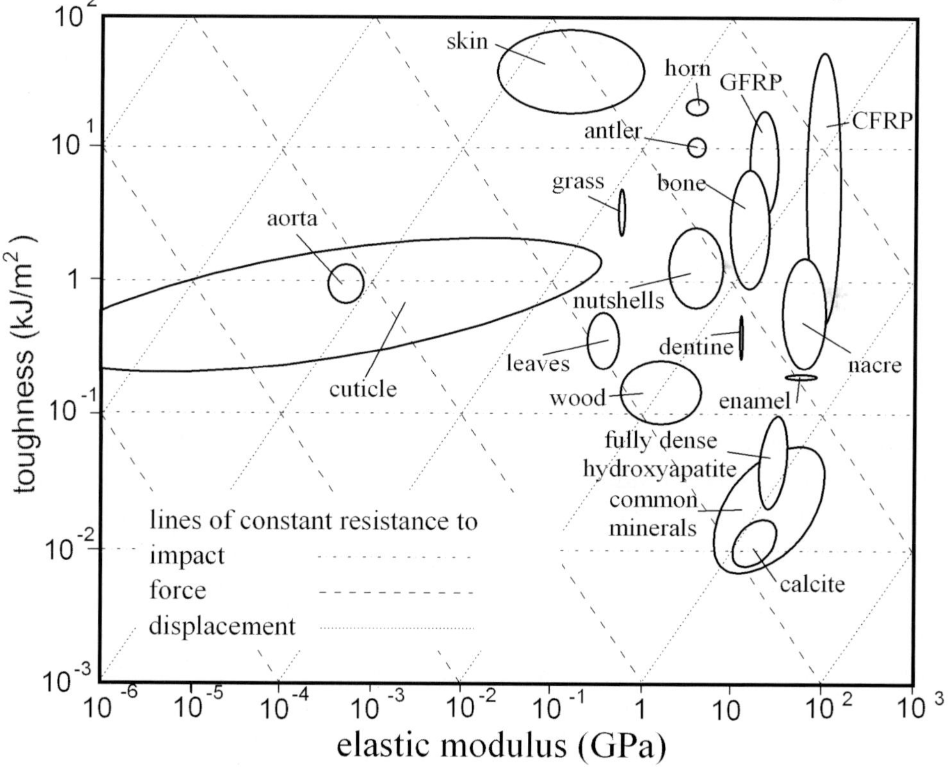

Fig. 3.2 Toughness of a range of biomaterials and some engineering composites as a function of elastic modulus for a range of materials (after Vincent and Wegst 2004).

Toughness and stiffness are the most important properties of materials biological or otherwise and maps showing the relationship between them have been constructed, such as that shown in Fig. 3.2. Lines of constant resistance to impact, force, and displacement controlled fracture are superimposed on the toughness-stiffness map.

3.2 Plant Tissues

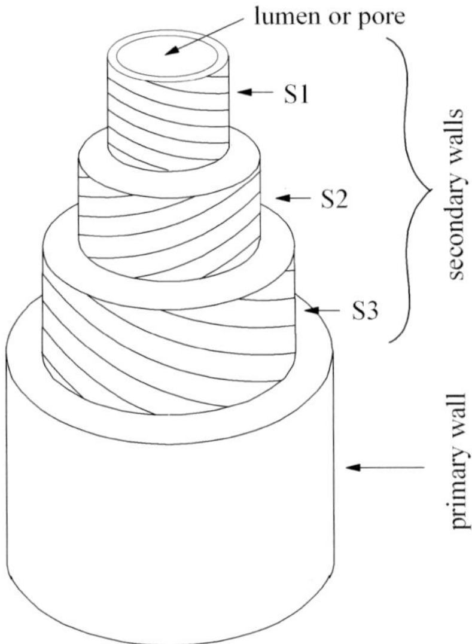

Fig. 3.3 Schematic cell structure of wood.

Both plants and animals have a cellular structure. In almost all plants the cell wall is composed of cellulose, a high polymer made from glucose, which crystallises into microfibrils of high stiffness and strength.[3] Animal cells generally lack cellulose and their cell walls are more pliant unless reinforced by mineral particles. The cellulose microfibrils make up about 50% of the cell wall in plants. The matrix of the cell wall is composed of lignin, an amorphous polymer similar to the epoxy in fibreglass, and hemicellulose, a semicrystalline polymer of glucose. The cell wall is not homogeneous but is composed of three main layers. The orientation of the cellulose microfibrils in these layers determines the mechanical properties of the cells. The main cells in woods are tracheids in hard woods and fibres in softwoods. The stiffness and strength in the grain direction comes from the alignment of the microfibrils. If all the microfibrils were aligned along the grain of the wood the toughness of wood to a crack propagating across the grain would be about an order of magnitude less than it is. However, the microfibrils are wound helically about the lumen or pore like a spring. In the first and third secondary walls (S1 and S3) of the cell, the microfibrils are wound as in a left-handed thread and in the S2 wall they are wound as in a right-handed thread as shown in Fig. 3.3.[4] The increase in toughness comes from the S2 layer which forms the bulk of the secondary wall of the cell wall where the microfibrils' angle to the cell axis, called the *microfibril angle*, is in the range 25–50°.[5] The spirally wound microfibrils are like springs bonded together. When a spring is stretched, the wire forming the spring is twisted. Similarly, when a cell is stretched the microfibrils split away from each other under the torsional shear. The splitting of the microfibrils absorbs energy and increases the toughness in itself. There is also a recovery mechanism whereby the amorphous matrix between the microfibrils within the cell wall is re-formed causing a slip-slip mechanism.[5] The split cell wall can also buckle inwards absorbing more energy.[6] The toughness of both soft and hardwoods is increased by these mechanisms to about 30 kJ/m^2.[7] Thus trees have evolved sufficient toughness to resist storms. If a tree is blown down in a storm it is usually up-rooted; rarely does the trunk fracture. Healthy branches of trees can of course break during a storm, but a tree can lose branches without dying so that branches did not have to evolve sufficient strength and toughness to prevent some of them fracturing in a storm. In any case, branches usually fail at their comparatively weak junction with another branch or the trunk which is also the point of most stress. Trees have also evolved another mechanism to strengthen their trunks and branches.

The strength of wood in compression is only about a third of its strength in tension. Under compression, wood buckles to form macroscopic creases, which lead to failure at much lower stresses than in tension.[8] Evolution has allowed for the lower strength in compression of trees by causing them to grow pre-stressed so that the outer layers of the trunk are under tension when the tree is under no load.[8] The tensile pre-stress can be more than 10 MPa. In all pre-stressed situations, since there is no applied load, there must be compressive stresses as well as tension for equilibrium and the core of the trunk is in compression. Under wind load the tension on the windward side of the trunk is greater than it would be if the trunk was not pre-stressed, but the compressive stress on the leeward side is some 10 MPa less. Since the tensile strength is so much more than the compressive strength, evolution has discovered how to compensate for the weakness in compression.

Grasses and leaves do not need to have particular toughness to resist fracture because they are thin and pliable. Bending a leaf back on itself induces a surface strain of 50% but, since the elastic modulus is low, the corresponding strain energy stored is small and the toughness does not need to be high to prevent tearing.[9] On release of the load, the leaf springs back into shape.

As has already been mentioned, leaf toughness can be a defence against herbivores, especially insects and other invertebrates. However, there are evolutionary forces working to improve tooth or mandible forms and jaw muscles, so that tougher plants can be exploited by herbivores as well as the evolution of tougher plants. Fruit is different, because evolution has designed fruit to be eaten so that its seeds can be dispersed. Flowering plants are divided into two groups: *monocotyledons* (monocots) whose leaf veins are mostly parallel and include grasses, palms and lilies, and the much bigger group, the *diocotyledons* (dicots), whose leaf veins reticulate from the stem. The grasses and other monocots are generally tougher than the herbaceous dicots, while the toughness of the leaves of woody plants (trees and shrubs) depends on whether they are new or mature. The relative toughness of some plant classes is shown in Table 3.1. There are two photosynthetic pathways C_3 and C_4 where the number refers to the number of atoms of carbon in the first product during the assimilation of carbon dioxide. The C_3 pathway is the more ancient of the two. Between about 7 and 5 million years ago during the late Miocene there was a global increase in the biomass of plants using the C_4 photosynthesis pathway, which may be related to a decrease in atmospheric carbon dioxide. Table 3.1 shows that the C_4 grasses, which also contain higher silica, are twice as tough as the C_3 grasses. Equids developed *hypsodonty*, high crowned teeth, in the late Miocene, which is probably partially an evolutionary change to cope with the higher toughness of grasses.[10]

Table 3.1 Toughness of leaves relative to herbaceous dicots (after Bernays 1991).

Plant Type	Relative toughness
Palms: expanded fronds	9.8
Woody plants: mature leaves	6.3
C_4 grasses	6.2
C_3 grasses	3.1
Woody plants: new leaves	1.7
Herbaceous dicots	1

3.2.1 The fracture toughness of plant tissue

The toughness of wood can be measured using standard engineering fracture toughness tests, but not the toughness of leaves or most other biological tissues. The toughness of plant tissues is most easily measured by cutting with a guillotine, microtome, and especially scissors, which have been pioneered by Peter Lucas and his collaborators.[11] The cutting of biological tissues is essential elastic apart from a small plastic zone or fracture process zone (FPZ) at the tip of the blade. Conventionally, the fracture toughness of essentially elastic fractures includes the irrecoverable work done in the plastic zone at the crack tip which, provided the zone is small compared with the specimen, is independent of the test geometry. Cutting operations are very nearly steady-state, that is, if one were to sit on the cutting edge one would observe no change in the deformation. Thus the entire work done in cutting goes directly to supply the necessary fracture work. The fracture toughness is therefore, the cutting work divided by the area cut. In practice the method is a little more difficult than described here. There is friction so that work is done in operating a pair of scissors even if nothing is cut, which must be subtracted from the total cutting work. Another complicating factor is that leaves and other biological tissues can be quite thin so that the plastic zone is not necessarily small compared with the thickness and the plastic work at the tip of the cut depends upon the thickness of the tissue being cut. In these circumstances it is necessary to separate the plastic work at the cutting edge from the total fracture work to get the essential fracture work. The separation can be accomplished by appealing to scaling arguments similar to those employed in the essential work of fracture method.[12] If the thickness, h, of the tissue being cut is small compared with the FPZ, the width of the FPZ will be the thickness of the tissue. When a length l is cut, the volume plastically deformed is proportional to lh^2 and the plastic work, W_p, is proportional to this volume. Thus

$$W_p = \beta l h^2, \tag{3.5}$$

where β is a plastic work density. The essential fracture work, W_e, on the other hand is proportional to the area cut, lh, and is given by

$$W_e = lhw_e, \tag{3.6}$$

where w_e is the specific essential work of fracture, that is the work that intimately is associated with the separation of the tissue by cutting. Thus the specific work of fracture, $R = \left(W_e + W_p\right)\big/lh$ can be written as

$$R = w_e + \beta h. \tag{3.7}$$

However, if the thickness of the tissue is larger than the FPZ then the width of the FPZ will be roughly equal to its length and independent of the tissue thickness. Hence in this case the specific work of fracture will also be independent of the tissue thickness and given by

$$R = w_e + \beta h^*,\tag{3.8}$$

where h^* is the width of a fully developed FPZ in a thick tissue.

(a)   (b)

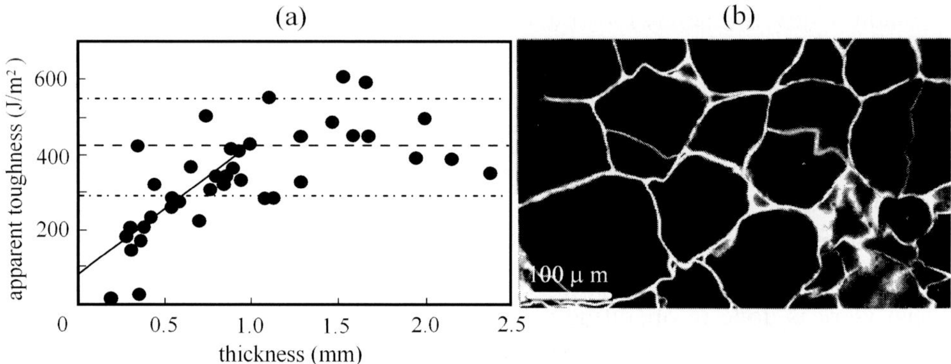

Fig. 3.4 (a) Specific work of fracture of green turnip sections as a function of the thickness; the full line is a linear regression for thickness less than 1 mm; the dotted line is the mean toughness obtained from a wedge test and the chain lines are two standard deviations from this mean. (b) Cell structure of green turnip (Lucas *et al.* 1995 with permission of the Royal Society).

In Fig. 3.4 (a) the specific work of fracture of sections of green turnip, measured by the scissors test, is plotted against the thickness. For sections thinner than about 1 mm, the specific work of fracture increases linearly with section thickness as predicted by Eq. (3.7). Above a thickness of 1 mm the specific work of fracture is roughly constant and so it can be assumed that the width of a fully developed FPZ at the cutting edge is about 1 mm. The plateau value of the specific work of fracture is very similar to the results of an independent wedge test[13] shown in Fig. 3.4 (a) by the dotted line. The intercept gives the specific essential work as 85 J/m^2. The plastic work comes mainly from the buckling of the cell wall and the slope of the line gives the factor $\beta = 341$ kPa. The cell wall fraction of green turnip (see Fig. 3.4 (b)) is very low at only 1.92% of the tissue and is the cause of the low toughness since the specific work of fracture of plant tissues depends almost entirely on the fraction of the tissue occupied by the cell wall. The specific essential work of fracture and the specific plastic work of fracture for a 1 mm thick section of a wide range of plant tissues are shown as a function of the cell wall fraction in Fig. 3.5.[14] The specific work of fracture is a function only of the cell wall volume fraction regardless of the tissue type. The specific essential work of fracture is a linear function of the cell wall volume fraction and the specific plastic work of fracture increases with cell wall volume fraction from zero to a maximum of about 30 kJ/m^2 at a volume fraction of 80–90%. At higher volume fractions the specific plastic work of fracture decreases sharply to become very small at 100%. The major part of the plastic work involved in cutting comes from the buckling of the spirally wound microfibrils in the S2 layer of the cell wall into lumen. If the cell wall is only a very small fraction of the tissue the specific plastic work is correspondingly very small, on the other hand if the cell wall occupies most of the tissue, then there is little room for buckling of the microfibrils and the specific plastic work is also very small.

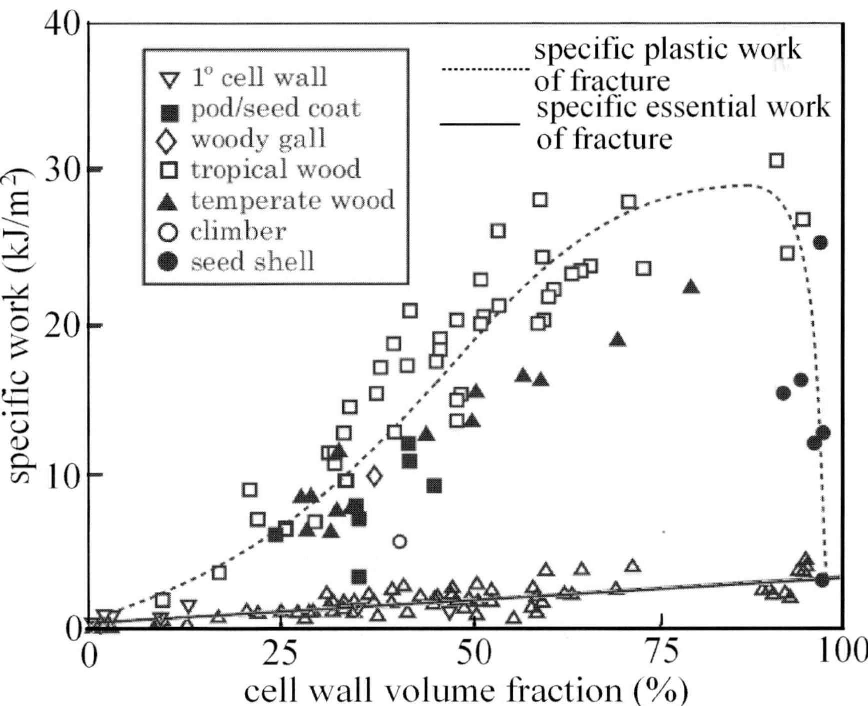

Fig. 3.5 The specific work of fracture of various plant tissues 1 mm thick as a function of cell wall volume fraction (Lucas *et al.* 2000 with permission OUP).

Seed cases have a high cell wall volume fraction and the contribution to their toughness from the plastic buckling of the cell wall is limited. However, seed cases do not rely on toughness alone to resist being opened by seed-eaters. The elastic modulus of plant tissues comes almost entirely from the cell wall and increases with cell wall volume fraction. Since seed cases have a high cell wall volume fraction (greater than 90%) they have a comparatively high elastic modulus of around 5 GPa, which can compensate for a moderate toughness since the parameter that controls the fracture force is $\sqrt{RE}$.

The harvesting of grass by vertebrates is somewhat different to leaves. In grasses and other monocots the strength comes from bundles called the *sclerenchyma*, which on maturity are highly lignified dead cells consisting of about 90% cell wall aligned along the grass blade. Mechanically, grass is like a bundle of fibres with the connecting tissue distributing the load between them. Consequently, when one of the *sclerenchyma* is cut the load is shared between the rest and the strength of the grass is only reduced in proportion to the number of *sclerenchyma* cut.[15] Such behaviour is in distinct contrast to that of a homogeneous brittle material where the strength depends on the stress intensity factor and for relatively small cuts is inversely proportional to the square root of the cut which causes a far more severe decrease in strength. In eating grass the larger animals, such as cattle, take a whole tiller of grass into their mouth at a time. Since there is little to be gained by notching the grass, the teeth of grazing animals serve primarily to grip the grass tillers[16] and cattle use their tongues to grip the grass not their teeth. The front teeth of deer rot with age, but that does not seem to affect them greatly in harvesting the grass provided they still have their back teeth for comminution of the grass.

3.3 Animal Tissues

Animal tissues can be divided into organic tissues, like skin, which are composed almost entirely of organic materials, and bioceramic tissues, like bone, which have a high mineral content. The organic tissues can vary widely in stiffness, and their stiffness and fracture toughness are also very much affected by the degree of hydration. The bioceramic tissues have high stiffness.

3.3.1 Organic tissues

Purely organic tissues vary in stiffness and hardness from the hard tissues based on chitin to the soft composites of collagen and elastin fibres and all have evolved to meet their particular function.

3.3.1.1 *Chitin fibres and cuticle*

Chitin is a polysaccharide polymer like cellulose, which was first discovered in the cell walls of mushrooms by Henri Braconnott (1781–1855) in 1811; Auguste Odier (1802–1870) named the material of insect cuticles *chitine* in 1823.[17] The chitin in cuticle is assembled into nanofibres about 3 nm in diameter and 300 nm long.[18] Although the elastic modulus of the nanofibres has not been measured, it has been inferred that they are stiffer than cellulose fibres and have an elastic modulus of at least 150 GPa.[18] Chitin fibres are the hard phase in the cuticle of arthropods and are also present in the shells of some molluscs. In cuticle the chitin is present in sheets about 6.5 nm thick in which the fibres are aligned. The orientation of the fibres in adjacent sheets changes by about 7–8 degrees so that there are about 25 sheets before the orientation repeats itself to form a kind of plywood. The sheets are bonded together by a protein that varies considerably between species of arthropods. Cuticle can have a very wide range in elastic modulus from 1 kPa up to 20 GPa.[19] There is still controversy on the mechanism responsible for the stiffening or scelerotisation of cuticle. The original suggestion made over sixty years ago is that the protein matrix is phenolically cross-linked in a process that is similar to the tanning of leather.[20] However, the dehydration of cuticle by secondary cross-linking, which was suggested shortly after the phenolically cross-linking hypothesis, is more likely to be the reason for scelerotisation.[21] When cuticle is fractured across the chitin sheets, the fracture path is deflected at each sheet into the interface absorbing more energy and the toughness of cuticle varies from 0.2 to 2 kJ/m^2. Judged on the basis of toughness alone cuticle is not particularly fracture resistant, but cuticle is good at resisting displacement controlled fracture since the controlling parameter is $\sqrt{R/E}$.

Herbivorous insects have little ability to digest plant cell walls and rely on the mechanical disruption of the cell wall to obtain nutrients. The 'teeth' of insects are integral with their mandibles, which are formed from cuticle. The form of the mandible depends upon the feeding habit, for example the caterpillars of the giant silkworm moths (family *Saturniidae*) simply snip off relatively large pieces of leaves, whereas the caterpillars of the hawk moth (family *Sphingidae*) tear away the leaf in much smaller pieces producing something that is closer to true chewing.[22] During the process of evolution the mandibles of the *Saturniidae* caterpillars have become relatively short with a broad base and no obvious 'teeth', whereas those of the *Sphingidae* caterpillars have become longer with narrower bases with two or three rows of sharp 'teeth'. Regardless of the shape of mandible they need to be hard.[23] The hardness of the 'teeth' on the mandible

                                    Fracture and Life

of a locust is about 360 MPa,[24] which is comparable to the hardness of the dentine in our own teeth, and is in the range 250–800 MPa,[25] which is certainly hard enough to cope with cutting leaves and grasses. The larvae of some beetle species chew into seeds. The hardness of seed coverings range from 170 to 270 MPa and it has been suggested that the hardness of the mandibles of these beetles is increased by high concentrations of metals such as zinc and manganese.[26]

3.3.1.2 Silk

Silk from the silk worm *Bombyx mori* is an important textile fibre and has great strength. Many other arthropods also produce silk as well as other caterpillars. However, only a few arthropods such as spiders make use of the high strength-to-weight ratio of the silk thread for their webs. For cocoon spinning arthropods the silk cocoon acts as a protection either during pupation as in the case of moths or the eggs as in the case of spiders. Since larvae are likely to be infected by parasites before they pupate, the protection of the cocoon is more against desiccation than parasitic attack, but the cocooned eggs of spiders are protected against burrowing larvae and the silk flocculent layer is a protection against ovipositors. Thus, except in the case of spiders, the high strength does not seem to be a result of direct evolution. As far as we humans are concerned the high strength of silk, not to mention its beauty, has been important to us since the third millennium BC.

Table 3.2 Mechanical properties of silk and some man-made fibres (after Elices *et al.* 2005).

Fibre		Density (kg/m^3)	Young's Modulus (GPa)	Tensile Strength (GPa)	Fracture Strain (%)	Specific Resilience (MJ/m^3)
Spider silk	Arigope trifasciata	1300	1–10	1.2	30	100
	Nephila clavipes	1300	1–10	1.8	30	130
Silkworm silk	Bombyx mori	1300	5	0.65	12	50
Man-made fibres	Nylon 66	1100	5	0.9	18	80
	Kevlar 49	1400	125	2.8	3	50
	E-glass	2560	76	1.4–2.5	2	15–25
	Carbon type 2	1750	250	2.7	2	25

Silk and spider web is formed mainly from strands of a protein called *fibroin* surrounded by an amorphous sheath of another protein called *sericin* which has the role of an adhesive.[27] Fibroin is a crystalline protein and during spinning the long linear polymer chains are aligned to give a stiff and strong crystalline thread. The mechanical properties of various silks are compared with made-made fibres in Table 3.2. The specific strength of silk, especially spider silk, is comparable to the strongest man-made fibres.[28] The resilience of silk, especially spider silk, is as good as or better than the best man-made fibres.

3.3.1.3 Tendon

Tendon, which connects the muscle to bone, needs high stiffness, so that the necessary muscle contraction is minimised, and high strength. The stiffness and strength come from collagen fibres which make up 70–80% of the dry weight of tendon. Collagen is a class of protein fibres that have a structure based on the helical arrangement which itself forms a second order helix to give a coiled-coil structure. The rope-like structure of collagen gives it high stiffness and strength. In unstressed tendons the collagen fibres are arranged in crimped bundles aligned along the tendon. A typical stress-strain curve for tendon, shown in Fig. 3.6, is similar to that of collagen. At low stress these crimps straighten and the stiffness of the tendon increase giving a J-shaped stress-strain curve. Once the fibres have straightened the stiffness remains constant until permanent non-elastic deformation occurs at strains greater than about 4%. Tendon needs to store large amounts of strain energy[29] to make running efficient. The kangaroo has an exceptional ability to store energy in its tendons, which enables it to reach speeds of up to 60 km/hr in its bounding locomotion. Although the collagen fibres are aligned to give maximum stiffness, the structure of tendon is in some ways similar to grass. The collagen fibres are connected by non-collagenous protein, which has a low stiffness. Consequently, tendon is notch insensitive. The notch insensitivity of tendon helped minimise the damage my wife did to the tendon in her little finger which she recently cut while preparing dinner. She partly severed the tendon but, because of the notch insensitivity, the cut in the tendon simply opened up rather than tore and the surgeon could suture it and she now has complete use of her finger again.

3.3.1.4 Skin

Skin is a composite of collagen and elastin fibres embedded in an amorphous polymer composed of proteins and polysaccharides in hydrated macromolecular complexes. Stress in the tissues is largely resisted by the deformation of the collagen and elastin fibres. Elastin is much more pliant than collagen with an elastic modulus of about 0.6 MPa and can be stretched elastically to more than twice its length. In skin the collagen fibres form a felt-like structure, but under strain they gradually become orientated in the direction of the strain. Before orientation of the collagen fibres, the stress is taken primarily by the low modulus elastin but at high strains the oriented stiffer collagen fibres increasingly take up the stress to give the distinctive J-curve for cat skin shown in Fig. 3.6, which makes skin difficult to tear.[8] Most animal skins have a well developed J-shaped stress-strain curve, but rhinoceros do not. The toughness of some animal skins is shown in Table 3.3.

Table 3.3 The toughness of some animal skins.

Animal	Skin location and direction	Test method	Toughness kJ/m^2	Reference
Human	Hand	Scissors	2.5	Pereira *et al.* (1997)
Rat	random direction	Scissors	0.6	Pereira *et al.* (1997)
	back longitudinally	Trouser	25	Purslow (1983)
	back circumferentially	Trouser	20	Purslow (1983)
	belly longitudinally	Trouser	15	Purslow (1983)
	belly circumferentially	Trouser	8	Purslow (1983)
Rabbit		Trouser	20	Atkins & Mai (1985)
White rhinoceros	back, deep	Trouser	43	Shadwick *et al.* (1992)
	Back, superficial	Trouser	78	Shadwick *et al.* (1992)
Wild ass	flank, deep	Trouser	33	Shadwick *et al.* (1992)

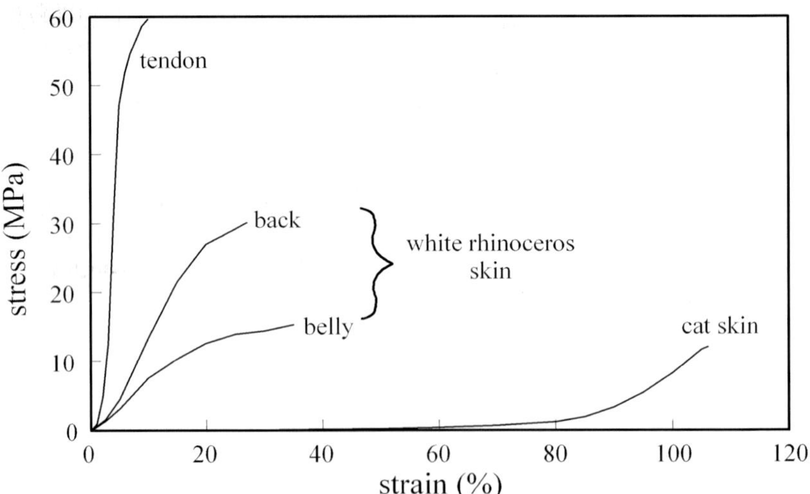

Fig. 3.6 Stress-strain curves for tendon, rhinoceros, and cat skin (data from Vincent 1990, Veronda and Westmann 1970, Shadwick *et al.* 1992).

The effect of the J-curve on the notch sensitivity can be estimated from an extension of fracture mechanics developed for rubber.[30] An approximate large strain expression for the energy release rate, G, for a crack or tear of deformed length $2a$ in an infinite sheet under uniaxial tension given by Graham Lake is[31,32]

$$G = 2\pi a U, \tag{3.9}$$

where U is the strain energy density away from the crack, which is simply the area under the tensile stress-strain curve. The deformed crack length is

$$2a \approx 2a_0 \lambda_x, \tag{3.10}$$

where $2a_0$ is the undeformed crack length and λ_x is the stretch ratio in the crack direction. Rubber is isotropic and deforms elastically with almost no change in volume so that the stretch ratio λ_x is given in terms of the stretch ratio normal to the crack, λ_y, by

$$\lambda_x = 1/\sqrt{\lambda_y}. \tag{3.11}$$

However, skin is anisotropic[33] and there is little change in thickness when it is stretched so that

$$\lambda_x = 1/\lambda_y. \tag{3.12}$$

The anisotropic nature of skin provides a small contribution to its tearing resistance. The critical tear length can be found by equating the energy release rate, G, to the fracture energy, R. The toughness of cat skin was not measured, but Table 3.3 gives the tear toughness of various other animal skins. The length of the critical tear length as a function of the strain has been calculated assuming that the toughness of cat skin is roughly the same as rat skin and is 25 kJ/m^2 and is shown in Fig. 3.7. Not until the strain is greater than about 80% will a tear of less than 50 mm propagate so hence the skin is very difficult to tear despite not having an exceptional toughness. If the cat skin deformed isotropically it would be marginally easier to tear. However, if the stress-strain relationship of cat skin did not have a J-curve, but was linear elastic and had the same strength at a strain of 100% as the real skin, then Fig. 3.7 shows that it could tear easily at strains as small as 30%. Your loved moggy would not survive his fights on the tiles at night if his skin was linear elastic. Skin can develop high strains and strength without storing a large strain energy which can provide the energy necessary for fracture and hence is difficult to tear.

It is also interesting to see from Table 3.3 that the toughness of the skin on the back of an animal is significantly greater than that on the belly. In fights an animal's belly is usually protected by a crouching posture – evolution is economic.

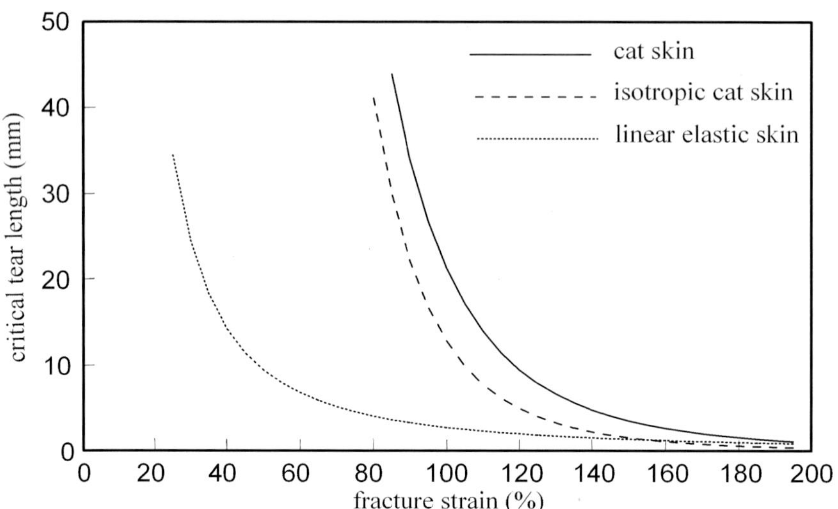

Fig. 3.7 Critical tear length in a cat's skin as a function of the applied strain.

The reader will notice that there is more than an order of magnitude difference between the toughness of rat skin measured by the scissors test and that obtained from a tear test shown in Table 3.3. The difference is so large that it cannot come from errors in determining the toughness of the skin. Surprisingly the large difference between the toughness measured by cutting and tearing has not received the attention it deserves. What is happening? It is claimed by fracture mechanists that toughness, measured correctly, is a material property but here are two very different values obtained correctly. Similar results are obtained for rubber: the toughness measured from cutting rubber is about 4 kJ/m^2, but the toughness measured by tearing is about 15 kJ/m^2.[34] A satisfactory explanation of the difference in toughness also has not been given for rubber. The clue lies in the fact that both skin and rubber are very extensible elastic materials. It is a necessary condition for fracture that the mechanical energy released is equal to or greater than the fracture energy required to form the new surfaces, but it is not a sufficient condition. The cohesive strength of the material must also be exceeded for fracture to occur. For most engineering materials the cohesive strength is exceeded once the energy condition is satisfied. However, for very extensible materials such as skin and rubber the tip of a crack becomes very blunt under load and the stresses at the crack tip are quite limited. Consequently it is difficult to develop stresses as large as the cohesive strength of the material and when fracture does occur, the energy released is much larger than the true fracture energy. In cutting, the sharpness of the cutting edge generates high stresses and the work done by the blade is equal to the true fracture energy. Hence the scissors test provides the more accurate estimation of the true toughness, but the tear energy is appropriate for calculations of the likelihood of tearing.

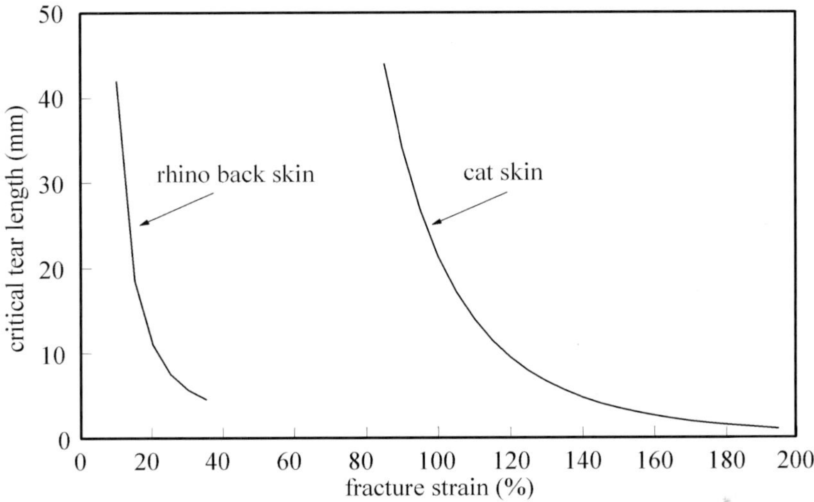

Fig. 3.8 Critical tear length in a rhinoceros' skin compared to cat's skin.

The rhinoceros has an armour skin averaging some 25 mm thick on the back and about 15 mm thick on the belly.[35] Rhinoceros skin is not much tougher than that of a cat, but the skin has to be penetrated before it can be torn. To prevent penetration a stiff as well as a tough skin is required and the rhinoceros skin has evolved to prevent penetration rather than to maximize tear resistance and for this reason a well developed J-shaped stress-strain curve is not necessary, but high stiffness is. Thus, once penetrated, strains as small as about 20% can tear a rhinoceros' skin, as is shown in Fig. 3.8. Evolution's strategy in the rhinoceros has been to develop penetration-resisting armour so that intraspecific combat is not too dangerous. The fracture mechanics of deep penetration of soft solids has been developed[36] and can be applied to the penetration of rhinoceros skin. The force necessary to penetrate the skin of a rhinoceros with a sharp indentor 40 mm in diameter representative of another rhinoceros' horn can be estimated using the paper of Oliver Shergold and Norman Fleck[36] to be about 25 kN (2.5 tonnes). While under impact during fighting, high impact forces can be achieved, the force necessary for penetration is impressive and it is easy to see how the expression 'hide like a rhinoceros' has arisen.

3.3.1.5 *Keratin*

Keratin is group of proteins cross-linked by sulphur that form the main fibres in hair, feathers, horn and hoof. As with all biological tissue the keratin fibres are used in hierarchical composite structures. The primary structure of the α-form of keratin is a three-strand rope or protofibril. The keratin fibrils are embedded in a protein matrix. In wool the proteins have high sulphur content and the elastic stress-strain curve has a sigmoidal form like vulcanised rubber. The small strain Young's modulus is about 4 GPa. Keratin is very sensitive to humidity. The compliant portion of keratin's sigmoidal stress-strain curve decreases by a factor of ten as the humidity increases from 0 to 100%, though the small strain Young's modulus is little affected. The effect of moisture on the stressed α-keratin is to unravel the helices transforming it into a β-sheet form. The protein matrix also absorbs moisture and is plasticised. On unloading there is a large hysteresis, and since the deformation is not immediately recovered on unloading, there is a large loss of energy during a loading-unloading cycle. The strength of hair is about 200 MPa and decreases with humidity.

The structure of the keratin in hooves is quite complex. The wall of a horse's hoof has to carry the load generated during galloping, especially over stony ground, and has two main keratin structures in roughly equal proportions. Running longitudinally in the wall there are cylindrical structures called *tubules* of about 200 μm in diameter which consist of helically wound layers of α-keratin around a central cavity. The fibres in the intertubular keratin run parallel to the sole of the hoof. The distal surface that contacts the ground is composed of dead cells which wear away. The proximal hoof wall grows much in the same way as our finger nails, which are also composed of keratin. Just as splits along the margins of our finger nails often lead to infection, so too do splits in the wall of horses' hooves if they penetrate to the proximal growth region. Thus the structure of the horse's hoof has evolved so that it is much more difficult to propagate a longitudinal fracture towards the proximal growth region than one in the comparatively unimportant transverse direction parallel to the sole along the intertubular fibre direction.[37] Even so, the fracture toughness in the transverse direction is relatively high and about 5 kJ/m^2 at 100% humidity.[38] The elastic modulus and yield strength in the transverse direction at 100% humidity are 180 MPa and 6.5 MPa respectively.[39] Hence the characteristic length, l_{ch}, of hoof keratin is about 20 mm, which means that even transverse cracks are unlikely to cause unstable fracture.

Male bovids use their horns to fight each other for the right of access to females. Unlike the antlers of deer, which are shed each year, horns are not renewed. The breaking of a horn is a serious event for a male bovid; certainly it will prevent him competing successfully for a female and may well lead to death from infection. Hence, evolution has ensured that horns are very tough. A bovid's horn has a spongy core of bone surrounded by a sheath of α-keratin. It is the outer sections of the horn that are the most highly stressed by bending during fighting and the bulk of the toughness of horns comes from the α-keratin sheath. Bovids prior to combat engage in a behaviour known as horning where they push their horns into wet mud or thrash them in wet bushes. While there is little firm evidence on the effect of moisture on the fracture energy of horn, notched tension tests on dry, fresh (water content 20%), and wet (water content 40%) gemsbok horn showed that dry horn was brittle and notch sensitive while the fresh and wet specimens showed considerable ductility and were not notch sensitive.[40] The total work of fracture of the fresh horn was about seven times that of the dry horn so it is very probable that the horning behaviour of bovids does increase the toughness of their horns and lessen the chance of fracture.

3.3.2 Bioceramic tissues

Bone is a biological ceramic containing a high percentage of calcium phosphate, as do teeth. The exoskeleton of molluscs and birds' egg shells are also bioceramics, but based on calcite instead of calcium phosphate. Bioceramics are utilised in animals mainly for their stiffness and hardness, but they need toughness to minimise fracture. Until comparatively recently, hard man-made ceramics have been brittle; remember your mother's favourite porcelain tea cup that you broke. Evolution found a way to produce hard materials that are tough. To obtain a stiff, hard material it is necessary to have a high volume fraction of the mineral constituent. The mineral phase in stiff animal tissue is in itself brittle. Toughness is achieved through the hierarchical structure of the composite, so that the fracture of the hard brittle phase entails considerable deformation and work in the softer phases.

3.3.2.1 *Mollusc shell structures and nacre*

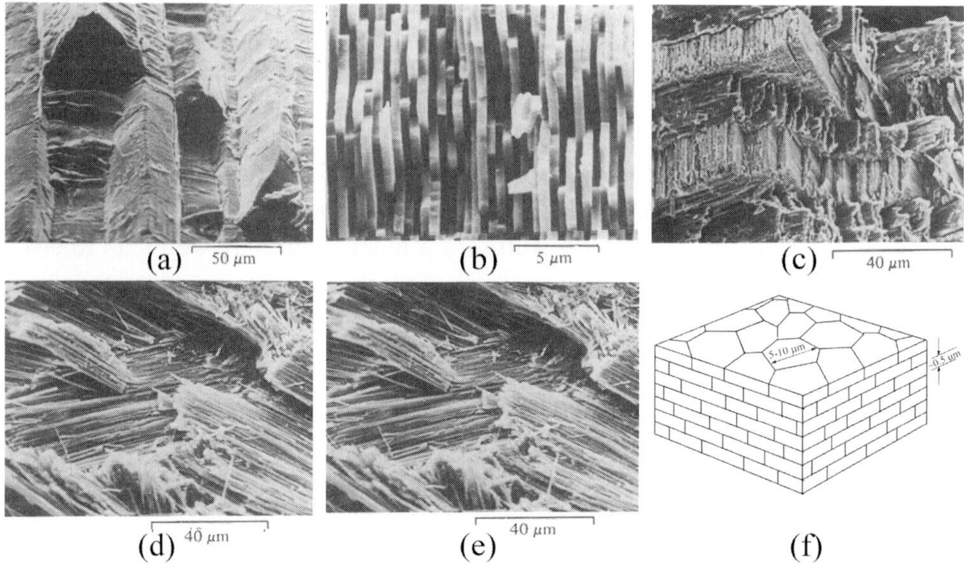

Fig. 3.9 (a–e) Scanning electron micrographs, at various magnifications, of the fracture surfaces of various mollusc shell structures (Currey 1980 with permission of the Society for Experimental Biology): (a) prismatic. (b) nacreous. (c) crosslamellar. (d) foliated. and (e) homogeneous. (f) schematic illustration of sheet nacre; the organic layers between the aragonite platelets are about 20 nm thick.

The structural compound in mollusc shell is chalk (calcium carbonate), not a material that one would immediately choose for hardness but of course one that is abundant in the sea. From this uninspiring material the mollusc grows a strong, tough shell. The chalk in mollusc shell is in the form of aragonite, the harder of its two crystalline structures,[41] with a volume fraction of the order of 95%. There are a number of different structures of aragonite in mollusc shells, which can occur in a single shell. The fracture surfaces of the different structures are shown in Fig. 3.9 (a–e). However, it is the plate-like structure of nacre (mother of pearl),[42] shown schematically in Fig. 3.9 (f), which is by far the strongest, and it is this structure that has created most interest. Nacre is the most primitive structure and found in those shells that evolved earliest, which is surprising because a strong material has been abandoned for a weaker one, but it is to be supposed that other advantages outweigh the decrease in strength. The nacre is in the form of platelets roughly 0.5 µm thick with sides of 5–10 µm cemented together with a thin (~20 nm) complex protein layer which can have chitin at the centre surrounded by a fibroin-like protein; an acidic protein is in contact with the aragonite platelets which are stacked like bricks. In columnar nacre the platelets, which are usually curved, are roughly aligned.

Up to strains of about 0.4% nacre is linear elastic with an elastic modulus of 60–100 GPa, which is similar to aluminium or glass. At higher stresses, hydrated nacre deforms plastically in the protein layers, though the protein is comparatively weak, the constraint of the aragonite platelets enables it to withstand stresses up to about 80 MPa and strains of the order of 1%. In contrast dry nacre behaves more like a monolithic ceramic with little plastic deformation and fails at strengths of the order of 100 MPa.[43] Inter-platelet fracture is comparatively easy; the fibrils of the fibroin-like protein draw out, similar to the crazing of polymers. Fracture across the platelets is more difficult. The aragonite platelets do not fracture and the crack runs around the platelets within the protein. The toughness of the nacre comes from the work done in pulling-out platelets from the highly ductile fibroin-like protein layers. There is a considerable crack growth resistance over a crack growth of some 500 μm[43] and the toughness at instability across the platelets is 350–450 J/m^2 in the dry state and 550–1250 J/m^2 in the wet state.[44] Nacre is a very successful composite because it has high stiffness, strength and toughness combined, which is needed for resisting force controlled fracture. Its success can be seen in the position of nacre in the toughness versus elastic modulus shown in Fig. 3.2.

3.3.2.2 Bone

John Currey has spent many years of research on the mechanical properties of bone and has recently published a new edition of his work, which provides an excellent insight into the behaviour of bone.[45] Bone has a complex hierarchical structure based on the two main constituents of protein, predominantly in the form of collagen, and an imperfect form of hydroxyapatite, in roughly equal volume fractions.[46] Plate-like mineral crystals, 30–45 nm wide, 4–6 nm thick and typically 100 nm long, grow and are aligned with the collagen fibrils. In mammalian bone the mineralised collagen is present in two types of bone: woven bone, where the collagen fibrils are arranged randomly, and lamellar bone, where the collagen fibrils are arranged into lamellae with aligned fibrils. Bone is also permeated by various specialised cells and blood channels. At a higher level the bone is organised in four different ways: primary lamellar bone, woven bone, and fibrolamellar bone, which consists of alternating sheets of lamellar and woven bone, and Haversian systems, where cylinders of lamellar bone have channels formed by osteoclasts that erode a cavity and osteoblasts that subsequently line the cavity with lamellar bone for blood vessels. At the highest structural level there is compact bone, which is essentially solid apart blood channels and a few cells and voids left by osteoclasts, and cancellous bone, with a high porosity visible to the eye, consisting of rods and plates which do not form closed cells. A schematic illustration of the hierarchical structure of mammalian bone at the different levels is shown in Fig. 3.10.

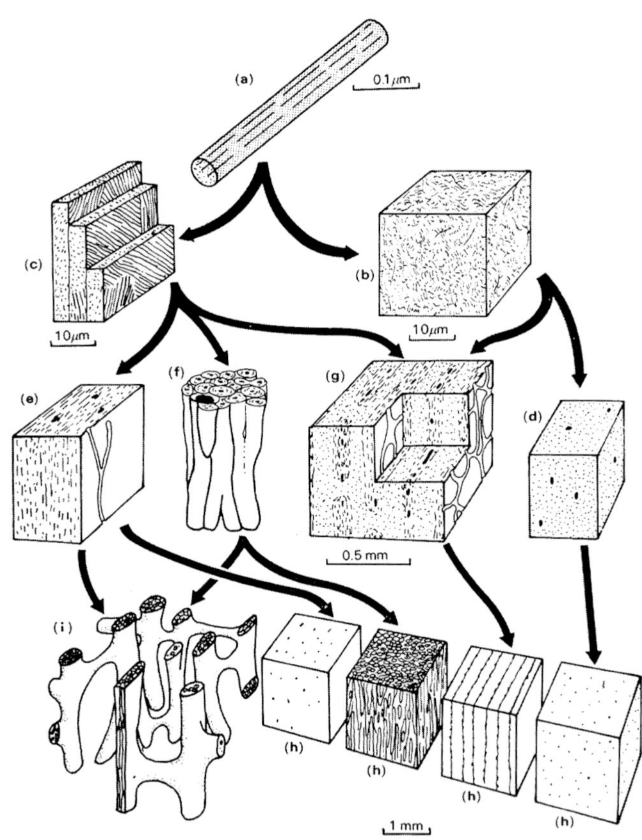

Fig. 3.10 The structure of mammalian bone at different levels with arrows showing what types may contribute to structures at higher levels, (a) Collagen fibril with associated mineral crystals, (b) Woven bone, (c) lamellar bone, (d) Woven bone; blood channels are shown as black spots, (e) Primary lamellar bone, (f) Harversian bone, (g) Lamellar bone, (h) Compact bone, (i) Cancerous bone, (Wainwright *et al.* 1982 with permission Princeton University Press).

In land mammals, the bone mass as a fraction of the total body mass increases with body mass from about 6% for the smallest mammals to about 27% for elephants; for humans the percentage is about 12%.[47] Galileo understood that scaling arguments show there is a limit to the size of animals because of the strength of bone,[48] and the bone mass as a percentage of the body mass increases in larger animals to compensate. Surprisingly the bone mass as a percentage of total mass for birds is not that different to mammals of the same weight, with the fraction varying from about 6% for the smallest birds to about 8% for the heaviest flying birds; for an ostrich, which can weigh up to 150 kg, the percentage is 9%.[47] Since the fracture of bones is not rare, evolution has found the bone mass to body mass ratios where agility and structural integrity are optimised. The bone mass as a percentage of body mass for marine mammals is similar to that for land mammals varying from about 8% for the smallest dolphin to 20% for the Blue Whale.[49] Since marine mammals need a density close to that of water to avoid buoyancy, there would not appear to be a need to keep the bone mass low and the similarity to the bone mass percentage of land mammals may be due to their earlier evolution as land animals. The bone mass of extant mammals fits to the power law

$$m = 0.061M^{1.09} , \qquad (3.13)$$

where m is the bone mass and M the body mass in kilogrammes.[47] However, the bone mass of the elephant is significantly higher than predicted by Eq. (3.13). The biggest dinosaurs were very much heavier than the elephant. The *Argentinosaurus huinculensis*[50] has been estimated to have weighed 73 tonnes.[51] If the bone/body-mass of the *Argentinosaurus huinculensis* was the same as the elephant, its bone mass would have been an incredible 20 tonnes, but it may not have been quite so heavy. A vertebra from the *Argentinosaurus huinculensis*, though 1.59 m long, is comparatively lightly built with deep weight-saving openings, or *pleurocoels*, in the vertebral body, or *centra*, and thin bony buttresses extending upward from the centra into the neural spine. Powerfully developed 'extra joints' (*hyposphenes* and *hypantra*) between adjacent vertebrae helped to stiffen and strengthen the spinal column, to support the tremendous weight. So evolution found a way of extending the size of animals.

The strength of bones comes from compact bone. However, compact bone has a comparatively high density of 2000 kg/m^3 and if bones were composed entirely of compact bone the strength-to-weight ratio would be too small. Cancellous bone, because it is porous, is much lighter and weighs from 1/10 to 1/2 the density of compact bone. Long bones like femurs are mainly stressed in bending. The main resistance to bending comes from the outside of the bone and, just as in bamboo, the strength of a hollow long bone is not much less than that of a solid bone. Hence evolution has designed long bones with a medullary canal along the centre, which are used as convenient storage vessels for marrow and blood cell production. Bones that have two dimensions much greater than the third, such as the vault of the skull, scapulae, and parts of the pelvis, have what Currey has called sandwich bones,[52] where low density cancellous bone is sandwiched between thin layers of compact bone. The compact bone takes the bending stresses and the cancellous bone provides the shear stiffness. Sandwich panels were used for the fuselage of the 1930s de Havilland Comet[53] and in the World War II fighter-bomber, the Mosquito had balsa wood sandwiched between layers of birch plywood.[8] Short bones are also lightened by having a thin shell of compact bone containing a cancellous core.

Bone has a unique property that is eagerly sought by engineers: it can repair itself and it can also remodel itself to account for changes in stress. If the stress on a bone decreases then the bone mass diminishes, for example, astronauts suffer bone loss during weightlessness in space. When bone is stressed, the higher strains developed around cavities or cracks are detected by osteocytes, which release chemicals that stimulate osteoblasts to form new bone. Osteoclasts dissolve the bone in under stressed regions. Bone likes to be evenly stressed. At the joints at the ends of long bones such as the femur, the bone needs to transmit the load from one bone to the next. The head of the femur articulates with the cup-like acetabulum in the pelvic girdle. The femur head is necessarily quite large and if it was composed entirely of compact bone the stresses would be quite low. Instead, the ends of long bones have a thin covering of compact bone supported by cancellous bone of lower stiffness which causes the stresses to be larger and prevents bone atrophy. Metal hip replacements are far from ideal because they take the load away from the bone and in time cause the bone to atrophy loosening the hip replacement. Modern ceramic hip replacements are designed to overcome this difficulty by matching more closely to the stiffness of the femur.

Bone has a quite impressive toughness, ranging from about 600 to 5,000 J/m^2, when compared with other biological materials (see Fig. 3.5) or the advanced ceramics which can have a toughness of up to about 500 J/m^2. The reason for the toughness of bone is its hierarchical composite structure. In lamellar bone the fibres are orientated in different directions in adjacent lamellar. Cracking starts in the lamella in the weakest direction whose fibres are normal to the stress direction. Since this lamella is reinforced by adjacent lamellae whose fibres are orientated parallel to the stress direction, many cracks can develop before cracking occurs in the strong direction.[52] The creation of multiple cracks is one of the contributions to the high toughness of bone. Cracking in the strong direction is made more difficult because of the pull-out of the collagen fibres. Another contribution to the toughness may be similar to a toughening effect achieved in advanced alumina ceramics where microcracking is triggered by the high stresses near a crack tip.[54] These ceramics exhibit crack growth resistance, that is the resistance to cracking increases as the crack grows and a microcracked zone is left in the wake of the crack tip. The microcracked wake zone shields the tip of the crack from stress and increases the apparent toughness. This mechanism can increase the toughness of alumina by a factor of five.

Table 3.4 Mechanical properties of three types of bone (after Currey 1979).

	Antler	Femur	Tympanic bulla
Mineral content (weight percentage)	59	67	86
Density (kg/m^3)	1860	2060	2470
Young's modulus (GPa)	7.4	13.5	33
Nominal work of fracture[55] (J/m^2)	6190	1710	200

Bones have evolved to perform a wide range of functions both within the same animal and across different animal species. Three examples of the different functions are provided by the femur of mammals, the antlers of deer, and the bone of the middle ear (*tympanic bulla*) of whales (see Table 3.4). The mammalian femurs perform what can be considered the classic bone function, that is, to be part of the endoskeleton which must support the body and its kinetics. The femur acts as a lever and it must have stiffness and strength. Since evolution is economic, it does not work on a high factor of safety and, though femurs must be reasonably tough, they are not so tough that they do not fracture under impacts as many of us know to our cost. On the other hand the tympanic bulla of whales does not need to be either strong or tough, but it must have high acoustic impedance so that sound is efficiently transmitted to the middle ear. Antlers have to be tough so that they can absorb impact during combat but, since they are renewed each year, they do not have to be as tough as horn.

Bones can also fracture by the accumulative effects of being stressed and unstressed over a number of cycles.[52] In engineering this type of failure, which is common in metals, is called fatigue. The medical community calls this type of fracture in bone *stress fracture*. In bone, fatigue fractures arise from the stressing of the bone during running or any other repetitive action. In humans the tibia is the bone most likely to suffer fatigue and in horses it is the first phalanx just below the fetlock. The mechanism of fatigue in bone is obviously very different to that in metals. In metals the fatigue strength is insensitive to the frequency of the load cycling and even what is termed as low cycle fatigue can take up to ten thousand cycles; in high cycle fatigue the life can be more than a million cycles. In fatigue experiments on bone the number of cycles are less than a hundred and more importantly the number of cycles to failure is very dependent on frequency. If the stress cycles are tensile it is the time of stressing rather than the number of cycles that is important.[56] Not surprisingly, the fatigue of bone is similar to that of oxide ceramics like alumina. Alumina like bone will fracture at smaller stresses than the short-term fracture strength if left under stress, especially in moist conditions, the time to fracture being a function of the applied stress; engineers call this failure static fatigue.[57] In fact for a time it was thought that ceramics were not susceptible to cyclic fatigue, but only susceptible to static fatigue.

3.3.2.3 Teeth

A recent book by Peter Lucas gives a fine review of properties of teeth.[58] There are four components in mammalian teeth as shown in Fig. 3.11. The main structural component of the tooth is dentine which is similar to bone. About 48% of dentine is mineral, with an organic matrix based on collagen. The odontoblasts, which form the dentine, retreat as it is formed into the pulp so that dentine grows from the inside and the pulp cavity gets smaller. The enamel forms a highly mineralised (97% by weight mostly apatite) hard cap to the exposed dentine that provides the hard surface for crushing or slicing of prey and food. The enamel is composed of mineral crystals 500 nm or longer, 25 nm thick, and about 100 nm wide, which are much larger than the crystals found in bone or dentine. The crystals form in bundles or prisms with almost no organic matter. The little organic matter (only about 1% by weight, the remaining 2% being water) forms the interprism boundaries. The cement, a modified bone, lines the roots of the tooth and can be remodelled during growth and movement of the teeth. The periodontal ligament allows the tooth to move so that the post canine teeth are aligned. Reptiles do not have a periodontal ligament because their teeth interdigitate and have no need to align accurately.

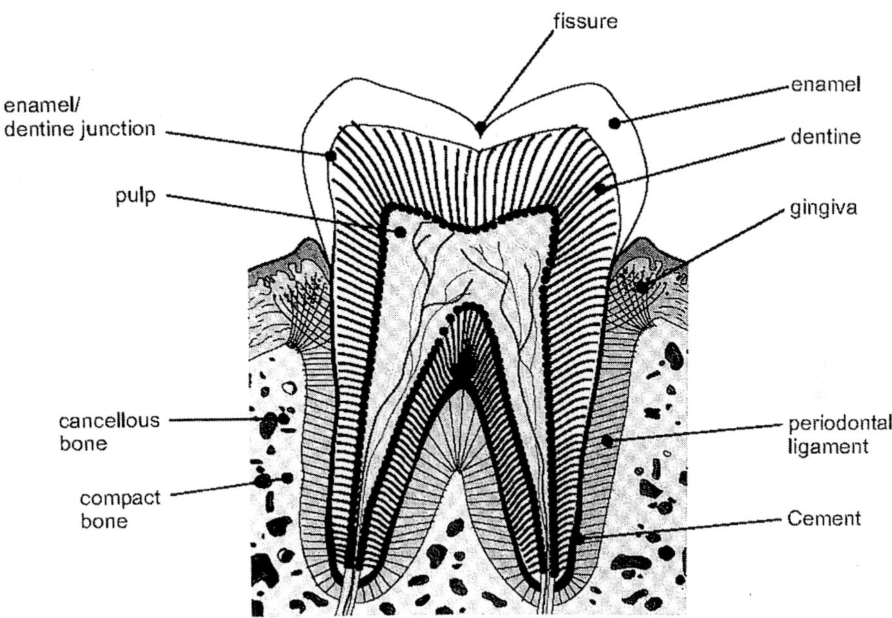

Fig. 3.11 Basic form of a mammalian tooth, illustrated by a human molar (Lucas 2004 with permission CUP).

Evolution has provided teeth shaped to accommodate the mechanical properties of the food eaten. Mammals began to evolve from reptiles through the *synapsids* about 300 million years ago. Their teeth had a single reptilian cusp and all the mammalian teeth evolved from this simple shape. The tearing canine teeth still have this primitive reptilian form. The post-canine molars have evolved multiple cusps to enable food to be chewed and fractured in bending. Carnivores also need blade like teeth, called carnassials, to shear their prey into small pieces. The anterior teeth have evolved into incisors which have multiple uses from a simple griping tool enabling leaves to be stripped from the stem, to removing flesh from a seed, or simple shearing of low toughness plants.

Fracture is not a significant problem with teeth. Even decay, a bane of the Western World, did not occur before humans started to include a high consumption of grain products in their diets about 10,000 years ago (now of course most decay is mainly due to sugars, not starches). However, abrasive wear, due to microfractures, is a problem. Not only are the teeth themselves a main source of wear, but also sand particles, which are almost inevitable in animal food and in human food as well until comparatively recent times. Since the efficiency of mammalian teeth depends on a reasonably accurate occlusion, crown wear is a potential problem. However, odontoblasts at the dentine pulp interface create dentine to accommodate the wear. Exposed dentine, which is softer than enamel, wears faster. In many mammalian herbivores and rodents this differential wear is utilised to produce efficient cutting edges. In these mammals the growth of the teeth is designed to expose areas of dentine that wear leaving sharp enamel ridges.[52]

3.4 Concluding Remarks

It would be foolish to claim that plants and animals evolved the way they have simply because of the restraints of fracture, but evolution has not been able to ignore fracture. Until recently man has exploited the materials he found to hand, such as wood and stone, even the first metals required only comparatively simple smelting. It is only now that we are designing materials almost from scratch. These new materials are almost entirely composites. In nature all materials are composites and have many different hierarchical levels and it is the mechanical interplay of these different levels that evolution has exploited to obtain unique fracture properties.

Evolution has the advantage over man in that there is no hurry. She learns from her mistakes, failures are left behind and we only see her successes. Of course the world is not static, the environment changes and the success of one era can be the failure of the next. Man too should understand that learning only comes from understanding our failures, not our successes.

3.5 Notes

[1]   Dobzhansky (1964).
[2]   Meyers *et al.* (2006).
[3]   The elastic modulus of cellulose fibrils glucose has been estimated to be about 130 GPa with strengths of up to 700 MPa.
[4]   Roelofsen (1959).
[5]   Keckes *et al.* (2003).
[6]   Gordon and Jeronimidis (1974).
[7]   Gordon and Jeronimidis (1980).

[8] Gordon (1978).

[9] The average specific essential work of fracture of the cell wall for plants is about 3 kJ/m^2 which is actually quite high (Lucas *et al.* 1995).

[10] Cerling *et al.* (1997).

[11] Lucas and Pereira (1990).

[12] Cotterell and Reddel (1977), Pardoen *et al.* (2004).

[13] Vincent *et al.* (1991).

[14] Lucas *et al.* (2000).

[15] Vincent (1982).

[16] The blades of grass springing from the bottom of the stalk.

[17] Rudall and Kenchington (1973).

[18] Vincent and Wegst (2004).

[19] The extensor apodeme, a ridge-like ingrowth of the exoskeleton that is the muscle attachment point, of a locust's hind leg has an elastic modulus of 20 GPa.

[20] Pryor, M.G.M. (1940).

[21] Vincent and Hillerton (1979).

[22] Bernays and Janzen (1988).

[23] Not so important for caterpillars which replace their mandibles at each moult as for beetles that have permanent mandibles.

[24] Hillerton *et al.* (1982).

[25] Marshall *et al.* (2001).

[26] Morgan *et al.* (2003).

[27] Dobb *et al.* (1967).

[28] Elices *et al.* (2005).

[29] The Roman torsion catapults, which could hurl boulders of up to 40 kg some hundreds of metres, used tendons for the torsion spring. Similarly the Parthians used reflexed composite bows with tendons on the tension side of a wood skeleton to halt the eastwards expansion of Imperial Rome (Cotterell and Kamminga 1990).

[30] Rivlin and Thomas (1953).

[31] Lake (1970). A finite element study by Lindley (1972) has shown this expression to be reasonably accurate. For small strains in a linear elastic material Eq. (3.9) is exact and identical to the expression for the classic Griffith crack given by Eq. (1.26) with $Y = 1$.

[32] Although Eq. (3.9) was derived for a nonlinear elastic material, it also gives the energy release rate at initiation if the deformation is nonelastic in just the same way as the J-integral (see §9.4.3), which was also derived for a nonlinear elastic material, gives the energy release rate at initiation for a plastically deforming material. Only on unloading can elastic and plastic deformation be easily distinguished.

[33] Veronda and Westmann (1992).

[34] Lake and Yeoh (1978).

[35] Shadwick *et al.* (1992).

[36] Shergold & Fleck (2004), see §11.3.1.

[37] Human finger nails are similarly more difficult to tear longitudinally than transversely especially over the central region of the nail (Farren *et al.* 2004) which is good news for those of us who bite their nails.

[38] The fracture toughness of the wall of a hoof is very dependent on the humidity and is a maximum at about 75% humidity. The stress-strain curve is even more affected by the humidity. The modulus of elasticity and strength of the hoof wall is a maximum in the dry state when the ductility is very small and maximum at 100% humidity when its ductility is the largest (Bertram and Gosline 1987).

[39] Kaspi and Gosline (1997).

[40] Kitchener (1987).

[41] The two crystalline forms of calcium carbonate are: aragonite with a Mohs hardness of 3.5–4 and calcite with a Mohs hardness of 3.

[42] The iridescent sheen of mother of pearl is caused by the interference of light because the thickness of the aragonite platelets is about 0.5 μm which of the order of the wave length of light.

[43] Barthelat and Espinosa (2007). The toughness along the platelets is lower: 330–440 J/m^2 dry, 790 J/m^2 wet.

[44] Jackson *et al.* (1988).

[45] Currey (2002).

[46] Wainwright *et al.* (1982).

[47] Prange *et al.* (1979).

[48] See §6.2.

[49] Prothero (1995).

[50] Bonaparte and Corio (1993).

[51] Mazzetta *et al.* (2004).

[52] Currey (2002).

[53] Not to be confused with the later 1950s de Havilland airplane of the same name.

[54] See §10.1.4.

[55] The total work to fracture of deeply notched bend specimens divided by the area of fracture.

[56] In compression cycling the number of cycles is more important.

[57] See §10.1.5.

[58] Lucas (2004).

Chapter 4

Human Evolution and Stone Tools

In the eighteenth-century, Benjamin Franklin called man 'a tool-making animal', but tool-making is not exclusive to man, or even primates. In 1835 Charles Darwin (1809–1882) visited the Galápagos Islands, during his famous voyage on the *Beagle,* and observed finches on the Galápagos Islands using twigs to pick insects out of tree-trunks. However, only man and his hominid ancestors substantially modified natural objects to fashion tools. Early man used wood, bone, and shell for tools as well as stone, but stone was the only material available that would take a sharp cutting edge. Since stone tools are enduring, they are our earliest evidence of tool-making. The oldest stone tools yet found are 2.6 million years old, and are from Gona, in Ethiopia (see Fig. 4.1).[1] The tool-makers have not been identified, but the hominids known to have been present in East Africa at this period are *Australopithecus garni* and *aethiopicus.* Fossilised animal bones with definite cut-markings were found in a 2.5 million-year-old site only 90 km to the south of Gona, but these bones were not associated with any stone artefacts.[1] Evidently, stone tools were used early in our evolution for the butchering of carcasses. The inclusion of meat in the hominid diet almost certainly led to increased body size and more importantly to encephalisation, or the increase in the ratio of brain to body weight. The mass of the modern human brain is about 2.5% of the total body mass and the human brain consumes some 20–25% of the resting metabolic rate compared to only about 8% for the great apes. Of course, humans can thrive on a vegetarian diet, but it must be well balanced and contain sufficient protein. The great apes, though largely vegetarian, do find the need to supplement their diet with meat and the diet of the early hominids is unlikely to have led to encephalisation if they had not included meat as a high energy source in their diet. The teeth and jaws of apes and hominids are not efficient at cutting flesh into small pieces to be swallowed. Apes, like carnivores, have incisors and canines that can grip prey at a wide gape angle. However, in carnivores at smaller gape angles blade-like teeth called carnassials occlude and are efficient at butchering the prey into pieces small enough to be swallowed. Apart from the canines, the teeth of apes and hominids occlude almost simultaneously and they lack efficient flesh-cutting carnassials. Stone tools made up for the lack of carnassials in the hominids an enabled them to obtain a more nutritional diet easily

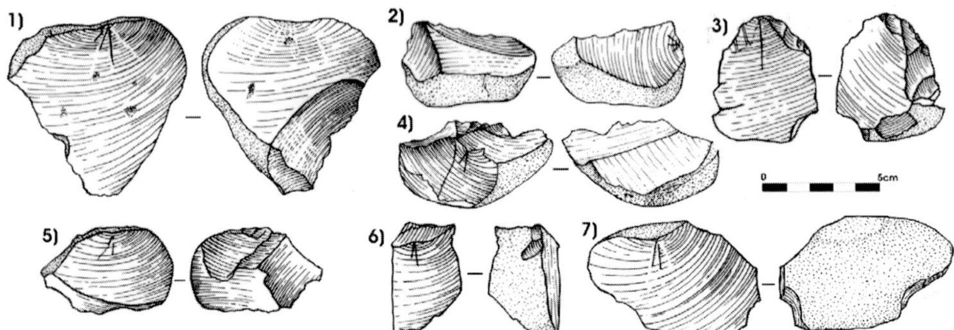

Fig. 4.1 Drawings of the stone artefacts excavated at one site in the Ounda Gona South area: sketches 1, 3, 5, 6 and 7 show flakes while 2 and 4 are heavily reduced cores from which flakes have been struck (Semaw *et al*. 2003 with permission Elsevier).

Tools can be fashioned from stone either by flaking or grinding. Since a sharp usable flake can be instantly fashioned from a stone nodule by striking it with a stone hammer but to grind a hatchet on a sandstone rock from a basalt cobble takes typically an hour or more, it is not surprising that flaking predates grinding. Hatchet grinding grooves on a sandstone rock in Kangaroo Valley, New South Wales, are shown in Plate 4. Ground stone tools have been found in northern Australia that probably date to about 30,000 years ago,[2] and the earliest ground stone tools in Japan date from 28,000 years ago.[3] In Europe, ground stone tools only date back to about 9,000 years ago.[4]

4.1 Modern Discovery of Stone Tools

The use of stone tools declined in Europe and the Middle East as metals came into use. Hammered native copper was used for some thousands of years prior to the exploitation of copper ores, but was soft and scarce and did not displace stone. However copper ores, which started to be exploited about 4,000 BC, were more abundant and occurred with impurities such as arsenic and antinomy, which automatically alloyed the copper, producing a much harder metal. Around 3,000 BC, copper started to be deliberately alloyed with tin to give bronze which could take a reasonable cutting edge and was much more durable than stone. As with any innovation, the first bronze tools would have been prestige goods just like the first mobile phones, and more important for giving status to the owner than functional use. However, as bronze came into more general use there was a decline in the level of control and formalism exercised in stone tool production. By the beginning of the first millennium BC in Europe, flakes were struck from stone for immediate use and then thrown away like today's disposable razors. By classical times stone tools had been forgotten, they were found of course, but were regarded as having superhuman origin with magical powers.

From classical times to the nineteenth-century stone tools were widely believed in Europe to be thunderbolts. Ulisse Aldrovani (1520–1605), who was a distinguished Italian naturalist of the Renaissance, described stone tools as 'a mixture of a certain exhalation of thunder and lightning with metallic matter, chiefly in dark clouds, which is coagulated and glutinated into a mass and subsequently hardened by heat like a brick'. In nineteenth-century Norway they were known as *tonderkilers* and the test that they were really thunderbolts was to tie a thread around them and place them on hot coals, if genuine the thread did not burn, but became moist. The origin and great antiquity of the stone tools began to be realised from the late seventeenth-century onwards. In 1679 the antiquarian apothecary, John Conyers (1633–1694), unearthed what is now termed a flint hand-axe, close to elephant bones from the Roman period, near Gray's Inn Lane in London, which he recognised as possibly an implement or weapon (see Plate 5). Conyers' flint hand-axe is now known to be about 350,000 years old and belongs to the Palaeolithic Age. The first definite recognition of stone tools and their great antiquity was made in 1797 by John Frere (1740–1807), a landowner with an interest in antiquities. Workers in a brick pit at Hoxne in Suffolk found a large quantity of flaked flint hand-axes at a depth of about 3.5 m. Frere mistakenly identified the hand-axes as weapons, but used the concept of stratigraphy, which in the eighteenth-century was just beginning to be used for relative dating, to argue that they came from 'a very remote period indeed; even beyond that of the present world'. In France, Jacques Boucher de Perthes (1788–1868), a geologist and antiquarian who was the customs house director at Abbeville, a town at the mouth of the Somme, discovered flint hand-axes and other stone stools in gravel pits, some *in situ* with the bones of extinct mammals. In 1838 he described his finds to a meeting of local intellectuals, again using arguments of stratigraphy he attributed the tools to antediluvian man, but he was disbelieved. The most authoritative confirmation of the great antiquity of the stone tools came in 1859 when Charles Lyell visited the excavated sites. We now know that the stone tools found by Boucher de Perthes date from the Palaeolithic Age and are some half a million years old.

4.1.1 The Brandon flintknappers

Since Neolithic times man has used flints to strike sparks from pyrites for fire-lighting. There is no need for a specialised form for the flint used to produce fire, but with the introduction of the flintlock gun in the seventeenth-century there came a need for shaped gunflints. There were a number of sites in England for the manufacture of gunflints; the longest-lasting was in Brandon, Suffolk, England, where there is still an inn called 'The Flintknappers' Arms'. The flintknappers had to learn to flake flint from scratch although there may have been a small industry making strike-a-lights before flints were needed for guns. The area around Brandon is rich in flint and there are extensive Neolithic flint pits nearby at Grimes Grave. The flint industry was at its height during the Napoleonic Wars when 200 flintknappers were employed. The introduction of the percussion cap, and later the modern cartridge, led to a decline in the industry during the nineteenth-century. Gunflints were knapped for export principally to Africa. In 1868, 20 flintknappers in Brandon were producing some 200,000 gunflints a week. By the beginning of the twentieth-century only a handful of flintknappers were left in Brandon. A few flintknappers still practiced their age old craft after World War II and the industry was finally killed in the 1960s when an arms embargo ended the trade to Africa. Although the modern flintknapper used steel knapping hammers, his technique was essentially the same as that used in the Upper Palaeolithic to produce blade flakes. It was from studying the flintknappers of Brandon in Suffolk that archaeologists began to understand how stone tools were produced.

4.1.2 The archaeological importance of stone tools

Stone tools are the most common artefacts found in early archaeological sites. We will see that definite tool types developed over time. Although archaeologists make great use of radiocarbon and potassium-argon dating, in unstratified Palaeolithic sites, stone tools are important for dating. Assemblages of stone tools are used by archaeologists to understand the technological development and cultural traditions of man and his hominid ancestors. Stone tools are durable but they wear during use, suffering characteristic damage in the form of microflakes or abrasion. Apart from mechanical damage, residues can be bonded chemically to the stone tool, giving insight into the use to which they have been put. We are used to forensic science identifying blood stains and the DNA from crime scenes. Blood residues are very strongly bonded to stone and are especially preserved in microfissures in the surface and can be detected after thousands of years. One of the oldest claimed recoveries of mammalian DNA is from the Middle Palaeolithic rock shelter site La Quina in the French Province of Poitou-Charentes whose archaeological deposits range from 35,000 to 65,000 years ago.[5] The three DNA sequences identified on the eight tools examined were boar or pig (100%), sheep (84%), and modern human (99%) where the figures in brackets give the percentage match.[6]

4.2 Stone Tool Types and Human Evolution

There are two main views of the differences in stone tool types. The earliest view, which was in vogue for the first two-thirds of the twentieth-century, is that the different stone tool types marked the evolution of man.[7] That culture should evolve similarly to species was the natural extension of Darwinism. However, from the 1930s some archaeologists argued that stone tool assemblages did not always follow a neat temporal sequence. In this view, the stone tool types are seen to represent functional or seasonal differences in site activities and could be interpreted as adaptive markers more than as markers for evolution. As in many explanations that seem to be polarised into two extremes the answer is probably some where in the middle. In addition another factor archaeologists tend to neglect is that the form of a stone tool is governed by the mechanics of flaking. There are only a limited number of ways that a flake being detached can be controlled: they are: the shape of the cobble or core from which the flake is being formed and the position, direction and magnitude of the percussive or pressure force being applied. The stone knapper has no control of the form of the flake once the fracture that separates it is initiated. Whether the flake is detached by percussion or pressure it is formed in a matter of milliseconds, because the flaking process is unstable. Of course, except in the simplest of flakes such as those shown in Fig. 4.1, the flake or core tool will have many flakes removed and the position and means of removal of those secondary flakes will have a profound effect on the final tool. However, the possible permutations are limited.

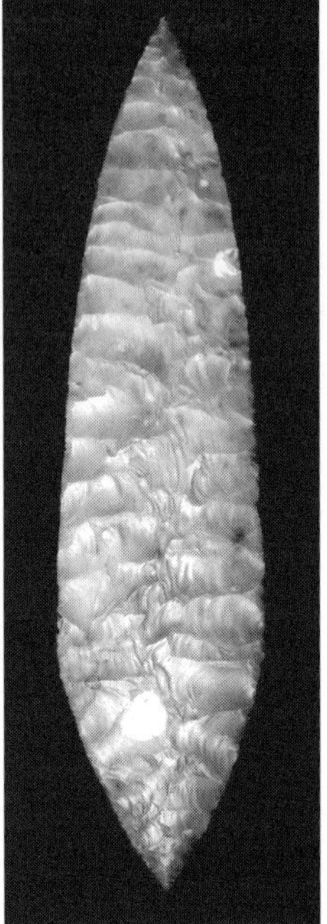

Fig. 4.2 Quartzite chopper from the Olduvai Gorge, Lower Palaeolithic   Fig. 4.3 Solutrean point
Collected by Louis Leakey (1903–1972) in 1934 (with permission BM).   South Western France,
Upper Palaeolithic
(with permission BM).

Having put forward the objections to the concept of stone tool evolution there is certainly still a very obvious difference in sophistication between the chopper tools from the Olduvai Gorge from the Lower Palaeolithic, dating from about 1.5 million years ago (shown in Fig. 4.2) and the beautiful Solutrean point from south-west France, dating from the Upper Palaeolithic and 14,000 to 18,000 years old (shown in Fig. 4.3). However, the simple 2.6 million-year-old percussion flakes shown in Fig. 4.1 could have equally have come from any time in the Stone Age, because there is no secondary working of the flakes and their form is controlled mainly by the mechanics of flaking. An example that similarity of form does not necessarily imply that there has to be a direct evolutionary link for the more developed flakes is provided by Clovis points from North America, shown in Fig. 4.4, which are similar in form to the Solutrean point from France except that there was usually characteristic fluting on Clovis points. Despite the Clovis point dating significantly later than the

Solutrean point to about 13,000 years ago, some archaeologists have mistakenly argued that there must be a link between the two, indicating possible contact between France and America which is extremely unlikely.

Fig. 4.4 Clovis points (courtesy Robert N. Converse).

Recently there has been an attempt at separating out the broad stone tool modes based on earlier work of the famous archaeologist Grahame Clark (1907–1995) and setting these against the backdrop of human evolution.[8] Clark defined the five basic modes of stone tools, shown in Fig. 4.5, which evolved with time. Mode 1 consists of the simplest of stone tools. In the Oldowan and Asian pebble and chopping tools, a few short flakes are removed from a pebble or cobble by percussion, with no preparation of the flaking platform to produce a chopping tool with sharp edges. The flakes removed would also have been used for cutting. Mode 1 is found throughout the 'old world', over the Pleistocene and into the Pliocene in sub-Saharan Africa. Mode 2, characterised by the Acheulian hand axe (see Plate 5), named after St Acheul, the suburb of Amiens in the north of France where such stone tools were first recognised. The flakes removed are longer compared with their thickness than those in Mode 1 and removed from both surfaces to produce a bifacial tool with some symmetry. There are a wide variety of hand axe types divided roughly into two series: one in which there is a

| Mode | Tool type | Description |

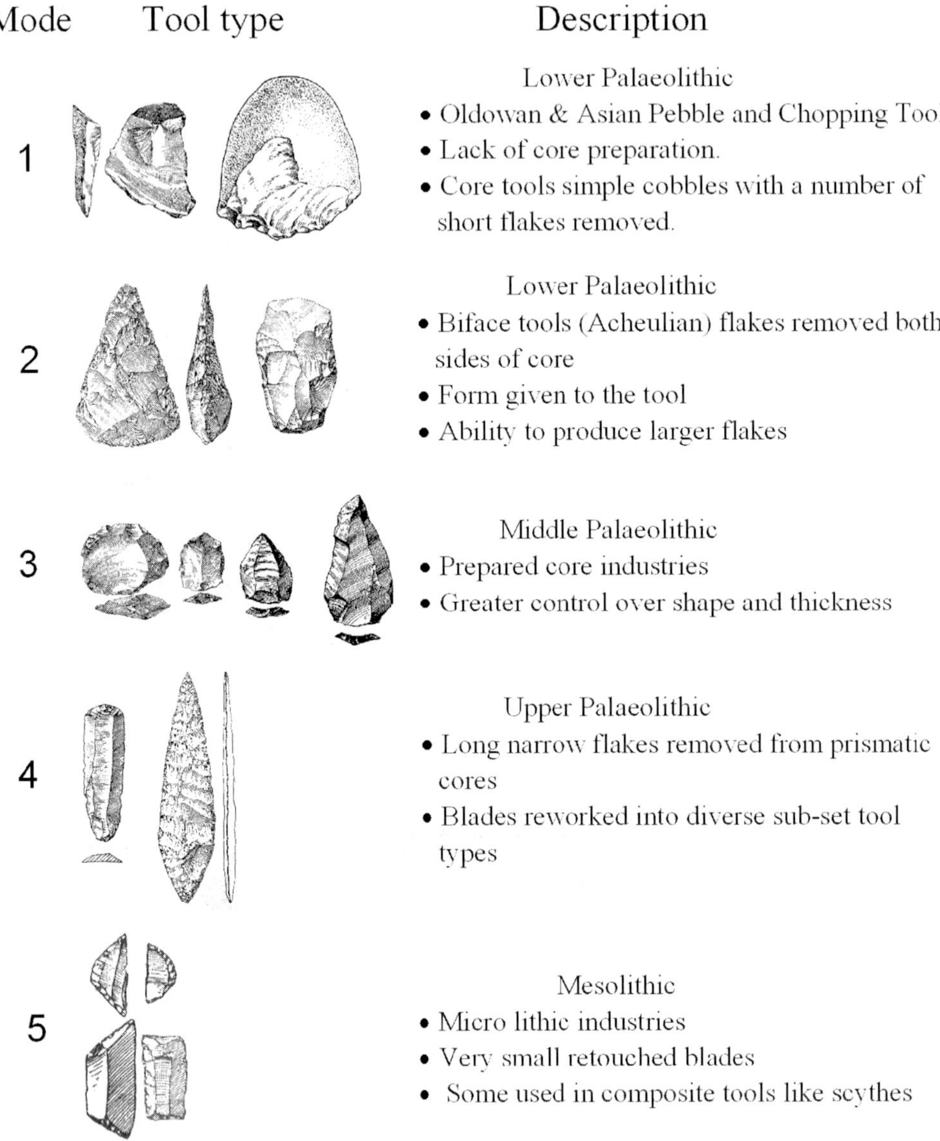

Mode 1

Lower Palaeolithic
- Oldowan & Asian Pebble and Chopping Tool
- Lack of core preparation.
- Core tools simple cobbles with a number of short flakes removed.

Mode 2

Lower Palaeolithic
- Biface tools (Acheulian) flakes removed both sides of core
- Form given to the tool
- Ability to produce larger flakes

Mode 3

Middle Palaeolithic
- Prepared core industries
- Greater control over shape and thickness

Mode 4

Upper Palaeolithic
- Long narrow flakes removed from prismatic cores
- Blades reworked into diverse sub-set tool types

Mode 5

Mesolithic
- Micro lithic industries
- Very small retouched blades
- Some used in composite tools like scythes

Fig. 4.5 Clark's stone tool modes (after Clark 1969).

butt and pointed end and the other more or less ovate in shape. In Mode 3 there is considerable preparation of a core before the removal of flakes, which enabled the shape of the flake to be controlled. These tools are more diverse than Mode 2. Long, thin, and narrow blade flakes characterise Mode 4. To remove these flakes carefully prepared cores were fashioned as shown in Fig. 4.6. These cores are very similar to those used by the flintknappers of Brandon to produce gunflints. It is also the first example of mass production. Once a prismatic blade core was prepared many blades could be struck from the core in quick succession. The Brandon flintknappers could produce about 200 gunflints an hour.[9] In the eighteenth-century Francisco Saverio Clavigero (1731–1787) observed a similarly impressive output of a hundred flakes per hour for the production of obsidian blade flakes by Mexican Indians. Some blade flakes were used just as removed from the core while others were reworked into beautiful symmetrical leaf shaped tools. The most beautiful delicate reworked blades would be liable to break if they were used and many are most likely prestige tools owned by a 'big man'. The final mode defined by Clark is the microliths of Mode 5. The microliths are very small flakes that are reworked into a variety of shapes. Microliths were often used in composite tools such as scythes.

Fig. 4.6 An obsidian core for blade flake production.

The evolution of the homo taxa for the early colonised regions proposed by Robert Foley and Marta Lahr[8] is shown in Plate 6 with the first appearance of the various stone tool modes superimposed. The evolution of *Homo ergaster* from *Homo rudolfensis* or *hablis* occurs without any technological change. However, around 1.4 million years ago *Homo ergaster* made a significant change to Mode

 Fracture and Life

2 stone tools. No technological changes were introduced by *Homo heidelbergensis*. In Africa the evolution of *Homo helmei* sees the new technology of Mode 3. Homo sapiens makes his first appearance in Africa about 130,000 years ago and Mode 3 and Mode 4 technologies developed comparatively quickly compared with the 2 million years taken to develop the first three technological modes. In other parts of the world the technological changes come somewhat later. The first technologies in the Middle East and Europe are Mode 1, but if *Homo sapiens* evolved in Africa and then spread to all other regions, as is most probable, the development of Mode 4 technology occurred not because it evolved in the region but as an import from Africa with the earliest *Homo sapiens*. Similarly Mode 3 technology in China, Java and Australia was imported by *Homo sapiens* and did not evolve from *Homo erectus*.

4.3 Stone Materials

What largely determined the evolution of a successful stone tool technology was the abundance of a material that could be easily worked and would take a sharp durable edge. Because many stones flake easily, natural flakes would have been found and used by the earliest hominids. Since it is not difficult to remove flakes from the right stone simply by hitting it against a rock the idea of producing flaked stone tools must have been reasonably common. The idea of grinding a stone to produce a tool is very much less obvious and it is not surprising that ground stone tools occurred only at the end of the Stone Age. The material properties suitable for grinding are also quite different from those for flaking.

4.3.1 Materials for flaked tools

For flaking, a hard homogeneous amorphous or fine-grained stone is needed so that the flaking is predictable and can take a sharp edge. The stone easiest to flake is also be the easiest to chip during use, so the best material for stone tool manufacture has to be a compromise between being easy to flake and difficult to chip. The more finely textured and dense siliceous stone has the right properties for flaking. Fortunately for the evolution of man, the siliceous stone family is large and well distributed.

Natural glasses are the most homogeneous of the silicates and have an amorphous structure. Obsidian, formed by the rapid cooling of siliceous lava, was the most highly prized stone for sharp-edged light duty cutting tools.

Obsidian is brittle like modern glass and easy to flake and gives an extremely sharp edge that is sharper than a surgeon's scalpel. When Don Crabtree (1912–1980) an American archaeologist who pioneered the replication of stone flaking techniques, had to have part of his lung removed his surgeon made half the incision with a normal surgical scalpel and half with an obsidian blade. That part of the incision made by the obsidian blade healed quicker than that made by the scalpel and left only a faint pink line compared to the usual highly visible scar. Obsidian blades are so sharp that they cause hardly any trauma to the tissue being cut and for this reason are used by some cosmetic surgeons who typically use the obsidian blade ten to twenty times before discarding it. The only thing holding back the more widespread use of this Palaeolithic technology today is the lack of expert knappers.

Obsidian was very desirable because of the sharpness of its flakes, but it was also one of the scarcest of the stone types suitable for flaking. Consequently, obsidian was traded over land and water for immense distances. The source of obsidian can be identified from trace elements, X-ray fluorescence, optical spectroscopy, or rock magnetic properties. In the first millennium BC obsidian from New Britain in Papua New Guinea was carried 3,800 km west to Sabah in northern Borneo and 3,300 km east to Fiji. In the Mediterranean there were a number of sources of obsidian such as Lipari in the Aoelian Islands which was settled some centuries before 4,000 BC, probably because of the obsidian which was exported to Liguria, Provence, and Dalmatia as well as to close by Sicily and Southern Italy bringing great wealth to the island.

Chert, another siliceous fine-grained sedimentary rock formed by impregnation of a rock by siliceous fluids or an accumulation of siliceous micro-organisms, was a favoured stone tool material though it does not give such a sharp edge as obsidian it is much more durable. Flint is a special form of chert formed through siliceous replacement into nodules or bands in limestone. Since flint is common in England and France where the first studies of stone tools were made, books on prehistory often used the term 'flint implements' as a generic term for stone tools. Jasper and chalcedony are other forms of chert. Like obsidian, chert was traded over large distances. In nineteenth-century Australia, the desert Aborigines discovered that the porcelain insulators used on the overland telegraph line were an excellent alternative to chert.[10] To prevent disruption to the telegraph, caches of glass bottles were placed at the foot of the telegraph poles, which the Aborigines found gave a sharper edge than the porcelain. An Aborigine flake fashioned from bottle-glass is shown in Fig. 4.7.

In Western Europe chert was relatively abundant, but in many parts of the world less suitable coarse-grained or crystalline materials like quartz, quartzite and silcrete had to be used. Quartz is a crystalline form of silica. In general, crystalline materials are not suitable for flaking because they have well-defined crystalline planes which are easily cleaved. What is needed for flaking is a material that is isotropic with the same properties in all directions. The flakes detached from a homogeneous isotropic stone can be controlled by preparation of the striking platform and the position and direction of the blow or pressure. Quartz can be flaked with some success because the cleavage planes are not well defined. Quartzite is quartz rich metamorphosed sandstone. Silcrete consists of quartz grains embedded in a matrix of amorphous silica. Coarse-grained silcretes produce an irregular blunter edge than fine-grained siliceous stone, but they can be preferable in activities like wood chopping because they are tough.

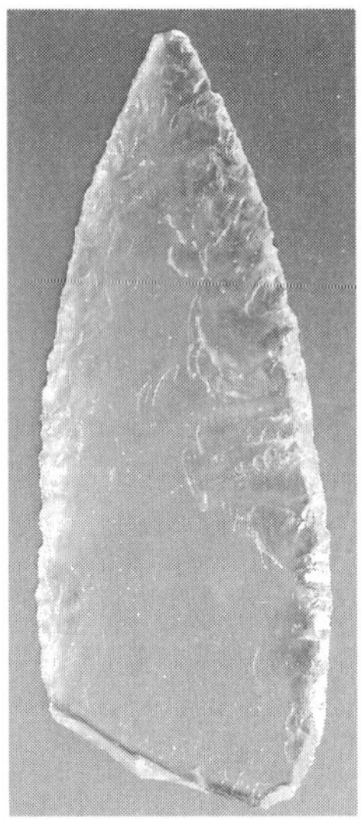

Fig. 4.7 An Aborigine bottle-glass flake from the nineteenth century.

4.3.1.1 Heat treatment of stone

Good flaking stone was reasonably abundant in Western Europe, but in other parts of the world less favourable stone had to be used. Heat treatment was used in comparatively recent times to improve the flaking properties of siliceous stone in Oceania, northern Australia, Bengal, southern Africa and parts of North America.[11] The fracture toughness of stone is reduced by heat treatment making it easier to flake, particularly to allow finer pressure flaking, but also makes it less durable. Even heating to 300°C reduces the fracture toughness of silcrete by about 17%.[11] The discovery of the benefit of heat treatment would have occurred when stones that had been in a camp fire were later flaked. Not only would the stone have been easier to flake but the difference in appearance of the flakes would have been noticed. The colour of flakes from heat treated stone is generally darker. For example, yellow-coloured siliceous stone contains goethite, a form of iron hydroxide named in honour of Goethe, under heat this hydroxide is reduced to an oxide haematite which is dark red.[11] Also the surfaces of a heated flake have a lustrous appearance. The reason for the lowering of the fracture toughness is not completely understood; certainly heat causes water loss in the stone and probably increases the bonding between the grains causing fractures to propagate transgranularly instead of intergranularly. A transgranular fracture is smoother and hence more lustrous than an intergranular fracture.

There are a number of ways that flakes can be tested to see whether the stone was heat treated. Siliceous stone is luminescent when heated, that is it gives off light and is thermoluminescent.[12] Natural ionising radiation falling on a siliceous stone causes the ejection of electrons from atoms leaving behind holes. The excited electrons move in the conduction band and holes in the valence band. Some of these are trapped at point defects in the crystal lattice. On heating, electrons are released, migrate to the trapped holes, and recombine with the emission of light. Heat treatment causes the atoms to give off photons so that in effect the clock is put back and subsequent thermoluminescence is much less than would be expected.

4.3.2 Materials for ground stone tools

Ground stone tools were made from tough igneous rock such as basalt, the material used for roads today, and greenstone. Igneous and metamorphic rocks are much more readily ground than chert and other flake industry materials. The earliest ground stone axe heads were made from water-worn pebbles of fine-grained rock. Quarrying igneous rock is not easy, requiring hammering with stone, prising with wooden levers, and perhaps the use of fire. The resultant lump of stone would be unlikely to have the form required for the ground stone tool. Since the igneous stones are difficult to flake, the quarried lumps would have to be dressed by hammering with a stone. The blows could not be too hard or the blank would fracture and so the dressing of a quarried lump of stone would have been slow and it is easy to see why pebbles of the right shape were eagerly sought.

4.4 Flaked Stone Tools

Describing in detail hominid and human flakes may seem somewhat esoteric, but it is necessary to be able to distinguish between flakes produced by humans and nature. In the late nineteenth-century a number of crudely shaped stones looking like tools were discovered in the Kent North Downs, England. These flakes, termed at the time *eoliths* (stones from the dawn of time) dated from the Pliocene. Subsequently these 'tools' were shown to be naturally flaked stones.

However, since then of course real stone tools have been discovered which do date to the Pliocene. As well as the necessity of being able to distinguish true stone tools, the tools chip and score during use and often their use can be determined from the flake markings

Bipolar flaking is perhaps the simplest and certainly one of the earliest techniques for flaking a stone pebble or core. In this technique an ovate pebble is placed upright on a rock anvil and struck at its pole by a heavy stone as one would crack a nut. One or more blows to the pebble results in the detachment of sharp compression flakes from the pebble either from the struck end of the pebble or where it rests on the anvil. After a number of blows the pebble takes on a battered appearance. The flake fractures run roughly in the direction of the blow. The bipolar technique is especially useful for the more difficult flaking materials like quartz and can be used on very small cores. It is a useful way of producing small flakes with sharp edges suitable for butchering small game or trimming wooden spear points. A bipolar flake is shown in Fig. 4.8.

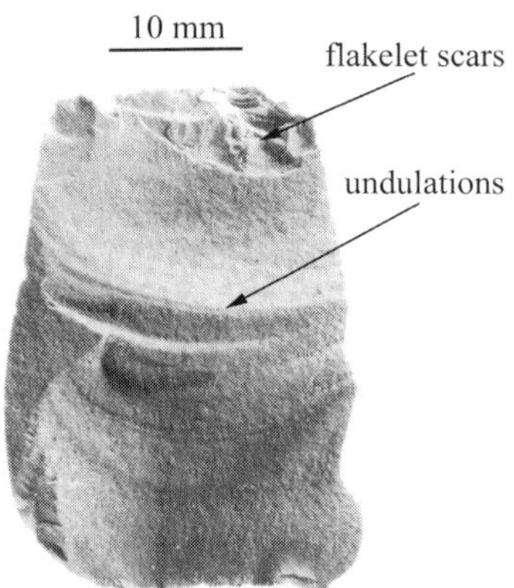

Fig. 4.8 A bipolar flake (Cotterell & Kamminga 1987 with permission Society for American Archaeology).

In bipolar flaking there is virtually no control over the process and the resulting flakes are very variable. In percussion flaking, a flake is struck from a core or nucleus with a hammer stone as shown schematically in Fig. 4.9. The resulting flake has a unionid shell-like appearance and is described as conchoidal (see Fig. 4.10). The core needs support, but the form of support can vary from a rock anvil to the knapper's thigh, depending on the size of flake that is being struck, and plays no part in the detachment of the flake other that to react against the blow. The flake has a characteristic shell-like swelling called the bulb of percussion at the struck surface; the rest of the surface is marked by gentle undulations similar to those seen on a bipolar flake and lance-like markings normal to the undulations. A reasonable degree of control over the flake being detached can be obtained in conchoidal flaking, firstly through the preparation of the edge of the core, in particularly the edge angle between the surface to be flaked and the striking platform as well as the plan view of the core's surface, and secondly by the placement of the point of percussion and the direction of the blow. The flakes that were removed to form the Acheulian hand axe shown in Plate 5 were conchoidal. The detachment of fine flakes is difficult with a hammer stone and for finer work a soft hammer of wood or bone was used. For even finer retouch work such as shown in the Solutrean point shown in Fig. 4.3 flakes would be removed by pressure with a bone point

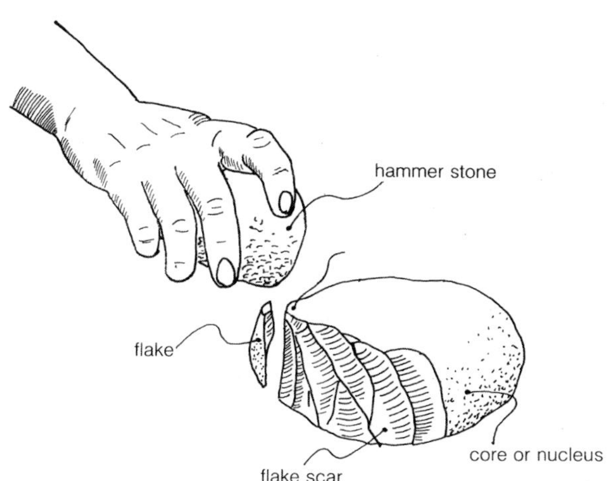

Fig. 4.9 Percussion flaking with a hammerstone (Cotterell & Kamminga 1990 with permission CUP).

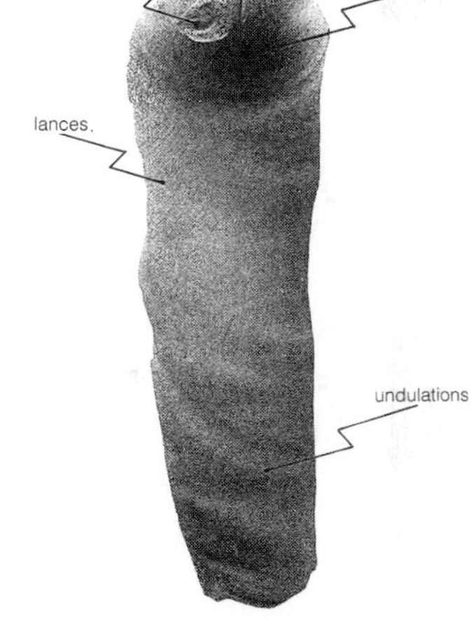

Fig. 4.10 A conchoidal obsidian flake sprayed with paint to prevent glare (Cotterell and Kamminga 1990 with permission CUP).

In conchoidal flaking the characteristic bulb of percussion is the remnant of the conical indentation fracture caused by the impact of a hammer or intense pressure under a bone point. The pressure at the point of contact, especially under pressure flaking, was often not intense enough to initiate a conical indentation fracture and a bending flake was removed instead of a conchoidal flake. A bending fracture initiates away from the point of application of the force under

                               Fracture and Life

bending stresses on acute edge angled cores. The bending flake has no bulb of percussion and has a typical waisted form when removed from a flat sided core as shown in the schematic illustration in Fig. 4.11.

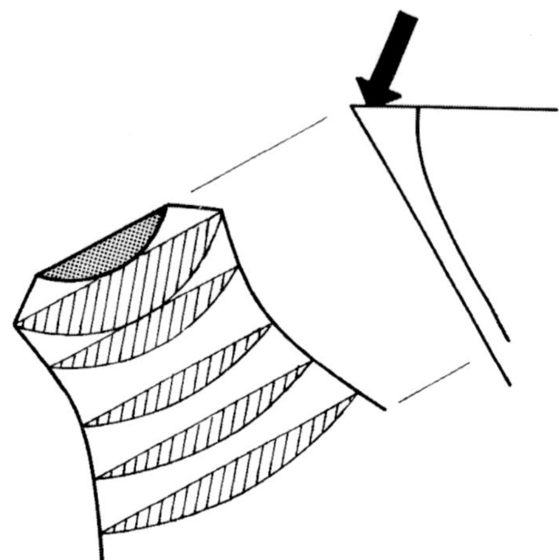

Fig. 4.11 Schematic illustration of a bending flake formed on a flat face of a core, cross-section indicted by hatching (Cotterell & Kamminga 1987 with permission Society for American Archaeology).

The bipolar, conchoidal and bending flakes are the three main types of flakes. To understand how the mechanics of fracture controlled the possible forms of stone flakes and tools, it is necessary to discuss the three phases in the detachment of a flake: initiation, propagation and termination. Within these three phases there are a number of possibilities which are schematically illustrated in Fig. 4.12.

4.4.1 Initiation phase

The three types of flakes discussed have been categorised by the method of flake initiation. In bipolar flaking, though there may be an element of initiation by indentation, the most likely form of initiation is by wedging. Repeated percussion blows will induce indentation cracks which will become filled with debris. Subsequent blows wedge open these indentation cracks and fracture initiates under the wedging forces just as a tree trunk can be split by steel wedges. Wedging initiation is more likely if the initiation takes place away from the edge of the core or if the edge angle is large, especially if it is obtuse.

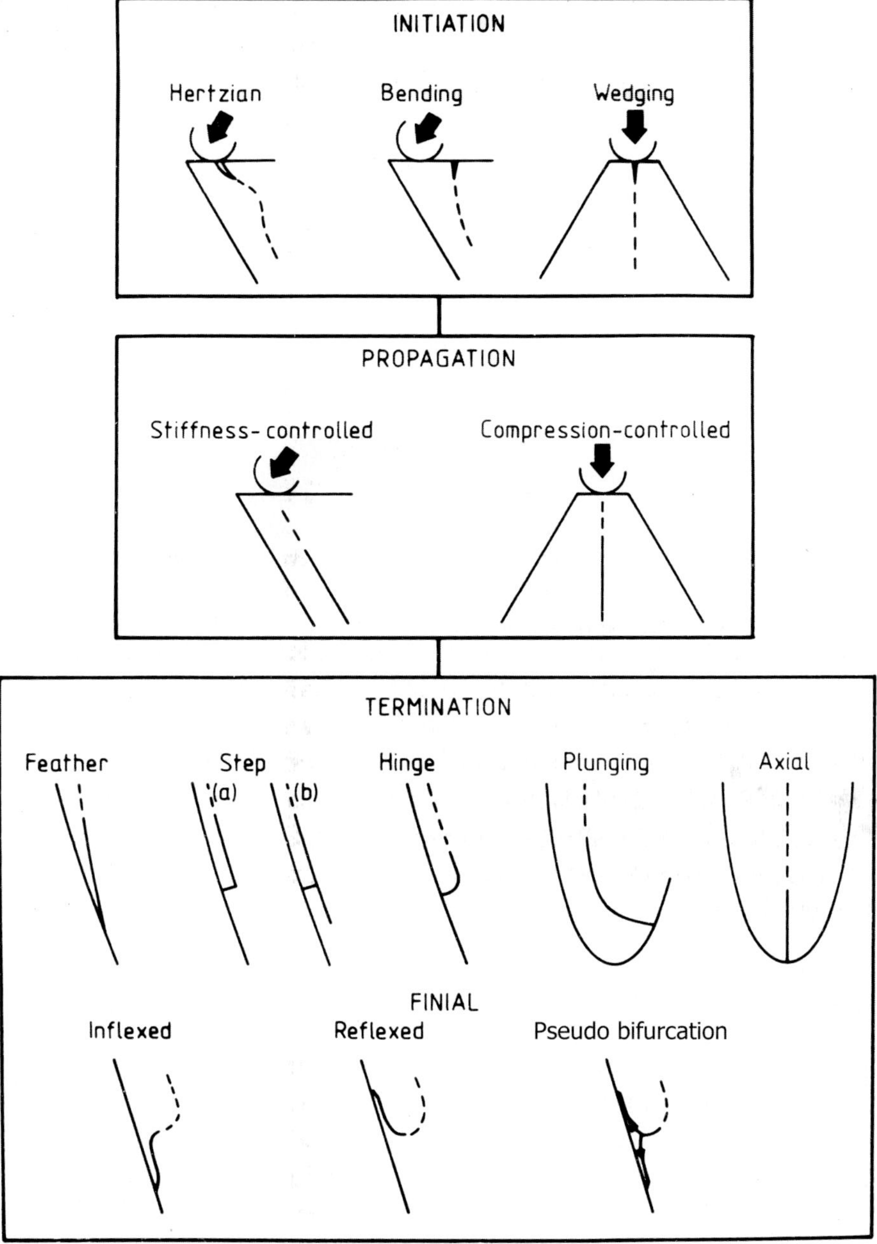

Fig. 4.12 The three phases in the formation of a flake (Cotterell & Kamminga 1987 with permission Society for American Archaeology).

The initiation of a conchoidal flake is due to indentation. Tensile stresses are needed to create a crack. It seems unlikely that indentation of a flat surface with a hard indenter can produce tensile stresses, but imagine indenting a rubber block with a spherical indenter. To accommodate the indentation the rubber around the contact area has to be stretched as the indenter sinks into the rubber and hence the stresses around the edge of the contact are in tension. Although it cannot be seen by eye the same phenomenon occurs when a piece of flint is struck by a hammerstone. If the force is near perpendicular to the surface and the blow is large enough, the tensile stresses initiate a conical fracture. A near perfect Hertzian cone formed in a flint core is shown in Fig. 4.13.

Fig. 4.13 A Hertzian cone formed in a flint core by hard hammer percussion (Cotterell and Kamminga 1987 with permission Society for American Archaeology).

The cone fracture, which is the starting point for a conchoidal flake, is named after Heinrich Hertz (1857–1894) who in 1880 first studied the conical fractures at elastic contacts between curved glass surfaces.[13] If the indenting force is at an oblique angle to the surface, as it is in conchoidal flaking, the tensile stresses on the trailing edge of the contact are much enhanced and those on the leading edge diminished and even replaced by compression. The tensile stresses are also enhanced if the indentation is made near the side face of the core. Consequently in conchoidal initiation, the fracture only forms around the trailing edge of the contact zone and only a partial cone fracture forms.

Both bipolar and conchoidal flakes require high contact stresses, if the flaking tool is too soft these high stresses cannot be generated and a bending initiation is likely. Bending fractures initiate away from the indenter under the action of tensile bending stresses and are easier if the edge angle of the core is acute. Bending flakes have no bulb of percussion, though the transition from initiation to propagation can be mistaken for a diffuse bulb of percussion. The initiation is cleaner than wedging or indentation and there are no secondary flakelets in the initiation region, but the absence of secondary flakelets does not necessarily indicate a bending flake. Stone tools suffer fracture during use and, on acute angled edges, bending initiation is common.

4.4.2 Propagation phase

The propagation phase is the most interesting phase in the formation of flakes. Why could long sharp flakes be detached easily from a core? Such flakes make possible the large variety of tools that man and his hominid ancestors made and in particular enabled him to have tools suited to the efficient butchering of his game without which the encephalisation of the brain may not have occurred and we may have evolved not into man but just another species of great ape with a small brain capacity. It is not skill that enables long flakes to be struck, but the mechanics of fracture. In homogeneous isotropic materials fractures propagate to maintain local symmetry in the stresses at the tip of the crack.[14] Such local symmetry ensures that the maximum energy is released for fracture. However, no fracture propagation is perfect and some local inhomogeneities will always deflect a crack from its idea path. If the crack path is stable, any deflections from the ideal path will be corrected and the fracture will return to its original path, but crack paths can be unstable so that once deflected from the ideal path, the crack continues to deviate. What determines whether a crack path is stable or not? Although it was stated in Chapter 1 that the stress field very near the tip of a naturally propagating crack is unique, the stress field loses that uniqueness the further one goes from the crack tip. Crack path stability depends on these non-unique stresses. If the general stress along the crack propagation direction is compressive the crack path is stable, but if it is tensile the crack path is unstable. It is fortunate for us that in flaking it is compressive and the crack path is stable.[14] However, establishing the stability of the crack path does not mean that long flakes can be detached, but only that whatever the flake form it will be predictable and reproducible.

                          Fracture and Life

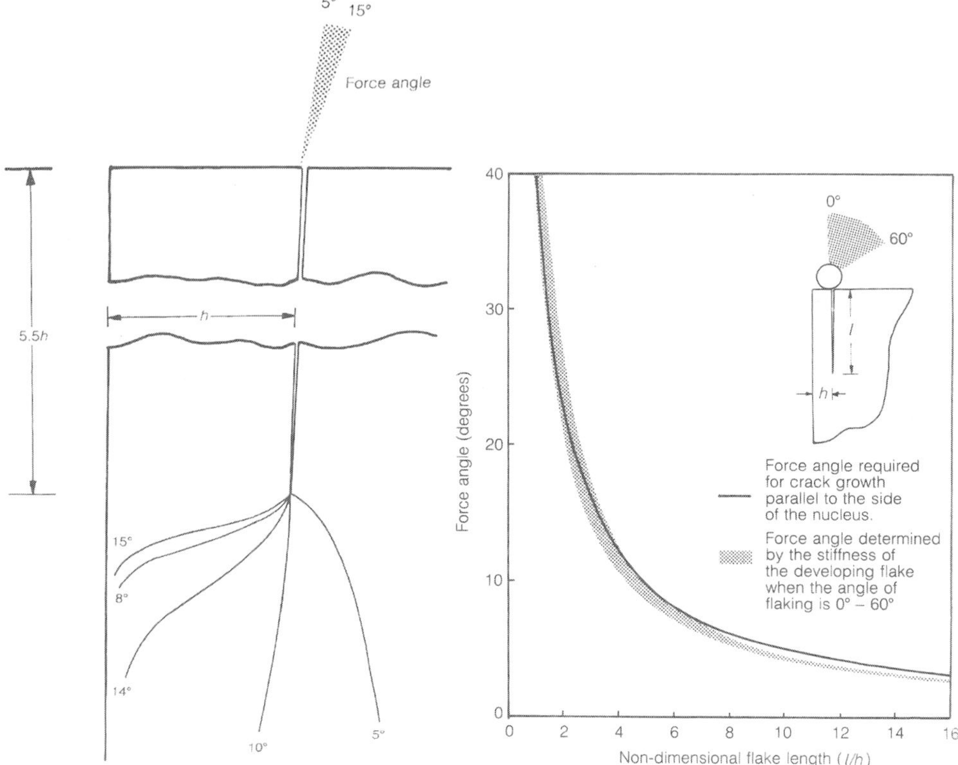

Fig. 4.14 Crack paths obtained from controlled force angles (Cotterell & Kamminga 1990 with permission CUP).

Fig. 4.15 Stiffness-controlled crack propagation (Cotterell & Kamminga 1990 with permission CUP).

Consider first a flake detached near a side face of a core, the mode of initiation is relatively unimportant. An idealised flake is considered where a fracture is propagating parallel to a flat side face of the core. It is assumed that the flaking force acts through the inner corner of the pupative flake. In a series of simulated flaking experiments a thick glass plate had a slit machined parallel to the side face to a depth 5.5 times its width with a natural crack at its tip, as shown in Fig. 4.14.[15] A force was applied at angles varying from 5° to 15° and the fracture paths shown in Fig. 4.14 were subsequently recorded. There is some inconsistency in the fracture paths caused by inaccuracies in forming the sharpened tip, but the general conclusion from this simulated flaking experiment is that the crack path is very sensitive to the force angle with the fracture hinging towards the side face of the core for large force angles and plunging into the core for small angles. For a $10°$ force angle the fracture only gently hinges towards the side face of the core. However, if the experiment is repeated with a longer initial slit length the force angle necessary to make the fracture run straight would be smaller. The necessary force angle, as a function of the length of the flake, to ensure that the fracture propagates parallel to the side face of the core has been calculated theoretically and is shown in Fig. 4.15. The force angle has to change rapidly with the propagation of the fracture to ensure that it continues to propagate parallel to a side face of the core. That is, if the stone knapper were controlling the force angle he would have to be continually changing the force angle to detach long flakes. A flake is detached from its core in a matter of milliseconds so that such control is impossible and long flakes could not be detached from the side of a core if it was the force angle that the knapper controls. But does the knapper really have direct control of the force angle? Consider percussion flaking. The stone knapper has a hammer stone weighing about 0.5 kg with which he hits the core. What the knapper controls is not the force angle but the direction of the blow, which is not the same as the force angle. The point on the flake under contact with the hammer stone follows closely the initial path of the blow since the hammer stone has a considerable inertia. As the fracture propagates so the stiffness of the detaching flake decreases. The result is that the force angle changes because of the change in stiffness of the partially detached flake. Superimposed on the force angle necessary to ensure the fracture propagates parallel to the side face of the core shown in Fig. 4.15 is the force angle caused by the stiffness of the detaching flake as the direction of the blow varies from $0°$ to $60°$.[16] The force angle determined by the stiffness remains in a narrow band despite the large variation in the direction of the blow and this band closely follows the force angle necessary to keep the fracture propagating parallel to the side face of the core. A high degree of skill is unnecessary for the detachment of long flakes. However, skill is necessary in the preparation of the core and placement of the blow to detach a predetermined flake. The mechanics of fracture happen to make the detachment of long flakes possible on a correctly prepared core. If flaking were not easy, stone tools would probably not have been made at all and man may not have evolved. Evolution is controlled by physics as well as by biology.

So far the discussion has concerned percussion flaking. Does the same apply for pressure flaking? Pressure flaking was predominantly used for the removal of thin flakes during retouch so that not so much inertia is needed to ensure that the direction of motion of the flaking tool remains sensibly constant. In this delicate flaking, the inertia of the hand and lower arm must be added to that of the light flaking tool. Large flakes were sometimes removed by pressure using a flaking point attached to the leg of a T-shaped piece of wood. The knapper would throw the weight of his chest on the bar of the T so that pressure flaking was really semi-percussion anyway.

The mechanics controlling the formation of bipolar flakes where a flake is initiated away from a side face of the core are somewhat different. In bipolar flaking the direction of the blow and the force are essentially in an axial direction. There is little bending and the main stress is compressive. Such a stress distribution produces a path stable compression-controlled fracture than tends to split the core into one or more pieces. The fracture is similar to that obtained in engineering compression tests on cylindrical brittle rock specimens.[17] The path is so strongly determined by the mechanics that even compression tests on single crystals of rock salt, orientated so that the compression force is at a significant angle to the highly preferred cleavage planes, fracture in the compression direction.[18] Compression-controlled propagation is unlikely to occur in flaking other than bipolar.

4.4.3 Termination phase

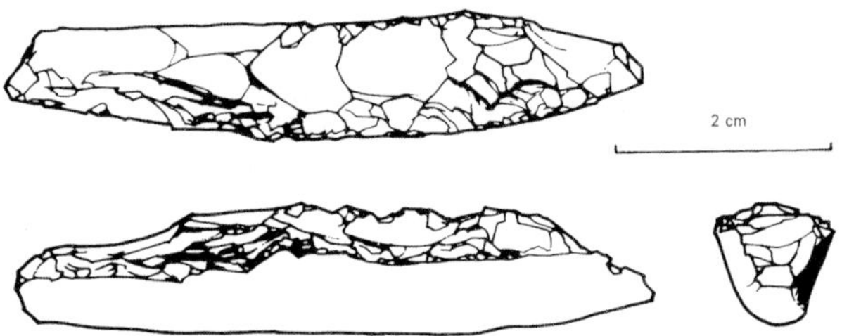

Fig. 4.16 Pirri graver made from a hinge flake, drawing by G. Happ (Kamminga 1985 with permission Aboriginal Studies Press).

The possible terminations of flakes are illustrated schematically in Fig. 4.12. The feather and the axial termination are simply the continuation of a stiffness-controlled and compression-controlled propagation. The fracture turns towards the edge face of the core to form a hinge termination if the force angle becomes too large to continue the fracture parallel to the edge face. Likewise a plunging termination results when the force angle is too small. Hinge terminations are usually undesirable, but the Australian Aborigines made deliberate use of a hinged termination to form a specialised tool, illustrated in Fig. 4.16, which was identified by Jo Kamminga and called the *pirri graver*.[19] The gouge-like tool, with the entire flake removed by retouch except the hinge termination, was used to cut decorative grooves on wooden artefacts such as boomerangs.

Step terminations are the result of an arrest of the flaking fracture. The fracture is reinitiated, either after a time or almost immediately, by a force with a large outward component causing an abrupt change in the fracture direction. There are two varieties: either (a) the flake detaches completely or (b) a small portion is left attached to the core. On semi-translucent stone, such as obsidian or flint, step terminations of type (b) can be seen highlighted by the reflection of light.

Fig. 4.17 (a) Acheulian hand-axe from Swanscombe: enlarged portions (b) an inflexion on a hinge termination, (c) a retroflexion on a hinge termination (Cotterell & Kamminga 1986 with permission Elsevier).

During the last stages in the formation of a hinge termination delicate finials can be formed on the most homogenous and fine-grained stone.[20] After the fracture has turned to propagate at right angles to the side face of the core, the general stress distribution is one of bending with tensile stresses parallel to the side face near the tip of the fracture and compressive stresses near the edge face itself. The tensile stresses make the fracture path unstable and the fracture either curves forward to run parallel to the edge face to form an inflexed finial or back on itself to form a retroflexed finial. These two types of finial can be seen on the enlarged portion of the Acheulian hand-axe from the British Palaeolithic site at Swanscombe shown in Fig. 4.17. Bifurcation is also possible where the fracture both inflexes and retroflexes detaching a small fragment.

4.4.4 Surface markings

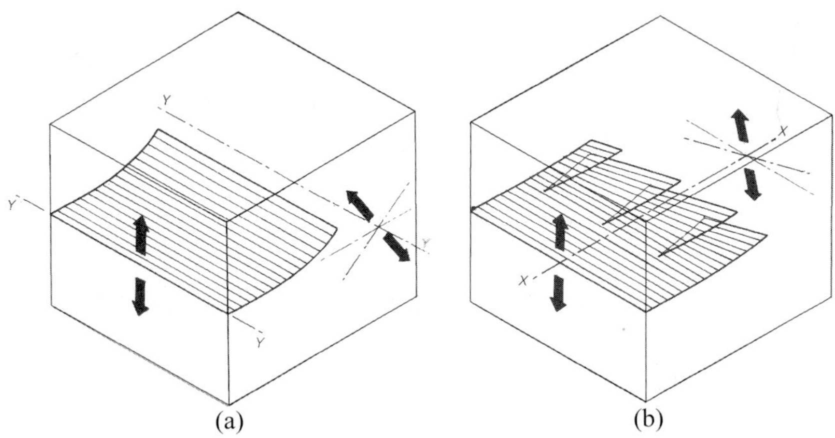

(a)   (b)

Fig. 4.18 (a) The formation of an undulation, (b) The formation of lances (Cotterell & Kamminga 1990 with permission CUP).

The surfaces of fine-grained stone tools are generally smooth apart from the features already mentioned, but they do have a number of finer features. Many of these features are commonly seen in the surface of glass fractures in ordinary life and were first described by Charles de Fréminville (1856–1936) and Frank Preston (1896–1948) in the early twentieth-century. The most common markings on the surface of a flake are undulations which can be seen in Figs. 4.8 and 4.11. Undulations result from a slight rotation of the principal stresses about the fracture front, as shown schematically in Fig. 4.18 (a). When the principal stress field rotates about the fracture front, the fracture surface can respond over its entire fracture front. However, if the principal stress field rotates about an axis perpendicular to the fracture front, as illustrated in Fig. 4.18 (b), the crack front cannot rotate as a whole and the crack front breaks up to form markings called lances because of their resemblance to medieval war lances. The surface of the conchoidal flaked shown in Fig. 4.11 has a number of lances in the direction of fracture propagation, but they are not very distinct. In unrelated experiments on the compression fracture of glass, I detached the blade-like flake shown in Fig. 4.19, which ended in a step termination and contained numerous very distinct lances. An érailure scar which is a common feature on the bulb of percussion (one can be seen on the conchoidal flake shown in Fig. 4.14) is a specialised lance where the tongue of the lance extends across the bulb of percussion. Often the érailure flake remains attached to the core.

Fig. 4.19 Compression glass flake showing numerous lances.

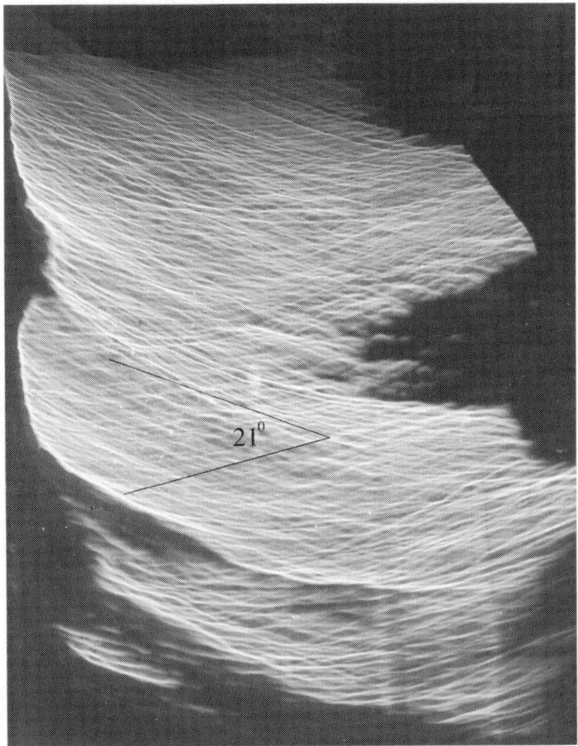

Fig. 4.20 The surface of a blade from Lipari in the Aoelian Islands showing Wallner lines.

Obsidian flakes often contain distinctive delicate surface markings called Wallner lines as can be seen on the obsidian blade from Lipari shown in Fig. 4.20. Wallner lines, which are discussed more fully in §9.1.4, are caused by the interaction of shear waves with the fracture front and the fracture velocity can be inferred from the angle of intersection of the lines, which form a record of the velocity of the fracture. Knowing the speed of propagation of shear waves in the stone material, the velocity of the fracture front can be calculated. The Wallner lines on the blade from Lipari shown in Fig. 4.20 are almost straight and the Wallner lines intersect at an angle of about 21°. Assuming that the velocity of shear waves in obsidian is similar to that in soda-lime glass, which is 3,500 m/s, Eq. (9.16) gives the speed of the fracture that formed it some 5,000 years ago as 630 m/s. It can be concluded that the flake was probably detached by percussion because its velocity is higher than that usually observed in pressure flaking. Being able to read the flaking velocity from the surface of the flake from Lipari is as exciting as being able to read the hieroglyphics written some 2,000 years later in Tutankhamen's tomb.

4.5 Ground Stone Tools

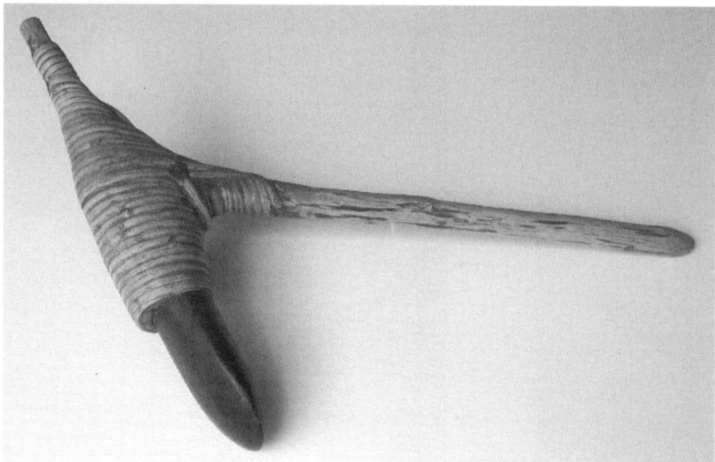

Fig. 4.21 A ground stone adze from Sulawesi.

A brief discussion of ground stone tools is included here because the abrasion of brittle materials is a fracture process, albeit on a small scale. Ground tools were primarily used for axe heads for chopping down small trees and the working of wood. A hafted stone adze from Sulawesi is shown in Fig. 4.21. Flaked stone tools are not efficient as ground axes for wood-working because they become quickly blunted by flaking during use, the tougher ground axes though not as sharp as flaked ones, are much more durable. The process of grinding a stone tool is simple but takes time. Friable sandstone, with clay bonding of the grains so that the grains break loose during the grinding process, is the best grindstone. Water is essential to wash away the blunted sandstone grains and to prevent over heating of the stone. The grinding was accomplished by pressure on the forward stroke with the pressure being relaxed on the return stroke. In time a groove develops in the grindstone, as seen in Plate 5. In Australia only the cutting edge of the tool was ground, but elsewhere they were often ground and polished on all the surfaces as for example the greenstone ceremonial axes of New Guinea and the stone imitations of the bronze axe in Europe.

4.5.1 *The mechanics of abrasion*

Abrasion is important to archaeologists for two reasons: firstly, the mechanics of grinding are of interest in understanding ground stone tools, but more importantly stone tools can wear by abrasion, as can millstones and stone mortars used in food processing. For effective abrasion the abrasive needs to be harder than the material being abraded, but to scratch a stone the particle does not need to be much harder. For minerals and stone the Mohs hardness scale is useful. In this scale, which is based on a mineral's ability to scratch another, mineral standards are arranged in order of their hardness from talc (hardness 1) to diamond (hardness 10). The quartz in sandstone has a Mohs hardness of 7 and basalt 5.

During the use of flaked stone tools abrasion can occur. Abrasion of flaked stone occurs by microfracture. The form of the microfracture depends upon the sharpness of the abrasive, whether it is free or fixed. The stress distribution under an indenter is basically compressive. Under a sharp particle the stone will yield plastically before cracking, whereas under a blunt particle the deformation is essentially elastic before cracking. The terms 'sharp' and 'blunt' are not absolute but depend upon the stone being abraded. The difference in the behaviour of sharp and blunt free and fixed particles is shown in Fig. 4.22.[21] Free blunt particles act like Hertzian indenters and produce isolated partial cone cracks with little material removal, as seen in Fig. 4.22 (a). If the blunt particles are fixed either a plastic furrow is ploughed across the glass if the load is comparatively

light, which in use-wear studies archaeologists call a sleek, or a series of partial
cone cracks around the trailing edge of the particles form under heavier loads, as
seen in Fig. 4.22 (b); again there is little material removed. With sharp indenters
the pattern is different. The glass under the indenter yields before the glass cracks
to form a median crack normal to the surface. Since the material has yielded, it
cannot return to its undeformed state on unloading and tensile stresses are created
that cause lateral cracks to form that vent to the surface with the detachment of
microchips. If the indenting particles are free, isolated median and lateral venting
microcracks occur, as can be seen in Fig. 4.22 (c). Abrasion under fixed sharp
particles causes more or less continuous median cracking followed by venting
with the removal of microchips, as shown in Fig. 4.22 (d). The glass cutter relies
on a diamond or carbide wheel tool to produce a median crack which he can then
propagate through the thickness of the glass by bending. If a cut piece of glass is
examined, the side scored will be seen to have many microchips which were
caused by the venting cracks.

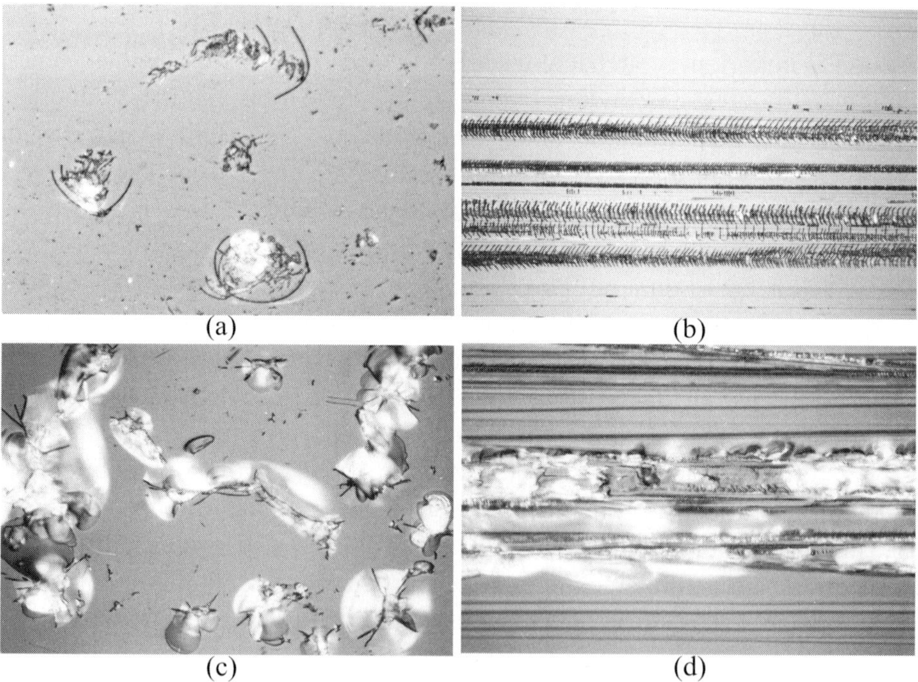

(a)   (b)

(c)   (d)

Fig. 4.22 Abrasive wear on glass caused by 220 grit silicon carbide particles: (a) free blunt
particles, (b) fixed blunt particles, (c) free sharp particles, (d) fixed sharp particles; width of field:
(a) and (c) 150 μm, (b) and (d) 300 μm (Cotterell & Kamminga 1990 with permission CUP).

The description given here of abrasion is for an ideal situation and these classic patterns are not necessarily distinguishable, but the microscopic abrasion and plastic flow patterns can reveal the function of a stone tool.

4.6 Use-wear on Stone Tools

The study of microchipping, scratching, fine abrasion polishing, and rounding of edges enables archaeologists to determine the possible use of stone tools. The recognition of use-wear is not new. In the nineteenth-century William Greenwell (1820–1918), a canon of Durham, observed that the working edges of some Palaeolithic end scrapers he had found on the Yorkshire Moors were smoothed and rounded.[22] Although Greenwell knew that this was likely to be use-wear he could not identify the possible use of the tools. It is now known that this use-wear was from the preparation of animal skins. When mammal skins are cleaned using stone scrapers, microchips are broken off which then abrade the tool's edge and it is this wear that Greenwell saw. Mammal skins are usually scraped immediately after the animal has been skinned and the use wear is due to free particles and similar to that shown in Figs. 4.22 (a) and (c). The Australian Aborigines processed marsupial skins differently; they dried the skin before scraping it and the use-wear is different.[23] When the marsupial skins were stretched and left to dry they were exposed to sand, ash and dust particles which adhered strongly. The use-wear seen on stone tools used by Aborigines to scrape skins is intense abrasion on the surfaces due to the fixed abrasive particles in the skin and akin to the damage shown in Figs. 4.22 (b) and (d).

Use-abrasion of stone tools occurs as described in the previous section but these classic patterns are difficult to identify on other than obsidian. Lances on a flake's surface can form a bench mark for assessing abrasion damage. Not all use-wear is by abrasion. Acute edges on tools readily chip during any cutting or sawing activity usually forming a bending microflake. Hard pieces embedded in the material being cut can also cause conchoidal microflakes. Again not all use-wear is fracture related. Stone tools can become polished during use. Silica polish due to the presence of opaline silica in grasses can be seen on sickle blades from the Middle East. Silica polish, also known as silica sheen, is strikingly lustrous and is the result of a combined chemical and mechanical process. Although it is essential to understand the mechanics of flaking and abrasion, use-wear analysis relies heavily on experimental studies.

Use-wear analysis has only been established as an archaeological discipline comparatively recently. However, it has already enabled a better interpretation of many stone tools. Without use-wear analysis it is not always possible to distinguish between a discarded flake and a flake tool or a discarded core and a core tool. For example, in Australia a distinctive domed discoid-shaped core usually about 0.5 kg in weight, called a horse-hoof core from it shape, was for long thought to be a pounder for a variety of uses varying from smashing shell fish to adzing wood until Jo Kamminga established by examining the so-called use-wear that they were in fact not tools, but discarded cores.[23] It is ironic that in the nineteenth-century British antiquarians correctly regarded similarly shaped European cores as just that – discarded cores, not tools. However, in 1877 Gillespie, in an address to the Royal Anthropological Institute, argued that the horse-hoof cores were tools.[24]

4.7 Concluding Remarks

Today fracture is usually studied so that it can be avoided. It is refreshing and humbling for someone who has studied fracture all their life to examine flaked stone tools because here, right at the beginning of human evolution, our hominid ancestors were unknowingly using the mechanics of fracture to make tools that enabled them to better exploit their environment and affect their evolution.

4.8 Notes

[1] Semaw *et al.* (2003).
[2] Mulvaney and Kamminga (1999).
[3] Oda and Keally (1973).
[4] Clark and Piggott (1970).
[5] Hardy and Raff (1997).
[6] Species of the same genus usually exhibit a 90–95% match; within the same family a 85–90% match; within the same order a 80–90% match.
[7] Clark (1968).
[8] Foley and Lahr (2003).
[9] Whittaker (2001).
[10] Spencer (1928).
[11] Domanski and Webb (1992).
[12] Dunnell *et al.* (1994).
[13] Hertz is better known for being the first to broadcast and receive radio waves in the laboratory and for showing that the properties of radio waves are the same as those for light and heat. The unit of frequency, the Hertz (one cycle per second), is named in

Hertz's honour. Hertz was one of the outstanding physicists of the nineteenth century and crammed more than one lifetime's work into a short 37 years.

[14] Cotterell and Rice (1980).

[15] Cotterell *et al.* (1985).

[16] Non-linear effects have been shown to be significant for long flakes (Chiu *et al.* 1998). However, though the detail is changed somewhat, the general conclusions remain the same.

[17] Bipolar flaking is similar to the splitting form of compression fracture shown in Fig. 2.6 (a).

[18] Gramberg (1965).

[19] Kamminga (1985).

[20] Cotterell and Kamminga (1986).

[21] Lawn and Wilshaw (1975).

[22] Greenwell (1865).

[23] Kamminga (1982).

[24] Gillespie (1877).

Chapter 5

Building in Stone and Concrete
in the Ancient World

The most visible remains of the ancient world are the stone buildings. Building in stone presented two fracture challenges: the quarrying and dressing of the stones and the design of stable buildings. Although less has been written on the extraction and dressing of stone than the building techniques, this challenge was the more difficult before iron came into general use. Nature after all had been building in stone since the world began. The stresses in geological formations are largely compressive. We saw in Chapter 2 that rocks are generally strong in compression, especially if the compressive hydrostatic stress is large. If we model Mount Everest as a pyramid with a height of 8,848 m, the average compressive stress at its base would be $8848\rho g/3$, where ρ is the average density of the rock and g is the acceleration due to gravity. The top of Mount Everest is composed of limestone laid down in a sea some 400 million years ago and rests on sedimentary and igneous rocks transformed into crystalline metamorphic rocks. Assuming a density of 2,700 kg/m^3, the average compressive stress at the base of Everest would be 78 MPa; considering that the hydrostatic stress would be similar, such a compressive stress is not high. The Great Pyramid of Cheops built out of limestone was originally 147 m tall. The density of limestone is 2,600 kg/m^3 thus the average compressive stress at the base of Cheops' Pyramid is only 0.4 MPa. A similar low stress can be estimated for stone columns. The Temple of Aphaia on the Greek Island of Aegina built in 490 BC has limestone columns with a diameter at the base of 0.98 m. The columns weigh some 8 tonnes and support a load estimated to be 20 tonnes,[1] giving the same low compressive stress at their base as Cheops' pyramid of 0.4 MPa.

The roofing of large buildings and the construction of bridges was more difficult, though even here some of the examples from nature are impressive. The impressive sandstone Landscape Arch in Utah shown in Plate 7 is the world's

116

largest natural sandstone arch, has a span of 88 m and is 1.8 m thick at its thinnest point.

5.1 Spanning Openings

The first and simplest method of spanning an opening was to use a lintel or architrave. Although not from the ancient world, the Late Middle Age clapper bridges of Dartmoor in England provide a classic example of the lintel construction. The term 'clapper bridge' is used on Dartmoor for a bridge which has one or more flat slabs of stones which rest on stone piers. The Postbridge spanning the East Dart River shown in Plate 8 is probably the oldest of the clapper bridges being built in the early fourteenth-century. The most efficient means of spanning an opening is with an arch, vault or dome.

5.1.1 Architraves

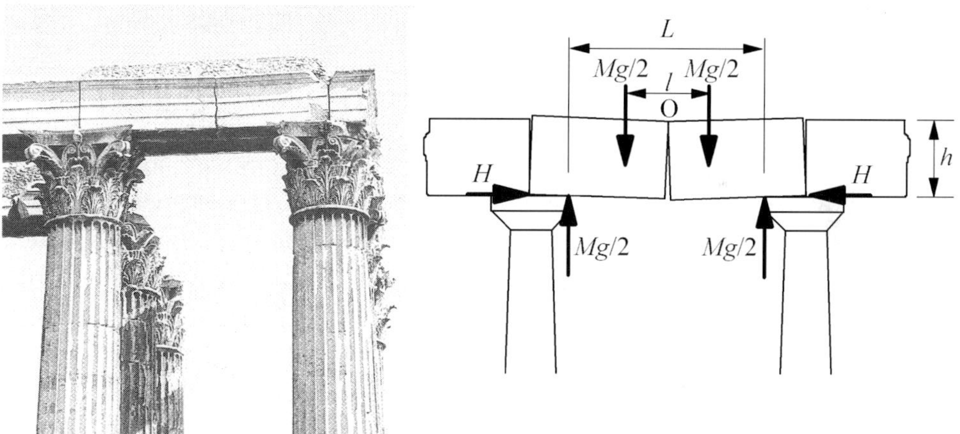

Fig. 5.1 Temple of Zeus, Athens   Fig. 5.2 Forces acting on a cracked yet stable architrave
(Cotterell and Kamminga 1990, with permission CUP).

The problem with stone lintel construction is that the lintel sustains a bending moment and the bottom of the lintel is in tension. Since stone has very limited tensile strength the opening that can be spanned by a lintel or architrave is quite small. However, even if the architrave cracks it is possible for it to remain stable provided there is sufficient horizontal thrust on it. Earthquakes have fractured many Greek architraves that have not fallen. A fracture in an architrave in the Temple of Zeus in Athens (shown in Fig. 5.1) has caused it to sag but not collapse.[2] The forces acting on an architrave that is cracked at its middle are illustrated in Fig. 5.2. In order for it to remain stable the horizontal thrust, H, must be large enough to ensure that the moment about O caused by the weight of the architrave and any superimposed load must be balanced by the moment of the horizontal thrust. Thus if M is the total mass of the architrave and the superimposed load then for the architrave to be stable

$$Hh \geq \frac{Mg}{2}\left(\frac{L-l}{2}\right),$$

$$\text{or } H \geq \frac{Mg(L-l)}{4h}.$$

(5.1)

The horizontal force, H, must be generated by the thrust from adjacent architraves. What is at first sight surprising is that if the halves of the architrave start slipping inwards, the frictional force destabilises the architrave and it must fall.

5.1.2 Arches

In architecture great distinction is made between what is called a *true arch,* where the joints between the brick or stone blocks (voussoirs) are normal to arch and the *false arch* which can be two stones leaning together at their apex or a corbelled arch, where the stones or bricks are laid horizontally, each course slightly overlapping the one below until they meet. True stone arches built from voussoirs are aesthetically pleasing, as is the Gate of Jupiter at Falerii Novi, shown in Fig. 5.3, built by the Romans in the style of the Etruscan Falerii Vertes and which had been destroyed in 241 BC, but structurally there is no real difference between a so-called true arch and any other. It is easier to build a corbelled stone arch without centring (temporary support) than a stone true arch. However, brick arches were built as far back as the first dynasty (3000–2600 BC) in Egypt. The controversy of the origin of the arch springs from the aesthetic distinction between true arches and corbelled ones.

The secret of designing an arch is to ensure that the thrust line is wholly within the masonry. The early arches were semicircular, but provided the horizontal thrust can be resisted, more elegant low arches such as the natural Landscape Arch shown in Plate 7 are stable. An arch of perfect shape would have its thrust line coinciding with the centre line of the stones with no tensile stresses.

Robert Hooke in 1675 was the first to understand the perfect shape but because of his quarrels with Isaac Newton (1642–1727), he gave the result as a Latin anagram, as he did for his law of elasticity; in 1705 his friend, Richard Waller (ca 1646–1715) published the solution in the posthumous papers of Hooke: *ut pendet continuum flexile, sic stabit contiguum rigidum inversum*, which, freely translated, reads 'as hangs a flexible cable so inverted stand the touching blocks'. For an arch of uniform thickness supporting just its own weight the perfect form is a catenary, the reverse of the shape of a hanging flexible cable. In modern limit theory applied to the analysis of arches or domes, it is assumed that masonry has no tensile strength and the so-called safe theorem states that an arch is stable if a thrust line that is in equilibrium with the external loads can be drawn that is wholly contained within the masonry.[3]

Fig. 5.3 The Gate of Jupiter at Falerii Novi (Adam 1994 with permission Indiana University Press).

5.1.3 Vaults and domes

Vaults are simply broad arches, and their stability is the same. Domes are even more stable than arches. Eggs, which can be viewed as two domes back to back, are very resistant to fracture from external loads and a chicken's egg can support a compressive force applied to its poles of about 30 N (3 kg load). However, the egg can easily be broken by the hatching chick pecking from the inside. A hemispherical dome has the advantage that there is no radial horizontal thrust at its base. However, the circumferential stresses do become tensile towards its base. For a hemispherical dome of constant thickness, the stresses become tensile at a height of approximately 0.31 times the diameter of the dome (51.8° from the crown).[4] The stresses are wholly compressive in a segmental dome whose height is less than about 0.19 times the diameter of the dome and can be very thin, but there is a radial horizontal thrust to be accommodated. Because of the circumferential tension, an unreinforced hemispherical dome is liable to vertical splitting and iron hoops and chains around the springing of some domes is testimony to the concern that engineers have felt about this liability. However, domes can develop vertical cracks around the springing and remain stable as will be discussed further in §6.5.

5.2 Ancient Egyptian Masonry[5]

True stone architecture starts quite suddenly ca. 2700 BC, during the reign of the third dynasty Pharaoh Djoser, with the building of the step pyramid and mortuary complex at Saqqara by Imhotep. Limestone sandstone and granite were required in enormous quantities for a period of 3,000 years, as were tools for extraction and dressing of the stones.

5.2.1 Building stone

Limestone was readily available along the Nile from Giza to Esna and quarries were developed, mainly on the east bank, along this long stretch of the Nile. From the third dynasty until the early new kingdom (ca. 1500 BC) limestone was the building stone of choice. Sandstone deposits lie in upper Egypt between Esna and Gebel Barakal and the most famous quarries were at Gebel el-Silsilia some 50 km north of Aswan. Sandstone was used from the eleventh dynasty (ca. 2000 BC) but did not become common until the until the beginning of the new kingdom when it began to be transported by the Nile some 600 km north to Memphis (about 16 km south of Cairo) and other towns in Lower Egypt. Most limestone and sandstone quarries were open cut, but such enormous quantities were needed that stone was also quarried deep inside mountains.

Limestone and sandstone are comparatively soft. Granite is considerably harder and more difficult to work with than the softer stones, especially before the advent of steel tools. In earlier periods surface boulders provided a source of granite. There are comparatively few deposits of granite in Egypt; the most famous granite quarry was just south of Aswan from which grey and rose granite was quarried. Granite was used as early as the first dynasty (ca. 3000 BC); though there was a peak in its use during the old kingdom (ca. 2600–2200 BC) most quarrying occurred during the eighteenth and nineteenth dynasties (ca. 1550–1200 BC). Other hard stones were also used. Reddish quartzite, from Gebel Ahmar just east of Cairo, was favoured by the sun kings of the fourth and fifth dynasties (ca. 2500–2400 BC) in their temples dedicated to the sun in the Abusir region just south of Cairo, and again in the eighteenth dynasty by Amenhotep III (1387–1350 BC) and Akhenaten (1350–1333 BC) at Luxor.[6] Dark green greywacke, a hard variety of sandstone, was quarried, at Wadi Hammamat some 50 km east of Karnak, from Predynastic period until the late Roman period (ca. 4000 BC –300 AD) as some 600 rock inscriptions testify.[6]

5.2.1.1 *Properties of building stone*

In discussing the working of Egyptian building stone it is usual to differentiate between soft stone: alabaster, limestone, the softer sandstones, and hard stone: the harder sandstones (such as greywacke), granite, basalts, diorites, and quartzite. A collection of the properties of these rocks is shown in Table 5.1. All rocks have sufficient compressive strength for use as columns, arches and domes. The tension strength and the fracture toughness are important when it comes to architraves or anywhere tension stresses are developed in a building. A better guide to the effective toughness of a stone comes from its characteristic length, l_{ch}, and Table 5.1 shows that the characteristic length of the hard rocks is much larger than that of the softer rocks.

When it comes to extracting and dressing the stone it is the hardness that largely controls the wear on the tools and the fracture toughness on the ease of extraction and dressing. The Mohs hardness obtained from a scratch test does give some indication of the ability of a stone pounder to pulverise a rock where obviously the pounder needs a higher Mohs hardness than the rock. However, the

Mohs scale only ranks the hardness. A better measure of the hardness is the diamond pyramid hardness (DPH) which is usually given in the non-SI units of kg/mm^2.

Table 5.1 Properties of Egyptian building stones (after Arnold 1991 and Whittaker *et al.* 1992).

Rock	Density (kg/m^3)	Hardness		Strength (MPa)		Fracture energy (J/m^2)	Fracture toughness (MPa√m)	l_{ch} (mm)
		Mohs	DPH (kg/mm^2)	Comp.	Tension			
Alabaster	2700	2–3	DPH (kg/mm^2)	–	–	–	–	–
Porous limestone	2650–2850	–	190	20–90	–	–	–	–
Dense Limestone	2650–2850	4	–	80–180	2.7–15	15–65	0.4–2.1	19–23
Sandstone	2000–2650	8	250	30–180	2.7–10	3–35	0.3–1.5	11–21
Quartzite	2600–2800	6–8	970	150–300	10–30	20–42	1.2–1.7	3–14
Granite	2600–3200	6–8	–	160–250	5–14	70–140	1.7–2.6	36–110
Diorite	2750–2870	5–6	850	170–300	15–30	21	3.2	11–45
Basalt	2800–3300	6–8	–	250–400	6–22	40–110	1.8–3.0	97–110

5.2.2 *Tools for extraction and dressing of stone*

The date when iron first came into general use in Egypt is disputed.[7] There was some doubt that hard stone such as granite could be quarried without iron tools and much has been made of the discovery of an iron plate in an airshaft in the Great Pyramid of Cheops in 1837, but its age has not been identified.[8] Nineteen iron objects were found in the tomb of Tutankhamen (1333–1323 BC) including a beautiful gold mounted iron knife alongside one with a gold blade which indicates the value of iron. However, from the time of Tutankhamen there is an increase in iron objects and by the twenty-sixth dynasty (ca. 660 BC) iron became as common as bronze.[7] Wrought iron in the annealed state has a DPH of about 120 kg/mm^2 which can be raised to about 200 kg/mm^2 by cold working. Flinders Petrie (1853–1942) found a score of iron tools at Thebes which he dated to the Assyrian occupation of ca 667 BC, including a steel mason's chisel that had been hardened by quenching to a DPH of 464 kg/mm^2,[9] a modern mason's chisel would have a DPH of 600–900 kg/mm^2. The implements found by Petrie were very likely brought to Egypt by the Assyrian armies, but finds of iron slag at Tel Defenneh and Naucratis, dating to the end of the seventh-century BC, show that by then some iron was being smelted in Egypt.[10]

Almost all copper in Egypt was obtained by smelting ores rather than from native copper.[7] However, Egypt does not have large copper ore deposits. Isolated copper deposits do exist in the eastern desert from about latitude 23°–29°N, with the most important at Um Seiuki about 50 km north west of modern day Ras Benas on the Red Sea coast. Sinai was a sparsely inhabited region exploited by Egypt for mining and also as military route between Egypt and the civilizations of the Fertile Crescent from the third millennium BC. There are copper ore deposits at Magharah and Serabit el Khadim in south-west of the Sinai Peninsula about 150 km from modern day Suez. Mining for turquoise started here in the third dynasty (2670–2600 BC) and continued until well after the twelfth dynasty (1994–1781 BC) when a temple dedicated to the goddess Hathor, the Mistress of Turquoise, was built. There is no proof that the copper ore was smelted down to copper here, but north-west in the Wadi Nasb which enters the Gulf of Aqaba at Dahab there is a large slag heap which indicates that some 5,000 tonnes of copper was extracted. Copper ores at Feinan and Timna in the rift valley between Jordan and Israel have been exploited from the mid-fourth millennium BC.[11] There is no evidence of Egyptian involvement at Feinan, but the ores at Timna were exploited during the nineteenth and twentieth dynasties (ca. 1290–1152 BC).

Copper is soft in the annealed state with a DPH of about 50 kg/mm^2 which can be doubled by cold working. The addition of arsenic increases the DPH of copper to 55–64 kg/mm^2 for arsenic contents of 4.2–7.92% and also significantly lowers the melting point. Cold working can increase the DPH of arsenical copper to between 195–224 kg/mm^2 which is comparable to cold worked wrought iron.[7] The addition of arsenic to copper was accidental and occurred because of impurities in the copper ore. The term bronze is now used for many alloys of copper, but what is known as bronze in the ancient world is an alloy of copper and tin. In ancient bronze the percentage of tin varied from 2 to 16%. The eastern desert has significant deposits of alluvial tin ore in the form of cassiterite.[12] Bronze with about 10% tin has a DPH of about 135 kg/mm^2 in the annealed state, after cold working the DPH increases to about 270 kg/mm^2 which is similar to the hardness of wrought iron. However, if the percentage of tin is greater than 4, cold worked bronze is brittle. Apart from the fact that wrought iron was not harder than bronze, one reason that iron masonry tools were not used until quite late in Egypt is simply supply. Copper and bronze tools are softer than stone and would wear comparatively quickly, hence vast quantities were needed for quarrying and dressing stone. It has been estimated that 690 tonnes of copper were needed for the tools to build Cheops' Pyramid.[13] The materials that were used to extract and dress the building stones of Egypt over some 2,000 years were stone, copper, and bronze.

It is doubtful that Egypt could satisfy her lust for copper locally. Frederick Wicker has suggested that one of the reasons for the expeditions to the fabled land of Punt was to exploit its copper.[13] Our knowledge of Punt is slight. It has long been suggested that Punt was somewhere on the Horn of Africa and was reached by a sea journey along the Red Sea. However, Wicker presents good arguments for Punt being on the south-east shores of Lake Albert and reached by the White Nile, he pinpoints Punt as being near the location of present day Buhuka. An abundant source of copper ore exists at Kilembe in the Great Rift Valley, about 240 km south east of Lake Albert, which is known to have been mined in ancient times. According to Wicker the likely ancestral home of the Egyptians was the area between the Ruwenzori range in Uganda and Nyanza in Tanzania, which is to the south of the location of Punt.[14] Some confirmation that the Egyptians knew the area of the Ruwenzori range comes from Herodotus who relates tales from Greeks that the annual flooding of the Nile could be due to three reasons, the third of which was that the flood came from the melting of snow.[15] He rejects this last reason, arguing that, since it gets hotter the further one goes south. How could there possibly be snow in the upper reaches of the Nile? The Ruwenzori Range has high mountains above 5,000 m which do have glaciers. It is true that the most of the annual Nile flood comes down the Blue Nile from the Ethiopian Mountains and not from the melting of these snows, but it indicates that the Ruwenzori range may have been known to the Egyptians.

5.2.3 Method of quarrying stone

The hardness of the softer stones is not much greater than that of cold worked arsenical copper or bronze and they were comparatively easily chipped with metal tools. The hard stones are much harder than the metal tools available to the Egyptians and were chipped with hafted hard stone picks of granite, chert, basalt or quartzite, or they were pulverized with spherical (diameter 15–30 cm; 4–7 kg) pounders.[16] Even Theophrastus (ca 370–287 BC), a student of Aristotle, wrote in his book *On Stones* that 'some stones…cannot be carved with iron tools, but only worked with other stones'.[17]

5.2.3.1 *Quarrying soft stone*

Fig. 5.4 The bottom of a large-scale limestone quarry north of the Chephren Pyramid shows the stumps of the removed blocks (Courtesy Dieter Arnold).

The Egyptians quarried their stones in a very regular fashion that required the minimum of subsequent dressing (see Fig. 5.4). For large limestone blocks such as those for Chephren's Pyramid trenches some 60 cm wide were cut with copper and stone chisels down to 30 to 40 cm below the depth of the blocks, a total depth of about 1 m.[16] The mechanism of chiselling was to chip flakes successively from a free edge in much the same way as flakes are detached from a core in making stone implements as described in Chapter 4, but building stones are less brittle. Unfinished trenches in the quarry for Mycerinus' Pyramid show that narrow slots were cut down each side of the trench with a copper tool to a depth determined by the length of the tool and then the projecting ridge was smashed with stone pounders.[18] Finally the block was detached from the rock by cutting slots and driving in wedges to split the rock. If the limestone had well-defined horizontal bedding planes, as in the case of the Tura quarries, whose brilliant white limestone was used to face the Great Pyramid, this final splitting of the block from its bed was comparatively easy.[19] However, the only wedges that have been discovered are those found in the Ramesseum (ca. 1250 BC). The famous English Egyptologist Flinders Petrie (1853–1942) dated these wedges to 800 BC,[20] but they could be of much later date. However Somers Clarke (1841–1926) and Rex Engelbach (1888–1946) stated that wedge slots could be seen in the sandstone quarries at Gebel el-Silsila used for the Ramesseum.[19]

5.2.3.2 The use of wooden wedges expanded by water

The American Egyptologist George Reisner (1867–1942) suggested that the limestone blocks for Mycerinus' Pyramid were split from the bedrock by cutting slots 160 x 160 mm into the sides of the trenches, packing the slots with wood and then flooding the trenches to cause the wood to swell and split the rock.[18] It is also stated that this technique was used in certain quarries until the eighteenth-century.[21] Spruce swells on wetting from between 6.5–9.5% and when constrained develops a maximum pressure of between 0.6 and 1.8 MPa depending on the condition of the wood before wetting.[22] Soft wood swells more than hard wood, but the maximum pressure developed when constrained depends largely on the Young's modulus. Since the Young's modulus of hard wood is greater than that of soft wood, the maximum constraint pressure is not likely to vary greatly with the type of wood. The maximum pressure develops after about an hour and then drops as the wood deforms viscously. Oven-dried wood not surprisingly develops the highest constrained pressure. Spruce that has been conditioned at 65% relative humidity develops a maximum constrained pressure of between 0.6 and 1.1 MPa.[22] The relative humidity in Egypt is low and perhaps the maximum constrained pressure could be nearly the same as for oven dried spruce which is 1.8 MPa. We can calculate the degree of splitting obtainable from the fracture toughness of rock given in Table 5.1 and the expression for the stress intensity factor, K, for a crack that is subject to uniform pressure over the depth of the slot which is packed with wood. The geometry is simplified to a semi-infinite rock with a total crack of depth, a, that has grown from an initial slot of depth, b and the stress intensity factor for this geometry is given approximately by[23]

$$K = 2p\sqrt{\frac{a}{\pi}}\sin^{-1}\left(\frac{b}{a}\right)\left[1.3 - 0.1785\left(\frac{b}{a}\right)\right], \tag{5.2}$$

where p is the maximum constrained pressure developed by the swelling wood. Obviously, this method would only be efficient if the initial slot that has to be cut is not too deep. Reisner believed that the 160 mm deep slots he observed in the

Mycerinus quarry were used to split the limestone blocks from the bed rock by wetted wedges and this is assumed to be the depth of the initiating slot.[18] The stress intensity factor is plotted against the crack depth, a, in Fig. 5.5 for three constrained pressures 0.6, 1.1 and 1.8 MPa, corresponding to the smallest and largest constrained pressure observed for conditioned spruce and the largest constrained pressure for oven-dried spruce. Superimposed on Fig. 5.5 are the range in fracture toughness observed for limestone and sandstone. Examining Fig. 5.5, it does not seem probable that simply packing the slots with wood and then wetting it would be an efficient method of splitting stone blocks from their bedrock. The method, which required the cutting of a slot 160 mm deep, is unlikely to have been used unless it was possible to cause a fracture of the order of half a metre long. For the toughest limestone and sandstone it would be impossible to even start a crack. If the largest constraint pressure for oven dried spruce was achieved, then a crack could propagate to a depth of half a metre providing that the fracture toughness of the rock was not more than 0.58 MPa√m, but the average fracture toughness of both limestone and sandstone is greater than that. A combination of hammering wooden wedges into a slot and then wetting them could might produce a high enough stress intensity factor to cause a long crack, but the contribution from the pressure due to wetting the wood would probably be the more minor. In the absence of any experimental evidence that soft rock can be split by wetting wooden wedges it must be concluded from the above analysis that the technique is unlikely to have been efficient.

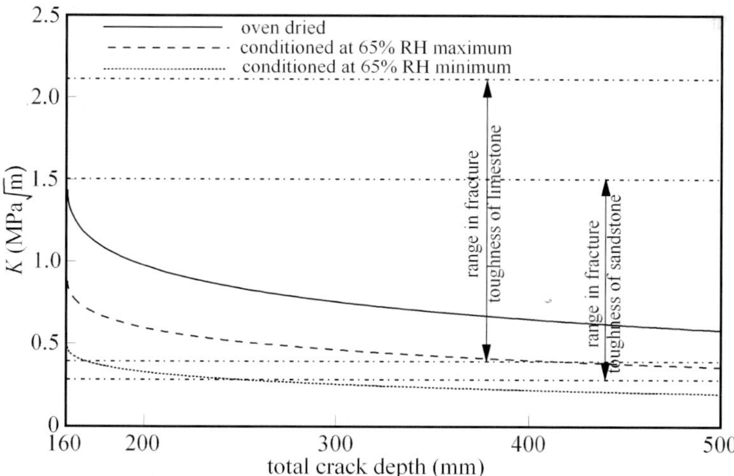

Fig. 5.5 The stress intensity factor induced by wetting a wood packed slot 160 mm deep.

5.2.3.3 Quarrying hard stone

The unfinished rose granite obelisk at Aswan shown in Fig. 5.6, which is 41.8 m long with a 4.2 x 4.2 m base and an estimated weight of 1190 tonnes, probably dates to the Thutmoside period 1479–1353 BC) and was the largest obelisk ever attempted, but it was abandoned because of the discovery of a large flaw.[24] However, it provides very valuable evidence as to how hard stone was quarried. A trench was cut around all sides, similar to the quarrying of large soft stone blocks, but copper or bronze tools were too soft for hard stone. Instead of chipping flakes, which is a comparatively efficient means of removing stone, spherical dolerite pounders were used to pulverise a trench, which in the unfinished Aswan Obelisk, had an average width of 75 cm. The sides of the trench show regular vertical flutes, as seen in Fig. 5.7, whose radii are comparable to that of the dolerite pounders. The Incas used a similar technique to hew their red granite blocks in the quarry at Kachiqhata leaving markings from their dolerite pounders almost identical to those at Aswan.[25] Engelbach surmised that labourers squatted in the trench at Aswan and pounded vertically, leaving the regular fluted pattern in the trench walls.[24] The progress of the labourers was monitored by lowering cubit rods and marking the upper end of the rod with red paint. A more efficient method of forming the trench once a small section had been formed, by using the pounders to chip flakes from its edge, would not leave the vertical fluting and was probably not used because it would limit the number of labourers who could work on the trench and the progress would have been slower even though more efficient on the basis of the amount of rock removed by each labourer. The basis for the pulverizing of rock with spherical pounders is the production of multiple Herzian cones. Engelbach experimented with dolerite pounders that were still scattered around the quarry and found that he could remove about 650 cm^3 in an hour.[24] From these experiments Engelbach estimated that it would take about seven months to complete the trench. Applying the same calculations to the extraction of the smaller obelisk of Queen Hatshepsut (1479–1458 BC), Engelbach obtained a time of about four-and-a-half months compared with the recorded time of seven months. Once the trenches around the obelisk were completed, it was detached from the bedrock by undercutting from both sides again using stone pounders.[24]

The red quartzite in the quarries at Gebel el-Ahmar, near Cairo, is harder than granite and the method used for granite seems to have been inadequate. The walls of the quarry are not as vertical as those in the granite quarries at Aswan and are inclined at about 10° to the vertical; also there are horizontal markings, which it is suggested were made by a sharp pointed stone pick.[19] The Gebel el-Ahmar quarries were still being worked in the twentieth-century, but were not greatly exploited in Pharaonic times probably because of the difficulty of quarrying this very hard rock.[7]

Fig. 5.6 The unfinished obelisk at Aswan (courtesy Tore Kjeilen, Looklex).

Fig. 5.7 The trench of the unfinished Aswan obelisk (Engelbach 1922).

5.2.3.4 Sawing and drilling stone[26,27]

Sawing and drilling using copper tools and sand as an abrasive were used to fashion monolithic sarcophagi from the time of the third dynasty (2670–2600 BC). The first sarcophagi were made in soft white limestone or alabaster but in the fourth dynasty, Cheops (ca. 2589–2566 BC) had a granite sarcophagus made. Quartz, the major component of sand, has a Mohs hardness of 7 (DPH 788 kg/mm2) and is one of the standard materials in the Mohs scale. A hint that the best abrasive sand was sought in early times is given by Pliny (AD 24–79) in his *Natural History* where he states that sand from a special shoal in the Adriatic was sought for sawing marble veneers.[17] The sawing and drilling using an abrasive leaves the characteristic striations discussed in §4.5.1.

It was estimated by Petrie that the saw used for the granite sarcophagus of Cheops was about 2.7 m long; there would have been no teeth and the thickness must have been about 5 mm.[28] In experiments, Denys Stocks used a copper saw 1.8 m long, 150 mm wide, and 6 mm thick to cut 30 mm deep slot, 950 mm long in 14 hours using dry sand in Aswan granite; only 7.5 mm was lost from the width of the saw in this process.[27]

The sarcophagi were hollowed out by drilling with tubular copper drills. The perimeter of the inside of the sarcophagi would have been drilled all round and some holes drilled along the centre. The stone between the holes would then have been smashed. The walls and floor were dressed. In Cheops' sarcophagus one of the drill marks was not completely removed by subsequent dressing and left a concave mark. Petrie measured this mark which was 2 mm deep and 33 mm wide.[29] From the dimensions of this mark, Stocks calculated that the drill used to hollow out Cheops' sarcophagus was 110 mm in diameter.[26] In experiments with an 80 mm diameter copper drill with a wall thickness of 1 mm, Stocks drilled a hole in Aswan granite 60 mm deep in 20 hours, but the drill lost some 90 mm in the process.[27] A thin drill is the most efficient at drilling the stone but, since copper is lost from the surfaces as well as the end of the drill the loss in copper is great. Drill holes for door jambs exist with the stub of the core remaining showing that the Egyptian copper drills has a thickness of about 5 mm similar to the thickness of their saws.[27]

Much of what is written on the quarrying of stone, including that in this chapter, comes from speculation rather than experience. It has been refreshing in this chapter to be able to draw on the experience of actual experiments performed by Rex Engelbach[24] and Denys Stocks.[26]

5.2.4 Building in stone

As already mentioned walls and columns presented no fracture challenge to the Egyptian architect, but architraves cannot take too large a load. Although the Egyptians used true brick arches in the first dynasty, there are very few examples of the true stone arch and the openings spanned by architraves were comparatively small. As Clarke and Engelbach remarked 'in the Great Hypostyle Hall at Karnak one cannot see the hall for the columns'.[19] Limestone was the principal building stone during the old and middle kingdoms (2600–1781 BC) and architraves using limestone from the Tura and Ma'sara quarries just south of modern Cairo were limited to about 3 m.[19] The introduction of sandstone enabled larger spans of up to 7 m or more. Prior to the exploitation of the great sandstone quarries, only granite could span such large openings. However, granite was not widely used for roofs because of the enormous labour needed in quarrying and dressing.[19]

Fig. 5.8 A section through the sepulchral chamber in Cheops' Pyramid (after Perrot and Chipiez 1882).

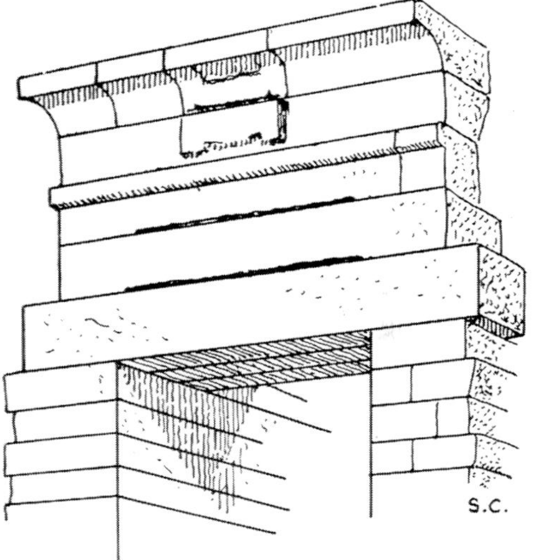

Fig. 5.9 Relieved architraves in the gateway of Nectanebo II at Karnak (Clarke and Engelbach 1930).

Architraves of any length can only carry so much load before they crack. The sepulchral chambers deep inside the pyramids had very high vertical and side loads. The Egyptians solved this problem by using huge stone blocks leaning against each other to form an arch to take the vertical load, and architraves to take the side thrust. The space between the inclined blocks and the architraves was unfilled so that no extra vertical load would bear on the architrave. Sections through the sepulchral chamber of Cheops' Pyramid are shown in Fig. 5.8; the span is 5.25 m. Limestone blocks were used at the very top of the chamber to take the vertical load with flat granite blocks 2 m thick taking the side thrust. There was not just one row of flat blocks over the chamber, but five with a space between them so that the weight of one block was not superimposed on the one below. As Clarke and Engelbach remark, really only one row of architraves was required to take the side thrust.[19] At the end of the Pharaonic era and during the

Greek period, the load on superimposed architraves was relieved, in a not-so-obvious fashion in their imposing gateways, by slightly hollowing out the architraves above the first as is shown in the sketch by Somers Clarke, shown in Fig. 5.9, of the gateway of Nectanebo II (360–342 BC) at Karnak.

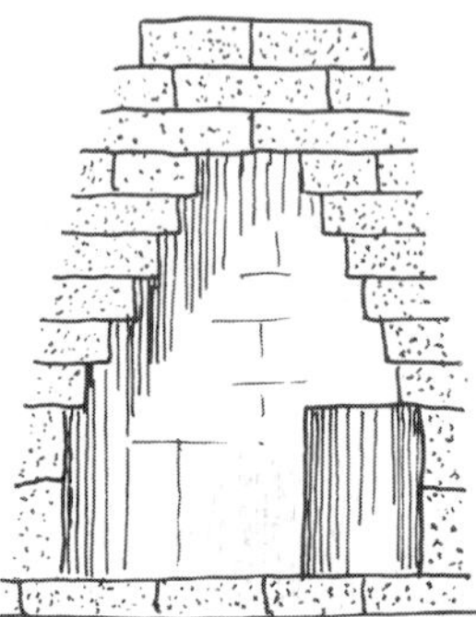

Fig. 5.10 Corbelled roof in the chamber of a twelfth dynasty mastaba at Dahshur (after De Morgan 1894).

Large corbelled chambers were constructed in the pyramids at Dahsur and Giza. In the middle kingdom (2040–1781 BC), corbelling was used in the mastaba tombs of nobles at Dahshur and Abusir. The corbelled roof over the 3.5 m wide chamber of a twelfth-dynasty (1884–1781 BC) mastaba at Dahshur is illustrated in Fig. 5.10. Stone barrel vaults with wedge shaped voussoirs were not used until about 750 BC during the twenty-fifth dynasty, when at least three were built.[16] The spans of these vaults, which were semicircular in form, ranged from 1.96 to 2.8 m.

5.3 Greek Masonry

Although arches and domes were hardly used in classical Greece, the Mycenaean tholos tomb built ca 1,250 BC, and fancifully called the 'Treasury of Atreus' by Heinrich Schliemann (1822–1896), had a corbelled stone conical dome 14.5 m in diameter and 13.2 m high (see Fig. 5.11). There was no larger diameter dome until Roman times. The Temple of Mercury at Baia north of Naples, with a diameter of 21.6 m, was built during the reign of Augustus (29 BC–AD 14). The stones of the corbelled dome of the Treasury of Atreus have been dressed to give a smooth surface so that visually its corbelled nature is unapparent. The entrance to the tomb, shown in Fig. 5.12, is through a magnificent doorway 5.4 m high by

2.4 m wide surmounted by a 120 tonne architrave with a relieving triangle in the corbelled dome above it, which would have been filled with a carved stone slab like the nearby Lion Gate. All the stone in the Treasury of Atreus would have been quarried and dressed with bronze tools. However, steel tools were used to fashion the stone of classical Greece.

Fig. 5.11 The Interior of the Treasury of Atreus, a drawing by Edward Dowdell, 1834.

Fig. 5.12 The entrance to the Treasury of Atreus (Cotterell & Kamminga with permission CUP).

Classical Greek architecture was essentially columns and lintels similar to that of ancient Egypt, but they did appreciate how to design the lintel to resist bending better. The aesthetic sensibilities of the Greeks were also more highly developed and proportions were of prime importance. The Greeks had also to contend with earthquakes, though they certainly could not take these into consideration in their architecture, they would have been conservative in design. The Greeks knew that the strength of a beam could be best increased by increasing its depth and, to satisfy both their aesthetic sensibilities and the requirements of strength, the Greeks occasionally left a hidden rib on a beam. The ceiling beams of the Hierion on the island of Samothrace, whose *cella* was completed in 325 BC, had their thickness tapered from the centre to the abutments showing that the Greeks appreciated the variation in bending moment (see Fig. 5.13).

 Fracture and Life

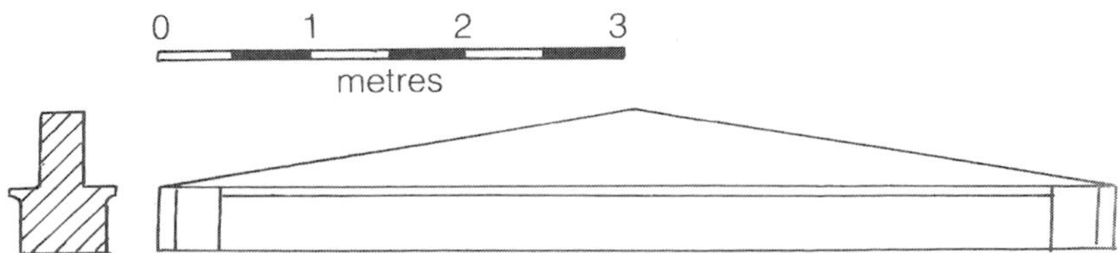

Fig. 5.13 A ceiling beam from the Hierion on Samothrace (Cotterell and Kamminga 1990, with permission CUP).

The Greeks used iron to bond their stone blocks together in three forms: iron clamps between blocks on the same course, dowels to fasten to courses below and iron braces to position stones while they were being dowelled. Iron was also occasionally used to strengthen architraves. The Propylaia in the Acropolis at Athens, built by the famous architect Mnesikles in about 437 BC, had architraves composed of two marble beams 850 m wide and 500 m high which supported the ceiling beams. The span of the architraves was not that large, 2.5 m, but Mnesikles obviously was not sure of their strength so he supported the ceiling beam at the centre of the architrave not directly on the architrave, but on iron bars about 75 mm wide and 115 mm deep which were in a groove cut in the top of the architrave the ends of which were some 25 mm less deep so that the load of the ceiling beams was transferred towards the abutments thus reducing the bending moment by about a half (see Fig. 5.14).[30] Of course the iron beams have long since rusted away and some of the architraves have been cracked by earthquakes.

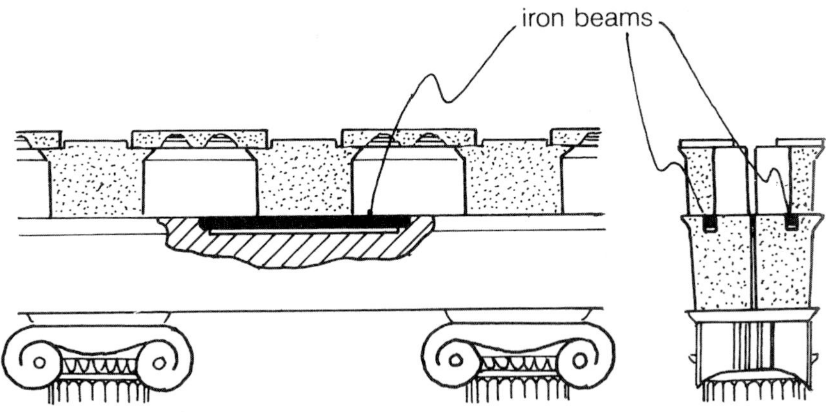

Fig. 5.14 Architrave of the Propylaia at Athens showing iron insets (Cotterell and Kamminga 1990, with permission CUP).

5.4 Roman Masonry and Concrete

The Romans introduced new elements in the design against fracture in their buildings: the wide use of the arch and dome, and the use of concrete. Their techniques for quarrying and dressing stone were little changed from those of the Egyptians, but of course they worked with steel tools. Plain saws with sand abrasives were used to cut the harder stones like granite, but for soft stone, serrated steel saws were used. The tools used by stone masons did not change from Roman times until the nineteenth-century and the introduction of the wire band saw.

The Egyptians used impure gypsum plaster as mortar. Slaked lime was used by the Greeks to create a non-hydraulic mortar, but it was the Romans who discovered that a hydraulic mortar could be made using slaked lime and the volcanic tuff called pozzolana after Pozzuoli the Italian town where a suitable volcanic tuff, containing aluminium silicate, is found. A hydraulic mortar will set under water, is more durable than non-hydraulic mortars, and can be used in the foundations of bridges. The Romans also made a hydraulic mortar by mixing crushed clay tiles with slaked lime and sand. The Romans started using lime mortar towards the end of the third-century BC. However concrete, using mortar and an aggregate of pebbles, stones or broken pieces of pottery, only started to be used towards the end of the second-century BC and was at first of poor quality. Even when Vitruvius wrote his *Ten Books on Architecture* during the reign of Augustus (29 BC – AD 14) and gave the different compositions for mortar, concrete was still an innovation being used sparingly. The quality of the concrete eventually developed by the Romans is evident in the still intact Pantheon in Rome, built by Agrippa in about 118–125, which has stood for nearly two millennia with only minor repairs.[31] Concrete made with pozzolana was important for the construction of the massive piers or *pilae* of Roman harbours like the one near the Villa Domitii Ahenobarbi at Santa Liberata, near the modern Porto Santo Stefano, which was built in the first-century BC. Recently, cores have been taken from these pilae to compare the composition and properties to concrete with a reproduction mortar using Vitruvius' proportions of two parts of pozzolana to one of lime.[32] Judging on composition, the actual mix used at Santa Liberata was probably nearer to 2.7 parts of pozzolana to one of lime. Unfortunately little detail is given on the aggregate, which was tuff or local limestone and sandstone in the approximate proportions or its proportion of 35:65 aggregate to mortar, in a modern concrete the proportion of aggregate would be much higher. The compressive strength of the cores taken from the *pilae* at Santa Liberata was 7.5 to 8.5 MPa which is much lower than modern concrete.[32]

It is unclear who originated the so-called 'true arch', constructed with voussoirs. Traditionally its invention has been attributed to the Etruscans but there is no firm evidence.[31] Roman examples of the voussoir arch, like the Gate of Jupiter (see Fig. 5.1) date to the third-century BC and there are a few Greek examples from the same period. However, regardless of who invented the voussoir arch, the Romans exploited it as an efficient and beautiful means of spanning space.

The shape of an arch is not a critical factor in its design, provided that it is reasonably thick. The Roman arches were usually semi-circular. Arches with rises of less than half the span, called segmental arches, create larger horizontal thrusts on abutments or piers than semicircular ones. The Romans were particularly conscious of the horizontal thrust of an arch and provided their bridges with massive piers of between a quarter and a third of the arch's span. What the Romans seemed to have failed to appreciate was that the horizontal thrust of adjacent piers balance each other. The Chinese made use of the segmental arch earlier than the West. The Zhao-Zhou Bridge, built by Li Chun in 610 over the river Jiao Shui in southern Hebei Province, is the world's oldest segmental bridge in the world.[33] Segmental bridges were not adopted by the West until the fourteenth-century. Fewer Chinese stone bridges have survived than Roman ones not because of their arch design, but because they did not possess a hydraulic cement for their foundations.[33]

The most impressive use of concrete by the Romans is in the Pantheon in Rome (the interior is shown in Plate 9) which has a dome 43.3 m in diameter. The dome, originally covered with bronze plates[34] which may have been gilded, would have presented an even more majestic sight when it was first built. Concrete enabled the Pantheon to be built in about seven years, whereas comparable buildings in stone could take up to century or more.[35] A dome of comparable size was not built again until the fifteenth-century dome of the Basilica di Santa Maria del Fiore, the Duomo of Florence, which is 42 m in diameter and has a double-walled brick design, the use of concrete having been long forgotten. The details of the concrete used in the Pantheon are indicated in the cross-section shown in Fig. 5.15. The Romans did not premix the aggregate with the mortar as is done now, but mixed the aggregate and mortar *in situ*. This practice may be the reason for the series of stepped rings on the lower portion of

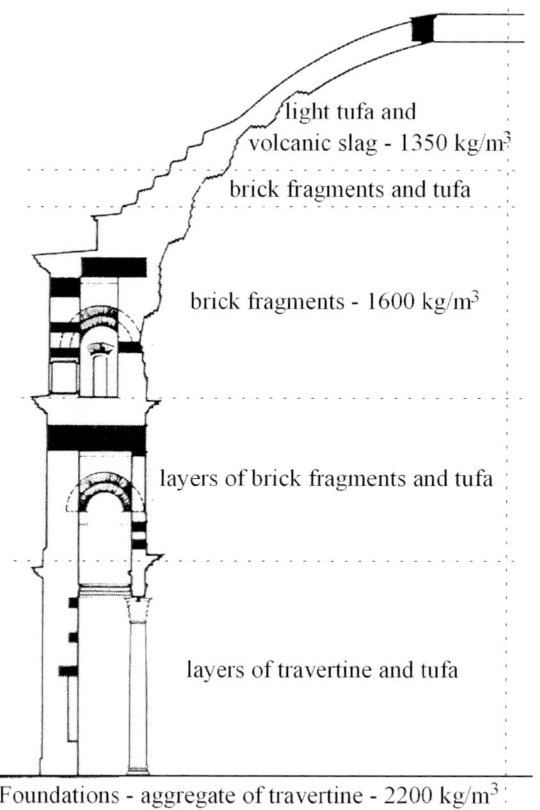

Fig. 5.15 A meridional section of the Pantheon (after MacDonald, 1982).

the lower portion of the Pantheon's dome which would have made the mixing of the concrete easier where the dome has a high slope.[36] The only light in the Pantheon comes from an oculus at the crown of the dome. Such an opening does not affect the stability of a dome, but does approximately double the maximum compressive stress. The weight of the dome was reduced by two means: the density of the concrete was graded as shown in Fig. 5.15 and the underside of the lower portion of the dome was coffered (see Plate 9). However, meridional tensile stresses in the lower part of the dome could not be avoided. In 1930, spalling caused small plaster fragments to fall and as a consequence scaffolding was erected for an inspection by Alberto Terenzio the Superintendent of the Monuments in the Latium.[37] Terenzio found vertical cracks in the dome which extended to about 33° from the crown, which is much higher than the transition from tension to compressive stress at 51.8° that occurs in a uniform thickness hemispherical dome. A finite element study of the Pantheon's dome where cracks were allowed to develop if there was tension predicts that cracks would extend up to 36° from the crown.[38] The distribution of the cracks around the dome generally corresponds to openings in the upper cylindrical wall as can be seen in Fig. 5.16. Cracks also were found in the cylindrical part of the Pantheon. Terenzio concluded that the cracks must have occurred very soon after construction because the bricks used for an earlier repair had the same stamps as those in the original portions of the Pantheon.[37] Provided the dome is thick enough the thrust line can be contained within the dome and stability is maintained even with vertical cracking as the Pantheon has demonstrated by standing for nearly two millennia.

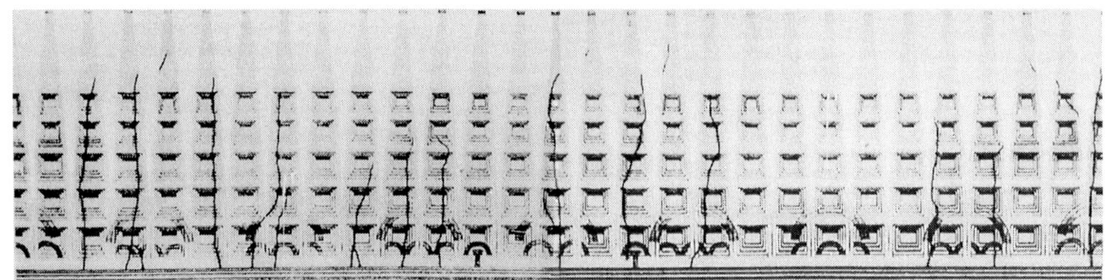

Fig. 5.16 Cracks observed in the interior of the dome (after Terenzio 1934).

5.5 Concluding Remarks

Many buildings from the ancient world are still standing to fill us with awe at their achievements. Perhaps even more awesome, but not so obvious, was their ability to cut the stone for the buildings especially before the use of iron and steel tools. The techniques of stone-cutting developed by Roman times were so efficient that they remained unchanged until the nineteenth-century. Before the Romans, the natural materials: wood, brick and stone were used for buildings. The Romans were always seeking cheaper and cost effective solutions to technical problems and developed a strong reliable concrete that could speed up construction.

Since the durable building materials of brick, stone, and concrete are comparatively weak in tension compared with their high strength in compression the design challenge in buildings is to span space causing the minimum tension in the structure. Column and lintels is an almost natural first choice but, because the lintel suffers bending stresses, entails considerable tension. Hence the Egyptians who mainly used this technique could only span relatively small openings and needed closely spaced columns. Although the Greeks largely used the same structural design, they increased the possible span slightly by shaping the stone so that the architrave is deeper where the bending moment is greatest and by some use of iron as reinforcement. It was the Romans who developed the arch and the dome which first enabled large building spans.

5.6 Notes

[1] Heyman (1972).
[2] Concrete reinforcement has been added to the architrave as a safety precaution.
[3] Heyman (1966, 1969). There is a whole family of thrust lines that satisfy equilibrium, but provided one of these is contained within the masonry, the arch or dome is stable.
[4] Heyman (1967).
[5] The study of masonry techniques used by the ancient Egyptians has been a comparatively neglected. The early classic work is Clarke and Engelbach (1930); a more recent work is Arnold (1991).
[6] Klemm and Klemm (2001).
[7] Lucas and Harris (1962).
[8] Sherby and Wadsworth (2001).
[9] Williams and Maxwell-Hyslop (1976).
[10] Snodgrass (1980).
[11] Weisgerber (2006).
[12] Muhly (1980).
[13] Wicker (1998).
[14] Wicker (1990).
[15] Herodotus, *The Histories* II,22.
[16] Arnold (1991).
[17] Humphrey *et al.* (1998).
[18] Reisner (1931).
[19] Clarke and Engelbach (1930).
[20] Petrie (1917).
[21] Noël (1965).
[22] Virta *et al.* (2006).
[23] Tada *et al.* (1985).
[24] (Engelbach 1922).
[25] Protzen (1985).
[26] Stocks (1999).
[27] Stocks (2001).
[28] Petrie (1883).
[29] Petrie (1884).
[30] Dinsmoor (1922).

[31] Adam (1994).
[32] Gotti *et al.* (2008).
[33] Needham *et al.* (1971).
[34] The bronze plates were replaced by lead in the Early Middle Ages.
[35] The Great Temple of Apollo at Didyma was abandoned still incomplete after 4.5 centuries.
[36] MacDonald (1982).
[37] Terenzio (1934)
[38] Mark and Hutchinson (1986).

Chapter 6

From the Renaissance to the Industrial Revolution

The approach to fracture in the ancient world was pragmatic though it was realised that the strength of materials was limited. The mechanics of the lever was known and extensively discussed in the *Mechanical Problems* written by the followers of Aristotle in about 280 BC. Problem 14 in the *Mechanical Problems* relates to the application of the mechanics of the lever to fracture. The question is asked 'Why is a piece of wood of equal size more easily broken over the knee if one holds it at equal distance far away from the knee to break it than if one holds it by the knee and quite close to it?…Is it because… the knee is the centre, but the further it is away from the centre the more easily is everything moved and what is being broken must necessarily be moved?' An attempt is also made at explaining fracture under impact in Problem 19: 'Why is it that if one puts a large axe on a block of wood and a heavy weight on top it does not cut the wood to any extent, but if struck splits in half even if the blow is light? Is it because all work is produced by movement?' However, the theory of fracture really begins during the Renaissance, though it was not seen as a discipline separate from mechanics in general until the twentieth-century.

6.1 Leonardo da Vinci (1452–1519)

The Greeks relied on logic, though obviously Problem 14 in the *Mechanical Problems* is based on everyday experience, their science was based on induction. The Renaissance saw the beginning of the use of observation in science. Georgius Agricola[1] (1494–1555) stated 'Those things which we see by means of our senses are more clearly to be demonstrated than if learned by means of reasoning',[2] and da Vinci described himself as a 'disciple of experience'[3]. A problem with da Vinci's writings, which is especially acute with his numerous inventions, is that it is difficult to distinguish between his ideas or designs and

141

what he actually tried experimentally or made. Since da Vinci probably did not start his Notebooks until 1489, when he was 37 years old, some of his notes must refer to earlier work and remembered experiments or inventions making the distinction even more difficult. For example, da Vinci took the Greek work on beams further. For beams of uniform cross-section he clearly understood that their strength varied inversely in proportion to their length stating that 'If a beam 2 braccia long supports 100 libbre, a beam one braccio long will support 200'.[2,4] Did da Vinci perform experiments to see if the lever-rule, he obviously used to estimate the predicted load carrying capacity of a beam, gave the correct prediction? Obviously the predicted carrying capacity he stated is based on an arbitrary 100 libbre for the 2 braccia long beam and experimentally he would not have obtained exactly double the carrying capacity for the one braccio long beam.

Da Vinci almost certainly performed tensile tests on iron wires. In his note: *Testing the strength of iron wires of various lengths* he described tests on wires of different length but the same diameter and illustrated the test method (see Fig. 6.1). The wire anchored to a support is attached at the other end to a basket into which sand is fed from a hopper. A spring is fixed so that the hole in the hopper is closed once the wire breaks. Da Vinci used this apparatus to test wires of different length starting with 2 braccia and repeating the test with wires each half of that used in the previous test and found that their strength varied inversely with their length. In earlier comments on this test it is assumed that da Vinci must have made a mistake because the strength of modern wires would vary only slightly with different lengths that were significantly longer than their diameter.[2] However, there are two reasons why da Vinci may have found that the strength of iron wire varied inversely with length. The first that is often mentioned is that wire drawing at the time of da Vinci would not produce nearly as uniform a wire as today. A contemporary illustration of an artisan wire drawing is shown in Fig. 6.2. The artisan sits on a swing and grips the wire with pincers and the swing is driven by a crank turned by a water wheel. When the swing reaches its limit the artisan lets go of the wire and swings back to grip wire once more. Today the wire is steadily drawn through a die by a capstan. The method illustrated in Fig. 6.2 may have produced significant defects

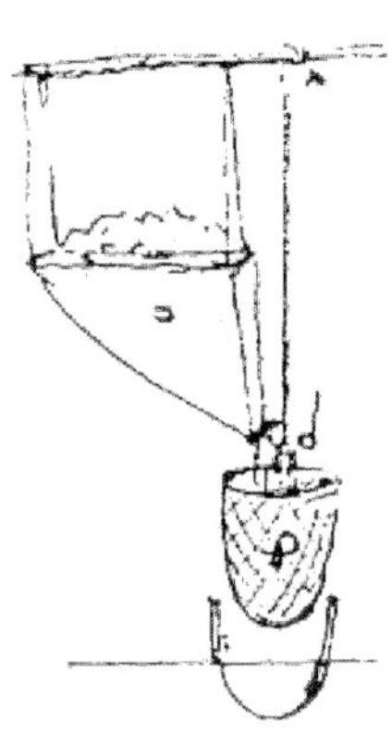

Fig. 6.1 Leonardo's wire testing apparatus.

along the wire. The second reason is that da Vinci specifically referred to iron wire, which would have been drawn from wrought iron. Residual weakening inclusions of slag are always present in wrought iron and these would be distributed along the length of the wire. Whether the defects were due to the drawing technique or the inclusions, the strength of the wire would have been controlled by the most severe defect in any particular length of wire. Since there is a greater probability of finding a weak link in a long chain, the mean strength of wire with significant defects varies inversely with length just as da Vinci reported.[5]

Fig. 6.2 An artisan drawing wire (Biringuccio 1540).

Elsewhere in his notes da Vinci discussed the strength of columns.[2] He noted that if there are two columns of equal length, one having a cross-section four times greater than the other, the larger one will support eight times the load of the smaller one. Of course, every first-year engineering undergraduate knows that da Vinci was wrong and that the strength of short columns is proportional to their cross-sectional area. Unfortunately da Vinci did not fully understand proportions based on a power of a factor.[2] He also considered columns of different length and the same cross-section and concluded that the strength of columns varies directly with the cross-section, but inversely with their length. The second conclusion could conceivably be made because the columns were long and buckled rather than fractured but, though it is probable that he actually performed tensile tests on wires, it is much less likely that he performed compression tests on columns and he was simply transferring his knowledge of wires under tension to columns under compression. Compression tests on columns are far more difficult than tension tests on thin wires. Even for a column of moderate cross-section the fracture load would be greater than could be achieved though direct load. The first recorded compression tests were made by Petrus van Musschenbroek in 1729 using direct load but were for studying buckling not fracture.

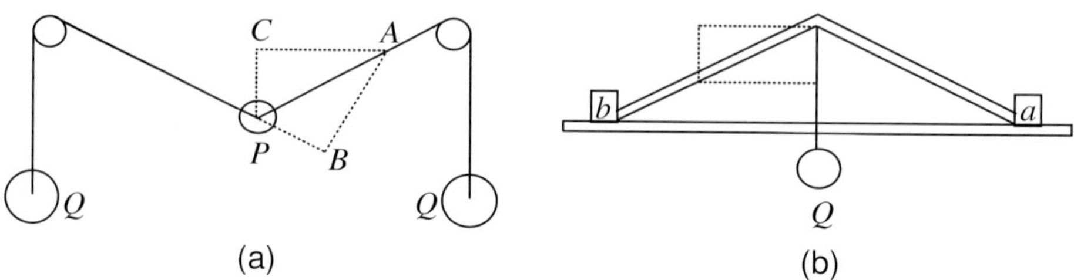

Fig. 6.3 (a) Leonardo's illustration of a string over two pulleys with the load *P* balanced by *Q*. (b) The horizontal thrust on a simple arch loaded at its apex.

Da Vinci understood the parallelogram of forces[6] very likely checking his analysis by an experiment using a strings and weights over pulleys, illustrated in Fig. 6.3 (a)[7]. The dotted lines in Fig. 6.3 (a) are da Vinci's construction to find the ratio *P:Q* which he correctly deduced was the same as the ratio *AB:AC*.[2] Da Vinci also discussed the horizontal thrust on the abutments at *a* and b of a simple arch consisting of two struts leaning together and supporting a weight, *Q*, illustrated in Fig. 6.3 (b). Although da Vinci did not refer to the dotted lines in his notes they are the parallelogram necessary to find the horizontal thrust if the weight of the struts is small compared with the load *Q*. Da Vinci used the thrust diagram of Fig. 6.3 (b) to discuss the fracture strength of an arch as illustrated in Fig. 6.4. In the analysis of an arch, da Vinci assumes that the thrust line is straight as it is in Fig. 6.3 (b). Although such a thrust line is not in equilibrium with the loads on the arch, it is not a bad approximation. Da Vinci states that 'an arch will not break if the chord of the outer arch [*abn*] does not touch the inner arch [*xby*]', which is very reminiscent of the modern *safe theorem* for an arch.[8] In his illustration the chord just touched the inner chord at *b* which is one of the fracture points, the others being at *a* and *n*.

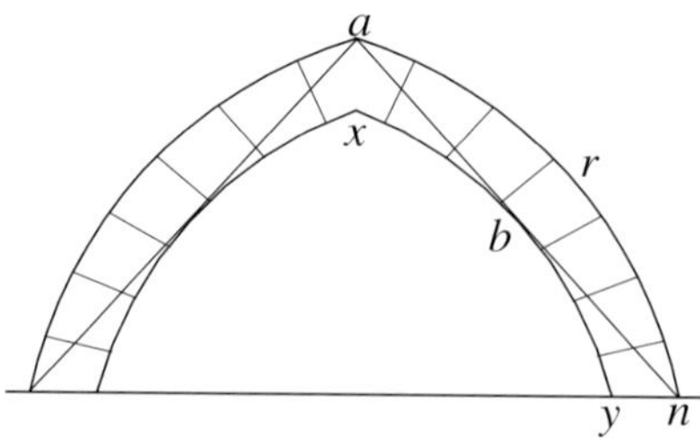

Fig. 6.4 Leonardo's analysis of an arch.

Leonardo's notes on fracture related mechanics constitute only a tiny fraction of his immense contribution to art, science, and engineering, but they are the first and important. However, because Leonardo did not publish his notebooks his work was little-known and did not affect the subsequent development of the mechanics of fracture.

6.2 Galileo Galilei (1564–1642)

In 1633, Galileo stood trial by the Holy Office for 'vehement suspicion of heresy' and was found guilty of having held and taught the Copernican theory of the solar system. He was ordered to recant and sentenced to imprisonment, but Pope Urban VIII (1568–1644) commuted the sentence to house arrest in his estate at Arcetri near Florence. Confined to his estate Galileo wrote the *Dialogues Concerning the Two New Sciences*, publishing it in 1638, in which he revisited the work of his youth.[9] Galileo considered this book to be 'superior to everything else of mine hitherto published'. The book consists of a hypothetical conversation between Salviati, Sagredo, and Simplico which supposedly took place over a period of four days. Part of the discussion on the first day and all of those in the second day are devoted to mechanics relating to fracture.

Galileo started his discourse on the strength of materials by considering the tensile fracture of 'a piece or wood or any other solid which coheres strongly'. He argued that the strength of bars under tension is proportional to their cross-sectional areas, reasoning that the number of 'fibres' binding the parts together is proportional to the area. Stone and metal which did not have a fibrous structure puzzled Galileo. He discounted the idea that some kind of glue held the parts together because no known glue could withstand the heat of a furnace and yet molten metals retained their strength when they resolidified. From experiments on copper wire Galileo deduced that, regardless of the diameter, the maximum length of wire that could be supported without breaking was 4,801 cubits (2770 m). He used this observation to discount the idea that solids fractured because a vacuum was produced. He had made tests that showed that water can only sustain a column 18 cubits (10.4 m) high and observed that if copper ruptured like water then it would only support a length of 2 cubits (1.16 m) because copper is nine times heavier than water. Galileo does not mention any dependence of strength on the length of wire. The copper wire used by Galileo would have been more ductile and would not contain such gross defects as the inclusions of slag in the iron wire used by da Vinci consequently there probably was not significant variation in strength with length.

Fig. 6.5 Galileo's cantilever.

On the second day Galileo discussed the fracture of a beam under bending. He observed that solids are far less capable of sustaining a transverse load than a direct longitudinal load and considered the fracture of the cantilever illustrated in Fig. 6.5. Galileo identified that fracture will occur at the plane *AB* and considered that the fracture will occur by the beam hinging about the point *B*. There are two lever arms: *BC* the arm through which the bending moment of the weight *E* is applied and the arm *BA* 'along which is located the resistance'. In this discussion Galileo faced the difficulty of describing the mechanical behaviour of solids without the concept of stress.[10] He assumed the resistance to fracture was uniform (equivalent to uniform stress) over the section *BA* and equated the moment of the load about his fulcrum at *B* to the moment of the resistance. In the absence of a concept of stress, Galileo defined the 'absolute resistance to fracture', *S*, as the tensile force that causes fracture in a bar of the same cross-section. Hence Galileo gives, in words, the equation

$$E = \frac{(BA/2)}{BC} S,$$
(6.1)

to determine the fracture load for a cantilever in terms of the tensile fracture load for the same bar. Using the concept of stress, Eq. (6.1) can be written as

$$\sigma = \frac{2l}{bh^2}Mg, \tag{6.2}$$

where $l = BC$, $h = BA$, b is the width of the cantilever, and Mg is the weight E. We now know that the stress is not uniform across BA and the maximum stress given by Eq. (6.2) is only a third of the actual stress if the cantilever remains elastic up until fracture.[11] Hence Galileo overestimated the fracture strength in bending by a factor of three.

Despite the error in the factor, Galileo's achievement in deducing Eq. (6.2) cannot be overestimated and it led him to grasp the importance of scaling in fracture. If the size of a beam is scaled up, its weight increases with the cube of its linear dimensions and the maximum moment by the fourth power, whereas its resistance to fracture only increases by the cube of its dimension. Galileo saw that this size effect placed a limit on the size of structures, both man-made and natural, which makes it impossible to build 'ships, palaces, or temples of enormous size in such a way that all their oars, yards, beams, iron bolts…will hold together; nor can nature produce trees of extraordinary size because their branches would break down under their own weight; so also it would be impossible to build up bony structures of men, horses or other animals so as to hold together…if these animals were to be increased in height enormously'.

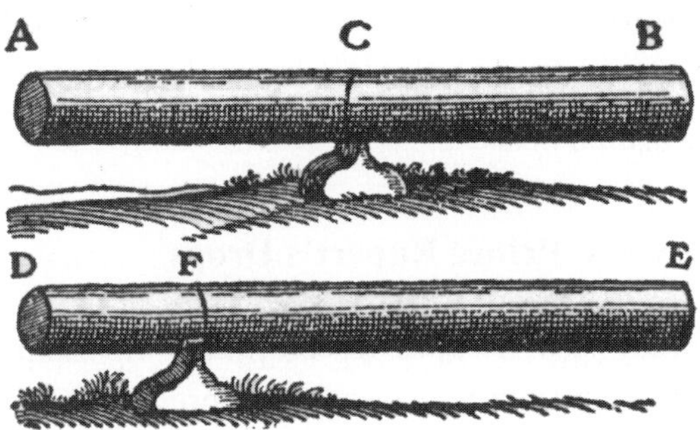

Fig. 6.6 Galileo's illustration for the Aristotelian Problem 14.

Galileo has an extensive section on various problems involving beams and discusses some of the Aristotelian *Mechanical Problems* including Problem 14 mentioned in §6.1 which he takes further to discover whether the force required to break a stick held at two fixed points and broken with the knee, is the same regardless of the position of the knee. The illustration that Galileo uses in his argument to show that the force necessary to break the stick is proportional to the product $(DF)(DE)$ and is a minimum when it is applied at the mid-point, is shown in Fig. 6.6.

Another concept of Galileo which was well ahead of his time is minimum weight design. He considers how the thickness of a cantilever should vary so that it is, in effect, equally stressed along its length (see Fig. 6.7) and therefore of minimum weight. At any section the resisting moment is proportional to $(CN)^2$ and the moment of the load is proportional to CB. Hence for minimum weight

$$\left(\frac{CN}{AF}\right)^2 = \frac{CB}{AB},\qquad(6.3)$$

which implies that FNB is a parabola. The parabola shown in Galileo's illustration (Fig. 6.7) is badly drawn. The vertex should be at B but appears to be at F. Galileo was not in control of the publication of the Dialogues and in any case had become blind by that date.[12]

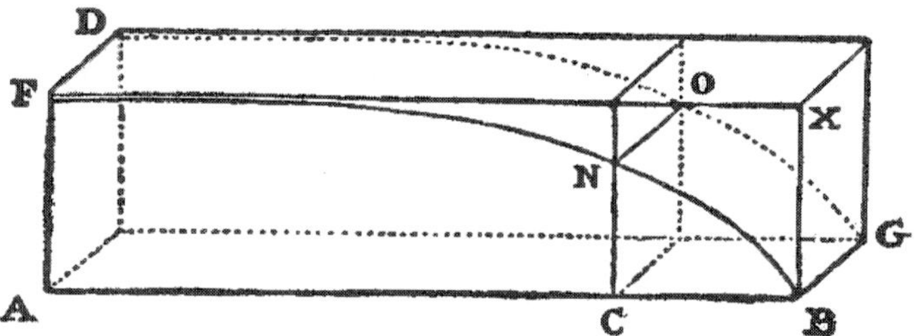

Fig. 6.7 Galileo's uniform strength cantilever.

6.3 The Royal Society and Prince Rupert's Drops[13]

The seventeenth-century saw a rapid development in mathematics, astronomy and the natural sciences and learned societies were formed in several European countries. The first society was in Italy where the *Accademia Secretorum Naturae* was founded in Naples in 1560. In England men of science were forming groups in the early seventeenth-century. John Wilkins (1614–1672), who was Chaplain to the Prince Elector Palatine, formed a group interested in natural philosophy and particularly what was being called the 'New or Experimental Philosophy' who met regularly from about the year 1645 in London. Some of the members of the group, including Wilkins, left London for Oxford, during the second of the English Civil Wars 1648–1649, and formed another group. However, the meetings continued in London and were later incorporated into the

Royal Society, which is usually considered to have been founded in 1660.[14] Among those first invited to become members of the Royal Society were Robert Boyle (1627–1691), the Irish natural philosopher, Christopher Wren (1632–1732), the great architect, and John Wallis (1616–1703) a mathematician who was one of the first members of the original London group.

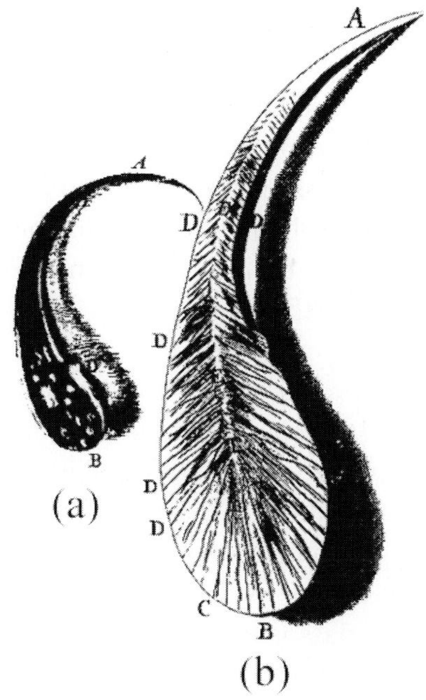

Fig. 6.8 Prince Rupert's Drops: (a) intact drop. (b) glue coated and fractured drop (Hooke 1665).

In March 1661[15] King Charles II sent the Royal Society 'five little glass bubbles … in order to have the judgement of the society concerning them'. The King had received the 'glass bubbles' from his nephew Prince Rupert who had recently returned from Germany with them. Rupert's drops, as they became known some thirty years later, are a paradoxical combination of strength and fragility. The spherical head of the drops, illustrated in Fig. 6.8, can be hammered on an anvil without breaking, but the tail can be broken off with finger pressure causing the drop to disintegrate into tiny fragments. Rupert's drops, made by dropping molten glass into water, created a minor sensation in the seventeenth-century both in England and in continental Europe – they were society's new toy.[16] The poet Samuel Butler (1612–1680) refers to the drops in his poem *Hudibras*:

> Honour is like that glassy bubble
> That finds philosophers such trouble,
> Whose least part crack'd, the whole does fly
> And wits are crack'd, to find out why.

The drops were also mentioned in a contemporary satirical poem about the Royal Society, possibly written by William Godolphin (1635–1696)[17] who was elected a fellow in 1663:

> And that which makes the Fame ring louder
> With much adoe they shew'd the King.
> To make glasse Buttons turne to powder
> If off the[m] their tayles you doe but ring,
> How this was done with soe small Force
> Did cost the Colledg a Month's discourse.

The phenomenon of Prince Rupert's Drops was understood by 1665 when Hooke published his *Micrographia*. The drops cool suddenly inwards from the outside when dropped into water so that the inside is at a higher temperature than the outside. When the centre of the drop becomes coherent the drop is stress free, during subsequent cooling the outside wants to shrink less than the inside, as a result, the outer skin of the drop is in compression whereas the inside is in tension. The low strength of glass is due to the presence of surface defects which are now under compression; the tail has a small diameter and, though its surface is also in compression, high bending stresses can be produced with a small finger pressure causing surface tensile stresses sufficient to cause the tail to snap. Once the crack reaches the centre of the tail, where there are high residual tensile stresses, the fracture propagates rapidly under the strain energy stored within the drop. The fracture surfaces revealed in Hooke's drop (see Fig. 6.8 (b)), which was coated with fish glue so that the glass fragments would not fly, are typical of a fast fracture running from the tail to the spherical bulb of the drop.

During the seventeenth-century Prince Rupert's drops were simply a scientific curiosity, but eventually the technique of strengthening[18] glass plate by quenching and producing compressive stresses in the surface became an established technology. François de la Bastie patented a process of strengthening glass plates by quenching them in oil in 1874. The technique worked but was a fire-hazard. Richardson invented a safer method of quenching using air blasts in 1888.[19]

6.4 Edme Mariotte (ca. 1620–1684)

Mariotte was one of the first members of the French Academy of Sciences, which was founded in 1666. He worked on a wide variety of physical sciences and in continental Europe the gas law that states that the volume of a gas varies inversely with pressure, which in English speaking countries is known as Boyle's Law, is attributed to him.[20] Mariotte's work related to the mechanics of fracture was not published until after his death in a paper on the motion of fluids.[21] Mariotte was the first to apply Hooke's law to the mechanics of solids. He was also an early proponent of experimental methods which he introduced to French science.

Galileo did not consider elongation in his mechanics of solids. When Mariotte did tension tests on bars he found that they elongated in proportion to the load as predicted by Hooke's Law and stated that fracture occurred when the elongation reached a critical value. He also did bending tests on the same bars and found that Galileo's formula overestimated the bending strength. Mariotte saw that for a beam to hinge about B in Fig. 6.5 bar would have to elongate in proportion to the vertical distance measured from B. Knowing that bars stretch in proportion to the load, Mariotte saw that Galileo's assumption of uniform resistance across the cross-section could not be correct, but that the resistance to fracture must vary as the elongation. Thus Mariotte obtained the expression,

$$E = \frac{BA}{3BC} S,$$
(6.4)

for the load that can be carried by the cantilever without fracture. He then went further to state that while the upper fibres in the beam are elongated, lower ones are compressed. Mariotte assumed the elongation in the top half was matched by the compression in the lower half so that what we now call the neutral axis was located at the mid-plane of the beam which is correct for a bean of symmetrical cross-section. However Mariotte now made an inadvertent mistake, he substituted half of the thickness, BA, of the beam into Eq. (6.4) to get the load supported by half of the beam when he should have also have substituted $S/2$ for S as well. The result was that when he added the load carrying contributions of both halves of the beam instead of obtaining the correct solution

$$E = \frac{BA}{6BC} S,$$
(6.5)

he obtained exactly the same formula as his first one given in Eq. (6.4). As a result of this error, Mariotte held the mistaken impression that the position of the neutral axis was immaterial;[22,23] Antoine Parent (1666–1716) gave the correct expression for strength of beams of rectangular cross-section in 1713.[24]

Mariotte's other contribution to mechanics of fracture arose from his work as an engineer to Louis XIV of France. The 'Sun King', Louis wanted elaborate fountains of great height in his palace, which was being built at Versailles. Mariotte experimented to find the necessary thickness in the pipes which were to supply these fountains. He attached long vertical pipes to a cylindrical pressure vessel and increased the water level in the pipes until they burst, the height of the water being in some experiments up to 30 m. From these tests Mariotte deuced that the required thickness of the pipe must be proportional to the internal pressure and the pipe's internal diameter.[22]

6.5 Dome of St Peter's and Giovanni Poleni (1683–1761) [25]

By 1743 St Peter's Basilica had developed alarming cracks running up from the drum into the dome (see Fig. 6.9) and Pope Benedict XIV (1675–1758) summoned Poleni to Rome in 1743 to report on the stability of the dome and make suggestions for its strengthening. The building of the present St Peter's was commissioned in 1451 by Pope Nicholas V (1397–1455), but it did not get much past the foundations until Pope Julius II (1443–1513) appointed Donato Bramante (1444–1514) as the architect. Bramante was not an engineer and designed a single shelled hemispherical brick dome with stepped rings like the Pantheon, not appreciating the advantages of the double shell that had been used on Basilica di Santa Maria del Fiore in Florence. After Bramante's death, Antonio da Sangallo (1484–1546), was appointed as the architect and redesigned the dome, changing the form to a segmental arc of revolution that increased the height from the spring to the lantern by about 10 m. In 1547 Michelangelo Buonarroti-Simoni (1475–1564) was appointed architect he changed the dome back to a hemispherical form. By the time of Michelangelo's death only the drum of the dome had been completed. The dome was redesigned by Giacomo della Porta (1533–1602), with the assistance of Domenico Fontana (1543–1607), as a double shelled paraboloid of inside diameter 42.3 m just slightly smaller than the Pantheon. The finial was finally placed on the lantern in 1593, but the dedication of the completed Basilica did take place until 1626.

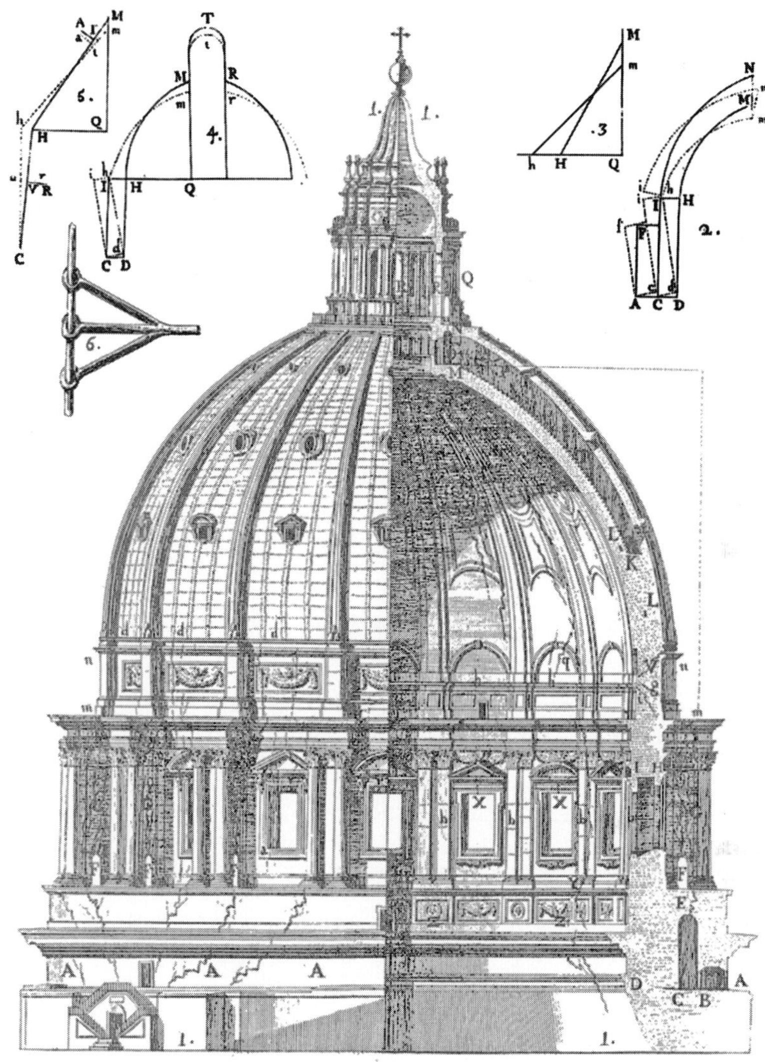

Fig. 6.9 Cracking in the Dome of St Peter's observed in 1742: Inserts 1–5 analytical drawings by Le Seur, Jacquier and Boscovich, Insert 6 drawing of tie bar connection.

The mechanics of the arch was well understood in 1743 through the works of Robert Hooke, Phillippe de la Hire (1640–1718), and Pierre Couplet (ca 1670–1743). Poleni knew that for an arch to be stable its thrust line must lie within the masonry. The meridional cracks had almost divided up the dome into lunes like orange slices. Thus Poleni hypothetically sliced the dome into fifty lunes and argued that if the sliced dome was stable so too would the dome be stable whether cracked or uncracked, which is in accord with modern limit theory. To test whether the thrust line in the lunes was within the masonry, Poleni made use Hooke's hanging chain analogy (see §5.1.2) and threaded a string with a series of beads whose weights were each in proportion to the weight of the lune at that point including a contribution for the weight of the lantern. The result of this experiment is shown in Fig. 6.10 and since the thrust line did indeed lie within the masonry, Poleni concluded that the dome of St Peter's was safe.

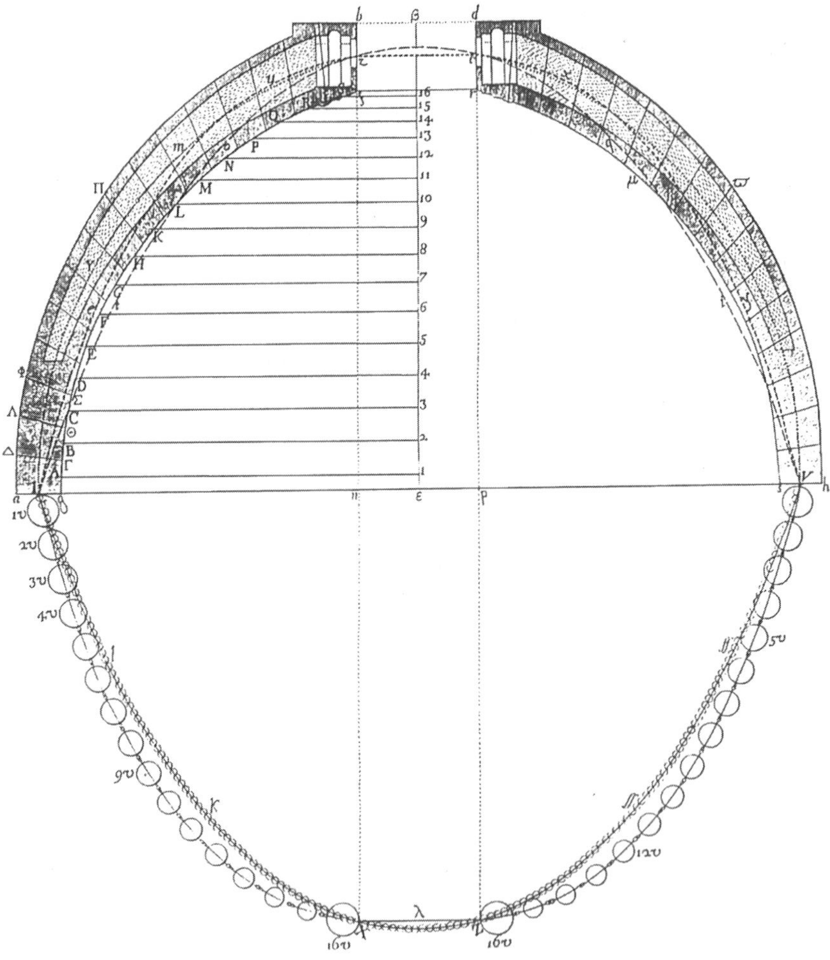

Fig. 6.10 The hanging chain analogy applied to St Peter's Dome (Poleni 1748).

Three mathematicians, Thomas Le Seur (1702–1771), François Jacquier (1711–1788), and Ruggero Boscovich (1711–1787) also made a study of the dome and concluded that extra ties were needed to counter the thrust of the cracked dome. Poleni agreed and estimated the tie force necessary from the horizontal pull of his hanging chain. Thus four chains were fitted between the double shells of the dome.

6.6 The Liberty Bell

Fig. 6.11 The Liberty Bell.

Bells were among the largest castings prior to modern times especially in China and Japan. Tōdai-ji's bronze temple bell weighing 26.3 tonnes was cast in 752.[26] The Chinese not only cast large bronze bells such the one in the Jueshing Temple in Beijing weighing 46.5 tonnes, dating to the Yongle period (1403–1424) of the Ming Dynasty, but also large cast iron ones of more than a tonne which date to the beginning of the first millennium.[27] In the West, large bronze bells weighing 15 tonnes or more were cast from the seventeenth-century. The largest bell in the world, the Tsar Kolokoi, weighing 196 tonnes was cast in Moscow in 1735, but has never been rung.[28] A fire in 1737 in the derrick erected to lift the bell from its casting pit caused uneven heating of the bell and during cooling residual stresses developed that caused an 11 tonne piece to crack away from its side. The bell was finally lifted and placed on a pedestal in the Kremlin in 1836.

The Liberty Bell, now in the Liberty Bell Center in Philadelphia, is not heavy, weighing just less than a tonne, but it is one of the most famous. Tradition has it that it was rung on the 8th July 1776 summoning the citizens of Philadelphia to hear the first public reading of the Declaration of Independence, but historians doubt this story. Originally it was called simply the State House Bell and it got its present name from the nineteenth-century abolitionists. The Bell was created to commemorate the 1701 Charter of Privileges, the original Constitution of Pennsylvania. Cast on the Bell is the quotation from Leviticus 25:10 'Proclaim Liberty throughout the land unto all the inhabitants thereof'. The purchase of the Bell was initiated 1751, and in 1752 an order was placed with the famous London Whitechapel Bell foundry which was established in the reign of Elizabeth I (1533–1603) and is still in business.[29] When it was hung in 1753 to try its tone, a stroke of the clapper cracked it. The cracking of bells was not uncommon and, as was the custom, the Liberty Bell was melted down and recast by two Philadelphia foundry workers, Pass and Stow. Like all other Western bells, the Liberty Bell was a high-tin bronze. The sonority of bells come from the composition of the bronze which typically has 20–22% tin and consists of islands of a eutectoid matrix and islands of a copper-rich solid solution. It is the eutectoid that gives the clarity of tone to the bell, but the eutectoid is brittle and the requirements of tone have to be balanced by the necessity of the bronze having sufficient toughness; the copper-rich solid solution provides the toughness. Pass and Stow realised that the Liberty bell probably had a too-high tin content and added an ounce and a half of copper to the pound of the old bell to make it tougher. Unfortunately, while the Bell was made tougher its tone suffered. Pass and Stow then experimented with small test bells and decided to add 0.25% silver, which together with virgins, was believed to sweeten the tone of a bell.[30] The second casting by Pass and Stow was no better than the first, but nevertheless was hung in the State House Steeple. A new bell was ordered from the Whitechapel foundry, but this proved no better than the Pass and Stow bell. The new Whitechapel bell was placed in the cupola on the State House to sound the hours and the Pass and Stow bell was retained in the State House Steeple. Apart from a period during the American War of Independence, when it was hidden to avoid it being melted down and used for cannons by the British, the Liberty bell was used to call the State Legislature into session and important events until the state and national governments were moved from Philadelphia in the early 1880s. However, it continued to be used for important events. It was during tolling for the funeral of the Chief Justice John Marshall (1755–1835) that the large crack visible in Fig. 6.11 appeared. In 1846 an attempt was made to enable the Liberty Bell to be used on George Washington's birthday in 1846 by drilling out of the crack, and fitting bolts with washers near the tip and start of the crack to prevent the lips of the crack rubbing together. However, when the Bell was tolled a new crack re-initiated from the end of the drilled-out original crack into the crown of the bell. Since then the Liberty Bell has only been rung by lightly tapping with a small mallet.

What were the causes of the cracking of the Liberty Bell? Primarily the problem was that the Bell had too high a tin content and was brittle. An analysis has been made on a small sample of the Bell was trepanned from the inner surface, about 100 mm from the lip and adjacent to the original crack, giving 73.1% copper and 24% tin with trace amounts of lead, zinc, silver, nickel and iron.[31] The high tin content would make the bronze brittle. The bell was tuned by chipping its edge with a chisel leaving a ragged edge, which would have provided initiation sites for a crack.[30] The original crack did not appear for some 80 years after the Liberty Bell was hung and the 1963 committee for the preservation of the bell has questioned whether it grew catastrophically.[31] Since the crack appeared during tolling of the bell for a funeral, it is unlikely that the clapper was swung particularly hard on that day. The probable scenario is that a small crack grew from the chipped edge of the bell with time until it became of critical length for the stress caused by clappering plus the residual stresses left from the casting process. The cause of the initial subcritical growth could have been either fatigue or stress corrosion. The drilling out of the crack in 1846 is reminiscent of the practice of drilling a hole at the tip of a crack to reduce the stress concentration and thus prevent further crack extension by fatigue. However, for this treatment to be effective it must be certain that all traces of the crack tip are removed. The bolting together of the crack surfaces would have done very little to protect the bell. The clappering of the bell on George Washington's birthday would not needed to have been particularly strong to reinitate a crack, particularly if the drilling had not completely removed the previous crack tip, because the starter crack was now very long.

6.7 Charles-Augustin de Coulomb (1736–1806)

Coulomb, a French engineering officer, was posted to Martinique in 1764. During his nine years there Coulomb wrote his famous paper that contains his work on fracture explaining that his aim was 'to determine, as far as a combination of mathematics and physics will permit, the influence of friction and cohesion in some problems of statics'.[32] Coulomb's work on compression fracture has already been discussed in §2.4.1; he also performed tension and shear tests on rock finding that the shear strength was equal to the tensile strength since the fracture of rock is largely determined by the maximum principal stress which is numerically equal to the shear stress.

Coulomb also considered bending. Unaware of the work of Parent, he independently obtained the correct expression for the bending of a rectangular wooden beam. The deformation in a wooden beam can be seen, but the deformation of stone beams is not so obvious and Coulomb assumed that for stone the beam is 'composed of stiff fibres that can be neither compressed nor extended'. He also assumed initially that the strength in compression is infinite and that the beam rotates about the bottom of the beam. Thus Coulomb obtained exactly the same expression for the strength of a stone beam as Galileo's Eqs. (6.1) and (6.2). However, Coulomb found that his expression overestimated the strength of the beams he measured and decided that the point of rotation must be higher than the bottom of the beam so that a portion of the beam is under compression. For beams of small cross-section, Coulomb was right in maintaining that the expression for the bending strength would be underestimated by Parent's expression. At high tensile stress stone, especially soft stone, strain softens due to microcracking and as a consequence the apparent nominal strength of such beams increases with decreasing cross-section.[33]

The assumption that in fracture a stone beam has infinite strength in compression and rotates about its top or bottom may not be correct for bending, but it does enable the stability of a stone arch to be predicted. A properly designed stone arch is under compression everywhere. Failure in an arch occurs not by the stone crushing but by a hinge occurring at either the intrados or extrados. Coulomb made use of this observation to calculate upper and lower bounds to the thrust on an arch, which seems to be the first statement that bounds could exist on the value of a structural quantity.[34] He considered a hinge occurring first at some point on the intrados and since the hinge converts a statically indeterminate problem into a static problem Coulomb could then calculate the horizontal thrust in terms of the weight of the arch and the position of the hinge along the arch. The position of the hinge is not arbitrary but occurs so that the horizontal thrust is maximised. Coulomb then stated that this maximum value would be the *minimum* thrust for a stable arch. A hinge in the extrados similarly gives rise to the maximum thrust for a stable arch.

6.8 Mechanical Testing in the Eighteenth-Century

The measurement of the strength of materials started during the Renaissance, but it was not until the eighteenth-century, when lever type testing machines started to be used, that mechanical testing of materials for their own sake developed. René Réaumur[35] (1683–1757) made tensile tests on iron wires in the 1720s during his study of heat treatment, but most testing in the eighteenth-century was on timber and stone. During the same period Petrus van Musschenbroek (1692–1761), professor of physics at Utrecht performed tension, compression, and bending tests, principally on wood, but he did test some metals.[22] Although his lever system gave him force magnification, the size of Musschenbroek's tensile specimens was very small (5 mm square) and his work was criticised. The choice of timbers by Musschenbroek also seems bizarre. Although he tested building timbers such as oak, fir, and cedar he also tested others, such as pomegranate, tamarind, orange, and quince, surely out of sheer curiosity. Georges-Louis de Buffon (1707–1788) performed a comprehensive series of tests from the late 1730s into the 1740s on the strength of wood beams which included tests to compare the strength of small and large specimens. After carefully testing more than 1,000 small specimens and being extremely careful to ensure that the specimens contained no knots or other defects, Buffon concluded that it was not possible to predict the properties of full-size timbers containing defects from tests of small specimens. Buffon then he began a series of bend tests on full-size structural members with cross-sections up to 200 x 200 mm and with spans of 8.5 m.[22] Not surprisingly Buffon criticised the small specimens used by Musschenbroek.

During the construction of the church of Sainte Geneviève in Paris during the 1770s there were differing opinions on the necessary size of the pillars. Émiland–Marie Gauthey (1732–1807) constructed a lever testing machine capable of testing the compression strength of 50 mm cubes of stone and compared his results with the compressive loads on existing buildings finding that there was a safety factor of about 10.[22]

Books on mechanics in the eighteenth-century often had a chapter on the strength of materials. One such book was that of William Emerson (1701–1782), who in 1754 published a book on mechanics with a section on the strength of timbers which contained results of his own experiments on some metals as well as timber.[22] The first book on the strength of materials was written at the close of the eighteenth-century by Pierre Girard (1765–1836) who was a member of the famous school for training engineers for construction of bridges and roads.[22]

6.9 Concluding Remarks

From the Renaissance onwards there was emphasis on the quantification of the strength of structures. This period also saw the birth of the experimental method.

Although technology is as old as man himself, the period between the Renaissance and the Industrial Revolution sees an exponential growth. Before the Renaissance, technology was largely the realm of the practical man, though there were notable exceptions like Archimedes who was both an intellectual and a practical engineer. Practical men were despised during classical times. This view of the engineer was expressed by Plato who wrote in *Gorgias* that 'though at times he has no less power to save lives than the general… do you place him in the same class as the advocate … You distain him and his craft and would call him engineer as a term of reproach and would refuse … to give your daughter to his son'.[36] Plutarch (ca. 46–119) claimed that Archimedes also held the practical arts in distain, 'regarding the work of an engineer and every art that ministers to the needs of life as ignoble and vulgar'.[37] However it is very unlikely that Archimedes himself held these views and it is simply Plutarch reflecting the accepted view of his time.

By the Industrial Revolution, engineering had become the driving force in the quest for understanding fracture.

6.10 Notes

[1]   Known as 'the father of mineralogy'.
[2]   Parsons (1939).
[3]   Clayton (2006).

[4] The libbra and braccio varied with Italian city. Leonardo spent time in Florence, Milan and Rome where the libbra was 0.327–0.3395 kg and the braccio was 0.584–0.670 m (Parsons 1939).

[5] The statistics of the strength of components controlled by the largest defect were given by Weibull (1939) see Chapter 8.

[6] The inaccessibility of Leonardo's Notebooks, unpublished until the nineteenth-century, has caused the parallelogram of forces to be credited to the Flemish mathematician and engineer Simon Stevinus (1548–1620).

[7] Experiments similar to that illustrated in Fig. 6.3 (a) used to be performed by first year engineering undergraduates.

[8] Heyman (1969).

[9] Galilei [1638].

[10] Stress was not defined until 1822 see §1.2.

[11] For a non-elastic cantilever the factor is within the limits of 2–3.

[12] (Heyman 1998).

[13] Brodsley *et al.* (1986).

[14] The First Charter was sealed on the 15th July 1662.

[15] March 1660 in the old style calendar.

[16] Prince Rupert's Drops were still being advertised in the *London Gazette* in 1695.

[17] Taylor (1947).

[18] It is customary to refer to this process as toughening, but the glass is not toughened, the increased strength simply comes from a favourable residual stress distribution.

[19] Richardson (1888).

[20] Boyle did not publish an explicit statement of his law. In 1660 he published a book, *New Experiments Physico-Mechanical Touching the Spring of Air and its Effects*, but it was in answering a critic Francis Linus that he made an explicit statement of the Law. Mariotte did not publish his work until 1676.

[21] Mariotte (1686).

[22] Timoshenko (1953).

[23] This mistaken impression about the position of the neutral axis not being important was repeated by Pierre-Simon Girard (1765–1836) in his book on the strength of materials *Traité Analytique de la Résistance des Solides*, published in 1798.

[24] The first expression for the bending stresses in beam with a general cross-section, symmetrical about a vertical axis, was given by Louis Navier (1785–1836) in 1826.

[25] Heyman (1967, 1988, 1998).

[26] Yoshimasa (1992).

[27] White cast iron, containing high sulphur and low silicon, with low damping capacity (Rostoker *et al.* 1984).

[28] The heaviest bell in use is the Mingun Bell in Mandalay, Myanmar which weighs 92 tonnes.

[29] The Whitechapel foundry also cast Big Ben, which weighs 13.8 tonnes and also cracked after two months in service. The crack was blamed on the use of a hammer twice the weight of that specified. The crack was not large and was chipped out, the bell given an eight of a turn and a lighter hammer fitted and has continued to chime the hours for nearly 150 years.

[30] Rosenfield (1976).
[31] Cornell *et al.* (1963).
[32] Coulomb (1776).
[33] Cotterell and Mai (1996).
[34] Heyman (1998).
[35] Réaumur is better known for the temperature scale.
[36] Plato, Gorgias 512b–e.
[37] Plutarch, Marcellus XVII, 3–4.

Chapter 7

From the Industrial Revolution to 1900

Before the Industrial Revolution, design against fracture was largely limited to buildings and bridges and the materials used were mainly timber and stone. The Industrial Revolution saw the development of improved spinning and weaving machines, James Watt's steam engine, and the railways. Consequently iron and steel became much more important. In 1740, Britain was the world's greatest metal producer but its iron output was less than 20,000 tonnes a year. The first cast iron bridge, spanning the Severn River in Coalbrookdale, was designed by Thomas Farnolds Pritchard (1723–1777) but built by Abraham Darby III (1750–1791)[1] because Pritchard died before in was completed in 1781. The bridge used 390 tonnes of cast iron. By 1840 Britain was producing 1.25 million tonnes of iron a year and in 1900 the production or iron and steel had risen to 9 million tonnes a year. These changes required development in the understanding of fracture. Concrete too started to be used again as a structural material. The first major concrete structure was the Grand Maître Aqueduct (1850–1865) which conveys water from the river Vanne to Paris on a long series of concrete arches. However, the approach to concrete was the same as for masonry – avoid tension. Attempts at reinforcing masonry with iron rods similar to their use by the Greeks were tried as early as the eighteenth-century by Jacques Scoufflot (1713–1781) in France and again by Isambard Kingdom Brunel (1806–1859) in the nineteenth-century, but moisture penetrating the mortar led to quick corrosion.[2] In the mid-nineteenth-century a French gardener, Joseph Monier (1823–1906) successfully made large tubs with a mesh of thin iron rods embedded in the concrete and William Boutland Wilkinson (1819–1902) suggested the use of old mining wire ropes to make reinforced beams by putting the ropes on the tension side.[2]

The foundations of the elastic theory of the mechanics of solids developed quickly once the concept of stress had been introduced by Cauchy in 1822. By 1900 all the basic equations of elasticity were known and solutions obtained to a large number of geometries and loading.[3] With the coming of the Industrial Revolution there was a need for information on the strength of iron and steel.

George Rennie (1791–1866), who constructed a lever testing machine capable of loads up to 11 tonnes and made a considerable number of tests on a wide variety of materials which were published in 1818, remarked about the lack of accurate mechanical data writing that 'The vague results on which the more refined calculations…are founded, have given rise…to a multiplicity of contradictory conclusions'.[4] The nineteenth-century saw an exponential growth in materials testing matching the growth in industry.

7.1 Emerson's Paradox

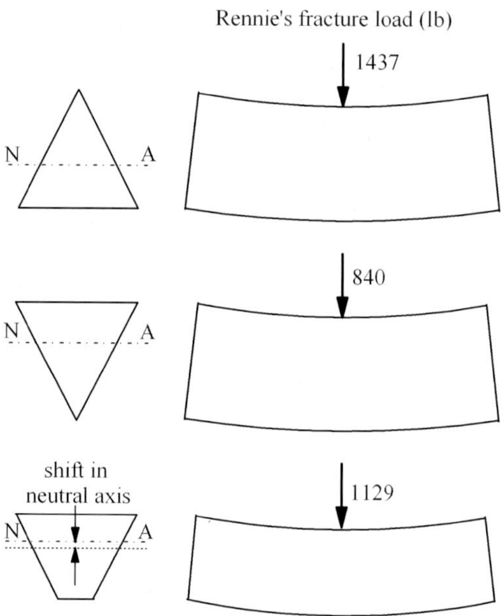

Fig. 7.1 Beams with triangular cross-sections giving fracture loads recorded by Rennie (1818).

Rennie made a number of bend tests on cast iron bars of different cross-section and also examined Emerson's Paradox which created considerable interest.[4] In his book on mechanics William Emerson (1701–1782) mentioned a curiosity in a beam whose cross-section is in the form of an equilateral triangle.[5] Emerson stated that if the beam is bent so that an apex is in tension then the beam is made stronger if the tip of the apex is removed; giving rise to the paradox that the part is stronger than the whole. No other details were given by Emerson. Although

Emerson's Paradox is in part simply a curiosity, it does raise fracture issues. Rennie performed experiments on simply supported cast iron beams with triangular cross-sections using a 2 ft 8 in (813 mm) span and found that when they were loaded so that the base of the triangle was in tension they fractured at 1,437 lb (6,392 N), when the apex was in tension they fractured at 840 lb (3,736 N), but when a portion of the apex was removed they fractured at 1,129 lb (5,022 N).[6] These tests are represented schematically in Fig. 7.1. In 1818 the position of the neutral axis was not known for a general cross-section and Rennie gave no explanation of his results.

In the year before Rennie published his paper, Peter Barlow (1776–1862) at the Royal Military Academy published a book on the strength of timbers.[7] Barlow too was interested in Emerson's Paradox and he quoted results of Benjamin Couch on Canadian oak that did not follow the 'commonly received notion', he repeated the experiments with fir and found that neither did these show any strengthening on removing the apex of the triangular cross-section. Barlow also stated that 'with regard to the situation of the neutral axis we have nothing to guide us in its determination but experiments'.

Thomas Tredgold (1788–1829) also examined Emerson's Paradox.[8] Despite scathing remarks about him by Isaac Todhunter (1820–1884), the English mathematician who wrote the classic nineteenth-century history of the theory of elasticity,[9] Tredgold did attempt to examine Emerson's Paradox theoretically. He failed because he did not position the neutral axis correctly, but nobody else knew its position for a section like a triangle at that time. Tredgold assumed that neutral axis is positioned so that the second moment of area for the two halves of the section should be the same. Thus for a triangular cross-section Tredgold positioned the neutral axis at a distance of 0.697 of the depth of the cross-section from the apex.[8] Louis Navier (1785–1836) correctly concluded that the neutral axis must pass through the centroid of the cross-section in 1826[10] and for a triangular cross-section that gives the neutral axis at a distance of 0.667 of the depth of the cross-section from the apex. For symmetrical cross-sections Tredgold correctly calculated what we now call the *section modulus*, $Z = y/I$, where I is the second moment of area of the cross-section and y is the distance to the furthermost fibre from the neutral axis.[8] Tredgold realised that, for a brittle material like cast iron, the beam fractures when a critical tensile stress is reached and the reduction in the second moment of area, when the tip of the triangular cross-section is removed, is more than balanced by the reduction in the distance to the furthermost fibre from the neutral axis. Tredgold calculated that if the depth of the cross-section is reduced by 10% that the strength is 1/37 (=2.7%) stronger. Eaton Hodgkinson (1789–1861) also gave an explanation of the Paradox in 1838.[11] The final word on Emerson's Paradox, as applied to a linearly elastic beam together with cases of square, circular, as well as triangular, was given by the great elastician Adhémar Barré Saint-Venant (1797–1886) in his copious notes to the third edition of Navier's *De la Résistance des Corps Solides* published in 1864.[12]

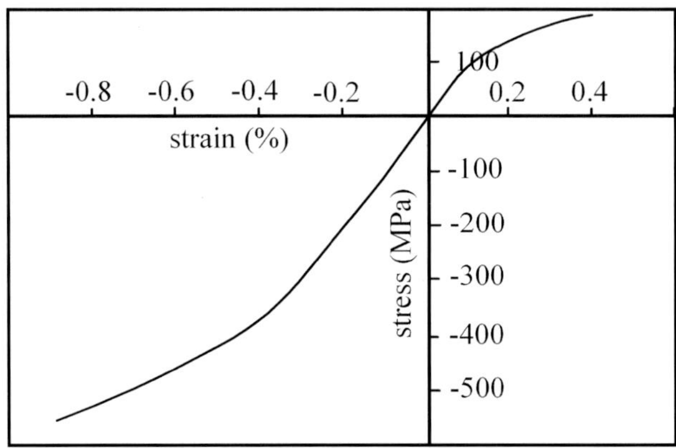

Fig. 7.2 Stress-strain curve for a typical cast iron (after Blanchard *et al.* 1997).

The degree of strengthening in a beam of triangular cross-section caused by cutting off the apex when the apex is in tension obviously depends upon the material. It also depends upon the strength in compression being considerably greater than that in tension. Timber is stronger in compression than tension which explains why Barlow did not observe the Paradox.[7] For Emerson's Paradox to be true the material must be brittle with a characteristic length small in comparison to the cross-section. Cast iron meets these conditions and exhibits the Paradox as Rennie found.[4] However, cast iron hardly exhibits a linear elastic stress-strain curve even at very small strains and the stress-strain behaviour in compression is different to that in tension whereas for most metals the behaviour is a mirror image. The stress-strain curve for a cast iron is shown in Fig. 7.2. Cast iron in the early nineteenth-century was not as strong as the cast iron shown in Fig. 7.2; Rennie's cast iron had a tensile strength of 132 MPa.[4] The non-linearity in the stress-strain relationship causes the degree of strengthening by cutting off the tip of the apex of the triangular cross-section to be underestimated by elastic theory. If Rennie's cast iron beams were truly linearly elastic he would have observed that the strength of a beam with a triangular cross-section tested with the apex in tension was a half of that when the base is in tension whereas his was 0.585 times the strength of the beam with the base in tension. For an elastic-brittle beam with a tensile strength less than half the compressive strength, 28% of the depth of a beam of triangular cross-section can be removed without weaken it when tested so that the apex is in tension. The maximum strengthening of 9.2% can be obtained when 13% of the depth of the cross-section is removed, whereas Rennie found that by removal of an unstated portion of the apex the beam was strengthen by 34.4%. Tredgold's estimate of the increase in strength of 2.7% by cutting off 10% of the depth of an elastic-brittle beam should be compared with the exact value of 8.86%.

7.2 Wrought Iron and Brittle Fracture

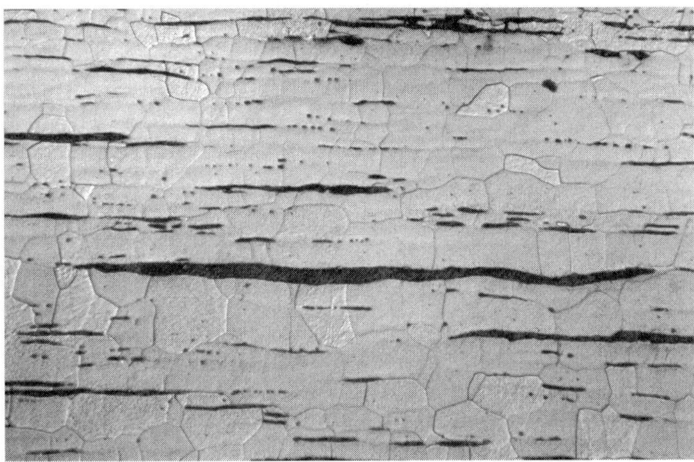

Fig. 7.3 Optical micrograph of wrought iron (courtesy Mike Meier).

Although there are generalities and similarities, fracture cannot be wholly considered independently of the material. During the Industrial Revolution and up until the middle of the nineteenth-century wrought iron was the main structural material. The early method of producing wrought iron consisted of extracting iron from a sponge consisting of a mixture of iron, iron carbide, and slag obtained by smelting iron ore with charcoal, and then repeatedly hammering and folding the sponge until the oxide scale and the iron carbide was reduced to almost pure iron (ferrite) with inclusions of slag, mainly iron silicate, in the form of stringers as a result of the repeated working. Since wrought iron is not now a common material a micrograph of its structure is shown in Fig. 7.3. During the Industrial Revolution the processes for manufacturing wrought iron were improved and production increased dramatically. The most successful was puddling, invented in 1784 by Henry Cort (1740–1800). The iron was tapped from the smelting furnace into a mould where most of the slag was removed resulting in what was known as *finers metal*, which had a similar composition to sponge iron. The finers were then used as the charge for a puddling furnace where the molten mass was stirred with iron oxide usually in the form of scale from the rolling mills. As the carbon was removed from the iron so the melting temperature rose and the *puddled balls* became pasty and could be removed from the furnace. The puddled balls were hammered or shingled and then rolled out into *muck bars*. The muck bars were *faggoted* together by rolling to produce *faggoted wrought iron*. The faggoted bars were themselves could be faggoted to produce *best wrought iron* and the process could be repeated to produce *best best wrought iron* or even *triple best wrought iron*.

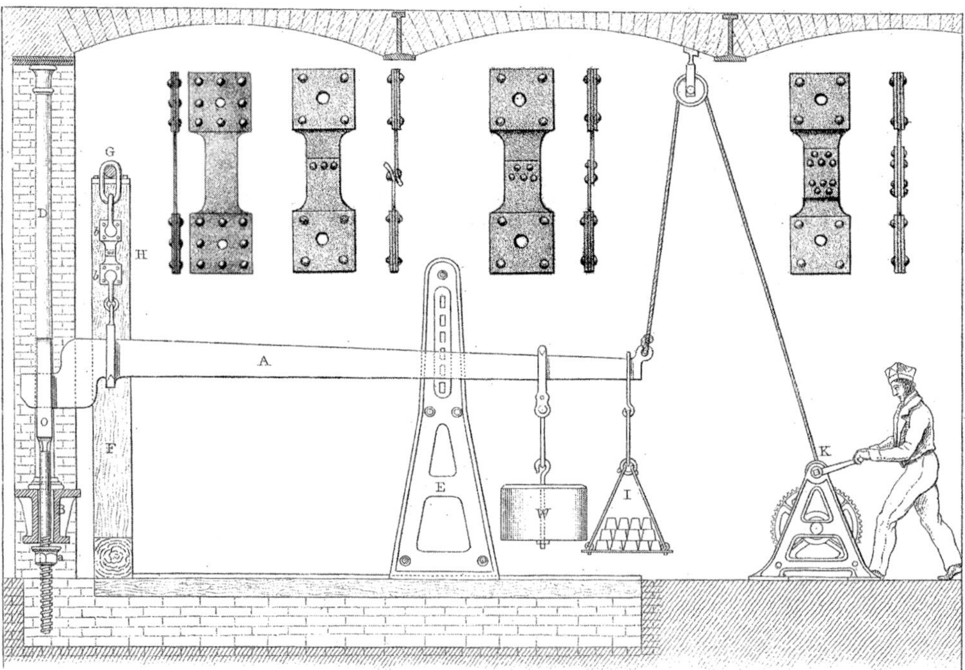

Fig. 7.4 Fairbairn's testing machine showing plate and riveted specimens as insets (Fairbairn 1850).

The price of wrought iron gradually fell after the introduction of puddling and the possibility of building ships with it became apparent. John Wilkinson (1728–1808), an ironmaster and strong advocate of the use of iron, built *The Trial*, an eight-tonne barge in 1787 to transport the 32 pounders, howitzers, swivel guns, mortars and shells he was manufacturing for the British Government; he also made an iron coffin for himself. However, the first ocean-going ships to use iron structurally were composites with iron ribs and wooden planks. There was resistance to using wrought iron for ships because it is heavier than water and it was thought to be a risk. William Fairbairn (1798–1874) was also an active advocate of wrought iron in construction and in 1838 embarked on a series of tests on the strength of plates and their riveted joints with the purpose of showing 'that the iron ship when properly constructed is not only more buoyant, but safer, and more durable than vessels built of the strongest English oak', which he published in 1850.[13] Fairbairn built the lever-type testing machine capable of testing up to loads of 30,000 lb (133 kN) shown in Fig. 7.4. With this machine he tested a variety of English wrought iron specimens, whose configuration is similar to that used today, cut both along and transverse to the direction of rolling, Fairbairn found little difference in strength between the two directions, but considerable difference between the different varieties. The average strength of the wrought irons was 22.8 ton/in^2 (352 MPa) or five times the strength of oak. Fairbairn gave drawings of the side-views and fracture surfaces showing that laminations in the plate were frequent and he described the fracture surfaces as fibrous with some areas of a crystalline appearance. These terms are still in use today for the ductile and cleavage regions in steel fractures, but during the first half of the nineteenth-century their significance caused a long debate. The fact that all metals were crystalline was not known and areas in a fracture surface, including fatigue fractures, which had a bright crystalline appearance, were thought to have been caused by the metal having crystallised for some reason in those areas.

John Robison (1739–1805), who wrote almost all the articles on the physical sciences in the 1801 supplement to the third edition of the *Encyclopaedia Britannica* and had a great influence on the ideas on the strength of materials in the early nineteenth-century, had implied that when a material was stretched it was damaged so that its subsequent strength would be impaired.[14] Although Fairbairn did no tests of his own to refute this notion, he quoted experiments at the Royal Dockyard in Woolwich, which showed that when an iron bar was fractured and the fractured parts themselves tested, the strength increased.

Fairbairn was an enthusiastic advocate of machine riveting and performed a range of tests on both riveted lap and butt joints. While Fairbairn was experimenting on the strength of wrought iron, Isambard Kingdom Brunel (1806–1859) was building in Bristol his famous iron steamship the *Great Britain*. Fairburn paid a visit to Bristol and described the double riveted joints of the ¾ in. (19 mm) plates used in the steamship's construction. In his quite comprehensive examination of the use of wrought iron in ships, Fairbairn also carried out tests on the resistance of iron plates to penetration and compared its resistance to that of oak.[13]

In mechanical testing the tensile strength has always been the easiest and most frequently measured mechanical property. However, the tensile strength of a material is not always a good guide to its fracture resistance. There were no quantitative toughness tests until the very end of the nineteenth-century.[15] In the absence of toughness tests the ductility in the form of the reduction in area in a tensile test is a better indication of the fracture properties of a material than the strength. David Kirkaldy (1820–1897) undertook a comprehensive series of tension tests from 1858 to 1861 using a lever-type testing machine while he was working for Napier and Sons in Glasgow. He wrote in his book giving the results of his tests that 'it seems most remarkable that an element of the highest importance should have been so long overlooked, namely, the Contraction of the specimen's area…The apparent mystery of a very inferior description of iron suspending, under a steady load, fully a third more than a superior kind, vanishes when we find that the former… [retained] its original area only slightly reduced; whilst the latter on breaking was reduced to very nearly a forth of its original area – the one hard and brittle, liable to snap suddenly under a jerk or blow, the other very soft and tough, impossible to break otherwise than by tearing slowly asunder'.[16] The toughness of wrought iron depends mainly on the phosphorus content. Phosphorus is present as an impurity in wrought iron and provided its percentage is less than about 0.2% does not affect the toughness very much.[17] However, phosphorus strongly hardens iron by forming a solid solution and frequently up to 0.6% was deliberately added to wrought iron to increase its strength. However, the increase in strength came at a price. In metals there is always a competition between flow and fracture. For cleavage fracture a critical stress over a critical distance is needed[18] and anything that increases the yield strength makes a brittle cleavage fracture in a body centred cubic metal more likely. A notch constrains yielding and increases the stress. The yield strength is also increased if the deformation takes place at high strain rates. Since the yield strength of iron increases as the temperature decreases, there is a transition from ductile fracture at high temperature to brittle cleavage fracture at low temperature. Because phosphorus increases the yield strength it raises the transition temperature. For steel an addition of 0.1% of phosphorus raises the transition temperature by about 70°C[19] and a similar increase would occur for wrought iron. Thus wrought iron with high phosphorus content had a high strength, but was brittle at room temperature and not suitable for structural use – hence the appropriateness of Kirkaldy's remarks. It was known in the eighteenth-century that phosphorus made wrought iron brittle or *cold short*[20] a term that is still in use today.[21,22] A series of letters was written to *The Times* in 1859 under the pseudonym of 'Amicus' commenting on the loss of the *Royal Charter* off the coast of Anglesey in Wales, warned of the dangers of using cheap cinder iron as boat plates.[23] Cinder iron was made by using slag left from the manufacture of wrought iron, which was as rich in iron as many iron ores, for furnace feedstock. The problem is that phosphorus is concentrated in the slag left from manufacturing wrought iron and cinder iron had a very high phosphorus content making it very brittle. Amicus describes such iron plates 'as brittle as glass'.

Kirkaldy has a long discussion on the fracture behaviour of wrought iron and steel.[16] He used three classifications for his wrought iron fractures: (i) fibrous, (ii) crystalline, and (iii) fibrous and crystalline. Kirkaldy gave a well-reasoned account of the causes of crystalline fracture stating that the change from fibrous to crystalline fracture can be explained without calling on the explanations that were current at the time namely vibration, percussion, heat and magnetism. He listed three ways wrought iron can be made brittle and crystalline:

(i)   By grooving a tensile bar.
(ii) By locally hardening the groove.
(iii) By increasing the strain rate.

Kirkaldy also quoted from an editorial in *The Engineer* of 1861 where it clearly stated that 'iron may be weakened both by percussion and by frost'.[24] What Kirkaldy objected to in this article was not the fact that percussion and frost made iron more brittle and the fracture surface has a crystalline appearance, but that frost and percussion are described as changing the iron from a fibrous state to a crystalline state. Despite this perceptive distinction made by Kirkaldy, the article contained a warning: the 'effect of percussion and frost upon iron… is one of the most important subjects that the engineers of the present day are called upon to investigate. The lives of many persons, and the property of many more, will be saved if the truth of the matter be discovered – lost if it not be'. Unfortunately it took some 90 years for the problem to be fully understood, during which time thousands lost their lives.

7.3 Steam Power and Bursting Boilers

The practical use of steam power began in the eighteenth-century. In Newcomen engines, which were used to pump water from mines from about 1711, the steam was condensed inside the cylinder and the power came from the vacuum produced. These engines ran at a very low pressure and the boilers presented little problem. James Watt (1736–1819) invented the separate condenser, which he patented in 1769, and made the cylinder double-acting, using steam pressure instead of just vacuum to return the piston. The pressure used in the early Watt engines was only about 5 lb/in^2 (35 kPa) and the boiler was constructed of wrought iron with the flat ends stayed by iron tie rods. With the introduction of the non-condensing steam engine by Richard Trevithick (1771–1833) in England and in the US by Oliver Evans (1755–1819) at the beginning of the nineteenth-century, the steam pressure rose. The Cornish boiler, introduced by Trevithick in 1811 usually produced a steam pressure of about 50 lb/in^2 (350 kPa). In the USA Evans was using an even higher pressure. Boilers began to burst. Enormous energy is stored in steam under pressure and the boilers burst explosively, causing great damage and all too often a large loss of life. In fact the boilers often burst so explosively that a myth arose that the steam under high pressure decomposed into hydrogen and oxygen, with the hydrogen exploding.[25] The first major boiler disasters occurred on steamships. Between 1817 and 1839 there were 23 major accidents in steamships due to bursting of boilers in British waters.[26] Even more disasters occurred in the USA, where steamships were more numerous. By 1818 30 people had been killed in seven separate boiler accidents in steamships and riverboats in the USA, a further 320 lives were lost between 1818 and 1830. The worst ever boiler accident occurred on the Mississippi steamboat *Sultana* in 1864 when 1,238 lives were lost, mainly from drowning rather than the explosion itself.[27] The causes of the bursting of the boilers were inadequate design, poor materials, loss of strength due to corrosion, overheating caused by an accumulation of sediment in the boiler, and deliberate tampering with the safety valves – tying down of the safety value was usual practice during the frequent steamboat races on the Mississippi.

The relationship between plate thickness, internal pressure, diameter, and stress for either a cylindrical or spherical pressure vessel, provided the thickness to diameter ratio is small, can be obtained from simple considerations of equilibrium and is fundamentally an easier problem than the bending of beams to solve. Hence in engineering the derivation of the stresses in thin walled cylindrical and spherical pressure vessels is usually taught before beam theory. However, until the advent of the Watts' steam engine, boilers worked at such low pressure that really no formal design was needed. Water was occasionally piped at pressure, such as for the fountains of Versailles, but even if these burst they did no damage because firstly they were of small diameter and secondly water is not very compressible and very little energy is stored. Thus it is understandable that, though Mariotte obtained a partial solution in the seventeenth-century,[28] the expression for the strength of cylindrical vessels was not formally obtained until the advent of boilers working at relatively high pressures. Oliver Evans published probably the first formula to calculate the thickness required for wrought iron boilers as a function of their diameter in 1805 and gave a table for the thickness required to sustain a pressure of 1,500 lb/in^2 apparently allowing for a factor of safety of 10 on the pressure.[29,30] Walter Johnson (1794–1852) also published a table of the plate thickness required in 1832 for a design pressure of 150 lb/in^2 which used a factor of safety of 5 on the pressure.[31] However, shell thickness frequently depended upon what was available and was often of inferior quality. Also with time the shell thickness was reduced by corrosion.[25] The formula for the required plate thickness for a cylindrical boiler adopted by the French Government in the 1850s was, in modern units,

$$t = 1.8 \text{ x } 10^{-5} pd + 3, \qquad (7.1)$$

where t is the thickness in mm, p is the pressure in kPa, and d is the diameter of the boiler.[32] This formula has a safety factor of about ten; the additional 3 mm is an allowance for corrosion.[33]

The large loss of life due to bursting boilers prompted governments for the first time to attempt to regulate the industry. Although select committees were set up in both Britain and the USA in 1817, the first country to legislate to control boilers was France, where a Royal Ordinance was issued in 1823.[25] An amendment in 1830 saw the establishment of a comprehensive boiler code giving design formulas and material strength values. Hemispherical heads were required for boilers working above 0.48 kPa and the boilers had to be tested at three times the design working pressure[34] both initially and yearly thereafter in case corrosion had weakened the boiler. This measure greatly reduced the number of boiler accidents in France.[25] At the request of the US Treasury Department, the

Franklin Institute set up a committee in 1830 under the direction of Alexander Bache (1806–1867) to report on the explosions of steam boilers and to carry out experiments. A subcommittee was responsible for the examination of the strength of materials used in boilers.[35] This subcommittee built the first testing machine in the USA and made hundreds of tests, including tests at high temperatures up to 700°C, from 1831 to 1836 reporting their results in a series of papers during 1837.[25] They also looked at the effect of rolling direction and found that the strength of wrought iron in the direction of rolling was 6% higher than in the direction transverse to rolling.[25] Despite this great effort the law enacted in 1838 was largely ineffectual and it was not until 1852 that an effective act was passed. In this act the maximum boiler pressure was set at 110 lb/in^2 (760 kPa) and boilers had to be tested at 1.5 times their working pressure every year.[25] However, even this act was not completely effective and between 1884 and 1892 there were 1,868 explosions in the USA killing 2,342 people and complaints about laxity in inspections were still being made.[36] Proof testing of boilers was resisted by some. At a legal inquiry in 1861, at the Guildhall in London into a boiler explosion on the London and North-Western Railway, strong opinions were expressed against the proof testing of boilers. An editorial in *The Engineer* in 1862 gave strong support to proof testing, but even here there were some reservations.[37] It was argued that 'the elasticity of iron and, therefore iron itself, is permanently injured by a strain somewhere between one-third and one-half of its breaking strength'; these arguments can be traced back to Robison.[38] We now know that a boiler or pressure vessel will not burst at a pressure less than its proof pressure unless corrosion has reduced its thickness. In the case of the explosion of the London and North-Western Railway boiler, its original thickness was 3/8 in [9.45 mm] and had been in service for more than nine years with no proof test at all and only one internal examination in 1857. After the explosion it was found that the plate was 'furrowed by corrosion for a total length of nearly 8 feet [2.4 m] along a seam of rivets, and in some places to the thickness of one-sixteenth of an inch [1.6 mm]'.[39]

7.4 Railways and Fatigue

Wilhelm Albert (1787–1846), a German mining engineer, was the first to study fatigue. In 1828 he performed repeated proof load tests with up to 100,000 cycles on iron mine-hoist chains.[40] However, the term *fatigue* was first used in 1839 in the lectures of Jean-Victor Poncelet (1788–1867) at the Metz military school.[41]

The French explorer Jules Dumont d'Urville (1790–1842), his wife and son were killed along with some fifty others in the first major railway accident. D'Urville had been attending inauguration ceremonies for the Paris to Saint-Germain railway and was on the unusually long returning train of 17 carriages, carrying some 1,500 people, drawn by two locomotives.[42] Near Versailles an axle on the leading locomotive fractured. The second locomotive crashed into the first and burst its boiler, and a fire started in the first six carriages. There was a large loss of life because the passengers were locked inside the carriages, as was then the usual practice. Fractures in the axles of locomotives or carriages after a few years of operation were not that uncommon and occurred by what we now call fatigue. The problem was recognised by the French Government who as a result of the Versailles accident issued an order within a few weeks requiring that state and length of service of railway axles be recorded and set up a committee to make experiments on the safe life of axles.[42] Shortly afterwards Charles Hood gave a paper to the Institution of Civil Engineers where he attributed the fracture of railway axles to the molecular change in the structure making the iron crystalline.[43] The reasons he gave for the molecular changes: vibration, magnetism, and heat are the same as those given for brittle cleavage fracture in wrought iron (see §7.2).

7.4.1 *The pragmatic approach to fatigue*

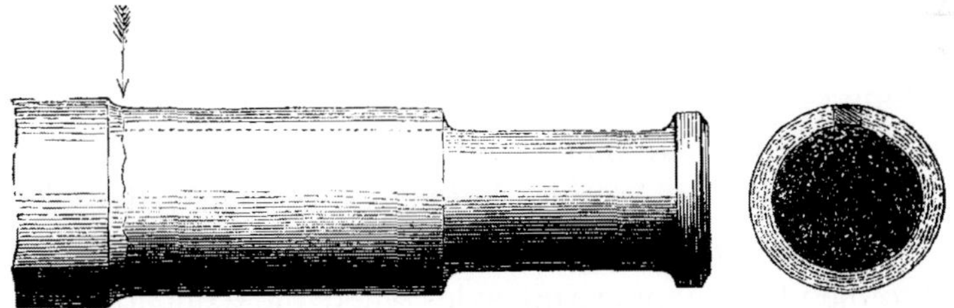

Fig. 7.5 Fractured railway axle, the fracture, indicted by the arrow, occurred at the end of the keyway shown by dotted lines (Glynn 1844).

At first there was no clear distinction between a brittle cleavage fracture and a fatigue fracture. Both were attributed to the iron having crystallised due to vibration, percussion, heat, cold, or magnetism. William John Macquorn Rankine (1820–1872) gave one of his first papers in 1843 on the subject of fatigue in railway axles when he was only 23 years old.[44] Rankine was sceptical of the notion that the iron changed its structure. He showed five axles which had fractured after two to four years and described the fractures as having started 'with a smooth, regularly-formed, minute fissure, extending all round the neck of the journal, and penetrating on an average to a depth of half and inch [12.7 mm]'. The axle finally broke when the 'iron in the centre became insufficient to support the shocks to which it was exposed'. There are no illustrations accompanying Rankine's paper, but in the same year Joseph Glynn did give an illustration of a fractured axle (see Fig. 7.5).[45] The fracture occurred at the tip of the transition radius where the axle was reduced to fit the wheel; the keyway also terminated at this point compounding the stress concentration. For about '½ in. [12.7 mm] in depth all round, there was a perfectly smooth cleft of a blue and purple colour; this annular part appeared to have been produced by a constant process; the central crystallized part being gradually reduced in diameter, until it was barely able to sustain the weight, and it broke on being exposed to a sudden strain'.[45] Railway axles frequently became overheated[46] and the 'smooth blue and purple cleft' was the slowly propagating fatigue crack coloured by oxidation due to the heat of the axle. An almost identical fatigue fracture in a railway axle occurred in a replica of Trevithick's 1804 *Pennydarren* locomotive in front of Rod Smith in 2004 and a railway enthusiast was quick to explain that the fracture had occurred because the metal had crystallised.[47] It is said that theories are not abandoned because they are proved wrong, but only when all those who believe them have died out – in this example it seems they can last for up to five generations! The extent of the problem of fatigue in railway axles during the 1840s can be judged by Glynn's comment 'that twice during the year 1843 [I] was placed in…danger from the breaking of the fore axle of a tender'. Glynn observed that the axles of tenders more frequently broke than those of other carriages, because they were more heavily loaded and made more journeys.[45] Rankine advised forming the reduction in axles before machining, which was a good observation, and he gave some support to the idea by comparing axles where the reduction in diameter was formed and machined to ones which were purely machined; 'the former resisted from five to eight blows of a hammer, while the latter were invariably broken by one blow'.[44] The paper given by James McConnell on railway axles in 1849 is most noteworthy for its discussion.[48] The chairman was Robert Stephenson[49] (1803–1859) who was not convinced by the structural change in iron theory of fatigue and said 'he was desirous to put members on their guard against being satisfied with less than incontestable evidence as to the molecular change in iron… He therefore wished the members of the Institution… to pause before they arrived at the conclusion that iron is a substance liable to crystallize or to a molecular change from vibration.' Following suggestions by Paul Hodge, Stephenson examined a piece of iron called crystalline and one called fibrous under a microscope and reported that 'it would probably surprise members to know that no real difference could be perceived'.[50] However, the belief that fatigue was due to crystallisation persisted.

Although the biggest problem from fatigue during the first half of the nineteenth-century was in railway axles, there was awareness that there might be problems elsewhere. In 1848 a Royal Commission on the *Application of Iron to Railway Structures* was set up. Henry James (1803–1877) and Douglas Galton (1822–1899) undertook an experimental investigation of the fatigue of iron bars under bending for the commission.[50] Two different cams, driven by a steam engine, were used to load the bars: one was rough so that in addition to producing the main load cycle each rotation, it also vibrated the bar at a higher frequency; the second cam was a smooth step cam and gave a simple stress cycle each rotation. It is difficult to assess the results with the first cam, so the remarks are limited to the tests using the second cam. Three cast iron bars bore 10,000 cycles that produced one-third of the static breaking load. One broke after 51,538 cycles and one bore 100,000 cycles, while three cast iron bars subjected to loads of a half of the static breaking load broke with 490,617 and 900 cycles. From these results it was concluded that iron bars would scarcely bear repeated loading to one-third of their static breaking load.[51]

The first use of the term *fatigue* in an English publication to describe the progressive failure under stress cycling was made in a paper by Frederick Braithwaite (1778–1865) in 1854.[52,53] This paper discussed the probable different instances of fatigue across a wide range of engineering applications. Braithwaite concluded that it was necessary to examine the strength 'of all bridges, machinery, and engines' where fatigue could cause fracture. Stephenson and Fairbairn had conceived the idea of building railway bridges using hollow wrought iron box beams and the first such tubular bridges, the Conway and the Britannia, across the Menai Strait were opened in 1849 and 1850 respectively. The tubular bridges were first designed so that their ultimate strength was six times the heaviest load, but subsequent considerations increased this figure to eight.[51] The Board of Trade specified that no wrought iron bridge with the heaviest load should exceed a stress of 5 tons/in^2 (77 MPa), which would be about 0.22 times the ultimate strength (or an ultimate strength 4.5 times the largest stress) but, as Fairbairn stated, on what principle this standard was

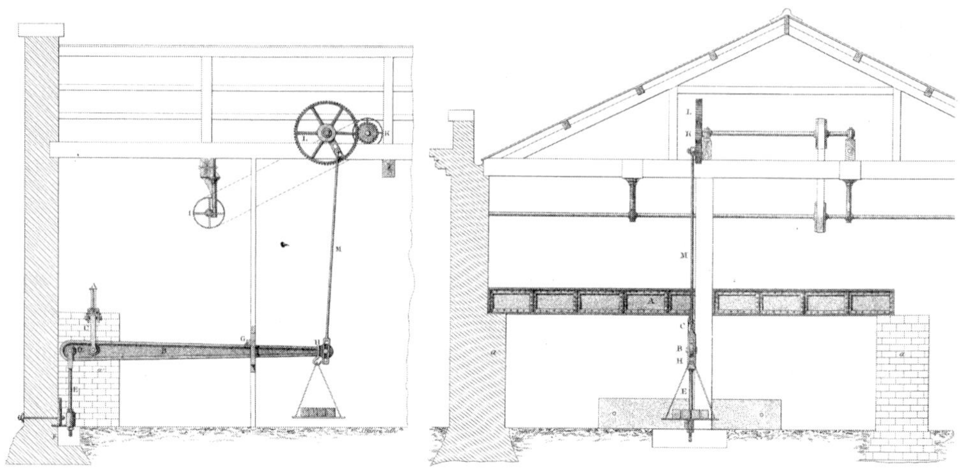

Fig. 7.6 Side and front views of Fairbairn's fatigue machine (Fairbairn 1864 with permission Roy. Soc.).

established is unclear. Fairbairn designed a fatigue testing machine capable of testing a riveted I-beam 22 ft x 16 in (6.7 m x 838 mm) deep, typical of the beams used in railway bridges, shown in Fig. 7.6. The beam was simply supported on two brick piers, and the load applied to the middle through a shackle. A dead load was applied to the lever and a connecting rod periodically lifted and applied the load on the beam – the spur-wheel being driven by a water-wheel through a pulley. This heroic apparatus could produce about eight load cycles a minute. In the first test a load of 6643 lb (29.54 kN), equivalent to a quarter of the ultimate load was applied. After two months the beam sustained half a million cycles without any visible damage. The load was then increased to two-sevenths of the static breaking load and the load cycling continued until the number of cycles reached a million. The load was further increased to 10,486 lb. (46.64 kN) or two-fifths of its ultimate strength and after a further 5,175 cycles the beam broke a short distance from the middle. The beam was now repaired by replacing the broken angle irons and putting a patch over the broken plate equal in area to the original plate. A load equivalent to a quarter of the ultimate strength was then applied. The beam then sustained a further 3,150,000 cycles without breaking again. At this point the load was increased to one-third of the ultimate strength and the beam broke after a further 313,000 cycles. The whole test took from the 21st March 1860 to the 9th January 1862. To put the test into perspective, using Fairburn's estimate of the number of train crossings on a bridge each day as 100, 3 million cycles would take about 80 years. Thus, Fairbairn concluded that the risk of failure with a load of one-quarter of the ultimate was low, but with a load of one-third was too risky and hence the Board of Trade requirement that the maximum stress be limited to 5 tons/in^2 (77 MPa) or about 0.22 of the ultimate load was very reasonable. Of course, Fairbairn's work could be criticized in hindsight as not taking into account the wide scatter in fatigue life; nevertheless it represented a good practical engineering approach to a difficult subject.

7.4.2 August Wöhler (1819–1914) and the systematic study of fatigue

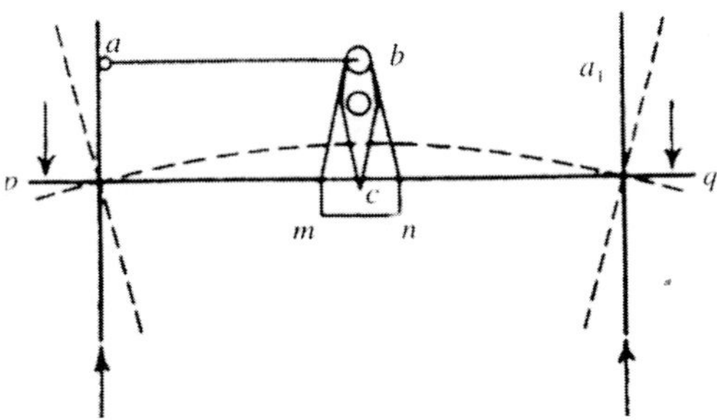

Fig. 7.7 Schematic illustration of Wöhler's device for recording maximum axle deflection.

Wöhler was put in charge of the rolling stock and the machine shop of the Niederschlesisch-Märhrische Railway at Frankfurt an der Oder in 1847 and he remained there for the next 23 years.[50] His main aim was to prevent the fatigue fracture of railway axles, but he approached his work in a systematic scientific manner. First Wöhler devised a method of measuring the maximum loads acting on the axles. In the schematic illustration of the device shown in Fig. 7.7, *pq* represents the axle and *aa₁* represents the wheels. The part *mnb* was fixed to the axle and a pointer *bc* was pinned to this plate, a hinged bar *ab* connected the pointer to the wheel and as the axle bent with the load, represented by the dotted line, the wheels rotated slightly causing the pointer to move relative to the part *mnb* and scratch a zinc plate at *mn*. The maximum axle load could be found from the length of the scratch. Wöhler found that the maximum dynamic load was 1.33 times the static load.[54] The maximum bending load on the axles occurs when the wagon goes over points and Wöhler considered that the number of maximum stress cycles did not exceed one per German mile (about 8 km) and assumed the

                     Fracture and Life

life of an axle as 200,000 German miles.[54] Hence Wöhler assumed that an axle was satisfactory if it could sustain 200,000 cycles of the maximum dynamic load. Naturally, the axle would experience many smaller load cycles, but he considered these to be unimportant.

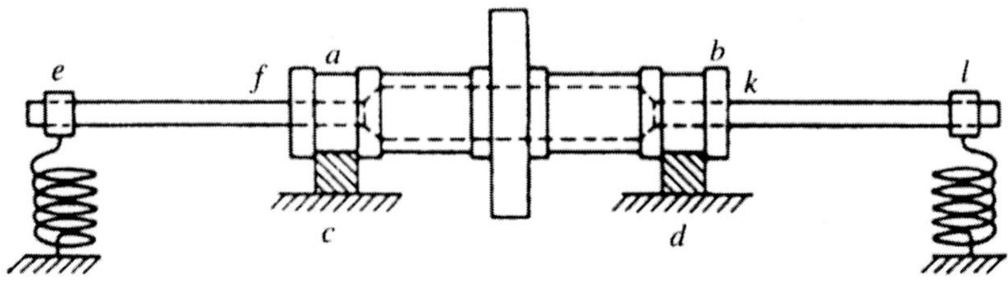

Fig. 7.8 Wöhler's fatigue testing machine.

To test the fatigue life of the axles Wöhler at first used full-sized axles, but they only ran at low speeds. Since Wöhler realised that a large number of tests were necessary to draw meaningful results, he used small axles machined from 38 mm bar with which the speed of revolution could be increased to about 40,000 revolutions per day, but he still retained an experimental configuration that was similar to that of railway axles (see Fig. 7.8). The two axles, *ef* and *kl*, were fixed in the rotor and bent by the springs at *e* and *l*. so that during each revolution the stresses passed through one cycle. With this arrangement, Wöhler could test specimens up to 10^7 cycles and established that there was a limiting stress range below which fracture did not occur. In other tests Wöhler compared the results for axles of uniform diameter with axles where the diameter either increased or decreased and found that sharp corners could reduce the limiting stress by between 25 to 33%.[50] Wöhler amassed an enormous amount of fatigue data, but the familiar fatigue logarithmic plots were first made by Basquin in 1910 who proposed the relationship

$$\sigma_a = CN^n, \tag{7.2}$$

where σ_a is the alternating stress, N the life at that alternating stress, and C and n are constants, which give a straight line in a logarithmic plot.[55] Using mostly Wöhler's 50-year-old data, Basquin gave a table of the values of C and n for many metals. For most metals the index n lies between −0.05 and −0.12.

The fatigue tests carried out with the apparatus shown in Fig. 7.8 entailed a complete reversal of stress. Wöhler also used rectangular specimens loaded by a cam so that he could vary the ratio between the maximum and minimum stress during each cycle. With this apparatus he showed that, though the *stress range* (maximum minus the minimum stress) had the biggest effect on the fatigue life, the life did decrease with increase in the maximum stress. In 1874, Heinrich Geber (1832–1912) proposed an empirical relationship between the limiting stress range, $\Delta\Sigma$, and the mean stress, σ_m, given by

$$\Delta\Sigma = \Delta\Sigma_a \left[1 - \left(\frac{\sigma_m}{\sigma_{UTS}} \right)^2 \right], \tag{7.3}$$

where $\Delta\Sigma_a$ is the limiting stress range for fully reversed alternating stress and σ_{UTS} is the ultimate tensile strength.[56] The fatigue strength of ductile metals matches the Gerber relationship reasonably well. A different relationship

$$\Delta\Sigma = \Delta\Sigma_a \left[1 - \left(\frac{\sigma_m}{\sigma_{UTS}} \right) \right], \tag{7.4}$$

was proposed in 1899 by John Goodman (1862–1935), who assumed that the limiting stress range, $\Delta\Sigma_a$, was two-thirds of the ultimate strength. Herbert Moore later modified the Goodman relationship using the actual limiting stress range.[57] The modified Goodman relationship matches the fatigue strength of brittle metals, but is conservative for ductile metals.[58] In 1939, Carl Soderberg (1895–1979) proposed an even more conservative relationship where the ultimate strength in Eq. (7.4) was replaced by the yield strength.[59]

7.5 The Coming of the Steel Age and Brittle Fracture

Until the middle of the nineteenth-century, steel was only produced in small quantities. Significant improvements had been made in the eighteenth-century when Benjamin Hunt found that he could produce steel by melting small quantities (about 40 kg) of wrought iron with carbon in a covered clay furnace, but the production of cheap steel in large quantities was first made possible by Henry Bessemer (1813–1898) and his converter, which was patented in 1855.[2] The open hearth process for producing steel patented by Carl Wilhelm von Siemens[60] (1823–1883) in 1861 fairly quickly became more important than the Bessemer process. From 1860 to 1890, wrought iron was gradually replaced by steel as a structural metal, because it was stronger and more reliable.[61] There were some early problems with Bessemer steel. The pig iron used by Bessemer in his original work was low in phosphorus and sulphur,[62] also the original converter was lined with clay or firebrick containing magnesia or dolomite which was later shown to remove phosphorus from the steel.[63] Bessemer sold the license to the Bessemer process to a number of ironmasters who used pig iron high in both the bothersome elements and lined their converters with siliceous materials such as ganister or sand which did not remove phosphorus. The steel that resulted from these ironmasters was often of unsatisfactory quality and the Bessemer process was vigorously criticised. Bessemer and his partner Robert Longsdon bought back the licences to the process and built their own steel works in Sheffield which started production in 1858.[63]

Fig. 7.9 Kirkaldy's hydraulic testing machine.

The 'new' material was rightly approached with caution and required extensive testing. Kirkaldy, already a mechanical testing authority, left Napier and Sons in 1861, moved to London and formed the first commercial testing house, which opened on New Year's Day in 1866, at just the right time.[64] He designed the very large horizontal hydraulic[65] testing machine shown in Fig. 7.9, built by Greenwood & Batley of Leeds, which was 14.5 m long, weighed 115 tonnes, and was capable of testing up to a load of 4.4 MN,[66] using a double lever system to measure the load.[67] Business was good and within a few weeks of Kirkaldy opening his testing house, even Krupp in Essen was shipping steel to be tested. A building designed especially to hold the large testing machine was built and opened at 99, Southwark Road., London in 1874. Kirkaldy was an outstanding experimentalist and his approach to strength of materials can be judged by the motto FACTS NOT OPINIONS which is carved in stone in the architrave above the door. The company started by David Kirkaldy was continued by his son, William George, and grandson, David. The family finally sold the testing house in 1965 and the new owners closed the business in 1974, the Kirkaldy testing machine having been in continual use for more than a hundred years. The Kirkaldy Testing Museum now occupies 99 Southwark Road and the Kirkaldy testing machine is still operational. Smaller versions of the Kirkaldy testing machine were built by Greenwood & Batley including one bought in 1884 by the first professor of engineering, William Henry Warren (1852–1926) at Sydney University in Australia. As a small footnote to history, I personally used the Sydney version of Kirkaldy's testing machine in 1964 to measure the velocity of crack propagation in large (1.8 x 1.2 m) sheets of polymethyl methacrylate (PMMA) since it was the most convenient machine to use at the time.[68]

New materials bring new problems. Wrought iron containing low phosphorus content is ductile and the transition temperature for brittle cleavage fracture, even in the presence of a notch and at high strain rates, is low and about −80°C. Steel is stronger than wrought iron and had an economic advantage when its cost dropped dramatically after the introduction of the Bessemer process. However, mild steel is more brittle than iron at low temperatures. Under impact, in the presence of a notch, the transition from ductile to brittle behaviour can be 20°C or higher yet the tensile fracture strength increases with decrease in temperature to below −100°C before the dropping dramatically when cleavage fracture is initiated at first yield even in a smooth tensile specimen. Since in the nineteenth-century, the tensile test reigned supreme brittle fracture problems that can occur with steel were masked and it is not surprising that the brittle behaviour of steel took a very long time to be understood.

7.5.1 *Brittle fracture opinions and tests*

In 1880, John Webster (1845–1914), a well-known bridge engineer, reviewed the state of knowledge of the behaviour of iron and steel at low temperature and presented some experimental results of his own.[69] As he stated 'from the earliest days of iron-bridge building, some forty years ago… the opinion of engineers as to the condition of iron and steel at temperatures below the freezing point of water has been much divided'. Webster stated that 'it is astonishing what strong opinions have been formed…on no stronger evidence…than perhaps breaking a few bars of iron with a sledge hammer in winter'. However, because there were no recognised tests that could assess the brittleness of steel, these 'opinions' were the best evidence available. Nathaniel Barnaby (1829–1915), Chief Naval Architect of the Royal Navy, earned Bessemer's enmity in 1879 by calling into question the suitability of steel for ship construction.[70] Barnaby gave some examples of fractures in steel angles, stating that 'recent cases have occurred of fracture in such bars, satisfying all the test conditions, without apparent cause, or from trifling blow or strain…nearly all [of which] took place during the late severe weather at Chatham'.[71] A vigorous discussion of Barnaby's paper led by Bessemer took place for two days with most speakers denouncing Barnaby. However Kirk, from Kirkaldy's old company Napier and Sons, did provide some support telling of 'a steel plate…[which] had been heated at one end and slightly thinned, and that plate, when cold, on being thrown down, split right up. Pieces cut from each side of the split stood all the Admiralty tests.' The tensile tests of Webster and those that he reviewed, not surprisingly, revealed no temperature effect.[69] However, Fairbairn, Sandberg[72], and Webster did perform impact tests, albeit without a notch, that did reveal some low temperature effects. Fairbairn reported impact test on cast iron to the 1849 Royal Commission into the *Application of Iron to Railway Structures* and stated that 'on the whole we may infer that cast iron of average quality, loses strength when heated beyond a mean temperature of 220°, and it becomes insecure at the freezing point, or under 32° Fahrenheit'.[69] Sandberg did impact tests in Sweden on wrought iron rails, as a consequence of three serious accidents on the railway between Stockholm and Gothenburg during the winter of 1865 when the temperature was about −25°C.[73] He found that 'the breaking strain[74], as tested by sudden blows or shocks is considerably influenced by cold; such iron exhibiting at 10°F [−12°C] only from one-third to one-fourth of the strength which it possess at 84°F [29°C]'.[69] Webster also performed impact tests on wrought iron, cast tool steel, malleable cast iron and grey cast iron, but found little difference in the fracture energy between tests at 5°F (−15°C) and 50°F (10°C) except in the case of grey cast iron where the fracture energy at 5°F was 21% less than that at 50°F. The fractures in the cast tool steel were interesting because the fracture bifurcated and the portion

DACBE shown in his figure (reproduced here as Fig. 7.10) detached and flew a considerable distance; similar fracture paths are observed in the terminations of stone flakes.

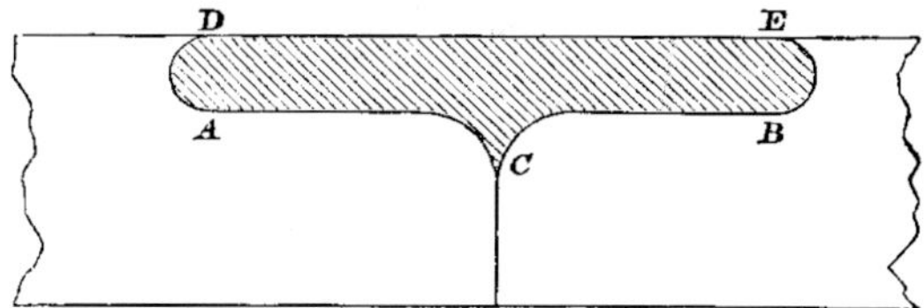

Fig. 7.10 Bifurcating fracture in impact test (Webster *et al.* 1880).

Despite the reservations about steel, there were no major brittle fractures in steel structures until the 1880s. High constraint is necessary for brittle fracture without high-impact loading and is unusual in steel plates thinner than 12 mm. Steel plate used in shipbuilding up until the end of the nineteenth-century was no thicker than about 12 mm and brittle fracture was unlikely. The Tay Bridge was constructed with cast iron columns and wrought iron. The disaster in 1879 was primarily due to the design wind pressure of 10 lb/ft^2 (9.93 MPa) being too small; at the time the French used a design pressure of 55 lb/ft^2 (54.6 MPa) and the Americans 50 lb/ft^2 (49.6 MPa).[75] The Tay Bridge was rebuilt for a design wind pressure of 56 lb/ft^2 (55.6 MPa) again using wrought iron. The Forth Bridge, which opened in 1890, was built of open-hearth Siemens-Martin steel. Different grades of steel were used for the tubular compression members and the braced tension girders both of which were of higher strength than usual. The tension steel had carbon content of 0.19%, an ultimate strength of 30–33 ton/in^2 (460–510 MPa), and a minimum elongation of 20%, whereas the compression steel had a carbon content of 0.23%, an ultimate strength of 34–37 ton/in^2 (525–570 MPa), and a minimum elongation of 17%.[76] The maximum thickness of the steel used in the tubular compression members was 1¼ in (31.8 mm), but the box lattice tension members were built up mainly from a number of ½ in (12.7 mm) thick steel plates. Benjamin Baker (1840–1907), the designer of the Forth Bridge, understood that comparatively high strength steel could be used providing the tension members had a laminated structure that used thin plates instead of thick plate and had reamed rivet holes.[77] Baker also guarded against fatigue by reducing the tensile working stress from 7.5 ton/in^2 (116 MPa), or a quarter of the ultimate strength, to 5 ton/in^2 (77 MPa) in the wind bracing where the stress alternated from tension to compression, and to 3.33 ton/in^2 where the stress alternated elsewhere.[78]

7.5.2 Major brittle fractures in the nineteenth-century

The brittle fracture of steel structures other than ships is well documented.[79] The earliest case of a major brittle fracture was in a water standpipe at Gravesend, Long Island, USA in on the 7th October 1886. The standpipe was 76.2 m high with a diameter of 4.87 m to a height of 18 m and then decreased conically to a diameter of 2.43 m at a height of 25.6 m. The thickness of the riveted plates decreased from 25.4 mm at the bottom to 6.4 mm at the top. Fracture occurred during the hydrostatic acceptance test, the temperature at the time was not recorded, but the average temperature in New York in October is 14°C. Water had been pumped to a height of 69 m when a vertical fracture running from the bottom of the stand pipe to a height of about 6 m. The whole tower then collapsed. The circumferential stress at the bottom of the standpipe at the time of the fracture would have been only about 65 MPa, less than about one-sixth of the probable ultimate strength of the plates.

The second major brittle fracture of the nineteenth-century was very similar to the first. The sealing tank of the Brooklyn, N.Y. Gasholder failed its hydrostatic acceptance test on the 23rd December 1898 (the average temperature in New York in December is 2.5°C).[79] The riveted tank was 54.3 m in diameter and 12.8 m high, the thickness at the bottom of the tank was 31.7 mm. There is no record of the height of the water when the fracture occurred but even if it were to the top the hoop stress at the bottom would only have been 108 MPa, about a quarter of the probable ultimate strength of the plates. An eye-witness described the fracture surface has having 'a rather coarse crystalline structure at the centre of the plate, shading off into a very fine grain at the surface, with here and there splinter edges much like a broken case-hardened material'. Another similar brittle fracture occurred in a water standpipe in Sanford, Maine just after the turn of the century in November 1904, though this standpipe had been in operation for seven years before it fractured.

7.5.3 Notch impact testing

In his 1818 paper, Rennie mentioned an experiment of Gaspard Prony (1755–1839) where it was found that a slight incision with a file on a bar reduced its tensile strength by a half.[4] No details of the material that was used by Prony were given, but Rennie repeated the experiment with a ¼ inch (6.35 mm) square bar of wrought iron with an incision of about 0.6 mm and found that the strength is only reduced by a sixth part.[4] With such a shallow notch and without a high strain rate wrought iron would not behave in a brittle fashion unless the phosphorus content was high. Kirkaldy recognised in 1862 that a notch and a high strain rate could make wrought iron brittle (see §7.2), but he did not do any impact tests. Un-notched impact tests were used by a number of researchers from the middle of the nineteenth-century,[69] but not having a notch did not produce fracture in the more ductile materials. Any complex structure will contain a notch-like feature either from its design or as a material defect and brittle behaviour can only be judged in the laboratory by a notch test. The first notch impact tests were those of André Le Chatelier in 1892 who found that some steels did not fracture when un-notched impact specimens were used, but fractured if there was a notch.[80]

Most early impact tests used the drop-weight test where the fracture energy could only be obtained from a number of tests that bounded non-fracture and fracture. The use of a pendulum to impact a specimen at its lowest point enables the fracture energy to be obtained directly from the difference in the height of the centre of gravity when the pendulum is released and the height to which it swings after fracturing the specimen. At the turn of the century, Bent Russell designed such a machine and performed over 700 impact tests on notched specimens examining the effect of the notch depth.[81] Russell only performed his tests at room temperature, but he produced a test that, when applied over a range of temperatures, would enable the transition behaviour of steel from ductile at high temperature to brittle at low temperature to be investigated.

7.6 Strength Theories in the Nineteenth-Century

Theoretical solid mechanics in the nineteenth-century was dominated by elasticity and there seems to have been some confusion over what should be called failure. Many theoreticians saw reaching the elastic limit as failure, whereas the practical men saw fracture as failure. Strength theories did not always clearly differentiate the different modes of failure. This confusion, at least as far as teaching is concerned, extended into the 1950s when I and all other engineering undergraduates were taught the five theories of failure[82] in a way that made little sense. Here we restrict the discussion to the theories that more or less suit fracture. Mohr's theory of fracture, primarily for compression, has already been discussed in §2.4.1.

The simplest strength theory to be suggested was that fracture occurred when the maximum principal stress reached a critical value. This theory was commonly associated with Rankine[83], but was also associated with Gabriel Lamé[84]

(1795–1870) and was the main strength theory used by engineers in the nineteenth-century. In unpublished lectures to the Sorbonne, Poncelet suggested that fracture occurred when the maximum principal strain reached a critical value which is a slightly different criterion.[85] Both of these criteria for fracture assume that the principle of similitude applies. In the nineteenth-century the first examples of the breakdown in the theory were found.

In the 1880s Hertz, studying indentation, described the cone-shaped crack that occurs when a hard sphere is pressed into the surface of a brittle solid like glass.[86] In 1891 Felix Auerbach (1856–1933) showed empirically that the critical indentation load to create a cone fracture in a flat surface was proportional to the diameter of the indenter.[87] If the principle of similitude applied one would expect that the critical load would vary with the square of the diameter. Thus similitude did not apply to indentation fractures. The theoretical explanation of Auerbach's law had to wait until 1967 when Charles Frank (1911–1998) and Brian Lawn published their paper on the theory of Hertzian fracture.[88]

7.7 Concluding Remarks

The Industrial Revolution saw wrought iron and steel become the materials for the development of the new technologies. The use of these materials on a very large scale brought new fracture problems that had to be solved. The new technologies caused the profession of engineer to become established. Prior to the eighteenth-century there were no professional engineers except for military men. The first civil engineering society was formed by John Smeaton in 1771, but it was quite informal and the first professional engineering society was the Institution of Civil Engineers founded in 1818. The term civil engineering was then only used to distinguish it from military engineering; a separate Institution of Mechanical Engineers was not formed until 1846.[89] In America the Franklin Institute of the State of Pennsylvania for the Promotion of the Mechanic Arts, to give it its original title, was founded in 1824. Although these institutes had an educational role, engineering schools where first established on the Continent. The Kaiserlich, Königlich Polytechnisches Institut, to become today the Technical University of Vienna, was established in 1815 along the original pattern of the Paris École Polytechnique formed in the eighteenth-century to train engineers. Similar polytechnics were created in Germany. In America the Rensselaer Polytechnic Institute was founded in 1824. Although engineering topics were introduced at Cambridge in the eighteenth-century, the first British school of engineering was established at King's College, London in 1838.

With the growth in engineering there grew a need for more standardised materials testing. Johann Bauschinger (1834–1893) established the first of a series of so-called Bauschinger Conferences in Munich in 1884 with a participation of international materials testing specialists. There followed three more conferences in Dresden, Berlin, and Vienna before his death in 1893. Due to the endeavours of the industrialist Werner von Siemens (1816–1892), brother of Carl, the Physikalisch-Technische Reichsanstalt, an institute designed to bring the research of scientists and industry together, was created in 1887 by the Reichstag. A fifth Bauschinger Conference was organised in 1895 in Zurich by Ludwig von Tetmajer (1850–1905) at which the International Association for Testing of Materials (IATM) was formed. The American Society for Testing and Materials (ASTM) was formed three years later in 1898.

7.8 Notes

[1] The third generation of Quaker iron founders all bearing the same name.

[2] Gordon (1968).

[3] The theory of elasticity is an essential tool for the full understanding of fracture. Its development is well documented in Todhunter (1886), Timoshenko (1953), and Love (1944).

[4] Rennie (1818).

[5] Emerson (1754).

[6] Rennie gives no details of the size of the cross-section or how much he removed from the apex.

[7] Barlow (1817).

[8] Tredgold (1824).

[9] Todhunter (1886) is very scathing about Tredgold largely because Tredgold wrote some silly things about 'Fluxions' (calculus) in the introduction to his book.

[10] Navier (1826).

[11] Hodgkinson (1838).

[12] Timoshenko (1940) used this problem as an example for his text book on the strength of materials.

[13] Fairbairn (1850).

[14] This notion of damage by preloading was also used as an argument against the proof testing of boilers (see §7.3).

[15] Tipper (1962).

[16] Kirkaldy (1862).

[17] In modern steels the phosphorus is usually kept to less than 0.04%.

[18] Ritchie *et al.* (1973).

[19] Dieter (1961).

[20] In the seventeenth-century cold short iron was termed coldshire-iron (Sturdie 1693).

[21] Pearson (1795).

[22] Brittleness in working at high temperature was termed red shortness in the eighteenth-century – now termed hot shortness. Hot shortness is caused by sulphur which forms the low melting point compound iron sulphide on the grain boundaries which is liquid at the hot working temperature. At the end of the eighteenth-century red shortness was thought to be caused by arsenic, but by the middle of the nineteenth-century it was starting to be correctly ascribed to sulphur. In the seventeenth-century red short iron was termed redshire-iron (Sturdie 1693).

[23] Boat plates were among the cheapest wrought iron plates available costing only about £9 a ton as compared with £25–£30 a ton for boiler plate (Amicus 1859).

[24] Editorial (1861).

[25] Burke (1966).

[26] Gordon (1978).

[27] Spence and Nash (2004).

[28] See §6.4.

[29] Evans (1805).

[30] Todhunter's (1886) only reference to thin walled vessels is to an anonymous paper published in 1818 which post dates Evan's paper.

[31] Johnson (1832).

[32] Todhunter (1886).

[33] The Prussian and Austrian formulae were very similar (Todhunter 1886).

[34] Cast iron boilers had to be tested at five times their working pressure.

[35] The members of this subcommittee were Alexander Bache, William Johnson, and Benjamin Reeves.

[36] Anon (1894).

[37] Editorial (1862a).

[38] See §7.2.

[39] Editorial (1862b).

[40] Albert (1838).

[41] Poncelet (1839).

[42] Smith (1990).

[43] Hood (1842).

[44] Rankine (1843).

[45] Glynn (1844).

[46] McConnell (1849) describes axles as 'nearly red hot from the want of proper lubrication'.

[47] Smith and Hillmansen (2004).

[48] McConnell (1849).

[49] The son of George Stephenson.

[50] Timoshenko (1953).

[51] Fairbairn (1864).

[52] Braithwaite (1854).

[53] Braithwaite states that the term fatigue was suggested by Field, perhaps Joshua Field (1757–1863).

[54] Schütz (1996).

[55] Basquin (1910).

[56] Geber (1874).

[57] Moore and Kommers (1927).

[58] Goodman (1899).

[59] Soderberg (1939).

[60] Siemens became a naturalised British subject.

[61] The quality of wrought iron was controlled largely by the puddler and shingler, the quality of steel was more controllable (Gordon 1988).

[62] Sulphur was not quite such a problem since Bessemer's friend, Robert Mushet, suggested that manganese in the form of an iron-manganese alloy, Spiegelseisen, be added to the steel (Street 1962). Manganese combines more readily with sulphur than iron and is not liquid at rolling temperatures.

[63] Bessemer (1905).

[64] Initially in rented premises at The Grove, Southwark, London.

[65] Water pressure from hydraulic mains, which started to become available in the 1860s, was used to supply the load. In 1871 the Wharves Warehouse Steam Power and Hydraulic Pressure Company was formed by an act of Parliament. This company, which later became the London Hydraulic Power Company, supplied water pressure for hydraulic lifts etc. through an extensive network of pipes from 1871 to 1977.

[66] The largest load actually used by Kirkaldy was 300 tons (3 GN). The hydraulic mains pressure was 75 lb/in^2 (517 kPa) which enabled a load of 50 tons (500 kN) to be achieved; a hydraulic pressure intensifier was used to increase the pressure for higher loads.

[67] Kirkaldy's testing machine was not the first hydraulic machine, this distinction goes to the testing machine Thomas Brunton (1793–1876) had constructed in 1814 by Fuller, which had a capacity of about 1 MN, and was used by Thomas Telford (1757–1834) in connection with his Menai Straits Bridge of 1826. However, the hydraulic pressure was used to measure the load which was about 25% high (Gordon 1988).

[68] Cotterell (1965a).

[69] Webster (1880).

[70] The enmity was so strong that in 1896 when Bessemer was writing his autobiography, he devoted the best part of a chapter refuting the opinions of Barnaby.

[71] Barnaby (1879).

[72] In an appendix to the Sandburg's translation of *Iron and Steel* by Knut Styffe (see Webster *et al.* 1880).

[73] Smedley (1981).

[74] In the nineteenth-century strain and strength were to a modern mind loosely used, here they undoubtedly mean energy absorption. Strain was also used to mean stress.

[75] Shipway (1989).

[76] Westhofen (1890).

[77] Wells (2000).

[78] Shipway (1990) Baker used the higher stress for the wind bracing where the load was due almost entirely to the wind, because the highest wind load would only be exceeded on a very small number of occasions.

                                    Fracture and Life

[79] Shank (1954).

[80] Le Chatelier (1892).

[81] Russell (1898).

[82] Maximum principal stress, principal strain, shear stress, total strain energy density, and distortional strain energy density.

[83] See Rankine (1888).

[84] See Lamé and Clapeyron (1831).

[85] Often wrongly attributed to St Venant who used it later (Timoshenko 1953).

[86] Hertz (1881, 1882).

[87] Auerbach (1891).

[88] Frank and Lawn (1967).

[89] The myth that the Institution of Mechanical Engineers was formed because George Stephenson was slighted by the Institution of Civil Engineers in an application for membership was not exposed until the 1950s.

Chapter 8

The First Half of the Twentieth-Century

By the beginning of the twentieth-century many of the tools needed for understanding fracture were in place. The mechanics of elasticity was fully understood and analytical solutions obtained for simple geometries and loadings. Thus the stage was set for the analysis of elastic fracture.

Much progress had been made in theory of the plastic deformation. Henri Tresca (1814–1885), on the basis of experiments on punching and extrusion had proposed that a metal yielded when the maximum shear stress reached a critical value.[1] Saint-Venant (1797–1886) had given the equations for plane strain isotropic plastic flow and had recognised that the total plastic strain depended upon the stress history. He also postulated that the direction of maximum shear strain-rate corresponded to the direction of maximum shear stress.[2] Maurice Lévy (1838–1910) had given the three dimensional relationships between stress and plastic strain rate.[3] During the first half of the twentieth-century plasticity theory developed to be comparable to elasticity theory and at the end of the period Rodney Hill published his seminal book: *The Mathematical Theory of Plasticity*.[4] However, analytical solutions to plasticity problems are only possible for the simplest geometries except for a plastic-rigid material, and the application of plastic theory to fracture had to wait for the development of the digital computer and the second half of the twentieth-century.

Metals and concrete were the dominant materials. Reinforced concrete was widely used.[5] Pre-stressed concrete, developed by Eugène Freyssinet (1879–1962) for the Plougastel Bridge over the Elorn River near Brest in France, which was opened in 1930, increased the efficiency of concrete. One of the major new metals of the first half of the twentieth-century was aluminium. Small quantities of aluminium had been produced in the early nineteenth-century but it was very expensive.[6] The independent discovery of the electrolytic aluminium process in the USA by Charles Hall (1863–1914) and in France by Paul Héroult (1863–1914) in 1886 made the use of aluminium as a structural material possible.[7] Pure aluminium is too soft for structural use and it needs to be alloyed. Alfred Wilm

(1869–1939) was researching into the alloying of aluminium with small quantities of copper and other metals for the possible use as cartridge cases when in 1909 he accidentally discovered the age-hardening of aluminium alloys. Wilm made tensile specimens of a quenched aluminium alloy, with 3.5% copper and 0.2% magnesium, some of which he tested on a Friday; during the next week he tested some more samples and found that they were very much stronger. Duralumin was born.[8] The early aeroplanes were constructed of wood and fabric with steel wire bracing. Hugo Junkers (1859–1935) designed the first all metal aircraft the J1 in steel, but it was not very successful because it was too heavy. In 1919 an aluminium (duralumin) aircraft, the J13, also designed by Junkers, flew for the first time.[9]

Plastics were slowly introduced during the first half of the twentieth-century. Celluloid had already been invented in the nineteenth-century and the Belgium born Leo Baekeland (1863–1944) invented Bakelite in 1907. Synthetic rubbers and the thermoplastics polyvinyl chloride (PVC) and polystyrene (PS) were developed in the 1930s. However, none of these plastics were used in critical applications. Polymethyl methacrylate (PMMA) more commonly known as *Perspex* developed by ICI in 1934 was an important material during World War II as it was used for aircraft cockpit canopies.

The period of this chapter saw two world wars which had a big effect on the development of fracture theory. They were both an impetus to the study of fracture and an impediment to the dissemination of its knowledge.

8.1 The Brittle Fracture of Steel

During the first half of the twentieth-century notch impact tests were by far the most usual test to determine the notch brittleness of steel. Notch impact tests were being performed over a range of temperatures at least from 1906 and the transition from ductile to brittle behaviour was known, but apparently not very widely because it did not prevent the brittle fracture of steel structures which, with the advent of electric welding, turned into an epidemic.

8.1.1 Notch impact tests

The pioneering work on the pendulum impact testing machine in America by Russell was continued by Augustin Georges Charpy[10] (1865–1945) in France and Edwin Izod[11] (1876–1946) in Britain. Charpy's first machine of 1901 was very similar in design to that of Russell, he used both vertically clamped and simply supported specimens preferring the later type of support which has become the standard.[10] His first machine was large with a 4 m pendulum and large specimens 30 x 30 x 120 mm. Charpy realised the importance of the sharpness of the notch being accurately controlled and used a keyhole notch formed by drilling a hole of 4 or 8 mm diameter at the bottom of the notch and sawing through the ligament; the depth of the notch was ⅓ or ½ of the depth of the specimen. Izod designed his pendulum impact test with a notched specimen clamped vertically in a vice as the result of being asked to test a burst shot-gun barrel against a sound one.[12] In his tests he clamped small notched test pieces from each barrel in a vice and hit them with a hammer; the test piece from the burst barrel snapped off but that from the sound barrel bent in two. Izod announced his pendulum impact testing machine, designed as a result of his tests on barrels, at a British Association meeting in 1903.[13] The tup of the Izod machine had 23 ft lb (31 J) of energy at impact. In the presentation of his machine, Izod remarked that the very early tests in the early eighteenth-century by purchasers of iron were crude impact tests, but since then the tensile test had practically replaced all others, except for the impact tests on rails and railway axles referred to in §7.5, yet 'fractures have occurred in steel which are inexplicable by known chemical or standard physical tests'.[13] Unfortunately, the clamping of the specimen in the Izod test makes it difficult to perform tests at other than the ambient temperature thus, though it was the test commonly adopted in Britain in the early years, its use declined. The simply-supported specimen used by Charpy can be easily cooled or heated outside of the machine and then quickly tested.

By 1905 Charpy had proposed an impact test that is essentially the same as the present-day standard and used a specimen 10 x 10 x 53.3 mm with either a keyhole notch 5 mm deep with a radius of ⅔ mm or a V-notch 2 mm deep with a 0.25 mm tip radius. The small Charpy machine for the 10 x 10 mm specimens had 30 kgm (294 J) of energy at impact; a 1916 version of such a machine bought by Imperial College and now in the Kirkaldy Testing Museum, Southwark, London is shown in Fig. 8.1. There was a large Charpy machine for the 30 x 30 mm specimens with impact energy of 300 kgm (2940J). The ASTM published their first standard E23-33T Tentative Methods of Impact Testing of Metallic Materials in 1933, where both the vertically clamped and simply supported specimens were described; the dimensions for the simply supported specimen with a V-notch, which is now known as the *Charpy Test*, were the same as those currently specified.[14]

                                    Fracture and Life

Fig. 8.1 The 1916 Charpy Impact Machine, now in the Kirkaldy Museum, London.

Charpy appears to have performed the first notch impact test over a range of temperatures (−80 to 600°C) on mild steel, nickel, and nickel steel in 1906.[15] James Dewar (1842–1923) and Robert Hadfield (1859–1940) had earlier, in 1904, made tensile tests on a wide range of metals at the temperature of liquid air (−182°C) and found that the ultimate strength of most iron and iron alloys increases markedly at low temperature, but that the ductility was reduced to practically nothing.[16] However, they discovered that nickel had a toughening effect on iron at low temperature. It was known at the time that nickel produced improvements in the properties of iron and steel, but the reason was not known. Nickel has a face centred cubic (fcc) crystal structure and fcc metals do not have a transition from ductile to brittle cleavage fracture. The idea that low temperature changes the structure of iron and steel was obviously still not completely discounted, because Dewar and Hadfield cooled wrought iron and steel specimens down to −182°C, allowed them to return to the ambient temperature and then performed a tensile tests that, of course, revealed the usual strength and ductility 'showing that the de-toughening or embrittling action is only temporary'.

Up until the beginning of the First World War there were six papers written on the behaviour of notch impact tests at other than ambient temperature.[17] During the First World War there was only one further paper published, but after the war there were 16 more papers published by the end of 1925.[17] There would almost certainly have been more work done on the notched impact test both by the Allies and the Germans, which was not published during the war. The present standard Charpy test piece was developed by the Aeronautical Inspection Directorate during the First World War.[18]

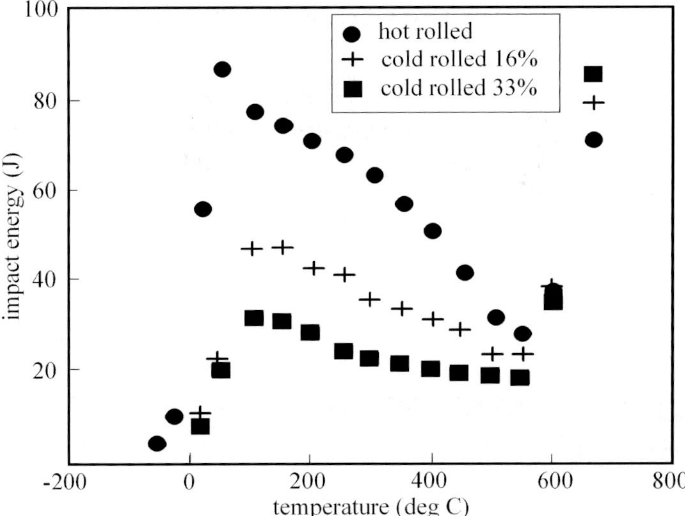

Fig. 8.2 Charpy impact energy for hot and cold rolled mild steel (after Greaves and Jones 1925).

Richard Greaves and Jones performed standard Charpy tests over a temperature range of −80 to 1000°C on a wide range of iron and steel.[17] The results for mild steel are shown in Fig. 8.2. Greaves and Jones were more interested in the maximum impact energy which was between −20 and 250°C and the minimum at 500 to 650°C due to blue-brittleness where ageing takes place almost instantaneously, than in the brittleness at low temperature even though an investigation of the embrittling action of cold was one of the aims stated in their introduction. It is surprising that no connection was made to the brittle fracture of structures that were occurring (see §8.1.3). The Hungarian born Egon Orowan[19]

(1901–1989) wrote in 1949 that 'since all the test methods to reveal brittleness in this period used impact loading, it became a widespread opinion, which had not completely disappeared, that brittle fracture in steel resulted from impact loading'.[20] Yet Greaves and Jones performed slow bend, albeit only for ambient and higher temperatures, as well as impact bend test on transverse specimens of Armco iron, a 0.45% carbon steel, a 3.77% nickel steel, and a nickel-chromium steel (3.72% Ni, 0.92% Cr); only the Armco iron showed no sign of embrittlement at ambient temperature.

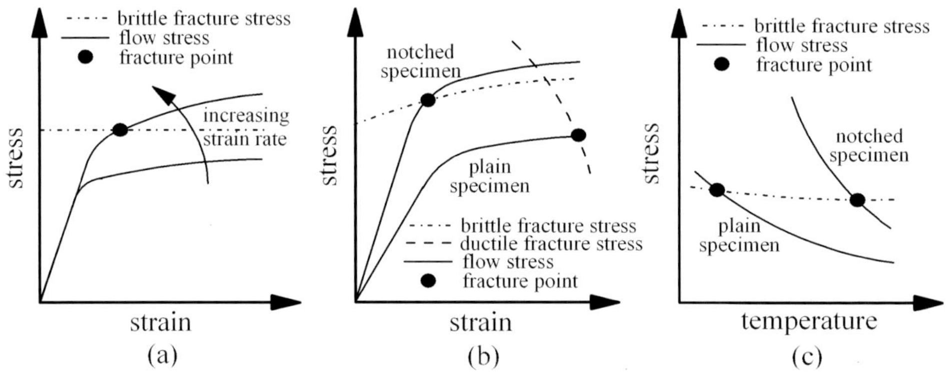

Fig. 8.3 Explanations of brittle and ductile fracture, (a) Ludwig's original concept. (b) Davidenkov's interpretation of Ludwig's concepts. (c) Orowan's transition concept.

8.1.2 Understanding notch brittleness and the ductile-brittle transition

In 1906 Augustin Mesnager (1862–1933) explained that triaxial tensile stresses would exist at the tip of a notch which, since plastic deformation is independent of the hydrostatic stress, enabled the maximum tensile stress at the notch to be much greater than that in a plain tensile specimen.[21] A few years later Paul Ludwik (1878–1934), working in the Technische Hochschule in Vienna, introduced a hypothesis to explain qualitatively brittle fracture in steel.[22] He assumed that the mechanical behaviour was determined by the stress-strain curve for plastic deformation as measured in a tension test, and a fracture stress which he assumed to be constant independent of the strain as shown schematically in Fig. 8.3 (a). According to this concept, brittle fracture at a low strain was possible if the flow stress-strain curve was elevated due to a high strain rate. Later, in 1923 Ludwik, in collaboration with Scheu, did consider the effect of a notch in increasing the maximum stress over that obtained in an un-notched specimen and thus favouring brittle fracture over plastic flow.[23,24] They also recognised that the fracture stress might not be constant and that a fracture curve could be obtained by varying the notches to change the constraint, fracture would then occur at different strains and stresses.[25] Ludwik explained his ideas more clearly in a 1927 article where the roles of strain rate and notches on increasing the yield strength are stated together with the effect of increasing the flow stress, but not the fracture stress, on the brittleness induced by low temperature.[26] Nikolai Davidenkov (1879–1962) summarised Ludwik's concepts and he also added a ductile fracture curve in an attempt to explain the two possible modes of the fracture of steel; he also gave a flow curve for notched specimens, as shown in Fig. 8.3 (b).[27] In 1949, Orowan published a figure similar to that shown in Fig. 8.3 (c) showing the effect of temperature and notching on brittle fracture, this is the figure often reproduced in textbooks.[20,28] Although Ludwik's hypothesis gave a qualitative picture of brittle cleavage fracture, it is not the whole story because brittle fracture does not only depend upon stress.

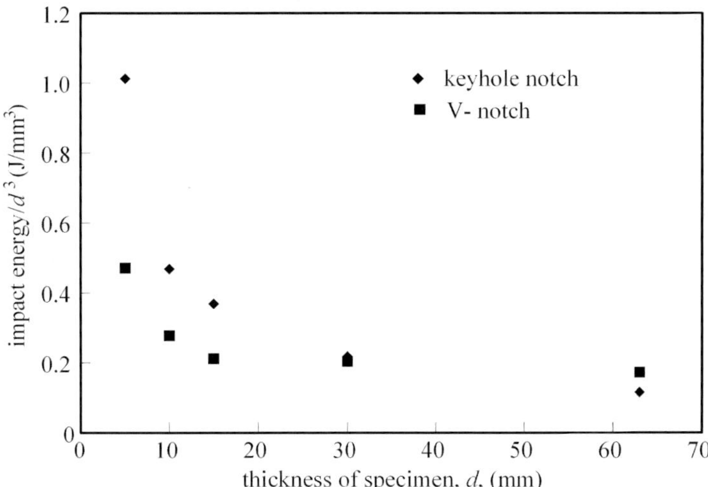

Fig. 8.4 Normalised impact energy as a function of specimen size (after Stanton and Batson 1920).

Since in Ludwik's hypothesis fracture depends upon stress, the laws of dimensional similitude should apply if it is true. For geometrically similar bend specimens with a characteristic dimension d, dimensional similitude would cause the force, P, to be proportional to d^2 to cause the same strain distribution at a deflection, Δ, proportional to d. Hence if P/d^2 is plotted against Δ/d one might expect that results from different sized bend specimens would all fall on the same curve. If brittle cleavage fracture occurs when the maximum stress reaches a critical fracture stress at a critical strain, then it might be expected that the work done to fracture the specimen would be proportional to d^3. The same result would be expected if the fracture was a stable ductile tear. Stanton and Batson performed notch impact tests on geometrically similar Charpy specimens with both V and keyhole notches in 1920.[18] All the fractures were crystalline except that for the smallest specimen with a V-notch and their results are shown in Fig. 8.4. Clearly, Ludwik's hypothesis did not hold over the range of specimens from 5 to 63 mm thick and there was a vigorous discussion to their paper. However, for specimens whose thickness was greater than 15 mm dimensional similitude is followed reasonably well. These results point to a general problem in assessing the fracture behaviour of large scale structures from small scale fracture tests: small scale tests tend to be non-conservative if applied to full scale structures using dimensional similitude.

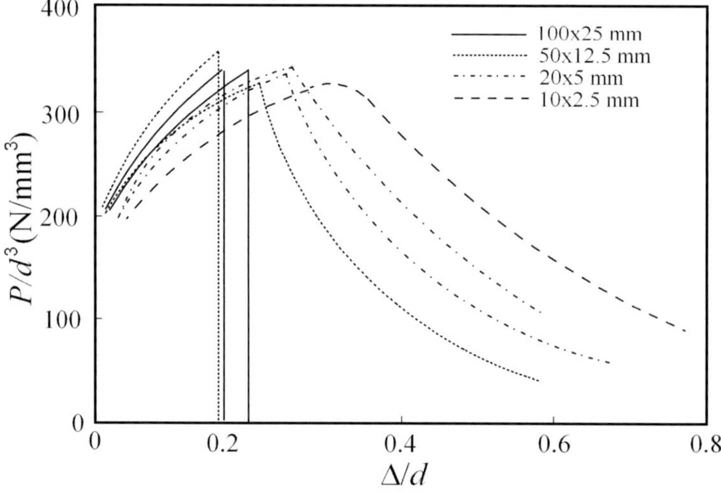

Fig. 8.5 Normalised load deflection curves for different sized notch bend specimens (after Docherty 1935).

For strict comparison the impact velocity in the tests by Stanton and Batson should have been proportional to the specimen size. In the 1930s Docherty repeated the experiments of Stanton and Batson using slow notch bend tests at different deflection rates.[29] They found that deflection rates in the range of 0.02 to 60 mm/s made little difference to the energy absorbed in fracture. The load deflection curves obtained by Docherty for mild steel, for specimens whose thickness ranged from 10 to 100 mm, are shown in Fig. 8.5. The total work of fracture varied much the same as those of Stanton and Batson, but here Fig. 8.5 shows that there was a distinct transition from ductile behaviour for specimens 20 mm or thinner to brittle cleavage for thicker specimens.

8.1.3 Brittle fracture of riveted steel structures

Brittle fractures in riveted steel structures and ships during winter were still occurring in the twentieth-century. Information on the fractures is more available for structures other than ships. As Edward Ellsberg wrote in 1925 'it is an unfortunate truth that no shipowner will publish the facts concerning structural failures of their vessels'.[30] However, more than a dozen riveted merchant ships have broken in two or have been lost in storms.[31] In May 1924 a fracture was discovered in the inboard section of C deck on the liner *Majestic*. Although C deck was the strength deck of the *Majestic*, the fracture was thought to be unimportant since this inboard section was not intended to take any significant load and repairs were delayed. In December 1924 during a rough weather passage from Southampton to the New York, a sound like a 'cannon shot' was heard and C deck fractured all the way from the starboard to the port side and down that side until it was arrested at a porthole. The *Majestic* sailed on to New York and despite the fracture, which was opening by about 12 mm, actually made the return trip to Southampton before being taken out of service for repairs. Another famous liner, the *Leviathan* suffered a similar fracture emanating from corner hatches in its strength deck in 1929.

The brittle fracture of riveted steel structures other than ships up to 1950 is summarised in Table 8.1.[32] The collapse of the large Boston Molasses Tank in 1919 was attributed, by witnesses for the defence in the ensuing lawsuit, to a bomb planted, by labour agitators.[32] It is not surprising that brittle fracture has often been credited to a bomb because the energy stored in a large structure is enormous and fracture releases it explosively. A two-tonne segment of a large thick-walled cylindrical pressure vessel of an ammonia plant which failed during its hydrostatic test was thrown 46 m in 1965.[33] The fracture of the Boston Tank in 1919 released 8.7 million litres of molasses, which drowned 12 people and injured 40 as well as drowning many horses, also a portion of the Boston Elevated Railway was demolished and many houses damaged.[32] The brittle fracture in this instance was due to design overstress as well as the brittleness of the steel. The failure of a crude oil storage tank in Ponca City, Oklahoma in 1925 was also at first attributed to an explosion. There was a large drop in temperature from 15°C on the day before the failure to −20°C. The failure was attributed to the large drop in temperature, but no thought was given of the possibility of brittle fracture. One crude oil storage tank in the Midwest of the USA suffered no fewer than five brittle fractures over a period from 1918 to 1933 all in the winter months.[32]

Table 8.1 Summary of brittle fractures in riveted steel structures (complied from Shank 1954).

Structure	Size (m)	Location	Plate thickness (mm)	Date	Temperature °C	Age Years
Water Standpipe	4.9 dia. 76 high	Long Island USA	25	Oct. 1886	–	On test
Sealing Tank of Gasholder	54 dia. 12 high	Brooklyn USA	32	Dec. 1898	–	On test
Water Standpipe	12 dia. 14 high	Maine USA	16	Nov. 1904	–	7
Molasses Tank	27 dia. 15 high	Boston USA	16	Jan. 1919	–	3
Crude Oil Tank	36 dia. 13 high	Oklahoma USA	25	Dec. 1925	−20	–
8 Storage Tanks	–	Midwest USA	–	1918–1937	−8 to −12	10–15
Oil Storage Tank	35 dia. 9 high	Middle West	12	Dec. 1943	−11	–

The last of the failures listed in Table 8.1 is interesting because, though the oil storage tank was of riveted construction, the tank had been leaking from the bottom and, in May 1943, a large equilateral triangular hole about 1.4 m on each side was cut in the bottom course of the plates to allow access for a wheelbarrow to remove the last of the oil. After the bottom was repaired a triangular patch was welded in the hole. A brittle fracture initiated in the weld of the patch on the 14th December 1943 when the temperature was −11°C and ran across the bottom plate. This fracture shows that welding can cause brittle fracture. The steel of the oil storage tank was clearly brittle at −11°C, but before the welding repair it had nevertheless not fractured though almost certainly experiencing as a low a temperature. The repair welds were of poor quality but, even if they had been of high quality, the tank may still have failed, because welding causes high residual stresses especially when a patch is welded into the high constraint provided by a solid plate.

The significant common feature of all the failures listed in Table 8.1 is that they all occurred in winter and presumably all at low temperatures.

8.1.4 Brittle fracture of welded steel structures

Electric arc welding did not create the problem of brittle fracture in steel structures, but it certainly aggravated it. Electric welding using covered metal electrodes was introduced at the turn of the century. The first all welded ship the 46 m long *Fullagar* was built in England in 1921. By the 1930s welding had become common in ships and other structures. In 1929 the first welded bridge, designed by Stefan Bryla (1886–1943) spanning the Sludwia river near Lowicz in Poland, opened. In 1937 a large ocean going tanker, the 11,860 tonne *J. W. Van Dyke*, was launched. By 1940 all ships under about 100 m in length were of welded construction.

Brittle fractures in welded structures started occurring in the 1930s when electric welding had become common. There are a number of reasons why welding makes steel structures more susceptible to catastrophic brittle fracture. The welding of two plates together essentially consists of laying a strip of molten steel between them. When the weld metal has cooled to become solid and capable of bearing a stress, there is still a large temperature difference between the weld and the plates. During subsequent cooling the shrinkage of the weld metal is resisted by the plates on either side, putting the weld into tension. The temperature difference is so large that, in massive structures, the weld metal and adjacent heat affected zone are plastically strained by about 2% in the temperature range of 200–400°C where ageing takes place almost instantaneously, lowering the transition temperature considerably. The structure thus contains high residual stresses of the order of the yield strength along its welds and heat affected zones which are embrittled because of plastic straining in the critical temperature range of 200–400°C. Welds are also a source of defects. In stick-welded thick plates the welds are laid down in a number of runs with the slag chipped away between the runs. If all the slag is not removed, then the slag from the previous run becomes a defect in the next weld. Electrodes with a rutile covering[34] are the easiest to use but deposit welds high in hydrogen and are prone to produce hot cracking during cooling. Low hydrogen electrodes are more difficult to use but are very much less likely to produce weld cracks.[35] Thus the chance of brittle fracture is enhanced by welding; there is the possibility of flaws

in material that has been embrittled and there are large residual stresses that add
to the stresses caused by the loads on the structure. In riveted structures, most
fractures stop at the edge of plate where there is a discontinuity, but a welded
structure is continuous and there is little hindrance to a propagating fracture.[36]

Fig. 8.6 T2 tanker, NBS ship No. 52 which fractured while in port (Tipper 1962 with permission
CUP).

Ships were vital in World War II and welding enabled them to be produced
quickly. Some 2,500 dry cargo welded Liberty ships were built, of these ships
145 suffered fractures where the ship was lost or so severely damaged that that it
was in danger of sinking.[37] Some 24% of fractures initiated at hatches which
originally had sharp corners; the later Liberty ships had reinforced rounded
corners.[31] A total of 20 welded ships built up to 1952 broke completely in two.
Riveted ships have only broken in two during storms whereas at least two welded
ships broke in two while tied-up in port. The T2 tanker SS *Schenectady* was still
in its fitting-out dock in January 1943, having been launched in October 1942,
when the air temperature fell from 3°C in the afternoon to −5°C at 11pm and the
ship broke in two with a loud report heard over a kilometre away. Another T2

tanker, NBS ship No. 52, shown in Fig. 8.6, also broke in two while in port, the air temperature being 1°C. To try to reduce the number of ships totally lost, a number of welded ships were fitted with *crack-arresters*, consisting of a slot over which a strap was riveted. However, despite this precaution two ships fitted with crack-arresters suffered complete fractures.[37] Riveting is not wholly benign. Cold sheared plates and punched holes are a source of embrittled material and cracks so that the incidence of service fractures is not that much less than in welded structures, but most fractures only propagate at most across a single plate.[38]

From the 1930s until the middle of the twentieth-century there were numerous brittle fractures in welded ships and other structures which cannot all be reviewed here.[39] The failure of a Vierendeel truss bridge over the Albert Canal in Hasselt, Belgium in March 1938 marks the first major brittle fracture of a welded structure other than a ship. Numerous other welded bridges failed by brittle fracture in Belgium and Germany during World War II, but the details were not known by the Allies until after the war. Brittle failures in welded bridges continued after the war. The Duplessis Bridge at Three Rivers, Quebec was completed in 1947. Fractures were found in February 1950 in the two girders, paint in the cracks indicating that the cracks had occurred before the girders had left the construction shop. Nearly a year later in January 1951, part of the West Crossing collapsed into the river when the temperature was −34°C.

Often it is found that structural accidents are caused partly because of misunderstanding or poor communications between the parties involved in their design and construction. In the case of the Duplessis Bridge the flange plate had been ordered to meet C.S.A. S-40 (ASTM A-7) specifications. Such thick structural plate is usually rolled from semi- or fully-killed steel, but in this case they were rolled from rimmed steel. The rimmed steel was also of poor quality containing high local concentrations of carbon and sulphur. Molten steel contains dissolved oxygen and other gases which need to be controlled during solidification. The amount of oxygen can be controlled by the addition of deoxidising agents such as aluminium. In fully-killed steels there is virtually no evolution of gas. With semi-killed steels a small amount of gas is allowed to evolve that benefits the ingot by minimising shrinkage. In rimmed steel the reaction of dissolved oxygen and carbon to form carbon oxide gasses is allowed to progress to form an ingot with blowholes of various sizes at its centre with a heavy rim of relatively void-free metal. During hot rolling the voids in the ingot are welded together. Segregation also occurs in rimmed steels with carbon, sulphur and phosphorus segregating in the centre and top of the ingot. Rimmed steels are cheapest because less of the ingot has to be discarded, but the transition temperature in rimmed steels can be high. Fully-killed steels have the lowest transition temperature. Welding becomes more problematic as the carbon equivalent[40] increases causing embrittlement of the steel and cracking unless measures such as pre-heating are taken. The Charpy impact values for the fractured girders of the Duplessis Bridge were only 4–8 J at 38°C.[41] Clearly the rimmed steel was far too brittle and almost guaranteed that the bridge would collapse.

Liquid natural gas containers need to operate at very low temperatures. A number of liquid gas pressure vessels operating at −162°C at a pressure of 34.5 kPa were built in Cleveland Ohio in the early 1940s.[32] In 1943 after the first pressure vessels had been operating for about two years an additional cylindrical-toroidal vessel 21 m in diameter and 13 m high was built in the same steel as the previous vessels. The steel was a nickel steel (0.08–0.12% C, 0.30–0.60% Mn, 3.25–3.75% Ni). A plate in the bottom of the vessel cracked when the vessel was first filled. This crack was repaired and the vessel put back into service. In 1944 after just over a year in service it cracked again releasing natural gas. An explosion followed as the gas ignited and the fire spread to an adjacent vessel and into the sewers. One hundred and twenty-eight people were killed. The 3.5% nickel steel used for the vessels is ferritic with a transition temperature about 30°C lower than similar steel without the nickel, but it is not low enough for operation at −162°C when at least 9% nickel steel is required. The problem is that nickel is about eight times more expensive than steel, so there is pressure to minimise its use. It was claimed after the accident that the Charpy impact value at −162°C was more than 20 J, but specimens taken from the plates that failed only gave values of 4–7 J. The designer of the pressure vessel in the course of the investigation stated that a sledge-hammer could be driven through the plates at −162°C, but did not think that such brittleness would obviate its use for construction purposes![32]

The brittle fracture of gas transmission lines in the USA during the period 1948 to 1951 are interesting because they illustrate the very high speed at which brittle fractures can propagate. When a fracture propagates, some of the energy released goes into the kinetic energy of the body. At the speed of Raleigh waves, which travel a little slower than the shear waves, all the energy goes into kinetic energy leaving none to create the fracture and is hence the theoretical limiting crack velocity. In practice the maximum velocity of crack propagation is of the order of one-third the velocity of shear waves or about 2,000 m/s in steel. The speed of sound in gas is about 300 m/s so that when a gas pipeline fractures, the unloading wave travels much slower than the fracture consequently the pipeline remains pressurised at the tip of the crack and the fracture can run for large distances. The length of fractures in the pipelines in the late 1940s varied from 50 m to nearly a kilometre.[32] The fracture of a 760 mm diameter pipeline is shown in Fig. 8.7. The sinusoidal fracture path seen in Fig. 8.7 is characteristic of fracture in pipes which was originally thought to be an instability caused by the high velocity of fracture propagation, but similar sinusoidal fracture paths have been observed where the crack velocity is less than a tenth of the velocity of shear waves.[42] The force of the gas discharge on the flaps which open on fracture cause a cyclic torsional loading that creates the sinusoidal crack path.[43]

Fig. 8.7 Fracture in a 760 mm gas transmission line showing a sinusoidal fracture (courtesy Lincoln Electric Arc).

8.1.5 Brittle fracture tests during the 1940s

Not surprisingly the brittle fracture of so many welded ships during War World II caused the formation of investigation committees in both America and Britain. In America the Secretary of the Navy, James Forrestal, established a Board in 1943 to inquire into the 'design and methods of construction of welded steel merchant vessels'. This board issued its report in 1946. The work was then continued by the Ship Structure Committee formed in 1946 under the chairmanship of Ellis Read-Hill. In Britain inquiries were instigated by the Admiralty in 1943. Whereas in America navy men were appointed to be chairmen of the committees, in Britain the Admiralty Ship Welding Committee, formed in 1946, was under the chairmanship of John Baker (1901–1985) at Cambridge University. Baker assigned the metallurgical investigation to Constance Tipper (1894–1995). Charpy impact bend tests were performed on plates taken from ship failures over the transition temperature by Constance Tipper and Morgan Williams in the USA in an attempt to determine the minimum impact value at which brittle fracture occurs. In general it is almost impossible to expect that an empirical test such as the Charpy test can be relied upon to determine the probability of brittle fracture which depends upon many factors. Having said that, the steels used in the ships built during World War II formed a fairly homogeneous group and correlations between a small empirical test and service behaviour were possible. The problem is that steels change with time so that the criteria developed from historical performance become outdated quickly.

The plates from ship failures were originally classified as source-plates that contained the fracture initiation, through-plates where the fracture propagated, and end-plates where the fracture arrested. In recognition that thicker plates present more constraint on fracture and hence behave in a more brittle fashion, the plates were also grouped according to thickness. It was found that there was a very significant difference in the impact energy of the source and end-plates. Only 10% of the source-plates had impact energy at the failure temperature greater than 10 ft lb (14 J) and none of had impact energy greater than 15 ft lb (20 J).[44] As a result of these investigations the post-war minimum Charpy impact requirement was set at 15 ft lb (20 J) but with the development of higher strength steels that figure has had to be revised upwards.

The effect of size on fracture was known in the 1940s and tests were made on 19 mm-thick steel-rimmed, semi-killed and fully killed plates up to 1,800 mm wide to try to understand brittle fracture.[45,46] Such tests required very large capacity testing machines. Large testing machines with tensile capacities of up to 3.5 MN were constructed in the USA at the end of the nineteenth-century for testing locomotive castings. In the 1930s The Baldwin-Southwark Company worked with A.H. Emery to produce testing machines with tensile capacities of up to 17 MN which were installed in the University of Illinois and Lehigh

University, the latter machine was subsequently moved to the University of California, Berkeley. The huge size of these testing machines can be appreciated from the 1965 photograph of the Berkeley machine shown in Plate 10.

Parallel tests were run at the University of California under Alexander Boodberg (1906–1952) and at the University of Illinois under Wilbur Wilson (1881–1958).[45,46] In these tests the plate width was varied from 75 mm to 1800 mm; the plates contained a central notch whose length was a quarter of the width of the plate in all except the narrowest plates which contained edge notches. The mean stress at fracture for all the plates was above the yield strength. There was only a slight decrease in fracture strength with increase in plate widths above 300 mm though the fracture strengths of the 75 mm wide plates were appreciably greater than those 330 mm wide. Obviously these tests did not provide an answer to the service brittle fractures.

The wide plate tests were expensive and a multitude of small full plate thickness tests were introduced. The US Navy Tear test was developed for a full thickness test with minimum specimen preparation.[47] The brittleness of the steel was judged from the energy required to propagate the fracture and the percentage crystallinity of the fracture surface. Similarly in Britain, Tipper developed what became known as the Tipper test in her work for the Admiralty Ship Welding Committee.[48] In this test, plates with a width twice the thickness were notched each side at the centre by 45° notches 3 mm deep. Tipper realised that reduction in thickness was a better measure of ductility than elongation. The transition from fibrous ductile tearing to cleavage in these tests was quite sharp and occurred over a temperature range of as little as 5°C. Again the percentage crystallinity of the fracture surface was also used as a measure of brittleness of the steel. Although these and other full plate thickness small scale tests more accurately duplicated the constraint that occurred in a service fracture they did not accurately reflect the behaviour of the later welded wide plate tests[49] for all steels.

8.2 The Beginning of Analytical Fracture Mechanics

The great advances in the theory of elasticity in the nineteenth-century made it almost certain that the first applications of mechanics to fracture should be for an elastic-brittle material. It was assumed almost without question in 1900 that fracture occurred when the maximum principal stress reached a critical value and that provided the stresses in a design were less than this value by a factor of safety then all was well. The idea of stress concentrations where the stress was much higher than the average was appreciated at the end of the nineteenth-century and Ernst Gustav Kirsch (1841–1901) published the stress distribution around a hole in tension in 1898.[50]

8.2.1 Wieghardt's pioneering work

Karl Wieghardt (1874–1924) when he was Professor of Mechanics at the Herzogliche Technische Hoschschule in Braunschweig wrote the first paper on the fracture mechanics of elastic bodies in 1907.[51,52] Although apparently the paper was well known at the time, it became forgotten and there were only two references to it after World War II until Peter Rossmanith translated it into English in 1995.[53] If this important paper had been developed, linear elastic fracture mechanics could have evolved in the early twentieth-century. Rossmanith believes that it was not developed because of lack of large-scale fracture failures in industry[54], but a more plausible explanation is that the crack geometry analysed had no direct application to the practical fracture problems coupled with the fact that notched bar testing had only just been developed and was not really understood, hence Wieghardt's paper was seen as just another theoretical problem in elasticity. When linear elastic fracture mechanics was finally developed the problem of the brittle fracture of steel was reasonably well understood and the race to build intercontinental rockets during the Cold War of the 1950s and 1960s brought problems in the brittle fracture of high strength steel which approximated to elastic-brittle behaviour.

Wieghardt solved the problem of a semi-infinite crack with either one force or two symmetrical forces applied at a distance from the crack tip either normal or tangentially to the crack surface. He derived expressions for the stress distribution near the crack tip, which have a square-root singularity of the same form as those given by Irwin 50 years later.[55] The exact expressions given by Wieghardt were independently derived again by Fazil Erdogan in 1962.[56]

The theoretical elastic stresses at the tip of a crack or sharp re-entrant corner are infinite even for the smallest force but, as Wieghardt stated, fracture cannot occur as the result of an arbitrarily small force. This observation led him to observe that elastic material does not rupture at a single point but rather fractures over a small area and that fracture did not occur when the stress at a single point reached a critical value, but only when the stress over a small area reached a critical value. Thus Wieghardt came close to formulating a criterion of fracture based on a critical stress over a critical area which was the criterion for cleavage fracture for steel proposed in 1973.[57]

8.2.2 *Inglis and the stresses due to cracks and sharp corners*

In 1913 Charles Edward Inglis (1875–1952), acting on a suggestion of Bertram Hopkinson (1874–1918) that the stresses around a crack should be analysed, solved the problem of an elliptical hole in an infinite plate under constant stress at infinity.[58] Inglis clearly wanted to promote the practical application of his paper, because he read it at a meeting of the Institution of Naval Architects and kept the mathematical part of his paper in the background. He anticipated fatigue crack growth and saw that under the high stresses at its tip a crack could grow under alternating loads so 'once fairly started, no amount of ductility [would] prevent it spreading'.

Inglis implicitly accepted the maximum tensile stress criterion and he concentrated his attention on the stress at the tip of an elliptical hole, unlike Wieghardt who saw the importance of the size of the zone at high stress. Consequently, since for an infinitely sharp crack the stress is infinite, Inglis did not examine the limiting case of a crack. He expressed the stress at the tip of a crack in terms of the root radius, ρ, and for and elliptical hole whose major semi-axis, a, is perpendicular to an applied stress, σ_0, obtained the expression for the maximum stress, σ, given by

$$\sigma = \sigma_0 \left[1 + 2\sqrt{\frac{a}{\rho}} \right].$$

(8.1)

Inglis also showed how the maximum stress can be approximated for a variety holes with round corners. Inglis's paper paved the way for Griffith to produce the most famous fracture paper of all time

8.2.3 *Griffith and the foundations of fracture mechanics*

Alan Arnold Griffith (1893–1963) graduated from Liverpool University at the beginning of World War I and after a year of research there, joined the band of gifted engineers and scientists at the Royal Aircraft Factory,[59] Farnborough in 1915. Probably the most gifted scientific member of the Royal Aircraft Factory was Geoffrey Ingram Taylor (1886–1975) who had joined the year before Griffith. One early task that Taylor undertook with Griffith was the stress analysis of the torsion of shafts with keyways.[60] Taylor thought of a soap film analogy to the torsion problem, apparently unaware that it had already been suggested by Ludwig Prandtl.[61] The soap film analogy enables the stress concentrations at the keyway corners to be estimated. This work earned Griffith the sobriquet of 'Bubble Griffith'.

Griffith was prompted to work on the effect of cracks and notches by Charles Jenkin (1865–1940), who was responsible for the preparation of specifications for aircraft materials during World War I, where the main concern was fatigue. The life of aircraft engines in World War I were found to be greatly increased by polishing the surfaces of the connecting rods.[62] Using the soap film analogy and the work of Inglis, Griffith estimated that the normal scratches observed on shafts caused an increase in the maximum stress by a factor of two to six which was practically independent of their depth. Griffith realised that the maximum tensile stress criterion thus predicted that 'the weakening of a shaft 1 inch in diameter, due to a scratch one ten-thousandth of an inch deep should be almost the same as that due to a groove of the same shape one-hundredth of an inch deep'. This conclusion he knew to be contrary to experience in fatigue. Griffith's famous 1920 paper was born.[63]

Having rejected the maximum stress criterion of fracture Griffith turned to examine energy. For an elastic body to be in equilibrium the potential energy of the whole system must be a minimum. Griffith took this theorem and obtained a new criterion of fracture by adding the corollary that an elastic system can only pass from an unfractured state to a fractured state 'by a process involving a continuous decrease in potential energy'. For the propagation of a crack, work must be done against the cohesive molecular forces to supply the energy of the new surface and this work comes from a decrease in the potential energy. Griffith concluded that the shape of the crack tip, and the intermolecular forces, remains the same during crack propagation so that the increase in the surface potential energy when the size of the crack increases in area by dA is $2\gamma dA$.

The classic Griffith's crack that he analysed to find the critical stress, σ_c, was a straight crack of length, $2a$, in an infinite plate under a constant normal stress, σ, perpendicular to the crack. Griffith calculated the strain energy per unit thickness, U, of a large circular region of radius R, where $R \gg a$, containing a crack from Inglis' analysis[58] and obtained

$$U = \frac{\sigma^2 \pi}{\bar{E}}\left[\left(1-\bar{\nu}\right)R^2 + \bar{\nu}a^2\right].\tag{8.2}$$

The first term in Eq. (8.2) is the strain energy stored in a circular region of radius R if there is no crack. Invoking Clapeyron's theorem[64] that the decrease in potential energy of the external forces from the unstrained to the strained state is twice the increase in the strain energy, Griffith assumed that the change in potential energy of the system, Π, due to a crack was given by

$$\Pi = 4a\gamma - \frac{\sigma^2 \overline{V} \pi a^2}{\overline{E}}.$$

(8.3)

So that the critical fracture stress is given by

$$\sigma_c = \sqrt{\frac{2\gamma \overline{E}}{\pi \overline{V} a}}.$$

(8.4)

However, Eq. (8.4) is incorrect. Griffith realised his mistake when he read the proofs of his paper and added a note at the end of the paper to this effect. In 1924 Griffith published another paper in which he gave the correct expression, but did not give any details of its derivation.[65] The correct expression for the critical fracture stress is

$$\sigma_c = \sqrt{\frac{2\gamma \overline{E}}{\pi a}}.$$

(8.5)

The original mistake made by Griffith was quite subtle. As he explained, the strain energy was erroneous, 'in that the expressions used for the stresses gave values at infinity differing from the postulated uniform stress at infinity by an amount which, though infinitesimal, yet made a finite contribution to the energy when integrated round the infinite boundary'.

The first published detailed derivation of Eq. (8.5) was given by Ian Sneddon (1919–2000) who in 1946 solved both the penny-shaped and the Griffith crack problem for cracks opened by uniform pressure on their surfaces.[66] Despite the difference in loading, the problem of a crack loaded by a uniform stress at infinity and one opened by a uniform pressure are virtually the same. Again it was war that caused Sneddon to work on fracture. In 1942 Sneddon joined the Ministry of Supply and was posted to the theoretical branch of the Armaments Research which was headed by Nevill Mott (1905–1996). The penny-shaped crack paper was the result of speculation by Mott that tank armour plate steel before rolling contained spherical gas bubbles which became flattened into penny-shaped cracks during rolling. Sneddon's job was to study these cracks.[67] The critical stress for a penny-shaped crack of radius a either opened by a uniform pressure or under a normal stress perpendicular to the crack is the same namely:

$$\sigma_c = \sqrt{\frac{\gamma \pi \overline{E}}{2a}}. \qquad (8.6)$$

That would be that, but for the fact that some people have believed that the original 1920 paper of Griffith was correct and the 1924 paper was incorrect even up until the late 1990s. Because of this ongoing controversy, at least two papers give the derivation of the critical stress for the Griffith crack since 1946 by different methods.[68]

Griffith attempted to verify his theory, choosing the glass as his model material. He estimated the surface energy of glass by extrapolating the surface tension of glass in the molten state to room temperature and obtained a value of 0.55 J/m^2. The fracture experiments were performed on tubes and spherical bulbs with cracks formed either with a glass-cutter's diamond or by scratching and tapping gently. Griffith expected to find that the fractures stress in these experiments could be predicted by his theory. In his first paper using Eq. (8.4) he did obtain good agreement by annealing the specimens at 450°C for an hour. In his second paper, Griffith argued that probably the cracks were blunted by the annealing at 450°C for an hour and shut off the furnace once the temperature reached 450°C. By this expedient he again got reasonable agreement by using Eq. (8.5). Although Griffith fiddled the results somewhat to obtain good agreement between the measured and the predicted fracture stress, the important result was that the fracture stress varied as the inverse of the square root of the crack length. We now know that the fracture energy of glass is about four times the surface energy.[69] In hindsight, we also know that the correction for the curvature of Griffith's specimens is significant and using these corrections, the value of the fracture energy calculated from Griffith's experiments is indeed about four times his measured surface energy.[70]

Griffith also estimated the cohesive strength of glass.[63] Assuming that the cracks were atomically sharp[71] so that the radius of curvature was of the same order as the molecular dimensions, he used Eq. (8.1) to obtain a cohesive strength of about 20 GPa. Realising that glass could not be elastic up to fracture as assumed by the use of Eq. (8.1), Griffith suggested that 7 GPa would be a more reasonable value for the cohesive strength. Since the Young's modulus of glass is 70 GPa, Griffith's estimate of the cohesive strength was $E/10$ which corresponds to modern estimates of the theoretical cohesive strength of solids.

The two papers of Griffith gave the first real insight into the mechanics of fracture and they will ever remain foundation stones for fracture research. Surprisingly, after 1924 he did no further work on fracture and soon began his work on gas turbines. Griffith's brilliant work in this field brought him many honours. Jim Gordon (1913–1998) quoted an anecdote told him by Ben Lockspeiser (1891–1990) that may partly account for Griffith's abrupt termination of his work on fracture. Lockspeiser was Griffith's assistant at the RAE and he told Gordon that one night he left a gas torch, used for drawing glass fibres, burning and a small fire was created in the laboratory. As a consequence Griffith was told by his superiors to stop his experiments.[72]

The brilliance of Griffith's two fracture papers, or perhaps because of it, has not stopped attempts at finding faults in the thermodynamics. Charles Gurney (1913–1997), who knew Griffith from the time he joined the RAE in 1937 and used to drink beer with him in the Mess, never discussed fracture with him, but felt it necessary in 1994 to give a thermodynamic derivation of Griffith's theory of fracture to refute the critics.[73]

8.2.4 Defects and the strength of brittle solids

In his first paper Griffith thought that molecular orientation would produce a glass surface layer that would have 'exceptional strength' since any flaws near the surface would be parallel to the surface.[63] However, in his second paper on the basis of the decrease in strength of fresh silica rods when lightly touching together he stated 'that the weakness [of glass] is due almost entirely to minute cracks in the surface, caused by abrasive actions to which the material has been subjected after manufacture' and noted that 'even atmospheric dust particles have an appreciable weakening effect'.[65] Griffith used the strength of his tubes and bulbs containing cracks to calculate the probable flaw size in glass as about 5 μm.

Many attempts were made after the publication of Griffith's papers to prove the existence of surface cracks. However, the separation across a surface flaw in glass is of the order of 50 nm, only about one-tenth of the wavelength of light, and the flaws are undetectable optically. Abraham Joffé (1880–1960), who made many contributions to the physics of crystals, showed indirectly in 1924 that the fracture of rock salt was due to surface cracks.[74] Rock salt fractures in a brittle fashion at a stress of 4.4 MPa when tested in air, whereas when the surface was being dissolved by water during testing, it failed at 1,600 MPa, not far from its theoretical strength of 2,000 MPa.[75] Joffé reported his findings on rock salt in

1924, attributing the increase in strength to the removal of surface cracks, to the First International Congress of Applied Mechanics at which Griffith gave his second paper on fracture. Even more convincing proof that the low strength of brittle materials was due to cracks was given by Orowan in 1933. The usual tensile strength of mica is between 200–300 MPa, but Orowan obtained strengths of more than 3 GPa by stressing only the central strip of a sheet of mica using grips that were much narrower than the sheet.[76] The small value of the usual tensile strength of mica is due to the presence of cracks at the edges of the sheet, which were not stressed in Orowan's experiments. Fred Ernsberger revealed the surface cracks on glass in 1960 by exchanging Na+ ions in glass with Li+ ions which are smaller, which induces local tension and causes the cracks present in the surface to propagate to a size large enough for them to be seen.[77]

Griffith also performed tensile tests on glass fibres.[63] He thought that the flaws would be restricted by the diameter of the fibre and tested fibres whose diameters ranged from 1 mm to 3 μm. Griffith estimated the strength of freshly made pristine fibres by bending them and measuring the radius of curvature at which they broke and obtained extremely high strengths. However, on exposure to the atmosphere of the laboratory the strength decreased, and he found a size effect in these aged fibres.[78] Extrapolating his results to zero diameter Griffith obtained a theoretical strength of 11 GPa, which was not much greater than the strength that he obtained for his pristine fibres. William Otto showed that if only the portion of the fibre up to the winder used to pull out the fibres, which was not damaged by contact, was tested, then the strength of the fibres over a diameter range of 5 to 15 μm did not vary significantly.[79] Otto also showed that the strength of glass fibres increased with the temperature at which they were drawn. These effects were explained by Dennis Holloway, who showed that the defects on pristine glass fibres were due to surface contaminants which dissolved at high temperatures.[80] Glass fibres drawn at high temperatures can retain their ultra-high strength if they are coated with a protective resin immediately after drawing. Today optical fibre technology produces fibres longer than 3 km with strengths greater than 1 GPa.

At the same time that Griffith was considering the effect of cracks and flaws on strength, the Hungarian born polymath Michael Polanyi (1891–1976) was considering the theoretical maximum possible cohesive strength of solids.[81] He argued that if the equilibrium interatomic spacing across a plane of atoms was a_0 and the cohesive strength was σ_c then the work done to part two surfaces of unit area would be or the order $\sigma_c a_0$. Since this work would reappear as the surface energy of the new surfaces

$$\sigma_c \approx \frac{2\gamma}{a_0}.$$

(8.7)

Others have made similar estimates of the theoretical cohesive strength. Orowan in 1934 assumed that the separation of the surfaces is elastic up to the cohesive stress thus obtained[82]

$$\sigma_c = 2\sqrt{\frac{\gamma E}{a_0}}.$$

(8.8)

Later Orowan realised that Eq. (8.8) gave an overestimate of the theoretical cohesive stress because there is a tail to the stress-displacement curve (see Fig. 1.1 (b)). Fitting a half-wave sine curve to the stress-displacement curve Orowan obtained the expression that is now usually quoted[83]

$$\sigma_c = \sqrt{\frac{\gamma E}{a_0}}.$$

(8.9)

8.2.5 *Obreimoff, stable fracture and its reversibility*

Muscovite mica from the White Sea region of Russia splits almost perfectly to produce near atomically smooth surfaces up to 100 mm^2 in area with virtually no damage so that if the two split surfaces are placed in contact they should adhere again spontaneously. In 1930 Obreimoff, working in the Physics and Technics Institute of Leningrad University, split thin strips from a mica sheet with a glass wedge.[84] The wedge-loaded cantilever specimen was stable and enabled fracture reversibility to be studied. The elastic energy released during crack propagation can be calculated simply from the engineers' theory of bending. Like Griffith, Obreimoff made a mistake in his calculation, but it was a very simple error of a factor of four.[85] The first experiments of Obreimoff were performed in air at ambient pressure; the crack propagated slowly for 10–15 seconds when wedged before stopping and the surface energy calculated from the equilibrium crack length was 0.38 J/m^2. As the pressure was reduced the time taken to reach equilibrium increased, taking four days under a vacuum of 2.7 10^{-4} bar. Propagation of the crack was often erratic and accompanied by a visible electric discharge especially at high vacuum when the glass vessel enclosing the specimen 'fluoresced like an X-ray bulb'. Under a vacuum of 10^{-9} bar, the surface energy increased to 5 J/m^2. The high surface energy under a vacuum was not immediate obtained when air prepared specimens were first tested, but only after the crack had penetrated about 1 mm into fresh mica. Thus Obreimoff had demonstrated that the fracture of mica was affected by the test environment and indicated the role of chemical kinetics in fracture.

When the wedge in a mica specimen tested under high vacuum was pulled back, the crack healed, but the surface energy of the healed mica was reduced. Mica is a material that heals readily, but other brittle solids such as glass and sapphire have also been shown to heal.[69]

8.2.6 *The extension of Griffith's theory to metals*

Until the 1940s Griffith's theory was thought to be inapplicable to metals. The surface energy of iron is of the order of 1 J/m^2 not much different than that of glass. The stress in iron or mild steel is limited by the yield strength to about 200 MPa thus, using a value of 200 GPa for Young's modulus, Griffith's equation (8.5) gives a critical flaw size ($2a$) for facture of about 12 μm. If Griffith's theory applied directly to iron or mild steel then their strength would be very sensitive to scratches or slight nicks which it is not. Orowan also showed that tin could sustain tensile stresses 5.5 times higher than the usual tensile strength if thin discs of tin were sandwiched between steel rods that inhibited their yielding.[83] This very high strength is more than 3 times the theoretical cohesive strength of tin obtained from Eq. (8.9). Although Orowan did not realise it at the time, his 1945 paper contained the explanation why a direct application of Griffith's equation using the true surface energy would not predict the brittle fracture strength of a metal. Tipper sent Orowan an example of a brittle fracture from a welded ship obtained during her work for the Admiralty. Orowan noted that usually a ductile tear preceded cleavage fracture in Charpy specimens, but in welded structures long brittle fractures were observed where there was no visible plastic deformation. However, when Orowan took back reflection X-ray photographs of the fracture surface he found that the reflections were blurred indicating that considerable plastic deformation had occurred. Orowan repeated his X-ray analysis after removing a layer 0.5 mm from the fracture surface and found that the reflections were sharp indicating that the plastically deformed layer was only about 0.5 mm deep.

On reading the proofs of his 1949 paper, Orowan realised that the thin layer of plastically deformed material that he had observed in 1945 provided the answer to why Griffith's equation did not apply directly to metals.[20] Orowan wrote in a footnote 'a thin layer of some 0.5 mm depth below the surface of brittle fracture in low carbon steel always shows a few percent plastic strain. The corresponding plastic work must be added to the surface energy in Griffith's equation; in fact, the order of magnitude of the [plastic work per unit area of surface, γ_p,] (10^3 to 10^4 J/m^2) is higher than that of the surface energy (1 J/m^2), so that the later can be neglected and, instead of Griffith's equation, the condition

$$\sigma_c = \sqrt{\frac{2\gamma_p E}{\pi a}} \tag{8.10}$$

used... The necessary crack length is [then] 1,000 to 10,000 times greater than the ordinary Griffith crack length'.[20] Hence the order of magnitude of the critical flaw size was 12–120 mm, not 12 µm. George Rankin Irwin (1907–1998) had presented the same idea two years earlier at the 1947 symposium of the American Society of Metals, also making use of Orowan's 1945 paper.[86]

The extension of Griffith's equation to metals is usually referred to as the *Irwin-Orowan* extension. The time was right in the 1940s for the extension of Griffith's equation. Many researchers were trying to explain the brittle fracture of steel and Griffith's equation was seen as a key to the quantifying of the fracture of metals. A year before Irwin's presentation of his ideas, Clarence Zener and John Holloman, who were working at the Watertown Arsenal, Massachusetts, wrote a review paper on the fracture of metals in which they discussed the application of Griffith's equation to metals.[87] They discussed the question of plasticity at the tip of a crack and speculated that perhaps, because the stress gradient was so high,[88] that the high stresses might not cause plastic deformation, but 'if plastic deformation does occur, the energy associated with it must be added to the surface energy in order that... [Eq. (8.5)] be applicable'. Eugene Merchant at about the same time was also making similar remarks in discussing metal cutting 'the energy required to separate the chip from the metal ... would of course be increased somewhat by ... the work expended in the local plastic flow of the metal adjacent to the newly created surfaces'.[89]

Since the fracture energy of glass is about four times the surface energy it might be expected that there is significant irreversible work performed in the fracture of glass and in which case a crack in glass would be blunted. However, there is evidence that the tip of a propagating crack is atomically sharp, but there is still controversy. Using an atom force microscope (AFM), Christian Marlière and his co-workers have claimed to identify voids ahead of a crack tip in glass indicating some plastic deformation, but Jean-Pierre Guin and Sheldon

Wiederhorn found no evidence of such voids in similar experiments where the two fracture surfaces matched one another to the limits of the accuracy of AFM.[90] Lattice trapping models where the discreteness of the atomic bonds is modelled have been proposed to account for the discrepancy between the fracture and surface energies of glass.[69]

8.3 The Statistics of Fracture

Nothing is completely certain, yet much of science and engineering is treated as if it were. In many cases the variation from a mean value is quite small and the assumption that one is dealing with deterministic quantities is justified. Some mechanical properties, such as Young's modulus, depend upon the average behaviour of the whole specimen and vary only slightly from one specimen to another. Even the yield strength of a metal specimen does not vary very much because, though the initiation of plastic flow depends on the point of local weakness, for yield to spread and be observed macroscopically a reasonably large volume of material is involved and the measured yield strength is the average of this volume. However, unstable fracture is different and most fractures are unstable. Fracture seeks the weak spot. From a design viewpoint it is not the average strength of components or structures that matters, but the strength that a very large fraction of them will exceed.

In 1939 Waloddi Weibull (1887–1979), working in the Royal Institute of Technology, Stockholm wrote a paper which has become the usual way of dealing with the statistics of brittle fracture.[91] In his original paper Weibull did not discuss the type of materials for which his theory was intended. The materials with which he illustrated his theory are mostly brittle, such as porcelain rods, but included others which are not so brittle, such as aluminium die castings. Weibull based his theory on the maximum stress criterion of fracture. He considered a body to be composed of infinitesimal elements and if one of these elements fractured then the whole body was fractured just as a chain is as strong as its weakest link.[92] He assumed that the risk of rupture in an elemental volume was dependent on the stress raised to a positive power m[93] which gave the probability of fracture $P(\sigma)$ under a uniform tension, σ, in a specimen of volume V

$$P(\sigma) = 1 - \exp - V \left(\frac{\sigma}{\sigma_0} \right)^m ,$$

(8.11)

where σ_0 is a reference stress. A size effect is predicted by Eq. (8.11) with the median strength being inversely proportional to $V^{1/m}$. Although Weibull did not mention Griffith's equation in his paper, his statistics has subsequently been interpreted in terms of crack-like flaws whose length is distributed according to the Pareto distribution.[94]

Ropes and composites behave differently to homogeneous materials in that individual components can fracture without fracture of the whole entity. The statistical theory of fibre bundles where the strength of each individual member has a particular distribution and the load is shared among the intact members of the bundle was derived in 1945 by Henry Daniels (1912–2000) who worked in the Wool Industries Research Association, Leeds.[95] For bundles containing a very large numbers of fibres, n, the deviation from the mean strength of the bundle becomes very small and the strength of the bundle can be visualised graphically. If a parameter w is introduced which is the inverse of the strength of any fibre and called the *weakness* by Daniels, and the probability of failure in any fibre is $P(w)$, the strength of the bundle or surviving fibres, Σ, is given by

$$\Sigma = \frac{n\left[1 - P(w)\right]}{w}. \tag{8.12}$$

The maximum strength of the bundle can be represented graphically, as shown in Fig. 8.10, by plotting $P(w)$ against w; the value of w that gives the strength of the bundle is obtained by finding w^* the value of w at which a straight line from $P(0)$ that just touches the curve. For smaller bundles of fibres the strength of the bundle is less simple.

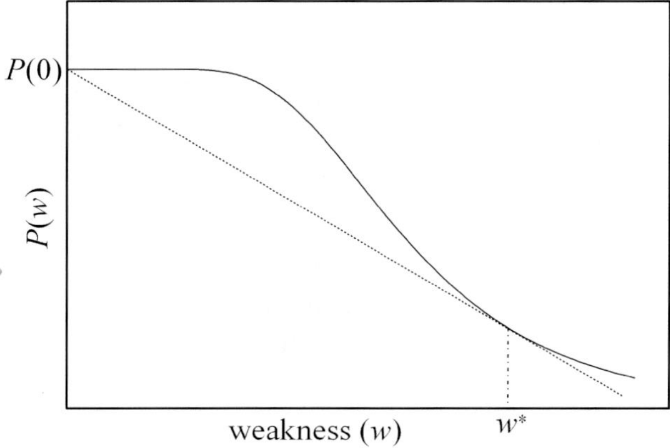

Fig. 8.8 Determination of the strength of a large bundle of fibres (after Daniels 1945).

8.4 Fatigue of Materials

Fatigue is a generic problem with metals. The railways' fatigue problems continued into the twentieth-century and the higher rotational speeds of machinery and engines brought more problems. Changes in materials also brought problems. By 1930, nearly all aircraft were constructed from aluminium alloys. One problem with age-hardening aluminium alloys is that, though they can have high strength, the fatigue limit as a fraction of the ultimate tensile stress is relatively low. Hence fatigue failures began to occur in aluminium aircraft. One of the first recorded fatigue accidents was in the inter-wing strut on a Curtiss Condor airliner over Germany in 1934 which caused failure of the lower wing and resulting in eleven deaths.[96] At least 20 Vickers-Armstrong Wellington bombers crashed due to fatigue of their wing spars during War World II. Alfred Pugsley (1903–1998), who was in the Airworthiness Department of the RAE during World War II, wrote that, despite the heavy losses of bomber aircraft to enemy action, airmen preferred the high risk of being shot down to the smaller risk of the aircraft breaking up due to structural reasons.[97] A high risk that is due to personal action seems to be far more acceptable to people than a low risk, such as fatigue failure, over which they have no control.

Aluminium alloys do not possess a true fatigue limit. The fatigue strength of duralumin is disappointing because, though its ultimate strength is similar to that of mild steel, its fatigue strength based on 2×10^7 cycles is only about half that of mild steel.

8.4.1 *Microstructural aspects of fatigue*

The notion that vibration caused metals to crystallise had not been completely dispelled by the beginning of the twentieth-century. In 1900 James Ewing (1855–1935) and Walter Rossenhain (1875–1934) had shown conclusively that the idea that metals became amorphous during cold working was wrong and identified the mechanisms of slip and twinning in plastic deformation.[98] Three years later Ewing and Humfrey observed a polished section of a fatigue specimen showing that the structure did not change during fatigue.[99] They also described how slip bands, formed in grains during the early stages, broadened during fatigue into what are now known as persistent slip bands (PSBs). They suggested that the broadening of the slip bands was 'due to a heaving-up of the surface' and gave the sketch of their conjecture shown in Fig. 8.11 (a). Finally, the slip bands developed into cracks. With optical microscopy it is impossible to see detail on the scale of their sketch, but with transmission electron microscopy such detail can be seen. Extrusions and intrusions in PSBs on the surface of fatigue specimens were first identified in copper by Alan Cottrell and Derek Hull in 1957 and are shown in Fig. 8.11 (b).[100] The usual schematic illustration of extrusions and intrusions is shown in Fig. 8.11 (c). The conjecture of Ewing and Humfrey looks very much like an extrusion-intrusion. Herbert Gough (1890–1965) also made early observations that showed that fatigue crack originated in PSBs.[101]

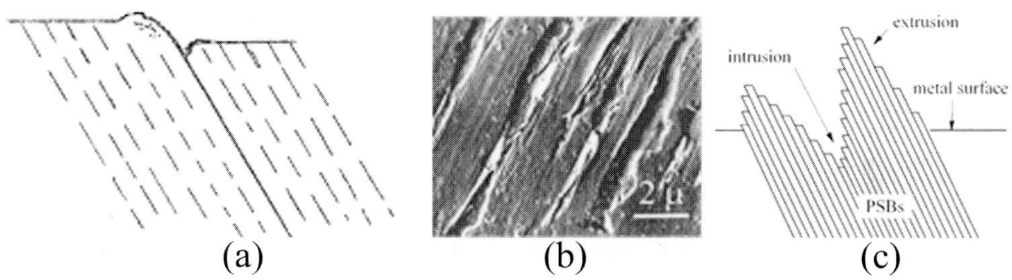

Fig. 8.9 Persistent slip bands (PSBs) (a) Ewing and Humfrey (1903). (b) Extrusions and intrusions in PSBs in copper (Cottrell and Hull 1957). (c) Schematic sketch of extrusions and intrusions; (a and b with permission of the Royal Society).

8.4.2 *Effect of frequency of stress cycling and corrosion fatigue*

The increasing speed of machinery led to investigations into the effect of the frequency of loading on loading. In 1911, Bertram Hopkinson performed fatigue tests on mild steel at 7,000 cycles per second which was three times faster than any others up to that time.[102] His tests seemed to show that the fatigue strength was markedly increased at high frequency, but he did not consider his tests conclusive. World War I intervened and he became involved in war work before he could do further experiments. Hopkinson was killed in a flying accident right at the end of the war, but Jenkin followed up his work and constructed a vibration fatigue machine that was capable of testing at up to 20,000 cycles per second.[103] Jenkin confirmed that the fatigue strength of a range of metals did increase by up to 60% when tested at 10,000 cycles per second. However, at 20,000 cycles per second there was a drop in strength of up to 9% in fatigue strength because at this high frequency there were problems with the specimens heating.

The first paper describing corrosion fatigue was published by Haigh in 1917 and records that the endurance of brass moistened by hydrochloric acid was greater than dry brass.[104] However, to determine whether there is a corrosion effect it is necessary to compare the results with specimens tested in a vacuum. The first such tests were performed by Gough and Sopwith in 1932, who tested a number of alloys representative of commercial aircraft materials.[105] They showed that fatigue life of carbon steel and duralumin is higher when tested in a vacuum compared with testing in air, but some other materials, such as stainless steel and a magnesium alloy, showed little difference. Usually the fatigue properties under a vacuum are superior to those in any other environment. Since there is less exposure time at high frequencies the fatigue strength is higher.

8.4.3 Cumulative damage

Most of the fatigue tests of this period were made with a constant alternating stress. The original fractures in railway axles, where fatigue first started to be understood as a different fracture mechanism, did occur under more or less constant load cycles as the bending stress due to the weight of the carriage alternated, but in most situations the alternating stress is not constant. Often as in the case of the loads on the spars of aircraft the alternating stress approaches the spectrum of white noise. The crash of a Lufthansa aircraft in 1927 with six fatalities led to load spectra being collected from the spars of Lufthansa aircraft in the 1930s using glass-scratch strain gauges.[106] The collection of such data continued in Germany during World War II. The first fatigue machine capable of variable-amplitude testing was constructed in Germany at the beginning of the war by Ernst Gassner (1908–1988). Gassner's machine was capable of applying programmed stress cycles simulating the mixture of high and low loads observed in service.[107] Random-amplitude fatigue machines were not built until the 1950s.[108]

In 1924 the Swede Arvid Palmgren (1890–1971) considered the endurance of ball bearings and proposed a linear cumulative damage law for fatigue.[109] This law considers the life of a specimen that is subjected to m different levels of alternating stress with n_i cycles at each level, where the number of cycles to failure at each level is N_i, predicts failure when

$$\sum_{i=1}^{m} \frac{n_i}{N_i} = 1. \tag{8.13}$$

This cumulative damage law was restated in 1945 by Miner and the cumulative damage law is usually known as the Palmgren-Miner law.[110] Fatigue damage depends upon the order in which the different alternating stresses are applied, especially in notched specimens or complex structures, hence the cumulative damage law does not always give a good estimate of fatigue life and in practice the right hand side of Eq. (8.13) can vary from about 0.2 to 20.

8.4.4 The effect of notches and size effect

In the nineteenth-century, Rankine had appreciated the effect of notches in the form of abrupt changes in section. Wöhler quantified notch effects in his early work, but they were not one of his major interests.[106] Gough, using theoretical values for the stress concentration factor obtained from photoelasticity found that the fatigue strength is not reduced as much as one might expect using a maximum stress criterion.[62] Rudolph Earl Peterson (1901–1982) at the Westinghouse Research Laboratories and August Thum (1881–1957) at the Technical University of Darmstadt contributed much to the effect of notches and size from the 1930s onwards. It was Peterson that introduced the notch sensitivity index q defined by

$$q = \frac{k_f - 1}{k_t - 1},\tag{8.14}$$

where k_f is fatigue stress concentration factor — ratio of the un-notched endurance stress to the notched average stress endurance, and k_t is the theoretical elastic stress concentration factor.[111,112] The notch sensitivity index, q, can vary from 0, when there is no notch sensitivity, to 1, where $k_f = k_t$. In grey cast iron the graphite flakes act like cracks and so dominate the initiation of fatigue so that Thum found the notch effect is small in this material.[113] Generally the higher the strength of the alloy, the greater is the notch sensitivity and the advantage of high static strength is lost when a complex component of structure is subject to fatigue. Gassner recognised this problem in 1941.[106]

In the fatigue testing of the first half of the twentieth-century there was little distinction between initiation and propagation of fatigue cracks. Usually the initiation phase occupies by far the largest portion of the total fatigue life of a plain un-notched sample. Since the initiation mechanism depends upon only a small volume of the material the fatigue life of plain specimens show little size effect, though there is considerable scatter in the lives. With notched specimens, especially where the stress concentration factor is high, the propagation phase of fatigue can be significant and dominate. The number of cycles to propagate a fatigue crack to a size that causes unstable fracture is very size dependent. Thus the notch sensitivity factor is size dependent. The effect of both strength and size on the notch sensitivity index can be seen in the results of Peterson and Wahl shown in Fig. 8.10.[114]

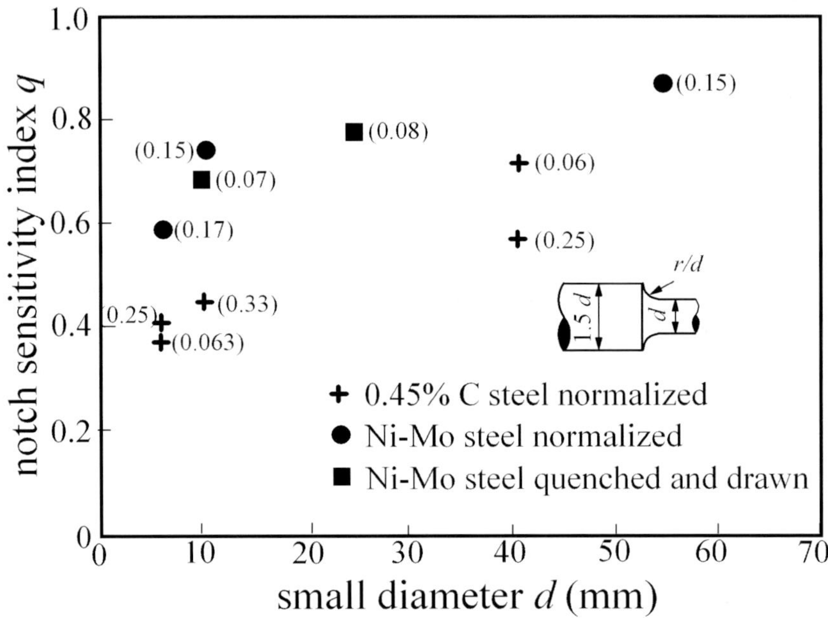

Fig. 8.10 Size effect in the notch sensitivity factor, r/d values shown in brackets (after Petersen and Wahl 1936).

8.4.5 *Component fatigue testing*

There are so many factors which determine the fatigue behaviour of a component or full-size structure that the prediction of the fatigue life from small scale laboratory tests is difficult. For this reason component or full-size fatigue testing has always existed to some extent complementing the more fundamental studies. The very first fatigue tests by Albert in 1837 were on the actual conveyor chains (see §7.4) and much of the early work continued to be at the component level. During World War I the Royal Aircraft Factory improved the fatigue life of air engine connecting rods by component testing. World War II saw the growth of aircraft component testing. In 1939 Gassner introduced block program fatigue tests that enabled more a realistic loading of aircraft components[115] with which Junkers performed a large number of fatigue tests on the wing spar of the Ju-88.[106] After World War II a vast number of surplus warplanes were a ready source of components for fatigue testing. In the late 1940s fatigue tests started on wings from the Mosquito and Mustang P-51 at the Aeronautical Research Laboratories[116] in Melbourne. Full size aircraft fatigue tests did not start until after the Comet disasters of 1954.

8.5 Concluding Remarks

The two major fracture problems of this period were the brittle fracture of steel and fatigue. By the end of the period enough was known about these two problems to reduce the frequency of their occurrence, but not to acceptable limits. Fracture and fatigue failures will never be completely eliminated, but their occurrence is being reduced. During this period it was new technologies rather than new materials that caused problems. Wrought iron practically disappeared and was almost completely replaced by mild steel. Aluminium was a new material but did not introduce major new fracture or fatigue problems during this period.

At the start of the period, the great elastician Augustus Love (1863–1940) could write 'the properties of rupture are but vaguely understood'.[117] The maximum stress criterion was still the main predictive fracture criterion. Many researchers saw that Griffith's theory marked the end of the dominance of the maximum stress theory for fracture, but it was not until the end of the period that it became clear how it could be applied to metals. However, the stress criterion remained the basis for discussion of fatigue, though it was realised that the effect of stress concentrations were not the same for all materials and there could be a size effect. The brittle fracture of steel was largely tackled by improving the detail design and by attempting to choose steels whose transition temperature was higher than the operating temperature. However, the necessary impact value to ensure ductile fracture was underestimated and is in any case really more of a guide to possible behaviour rather than a means of ensuring ductile behaviour. Henri Schnadt remarked in criticism that trying to measure the transition temperature from a Charpy test was like trying to weigh a person by measuring his height.[118]

8.6 Notes

[1] Tresca (1869).

[2] Saint-Venant (1871).

[3] Levy (1871).

[4] Hill (1950).

[5] Reinforced concrete was even considered for ship construction in the 1920s.

[6] Napoleon III had an aluminium dinner service which was used for the favoured few at his banquets while the nobility dinned from mere gold plates (Street 1962).

[7] In 1889 Hall, with the financial backing of Alfred Hunt formed the Pittsburg Reduction Company which in 1907 was renamed the Aluminum Company of America (Alcoa).

[8] Street (1962).

[9] In the USA this aircraft, which was used by the US Postal Department, was known as the JL-6.

[10] Charpy (1901).

[11] Izod did little to develop his discovery apart from telling a few friends and had the misfortune to have the machine known after him mispronounced. The machine is usually pronounced as Ī-zod whereas his name is pronounced Ĭzzod, Anon (1946).

[12] Anon (1946).

[13] Izod (1903).

[14] Siewert *et al.* (2000).

[15] Charpy (1906).

[16] Dewar and Hadfield (1904).

[17] Greaves and Jones (1925).

[18] Stanton and Batson (1920).

[19] Egon Orowan and Geoffrey Taylor independently developed the dislocation concept to explain metal plasticity.

[20] Orowan (1949).

[21] Mesnager (1906).

[22] Ludwik (1909).

[23] Ludwik and Scheu (1923).

[24] In the same year Prandtl (1923) showed rigid-plastic material that the maximum tensile stress for a notch under plane strain was 2.57 times the yield strength.

[25] Kuntze (1928; 1932) attempted to obtain a unique fracture stress/strain curve but the assumptions on which his method was based were untenable Orowan (1949).

[26] Ludwik (1927).

[27] Davidenkov (1936). Orowan (1949) rejected Davidenkov's ductile fracture curve on the grounds that it is not simply a function of strain.

[28] Orowan did not make clear in his paper what his contribution was and the subsequent literature has often been wrong in its attribution of the development of the theory. I am very grateful to Finnie and Mayville (1990) giving the historical development.

[29] Docherty (1932, 1935).

[30] Ellsberg (1925).

[31] Biggs (1960).

[32] Shank (1954).

[33] Boyd (1970).

[34] Titanium dioxide, quartz, kaolin, mica, and feldspar.

[35] Low hydrogen electrodes must be dried and protected from absorbing water by measures such as keeping them in a hot box until immediately prior to welding. Poorly handled low hydrogen electrodes can be as prone to weld cracking as rutile electrodes.

[36] In some cases a fracture can continue from one plate to another in a riveted structure, such as in the case of the fracture in the Gravesend Water Standpipe in 1886 discussed in §7.5.2 (Shank 1954).

[37] Tipper (1962).

[38] Comparisons of the incidence of brittle fracture in riveted and welded structures are difficult, but Hodgson and Boyd (1958) estimated that welded ships were 18% more likely to have brittle fractures. However, the fractures were much more extensive in welded ships.

[39] See Williams and Ellinger (1952), Shank (1954), Hodgson and Boyd (1958), and Boyd (1970).

[40] The carbon percentage that causes equivalent hardening; $CE = C + (Mn+Si)/6 + (Cr+Mo+V)/5 + (Cu+Ni)/15$.

[41] A common Charpy impact requirement at the design temperature is 27 J.

[42] Fujimoto and Shioya (2005).

[43] Wells (1973).

[44] Williams and Ellinger (1953).

[45] Boodberg *et al.* (1948).

[46] Wilson *et al.* (1948).

[47] Kahn and Imbembo (1948).

[48] Tipper (1948).

[49] See §9.2.4.

[50] Kirsch (1898).

[51] Wieghardt (1907).

[52] A short biography of Wieghardt is given by Rossmanith (1999).

[53] Rossmanith (1995).

[54] Rossmanith (1999).

[55] Irwin (1957).

[56] Erdogan (1962).

[57] Ritchie *et al.* (1973).

[58] Inglis (1913).

[59] The Royal Aircraft Factory became the Royal Aircraft Establishment (RAE) in 1918. The Farnborough site of the RAE is now occupied by the Defence Science and Technology Laboratory (DSTL), QinetiQ, the Air Accidents Investigation Branch and the British National Space Centre.

[60] Griffith and Taylor (1917).

[61] Prandtl (1903).

[62] Gough (1924).

[63] Griffith (1920).

[64] Griffith references Love (1906) for this theorem, but it is due to Claperyon see Timoshenko (1953).

[65] Griffith (1924).

[66] Sneddon (1946), Sneddon and Elliot (1946).

[67] Chadwick (2002).

[68] Spencer (1965), Cotterell (1997).

[69] Lawn (1993).

[70] Cotterell (1997).

[71] Modern studies have shown that crack tips in glass are atomically sharp (Guin and Wiederhorn 2005).

[72] Gordon (1968).

[73] Gurney (1994).

[74] Joffé *et al.* (1924).

[75] The fracture strength of the rock salt in air was independent of temperature over the temperature range -200 to $650°C$.

[76] Orowan (1933a).

[77] Ernsberger (1960).

[78] The strength did not vary inversely as the diameter to the power of 0.5 as would be expected if the flaws where proportional to the fibre diameter but to the power 0.73.

[79] Otto (1955).

[80] Holloway (1959); Holloway and Hastilow (1961).

[81] Polanyi (1921).

[82] Orowan (1934).

[83] Orowan (1945).

[84] Obreimoff (1930).

[85] This mistake was found by Orowan in 1933. All the values for the surface energies quoted here have been corrected to account for the error.

[86] Irwin (1948).

[87] Holloman and Zener (1946).

[88] We now know that conventional plasticity theories are inadequate to model the stress-strain behaviour at the micrometre and explain atomic decohesion, but they are adequate for the plastic zone at the tip of a crack as a whole (Xia and Hutchinson 1996).

[89] Merchant (1945a).

[90] Célarié (2003); Guin and Wiederhorn (2006).

[91] Weibull (1939).

[92] It should be noted that the weakest link concept does not apply to less homogeneous brittle materials such as mortar and concrete.

[93] A typical value of the Weibull index m is 10.

[94] Hunt and McCartney (1979).

[95] Daniels (1945).

[96] Campbell (1981).

[97] Pugsley (1966).

[98] Ewing and Rosenhain (1900).

[99] Ewing and Humfrey (1903).

[100] Cottrell and Hull (1957).
[101] Gough (1933).
[102] Hopkinson (1912a).
[103] Jenkin (1925), Jenkin and Lehmann (1929).
[104] Haigh (1917).
[105] Gough and Sopwith (1932).
[106] Schütz (1996), this history of fatigue gives details of German research that is neglected in other works.
[107] Gassner (1941).
[108] Freudenthal (1953), Head and Hooke (1956).
[109] Palmgren (1924).
[110] Miner (1945).
[111] Peterson (1933).
[112] Later generalized for finite fatigue lives.
[113] Thum and Ude (1930).
[114] Peterson and Wahl (1936).
[115] Gassner (1939).
[116] Now the Defence Science and Technology Organisation (DSTO).
[117] Love (1906).
[118] Unfortunately the test which he proposed was no better at measuring a person's weight it just used a different height gauge Schnadt (1944).

Chapter 9

Fundamentals of Fracture and Metal Fracture from 1950 to the Present

The second half of the twentieth-century saw fracture mechanics emerge as an engineering discipline and come to maturity. The development of linear elastic fracture mechanics (LEFM) in the 1950s was the first practical move away from a maximum stress fracture criterion in engineering.

At the beginning of the period the problem of the brittle fracture of steel had not been solved. The solution lay mainly in determining the effective transition temperature and in metallurgical improvements to produce steels of lower transition temperatures. However, for the assessment of the significance of defects needed elasto-plastic fracture mechanics (EPFM).

The growth of computers during this period enabled more realistic models of deformation and fracture to be implemented. At the end of the period computing power had grown enormously, enabling fracture at the atomic scale to be simulated using molecular dynamics which will be discussed in Chapter 12.

9.1 Linear Elastic Fracture Mechanics (LEFM)

George Rankin Irwin (1907–1998) was working on armour materials in the in the Ballistics Branch of the Mechanics Division of the Naval Research Laboratory (NRL) during War World II. Evidence of a fracture size effect led him to obtain a research contract for the study of fracture at the University of South Carolina from 1941 to 1948.[1] Irwin was joined at NRL in 1948 by Joe Kies (1906–1975). The initial fracture experiments of Irwin and Kies were on polymethyl methacrylate (PMMA) and cellulose acetate sheets loaded to different tensile stresses and fractured by firing a bullet into them, where the stress necessary to cause complete fracture was size dependent.[2] In 1952 Irwin and Kies used

232

Griffith's theory to estimate the brittle fracture strength of a steel plate containing a large crack.[3] Instead of using Griffith's equation directly they obtain the energy release rate from an empirical expression for the stiffness of a plate of finite width and length containing an elliptical hole.[4] From this work Irwin and Kies later developed the compliance method of calculating the energy release rate.[5] Meanwhile on the opposite side of the Atlantic, Alan Wells (1924–2005), a brilliant innovative researcher who had joined the British Welding Research Association (BWRA) at Abington in 1950, had the idea of directly measuring fracture energy using a method that did not rely on any mechanical analysis.[6] Geoffrey Taylor, together with Harold Quinney, had shown that almost all plastic work is converted into heat.[7] Brittle fractures propagate at high speed, almost instantaneously as compared with the conduction of heat, so Wells realised that the heat generated by a fracture would propagate as a plane wave away from the fracture surface and the heat generated could be measured from the maximum temperature recorded by a thermocouple positioned a short distance from the pupative fracture plane. Wells used an automobile spring steel as his test material and calculated that a fracture energy of 70 kJ/m^2 would cause a temperature wave of 1°C at a distance of 2.8 mm from the fracture surface, which would arrive 0.48 s after the fracture, making the test feasible. Specimens with different notch depths were tested and calculated that the average fracture energy of those with a 100% cleavage fracture was 49 kJ/m^2. Wells' measurement of the fracture energy for spring steel from the heat generated is the only direct measurement that has been made. This innovative work of Wells caught the attention of Irwin and led to a close relationship between them.

At about the same time Ronald Rivlin (1915–2005), who was working with the British Rubber Producers' Research Association[8] at Welwyn Garden City some 45 km south west of the BWRA, was applying the Griffith concept to the non-linear elastic fracture of rubber in collaboration with Alan Thomas.[9] Rivlin and Thomas used the two test geometries shown in Fig. 9.1: (a) simple extension, and (b) short tension.[10] Rivlin and Thomas showed that the fracture energy for these two geometries is given by

$$R = 2\lambda\frac{F}{h} - UW \quad \text{for simple extension specimen,}$$

$$R = 2UL \quad\quad\quad\quad \text{for short tension specimen,}$$

$$\tag{9.1}$$

where λ is the stretch ratio, F is the force, U is the strain energy density well away from the crack tip in the arms for the simple extension specimen and in the ligament for the short tension specimen, h is the thickness of the specimen, and W is the half width of the simple extension specimen. The value of the fracture energy obtained from these specimens for three different vulcanized rubbers varied from 2.3 to 6.9 kJ/m^2.

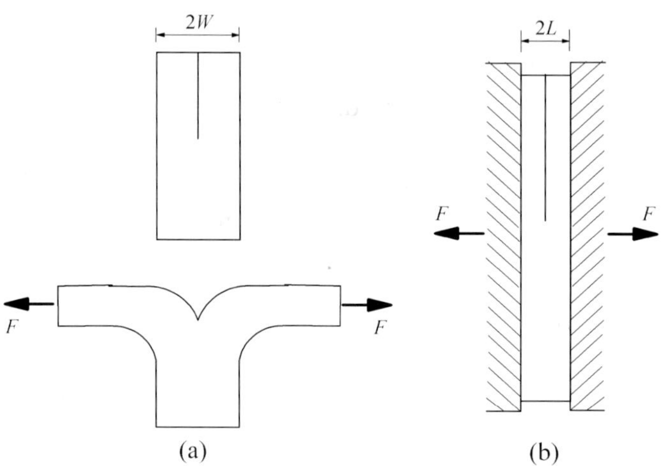

Fig. 9.1 (a) Simple extension specimen, (b) Short tension specimen (after Rivlin and Thomas 1953).

Back at the NRL Irwin's ideas on the energetics of linear elastic fracture mechanics (LEFM) had crystallised by 1955.[11] Irwin named the energy release rate the *crack extension force* and gave it the symbol G in honour of Griffith. Irwin probably had in mind an analogy to the force on a dislocation when he called the energy release rate the crack extension force. James Koehler (1914–2006), Nevill Mott (1905–1996), and Frank Nabarro (1916–2006) had introduced the useful concept of a force per unit length, F, acting on a dislocation from the rate of increase in potential energy, Π, when a dislocation is moved[12] or

$$F = -\frac{\partial \Pi}{dx}. \tag{9.2}$$

Similarly G can formally be represented as a force per unit length of crack in the direction of crack propagation. Irwin gave no specific description of the fracture energy in his 1956 paper simply calling it, G_c, the *critical crack extension force*. Frederick Forscher at Westinghouse Electric Corporation called the critical crack extension force *fracture toughness*.[13] However, since this term subsequently became used for the critical stress intensity factor, the term *fracture energy* is used in this book to avoid confusion though many, who favour the energy approach to fracture, still use the term fracture toughness. Irwin gave a table of the fracture energies for a variety of materials including the values for vulcanised rubber obtained by Rivlin.[11] It is interesting that Irwin gave estimates of the fracture energy for steel ship plate showing that Irwin originally saw the possible application of LEFM to the brittle fracture of steel.

Irwin made use of the toughness measurements of aluminium alloys to estimate the length of the fatigue crack growing from the automatic direction finding (ADF) window in the Comet airliners that caused the crashes in 1954. Using toughness data from a similar aluminium alloy, Irwin calculated that a crack 125 mm long would be unstable at the stresses in the skin of the pressurized cabin. From the analysis of a circular hole with radiating cracks by Oscar Bowie (1921–1995)[14], Irwin knew that the effective crack length for a crack from the ADF window was the size of the window plus any fatigue crack, once it was more than about one eighth of the size of the window and so only a short fatigue crack could have caused unstable fracture. The wreckage from Comet G-ALYP, which crashed half an hour out of Rome bound for London on the 8th January 1954, was reconstructed and the fatigue crack that caused the crash was 25 mm long. Wells, who worked at NRL with Daniel Post for a year in 1954 while his wide plate testing facility was being constructed, made similar estimates of the critical length of a fatigue crack in the Comet.[15] As a result of the analysis of the Comet disaster, Kies and Irwin discussed the application of fracture mechanics to fracture-safe design with West Coast aircraft companies. These discussions led to the adoption of materials with higher fracture toughness and crack-arrest features, which probably prevented some failures.

Kies had the idea that hot stretching PMMA, an important material for aircraft canopies but quite brittle, might increase its toughness.[16] The critical crack extension force, G_c, of the stretched PMMA could be measured using a precracked tensile specimen, but required the Young's modulus, E, for evaluation whose value was somewhat uncertain because of its sensitivity to strain rate. However the critical stress depended only upon EG_c, which could be calculated directly from the critical stress and crack length without knowing E and could be used to assess the toughening action of hot stretching. As a consequence of Kies' observation, the West Coast aeronautical engineers expressed their fracture tests in terms of $(EG)^{0.5}$ which they called K_c after Kies.[1] In his 1956 paper Irwin gave the expression for the stresses in the vicinity of a crack tip noting that $(EG/\pi)^{0.5}$

measures the intensity of the stress for a Griffith crack. In later work Irwin changed his definition of K the stress intensity factor so that the normal stress acting across the prolongation of a symmetrically loaded crack (mode I) is given by

$$\sigma_y = \frac{K}{\sqrt{2\pi x}},$$

(9.3)

where x is the distance from the crack tip. With this definition of the stress intensity factor

$$K = \sqrt{\bar{E}G},$$

(9.4)

but it was not known whether this expression was a general result. The proof of the generality of Eq. (9.4) was given by Irwin in 1957.[17]

During the first years of LEFM only cracks with local symmetry in the stresses at their tips (mode I), which is the natural growth for cracks in isotropic solids, were considered. However, Irwin realised that three archetypal stress distributions were possible at the tip of a crack, mode I, a pure opening mode, mode II, an in plane shear mode, and mode III, an out of plane shear mode as illustrated in Fig. 9.2. Irwin published the full crack tip stresses for these three modes in 1958.[18]

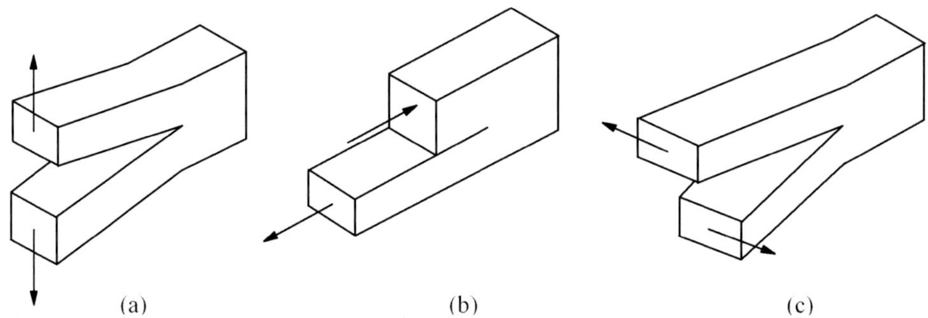

Fig. 9.2 The three archetypical modes of fracture (a) opening, mode I, (b) in plane shear, mode II, (c) out of plane shear, mode III.

Paul Paris observed a thickness effect in the critical stress intensity factor during the summer of 1955[19] and the plane strain critical crack extension force, G_{Ic} and plane strain critical stress intensity factor, K_{Ic}, which are the minimum toughness values, were first described in 1958.[20] The critical value of the stress intensity factor was not called the *fracture toughness* until the first report of the special ASTM committee in 1960.[21]

In 1952 Max Williams published a paper on the stress distribution near a sharp re-entrant corner with a finite included angle, which he used in 1957 to give the polar stress distribution near the tip of a sharp crack under mode I and II loadings.[22] This paper is important because it not only gives the singular terms, but also the higher order non-singular terms. It was later shown that the second term in the series expansion of the stress at the tip of crack, which does not depend upon the distance, r, from the crack tip, determines the crack path stability.[23] Williams also showed that for mode I loading that the maximum principal stress at a fixed distance r close to the crack tip is a maximum not along the prolongation of the crack but at an angle of 60° to the prolongation, where it is 30% higher than along the crack line.

Two major engineering problems of the time whose analyses were amenable to LEFM demonstrated the importance of the fracture work at NRL. In the 1950s General Electric introduced new two-pole generators that ran at 3,600 RPM, twice as fast as the previous four-pole generators. The diameters of the rotors were decreased to around 1 m, but that was not enough to compensate for the higher speed. Three of the rotors of these new generators burst in 1954–1956 during their commissioning trials. The steel of these rotors had a yield strength of around 700 MPa and they burst at tangential bore stresses of about half the yield strength. The rotors operated below the Charpy appearance transition temperature where the impact energy was from about 6 to 35 J. The first disk burst tests on sound and cracked material taken from the failed rotors were only 9.5 mm thick, which was far too thin to enable the plane strain conditions in the actual rotors to be simulated.[24] Boris Wundt, the secretary of the General Electric team investigating the problem of rotor failures, was impressed by the work of Irwin and, in collaboration with David Winnie, he applied Irwin's theory to the bursting of thick (50 and 150 mm) disks and notch bend tests on beams 50 x 50 mm.[25] At this early stage of LEFM the notch tips were machined and had a tip radius of 0.125 or 0.25 mm. The fracture energies for Ni-Mo-V steel, with a 50% crystalline Charpy transition temperature of 107°C, obtained from the burst disk tests had an average of 32.6 kJ/m^2 and those obtained from the notch bend tests had an average of 44.3 kJ/m^2. There were no minimum thickness requirements for plane strain fracture at the time, but the modern minimum thickness for valid plane strain fracture toughness measurement would have been 65 mm so, though the fracture energy values are probably high because the notch tips were machined, they are representative of plane strain conditions. Fracture energies were also calculated from burst disk tests on other steels with lower transition temperatures which had much larger fracture energies. This paper, which created great interest at the time, considerably helped the acceptance of Irwin's new approach to fracture.

The second of the engineering problems that needed fracture mechanics for their solution were the Polaris and Vanguard missile programs of the Cold War. These programmes were delayed by failures during hydrostatic tests of the high strength steel rocket chambers and the fracture mechanics group at NRL was asked for assistance. In 1959 the US Department of Defense asked the ASTM to organise a special committee to review testing methods for high strength sheet materials and to recommend a standard method of evaluating their resistance to brittle fracture. The reports of this committee from 1960–1964 were the beginning of the codification of LEFM. In 1965 the ASTM Committee E-24[26] on Fracture Testing of Metals was formed and their first standard: E399 Test Methods for Plane-Strain Fracture Toughness of Metallic Materials was published in 1970. It is interesting to note that the first application of LEFM in a structural code occurred in the Australian Timber Engineering Code AS CA65 of 1972 as a result of the work of Bob Leicester in the CSIRO Division of Building Research.[27]

In 1965 Takeo Yokobori founded the International Congress on Fracture and instigated the first International Conference on Fracture (ICF) held in Sendai, Japan. The ICF is held every four years and is the premier conference on fracture; ICF-1 attracted about 500 participants from 19 countries showing how the interest in fracture had blossomed since the end of World War II. Also in 1965 Max Williams founded the *International Journal of Fracture Mechanics* (now the *International Journal of Fracture*), the first journal devoted to fracture. Fracture mechanics had become a separate discipline.

9.1.1 *Fracture of high strength metals*

The first developments in LEFM were for application to high-strength metals. A plastic zone forms at the tip of a sharp notch or crack before fracture initiation and inside this zone the stress is not given by the elastic solution. However, Irwin realised that if the plastic zone was small enough, then the stress outside of it would still be approximately given by Eq. (9.3). With high strength thin sheet materials the stresses near the crack tip is in a state of plane stress. Irwin, assuming that the stresses outside of the plastic zone are given by Eq. (9.3), used a simple equilibrium argument to calculate the size of the plane stress plastic zone, d_0, as

$$d_0 = \frac{1}{\pi}\frac{K^2}{\sigma_Y^2},$$

(9.5)

and the effective length of a crack as $a + d_0/2$.[28] The plane strain size of the plastic zone, D_0, was estimated on the basis of the experimental values of the net section stress of circumferentially notched round bar specimens[29,30] and the accepted value has become

$$D_0 = \frac{d_0}{3} = \frac{1}{3\pi}\left(\frac{K}{\sigma_Y}\right)^2.$$

(9.6)

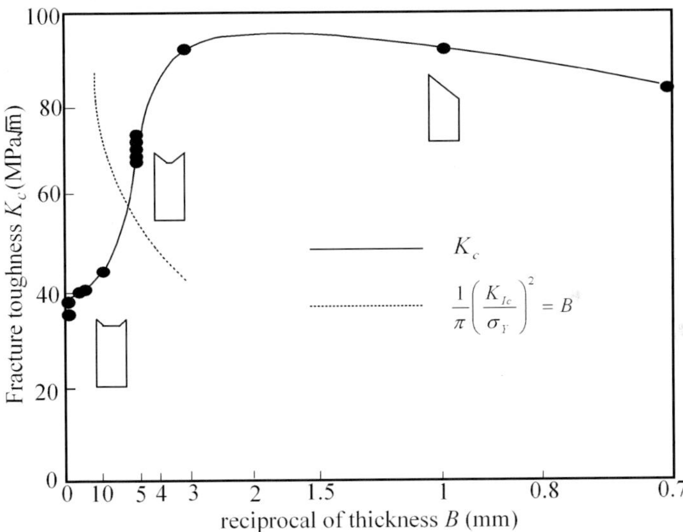

Fig. 9.3 The fracture toughness of aluminium 7075-T6 as a function of sheet thickness (after Irwin and Srawley 1962).

Although the effect of thickness was first noted in 1955, the data for the high-strength aluminium alloys 7075-T6 and 2024-T3 were not presented until 1958.[18] In Fig. 9.3 the fracture toughness of the aluminium alloy 7075-T6 is plotted against the reciprocal of the sheet thickness, B; the dotted line gives the locus of $d_0 = B$. When $d_0 < B$ the fracture surface is mainly transverse with small shear lips, if $d_0 \ll B$ the shear lips are very small compared with the thickness of the

                                          Fracture and Life

sheet and $K_c \rightarrow K_{Ic}$. With thin plates where $d_0 >> B$ there is a slant shear fracture surface and the fracture toughness decreases with decrease in the sheet thickness, because most of the fracture work per unit of crack growth is associated with the plastic work which is roughly proportional to B^2 and hence the fracture energy is proportional to B. This relationship between sheet thickness and fracture energy is the reason why gold, the most ductile of the metals, is nevertheless very brittle in the form of gold leaf which is only about 100 nm thick. Irwin introduced a new parameter β defined by

$$\beta = \frac{1}{B}\left(\frac{EG_c}{\sigma_Y^2}\right) = \frac{1}{B}\left(\frac{K_c}{\sigma_Y}\right)^2 , \qquad (9.7)$$

and stated that plane strain conditions prevailed when $\beta < 1.9$.[18] Later it was found that the plastic zone size needed to be smaller to ensure plane strain and the ASTM Standard of 1970 defined plane strain conditions as[31]

$$B \geq 2.5\left(\frac{K_c}{\sigma_Y}\right)^2 . \qquad (9.8)$$

For LEFM to be applicable the plastic zone size must be small compared with the dimensions of the specimen and the crack size, a, and the remaining ligament, l, are both required to satisfy the same inequality as the thickness.

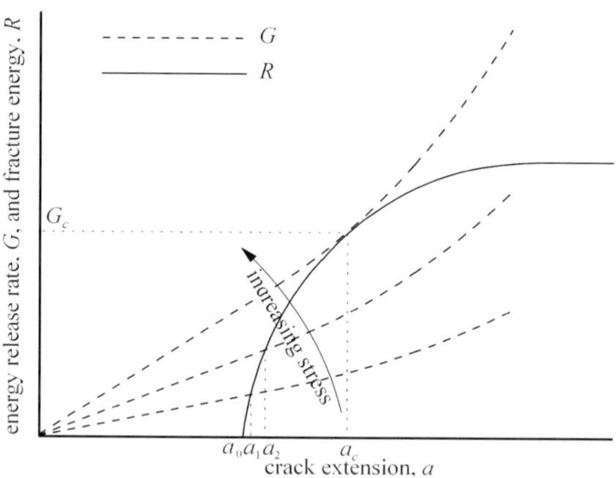

Fig. 9.4 The concept of crack growth resistance.

The initiation of a fracture from a straight notch occurs under more or less plane strain conditions and shear lips or slant fracture only develop with propagation. Thus the fracture toughness increases with crack propagation and the fracture does not become unstable until the crack has grown to a critical length.[32] This phenomenon was briefly mentioned in the first report of the special ASTM committee,[21] but was first described in detail by Joe Krafft, Manevelette Sullivan, and Bob Boyle in 1961.[33] Unless the thickness of the sheet is considerably greater than the plastic zone at the crack tip the fracture energy depends upon the sheet thickness and increases with crack extension. The fracture toughness also increases with crack extension giving rise to a K_R-curve. Thus there was a change in the concept of fracture instability, which is illustrated in Fig. 9.4. In 1954, the fracture energy R was assumed to be a constant and unstable fracture occurred when $G = R$, but in the 1959 concept unstable fracture occurred when

$$\frac{dG}{da} = \frac{dR}{da}.$$

(9.9)

9.1.2 The fracture process zone (FPZ)

A paradox of Griffith's theory is that the stress is infinite at the crack tip. Griffith realised this paradox and tried to answer it by arguing that the radius of curvature at the tip of the crack was of the same order as the molecular dimension, which implied a maximum stress of the order of 20 GPa supposing that Hooke's law applied at such high stresses, which Griffith doubted.[34] Grigory Barenblatt in the USSR addressed this paradox in 1959 and formulated the concept of a FPZ.[35] He assumed that cohesive forces[36] would act across the crack faces near the crack tip which would induce a negative stress intensity factor. The stresses at the crack tip can only be finite if the total stress intensity factor is zero and hence the total stress intensity factor, K, must satisfy

$$K = K_{app} + K_{coh} = 0,$$

(9.10)

where K_{coh} is the stress intensity factor due to the cohesive forces and K_{app} is the stress intensity factor due to the applied forces. In Barenblatt's concept there is no energy released during crack propagation because the cohesive forces, whose work of separation is the surface energy, are modelled explicitly. The relationship between K_{coh} and the surface energy, γ, for an idea elastic-brittle material is given by

$$\frac{K_{coh}^2}{E} = 2\gamma.$$

(9.11)

In Barenblatt's concept the FPZ is an infinitesimally thin line along the crack. Naturally in practice the FPZ must have some width, but on that part of a FPZ where the cohesive stress decreases with separation of the atoms, or *strain softens*, the width of the zone must collapse. It is only the discreteness of the microstructure that prevents it collapsing into a line. Just as the concept of surface energy has been generalised to account for plastic deformation so the concept of a FPZ has been extended to the non-elastic region at the tip of a crack.

The FPZ can be modelled as a line provided there is strain softening. In thin ductile metal sheets deforming under essentially plane stress conditions the sheet will thin ahead of the crack and effectively strain soften. Donald Dugdale[37] in Britain and Volodymyr Panasyuk[38] in the Ukraine, independently introduced strip yield models for crack tip plasticity in 1960. In the strip yield model the plastic zone is replaced by a fictitious extension of the crack across which act the yield stress, σ_Y, holding the faces of the crack together. Using the condition that the effective stress intensity factor at the tip of the fictitious crack is zero, it was shown that the length of the plastic zone, d_0, for a Griffith crack loaded by a remote stress σ, is given by

$$\frac{d_0}{a} = \sec\left(\frac{\pi\sigma}{2\sigma_Y}\right) - 1. \tag{9.12}$$

For small scale yielding Eq. (9.12) becomes

$$\frac{d_0}{a} \approx \frac{\pi^2}{8}\left(\frac{\sigma}{\sigma_Y}\right)^2 = 1.23\left(\frac{\sigma}{\sigma_Y}\right)^2. \tag{9.13}$$

Eq. (9.13) is very similar to Eq. (9.5) derived by Irwin. In 1964, Hahn and Rosenfield showed that Dugdale model was a good approximation for the local yielding at the crack tip in thin metal sheets.[39]

The crack tip blunts before fracture causing an opening displacement at the crack tip. Dugdale did not calculate the crack tip opening displacement (CTOD), but Panasyuk did. The first exact expression for the CTOD, δ, of a Griffith crack with a FPZ

$$\delta = \frac{8\sigma_Y a}{\pi E}\ln\left[\sec\left(\frac{\pi\sigma}{2\sigma_Y}\right)\right]$$

$$\approx \frac{1}{\sigma_Y}\left(\frac{\sigma^2\pi a}{E}\right) = \frac{G}{\sigma_Y}\ \text{ for } \sigma \ll \sigma_Y, \tag{9.14}$$

was given in the West in 1962.[40,41] Panasyuk was the first to recognise and state that the fracture energy, R, was related to the yield strength, σ_Y, and the critical CTOD, δ_c, by

$$R = \sigma_Y \delta_c,$$ (9.15)

This expression is exact for infinitesimally small FPZs or geometries that allow steady state crack propagation.[42] Alan Wells, using the Irwin plasticity model, gave an approximate expression for the CTOD for small scale yielding in 1961, which is similar to the approximation given in Eq. (9.14).[43]

Alan Cottrell also saw the need for the analysis of a crack with plastic zones in the early 1960s. Being unaware of either Dugdale's or Panasyuk's work he saw that the spread of yielding from a crack could be modelled by a distribution of dislocations. He approached his mathematician friend Bruce Bilby, to collaborate on the analysis, which was completed with the help of Swinden, one of Bilby's students.[44]

9.1.3 Crack paths in low velocity elastic fractures

Cracks do not always do what is wanted. Crystals have cleavage planes, usually the close-packed planes, where the surface energy is low. Diamond can be cleaved on octahedral planes but it requires great skill to identify the cleavage planes and avoid ruining the stone. King Edward VII wanted the famous 3,106 carat Cullinan Diamond cut into a number of smaller stones and Joseph Asscher of Amsterdam took two months to study the diamond before he placed his cleaving knife on the diamond and struck his blow. Asscher cleaved the stone perfectly, but the tension caused him to fall to the floor in a faint.

The wedging experiments of Obreimoff to measure the surface energy of mica were possible because mica has a silicate sheet crystalline structure where the bonding between the sheets is by weak van der Waals forces and the surface energy is small.[45] In 1957, John Benbow and Frank Roesler wanted to measure the fracture energy of the isotropic polymers PMMA and polystyrene using a double cantilever beam (DCB) specimen similar to that of Obreimoff which has the advantage over the Griffith central crack loaded in tension that, under displacement control, crack propagation is stable.[46] They found that while the crack propagation is stable, the crack path, instead of following the line of symmetry, was unstable and veered off to one side. However, the crack could be made to follow the line of symmetry if a sufficiently high compression force was applied parallel to the crack. Later it was found that side-grooves along the line of the crack could also be used to control the crack path.[47]

In an ideal elastic-brittle isotropic solid a crack would take the path that released the maximum energy which implies a pure mode I stress field,[48] called the condition of local symmetry by Robert Gol'dstein and Rafael Salganik,[49] is maintained at the crack tip. However, no material is perfectly isotropic, and the crack path will deviate from the ideal. Whether such deviations are self-correcting so that the crack path is stable and on average the ideal path is followed, or the crack continues to deviate from the ideal path so that the crack path is at least locally unstable depends upon the second term in the Williams' series expansion[22] for the stress at the tip of a crack as shown by Jim Rice and me.[23,50] The first two terms in the series expansion for the stress acting along the tip of a mode I crack can be written as

$$\sigma_x = \frac{K}{\sqrt{2\pi x}} + T. \tag{9.16}$$

The crack path after a small deviation from the ideal mode I crack path in an isotropic elastic material is shown schematically in Fig. 9.5. If the T-stress is compressive the propagation of the crack is stable and after a small deviation the crack grows so that it returns towards to the ideal path. However, if the T-stress is tensile then after a deviation from the ideal crack path, further crack growth increases the deviation. In the case of the classic central Griffith crack under a simple tension stress, σ, $T = -\sigma$ and the crack path is predictable and almost exactly at right angles to the applied stress. However, in the DCB specimen, as used by Benbow and Roesler,[46] $T > 0$ and the crack path is unstable unless a sufficiently large compressive force is applied parallel to the crack line to make the T-stress compressive, or sufficiently deep grooves are provided along the crack path.

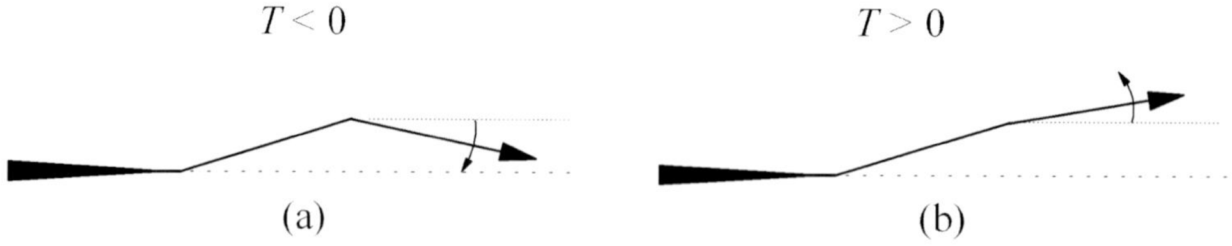

Fig. 9.5 Crack path after a small deviation from the mode I crack direction, (a) compressive T-stress, (b) tensile T-stress.

This simple condition for the local stability of the crack path cannot determine whether the crack path is globally stable or not.[51,52] For example, once a crack has deviated away from its ideal path in the DCB specimen isochromatic photoelastic fringes near the crack tip show that $T < 0$ and the crack path is stable and predictable.[48] On the other hand, a central crack under a remote biaxial normal system is stable globally even when initially T is slightly positive provided $T < K(1 - \pi/4)/\sqrt{\pi a}$.[51]

9.1.4 Dynamic crack propagation

Just before World War II, Adolf Smekal (1895–1959) at the Martin Luther University in Halle and Hubert Schardin (1902–1965) at the University of Freiberg were measuring the fracture velocity of glass. Schardin developed the multiple-spark high speed camera with frame rates of around 100 per second that could catch the extremely high velocities of cracks. In 1937, Schardin and Struth found that the maximum velocity of cracks in glass is a constant which depends upon the type of glass.[53] The maximum crack velocity is 1,500 m/s in soda lime glass which is about 0.43 times the velocity of shear waves. Shortly after the publication of the results of Schardin and Struth, Frederick Bastow and Harold Edgerton at the Massachusetts Institute of Technology measured almost the same fracture velocity for glass using a photograph taken with two successive sparks.[54] Hermult Wallner, an assistant of Smekal, published his description of the patterns produced on the surface of glass by the interaction of shear waves reflected from the edges of the specimen with the crack front, which are now known as Wallner lines, just before World War II.[55] Although their significance was not fully appreciated at the time, Wallner lines had been discovered by Charles de Fréminville (1856–1936) in 1907.[56] An obsidian blade flake with Wallner lines on its surface is shown in Fig. 4.20. Wallner lines are usually only visible in materials that have glassy smooth fracture surfaces such as glass, glass ceramics, glassy polymers such as epoxy, diamond, and tungsten. As the fracture surface of all materials roughen when approaching the maximum velocity, Wallner lines cannot be used to measure the maximum crack velocity. There are various methods of calculating the crack velocity from Wallner lines which have been summarised by Smekal.[57] Only the simplest method is given here.

The mechanism of the formation of Wallner lines on the fracture surface of a plate is illustrated in Fig. 9.6 The shear waves travelling at a velocity, c_2 are generated at O on the edge of specimen and travel faster than the crack, whose velocity is v_f, and intersect the crack front at D. The disturbance to the stress field by the interaction of the stress waves and the crack front causes a slight ripple in the fracture surface leaving a Wallner line. When the shear wave meets the opposite edge of the flake at C' it is again reflected to form a complementary Wallner line. In this way a series of intersecting lines are left on the surface. During a small interval of time dt the crack front advances from AA' to BB'; thus $DE = v_f dt$. The shear wave advances from aa' to bb' in the same time and $DG = c_2 dt$. From the geometry of the triangles DFG and DEG

$$v_f = \frac{c_2 \sin \theta}{\cos \beta}. \tag{9.17}$$

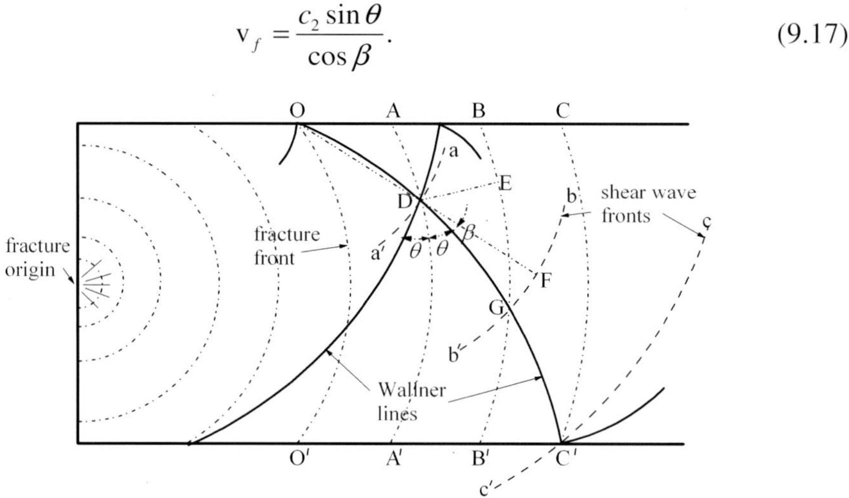

Fig. 9.6 Schematic development of Wallner lines on a plate.

In 1953, Frank Kerkhof at the University of Freiburg, developed a method of modulating the fracture with shear waves generated from an ultrasonic transducer which interact with the crack front to produce controlled ripples.[58]

The fracture surface in amorphous brittle materials such as glass or PMMA is optically smooth, with a mirror-like appearance at low velocity and it is in this phase of crack propagation that Wallner lines can be seen. In glass the fracture surface passes through four phases as the crack velocity increases: a smooth mirror phase at low velocities, a mist-like region[59], followed quickly by a very rough hackle phase where there are macro attempts at crack branching, and possibly finally branching of the fracture into two at a critical crack velocity. Branching can occur many times as many boys have found when they have hit a cricket ball though a window. Unlike the crack path instability at low crack velocity, at high crack velocities the crack does not simply deviate but always branches. The development of the fracture zones on the fracture surface of the stem of a wine glass that I accidentally broke are shown in Fig. 9.7.

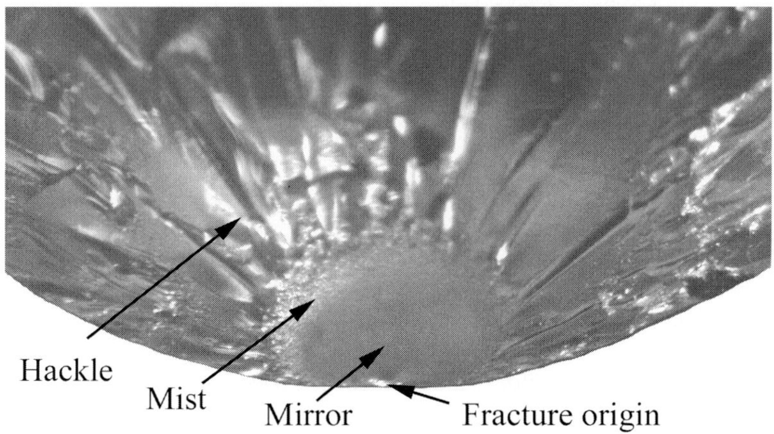

Fig. 9.7 Fracture surface of the 8 mm diameter stem of a wine glass.

Similar fracture markings caused by high crack velocity occur in other amorphous isotropic brittle materials such as glassy polymers and were first described by Joe Kies, Manevelette Sullivan and George Irwin.[60] In glassy polymers, crazes develop at a crack tip where the polymer chains become orientated. In PMMA at low crack velocities the fracture surface is glassy smooth and exhibits a brilliant rainbow effect because the thickness of the craze material, 0.5 μm, is of the same order as the wavelength of light and the craze material has a lower refractive index.[61] At crack velocities above about 150 m/s parabolic markings appear in the fracture surface caused by microcracks initiating within the craze layer ahead of the main crack front and leaving a slight step where the two cracks converge as can be seen in Fig. 9.8.[60,62,63] Above crack velocities of about 340 m/s there are micro attempts at branching with microcracks visible below the fracture surface.[63] At crack velocities above about 500 m/s the fracture surface becomes very rough with subsurface periodic markings, which have been described as ribs, which are spaced at about 1 mm.[62,63] This velocity phase is akin to hackle in glass. The limiting crack velocity for PMMA, when the crack divides into two distinct branches, is about 670 m/s ($\sim 0.6\ c_2$).[62] Cracks can be prevented from branching by scoring grooves in the surface to guide the fracture and can then propagate at higher velocities.[62]

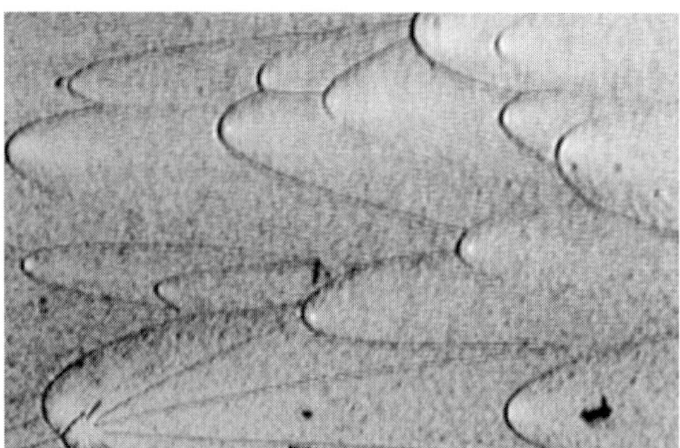

Fig. 9.8 Parabolic markings on a fracture surface in PMMA, the direction of fracture propagation is from left to right (Fineberg and Mander 1999 with permission Elsevier).

Cleavage fractures in steel also become rougher as the crack velocity increases before the fracture finally branches, but the different phases are not distinct. Brittle fracture surfaces are relatively smooth and featureless at low crack velocities, but as the crack velocity increases the fracture surface roughens and chevron markings normal to the parabolic tunnelling fracture front appear that are very characteristic of cleavage fracture in steel and always point back towards the site of initiation. The fracture surface in the high velocity region of a cleavage fracture in a 1.8 m wide rimmed steel, 19 mm thick, plate tested at 26°C by Robert Lazar and William Hall in the huge 13 MN testing machine at the University of Illinois in 1959 is shown in Fig. 9.9.[64]

Fig. 9.9 High velocity region of a cleavage fracture in 19 mm thick rimmed steel (Hall & Barton 1963, courtesy the Ship Structure Committee).

9.1.4.1 Analysis of dynamic fracture

In unstable fracture from a classic Griffith crack of length $2a_0$ in a large plate specimen under tensile stress, the elastic energy released increases with the crack length so that during crack propagation there is an excess of energy available over that needed to create the fracture. As a consequence the crack accelerates and the excess energy becomes the kinetic energy of the plate as it moves away from the crack path. Mott was the first to discuss the kinetics of crack propagation at a conference on the brittle fracture of mild steel plates in 1945.[65] He showed on dimensional grounds that if the fracture energy is independent of the crack velocity, the crack velocity v_f must be given by

$$v_f = kc_0 \sqrt{\left(1 - \frac{a_0}{a}\right)}, \qquad (9.18)$$

where a is the current half crack length, $c_0 = \sqrt{(E/\rho)}$ is the velocity of longitudinal waves in a prismatic bar, ρ is the density, and k is a constant. Hence Eq. (9.18) indicated that there should be a limiting velocity to crack propagation.

The first exact dynamic solution to a propagating crack was made by Elizabeth Yoffe, an Australian who moved to Cambridge in 1946.[66] Orowan suggested to Yoffe that she extend the work of Jock Eshelby (1916–1981) on the propagation of dislocations[67] to cracks. Thus Yoffe in 1951 solved the problem of a crack of constant length traversing a plate by opening up at one end and closing at the other.[68] Despite the unrealistic constant stress intensity factor at the crack tip, the polar variation in the stress at the tip of a crack obtained by Yoffe gave the unique velocity dependent distribution for a mode I propagating crack. The main differences between the dynamic and the static stress distributions at the tip of a crack are: in the dynamic case the ratio of the stress, σ_y, acting across the crack direction to the stress, σ_x, acting along it decreases with crack velocity becoming zero at the velocity of Rayleigh waves, and above a critical crack velocity the polar stress, σ_θ, is a maximum away from the crack line. Yoffe suggested that crack branching, which occurs at a definite fraction of the velocity of Raleigh waves, was due to the maximum polar stress, σ_θ, shifting from the crack line. This explanation of branching was accepted for many years but it has subsequently been shown to be incorrect.

In 1960 Bertram Broberg analysed the more realistic problem of a Griffith crack growing at both crack tips under a uniform stress.[69] The variation of the stress with angle from the crack line is the same as that obtained by Yoffe, but the energy release rate now becomes zero at the velocity of Raleigh waves (c_R). The energy release rate can be written as

$$G = G_s F\left(v/c_R\right), \qquad (9.19)$$

where G_s is the static energy release rate and the function, F, depends weakly on the Poisson's ratio as well as the normalised crack velocity. The function, F, is shown in Fig. 9.10 for a Poisson's ratio of 0.35. In 1970 Eshelby recognising that

                         Fracture and Life

the function, *F*, is almost a linear function of the crack velocity postulated that the critical crack velocity for branching must be about $0.5c_R$ since after branching twice the fracture energy is required.[70]

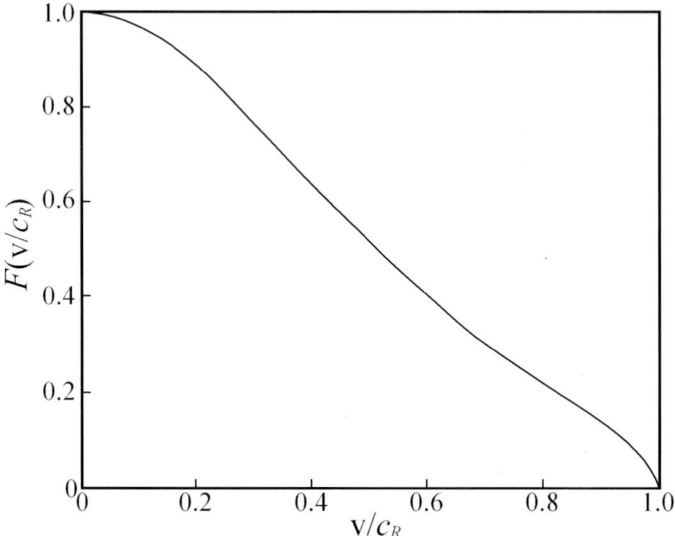

Fig. 9.10 Variation in the energy release rate with crack velocity for a Griffith crack ($\nu = 0.3$).

Only at low velocities is the dynamic fracture energy a predictable function of crack velocity, causing a crack to accelerate smoothly. At higher velocities where the fracture surface is very rough the crack velocity fluctuates considerably. Jay Fineburg and his co-researchers have developed an accurate method of measuring the crack velocity in glassy polymers by using the resistance of a thin layer of aluminium evaporated on the surface which enables the crack velocity to be resolved to 10 m/s.[63] Cracks in PMMA accelerate more or less smoothly up to crack velocities of about 340 m/s, but at this velocity microscopic branches form beneath the fracture surface and the crack velocity oscillates violently. The crack velocity decreases sharply when the micro branches form and more energy is needed for the larger fracture surface area. When the micro branches arrest and the fracture once more becomes a single crack, there is a sudden increase in velocity. The fracture energy derived from the energy release rate given by Eq. (9.19) for the peak velocities, when there is a single crack, it is almost constant at about 90% of its value at the critical velocity of 340 m/s.

The main features of dynamic fracture are very similar in all brittle materials. At low velocities the surface is smooth becomes rougher with increasing velocity and finally branches into two separate cracks which, if the specimen is large enough, can themselves branch again. The prime reason for this similitude is that all materials behave essentially elastically at high strain rates. However, the branching velocity is material dependent and, as Krishnaswamy Ravi-Chandar points out, this material dependence seems to rule out a purely mechanically based theory of branching.[71] It has been suggested that in glassy polymers the FPZ, which takes the form of a craze, has a significant control over the phenomenon of branching.[71,72] Certainly the craze at the tip of a crack, which in glassy polymers can be relatively large, has an effect on fracture, but in inorganic glasses the crack tip has atomic sharpness and the FPZ is extremely small hence, though a FPZ may explain why the critical branching velocity is not a constant fraction of the velocity of Raleigh waves, it cannot explain why branching occurs. The energy balance explains how branching is possible, but not why it occurs. A complete explanation of the features of dynamic fracture propagation is yet to come.

9.2 The Brittle Fracture of Steel

The advent of nuclear power in the 1950s in an era when the incidence of brittle fractures of steel structures, though much less than before 1950 was still appreciable, presented a challenge to engineers and made the understanding and control of brittle fracture imperative. It is testimony to the expertise of men such as Roy Nichols (1923–1999) who had a long and distinguished career in the UK Atomic Energy Authority, that there has been no serious fracture in any nuclear power plant despite there being over 10,000 reactor years of operation worldwide. The two notorious major nuclear accidents at Three Mile Island, USA in 1979 and Chernobyl, Ukraine in 1986 were not primarily due to fracture. Nuclear pressure vessels are constructed from very thick steel which suffers damage from neutron radiation in the belt-line that section of the vessel wall closest to the reactor fuel. The primary damage is lattice defects in the form of vacancies. Self-interstitials diffuse causing solute clusters and distinct phases, which pin dislocations and cause hardening which can increase the brittle-ductile transition temperature by more than 100°C.[73] Thus allowance must be made for the ageing of nuclear pressure vessels. To check the ageing process test coupons are located on the inside of the reactor pressure vessel, where the neutron flux is several times higher than in the vessel itself, which can be retrieved for periodic fracture tests.

At the beginning of this period advances were made in understanding the mechanisms whereby cleavage fractures initiated and propagated. Engineers still largely saw the way to avoid brittle fracture was simply to ensure that the brittle/ductile transition temperature was high enough to prevent cleavage fracture, but it was appreciated that the transition behaviour of a structure could not be assessed accurately from a small-scale test such as the Charpy test. However, gradually fracture mechanics was seen to have a role to play in avoiding brittle fracture.

9.2.1 *Theory of cleavage initiation and propagation*

In the late 1940s Norman Petch (1917–1992) and a New Zealander, Eric Hall working in parallel groups in the Cavendish Laboratories in Cambridge on the yield and fracture of steel achieved results that greatly influenced the development of strong tough steels. Together they showed that the lower yield strength of steel, σ_Y, depended upon the grain size through the relationship

$$\sigma_Y = \sigma_i + k_Y d^{-\frac{1}{2}}, \tag{9.20}$$

where σ_i and k_Y are constants. The papers containing the Hall–Petch relationship were not published until the early 1950s.[74] Petch's experiments were made at the temperature of liquid nitrogen (77°K), where yielding was followed by cleavage fracture and he found that fracture followed an equation of the same form. Orowan had deduced a similar relationship, without the constant term, for fracture from Griffith's theory by assuming that the crack size was equal to the grain diameter.[75] Alan Stroh (1926–1962) showed that transition temperature increases with grain size, because the cleavage strength is almost independent of the temperature and is given by[76]

$$\frac{1}{T} = A \log d + B, \tag{9.21}$$

where T is the absolute transition temperature.

The idea that cleavage microcracks form at the tip of a slip band blocked at a grain boundary by a pile-up of dislocations was proposed by many in the 1950s, but Cottrell pointed out that this idea posed a problem.[77] Brittle fracture depends strongly on the hydrostatic stress, yet if the nucleation was due to a pile-up of dislocations on a slip band it would be independent of the hydrostatic stress. This fact led Alan Cottrell to realise that the propagation of a microcrack was more difficult than its initiation. There are three distinct stages to brittle fracture: the nucleation of slip bands at the yield stress, the nucleation of microcracks, and the propagation of the cracks. Cottrell proposed an easier mechanism of crack nucleation than slip blocked at a grain boundary. A microcrack can be formed in body centred cubic (bcc) iron on a cleavage plane (001) by the intersecting slip planes on (101) and $(10\bar{1})$ which form a new dislocation with a release in energy as is shown schematically in Fig. 9.11. The corresponding dislocation reaction for the fcc lattice does not release energy and so cleavage is unlikely to develop. Cottrell calculated the condition necessary for a crack to grow as

$$\sigma_Y k_Y d^{1/2} > \beta\mu\gamma \tag{9.22}$$

where γ is the surface energy and $\beta \approx 1$ for tension and $\beta \approx 1/3$ under high constraint at the tip of a notch.[77] The constant k_Y decreases rapidly with temperature and, though microcracks may be initiated at higher temperature, they cannot grow. Cottrell's insight into fracture showed that many effects of alloying and metallurgical treatment could be rationalised. The grain size can be reduced in steel by finishing rolling at a low temperature thus reducing brittleness. Aluminium also refines the grain size. Manganese reduces both the grain size and k_Y and is thus doubly beneficial

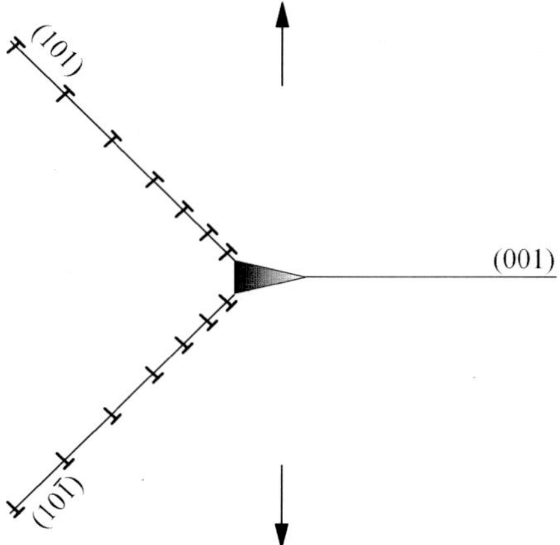

Fig. 9.11 Coalescence of two slip bands to form a cleavage crack (after Cottrell 1958).

Crystalline materials can be intrinsically ductile or brittle. Face-centred cubic (fcc) metals such as aluminium always fail by ductile tearing, while diamond and common salt are intrinsically brittle and fail by cleavage. Many crystalline materials, such as the bcc metals like iron, intermetallics like magnesia and other materials such as silicon and sapphire, exhibit a transition from brittle to ductile with temperature. It is unfortunate that steel has such an abrupt change in behaviour that was not properly understood until the middle of the twentieth-century. A milestone in the understanding of why some crystals are intrinsically ductile and others intrinsically brittle was the 1967 paper by Anthony Kelly, Bill Tyson and Alan Cottrell.[78] They considered a cleavage crack in a crystal and calculated whether it would blunt by shear or propagate as a sharp cleavage crack. The criterion that they deduced was that a crystal was intrinsically ductile if the ratio of the theoretical cohesive strength to shear strength was large (greater than about ten) as it is for the face-centred metals. Kelly and his colleagues realised that their criterion was a simplification of the real situation, but it did separate most crystals into the two camps. Jim Rice and Robb Thomson suggested that it would be better to examine whether a spontaneous emission of dislocations from an atomically sharp crack could occur.[79] A shear stress equal to the theoretical value can only cause atoms to shear past each other if it is constant along the whole shear plane, but near a crack tip the shear stresses are not constant. Rice and Thomson showed that if $\mu b/\gamma$, where b is the Burgers vector, is greater than 7.5 to 10 then a sharp crack will remain sharp. Ali Argon has reviewed brittle-ductile transitions.[80] There are two types of brittle-ductile transformations those in bcc transition metals such as iron with high dislocation mobility and most other intrinsically brittle crystalline solids which have sluggish dislocation mobility such as silicon. In the bcc metals there is a single energy barrier to the motion of a dislocation from one lattice to the next and the dislocations can readily multiple to produce large plastic strains. In materials such as silicon, the ductile transition is governed by the mobility of the dislocations away from the crack tip. Mechanistic models agree with experimental data on the transitions. Molecular dynamics has also been used to model the transitions.[81]

9.2.2 Propagation tests

The philosophy in these tests is to ensure that the service temperature is high enough so that even if a crack is initiated it cannot propagate. In the 1950s William Hall and his colleagues at the University of Illinois carried on the earlier research of Wilson[82] on 1.8 m wide, 19 mm thick, steel plates.[83] In their tests the plates had small edge notches and fracture was initiated by wedge impact on one of the notches while the plates were loaded to stresses representative of service conditions. The effect of residual stresses on fracture propagation was studied by welding up tapered edge slots cut above and below the fracture plane. Brittle fractures were successfully initiated by the wedge impact in such plates even if there was no applied external load.

Since wide plate tests were expensive, small-scale tests were devised to determine the temperature at which a crack initiated in a brittle weld could not propagate. In the 1950s William Pellini (1917–1987) and his colleagues developed the explosive bulge and drop weight tests at the NRL in Washington in the 1950s, which were designed to determine the lowest temperature at which a crack initiated in a brittle weld could not propagate.[84,85] In the explosion bulge test a plate 14 in. (356 mm) square was supported on a heavy die with a 12 in. (305 mm) opening and an explosive wafer was detonated at a fixed stand-off distance from the plate. A short brittle weld bead was deposited across the centre of the plate and a notch ground in it down to the plate's surface to provide easy crack initiation. Above the transition temperature a limited amount of dishing occurs and a crack does not propagate into the plate. As the temperature is reduced a critical temperature is reached at which cracks do propagate into the plate, but do not run to its edges. At lower temperatures the cracks run right to the edge of the plate. The NRL drop weight test, also known as the Pellini test, was similar to the explosion test but easier to control than the explosion test. A rectangular plate has a brittle weld bead laid at the centre of the plate parallel to the long axis of the plate and again has a notch ground in it at the centre down to the plate's surface. The plate is supported near the two short edges and impacted at the centre by a falling weight. A stop at the centre limits the deflection to a 5° bend angle. The brittle weld cracks when the angle of bend is about 3°, but above the transition temperature fails to propagate through the plate. As the temperature is reduced the nil ductility transition temperature (NDTT) is reached at which a crack will propagate right across the plate. The nil ductility test has retained popularity especially in the USA where it is an ASTM standard.[86]

9.2.3 Crack arrest tests

The philosophy in crack arrest tests is to ensure that the toughness of the steel is high enough to arrest a propagating crack. In the early crack arrest tests it was considered that a crack arrested if the stress was below a critical value depending upon the temperature. However, there are two different arrest regimes. At temperatures above the transition range from cleavage to ductile behaviour the toughness is high enough to ensure that a crack is arrested almost regardless of its length for applied stresses less than the yield. Cottrell suggested that a crack arrested when at least one third of the grains failed by ductile tearing.[87] In this case, since the plastic constraint at the tip of a crack causes an effective three-fold increase in the yield strength, the load carried by the ductile ligaments exceeds the yield strength of the unconstrained regions. Below the transition range arrest depends upon the stress intensity factor.

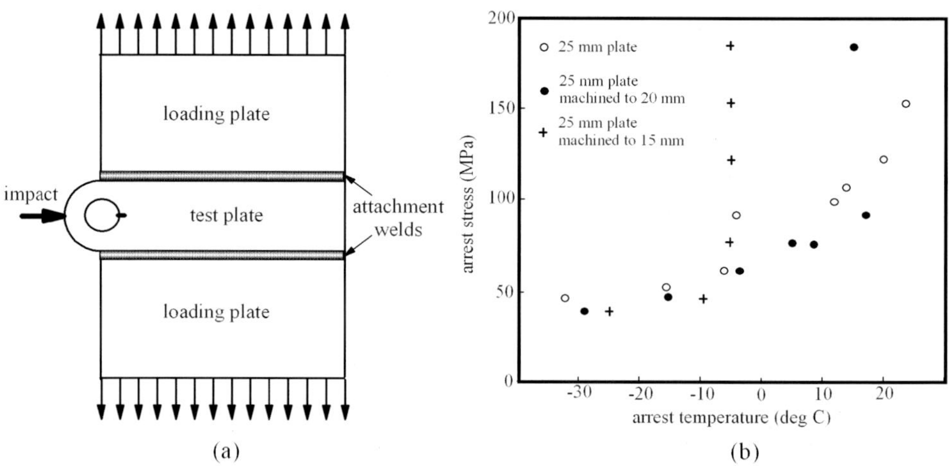

Fig. 9.12 The Robertson crack arrest test, (a) specimen. (b) arrest stress as a function of arrest temperature (after Robinson 1953).

Robertson, working at the Naval Construction Research Establishment, Dunfermline, Scotland devised the first crack arrest test in 1951.[88] The Robertson crack arrest specimen consisted of a test coupon shown in Fig. 9.12 (a) which was welded to loading plates typically thinner than the test plate and designed to yield to provide more or less a constant stress across the plate's width. An initiating saw cut was made on the inner surface of a hole cut into a protrusion at the end of the coupon. This protrusion was cooled with liquid nitrogen while the opposite end of the coupon was heated. Thus a temperature gradient was set up across the coupon. The specimen was loaded to a given stress level in a simple, purpose-built, hydraulic testing machine and a crack initiated by firing a bolt gun at the protrusion. The temperature at the arrest point was taken as the crack arrest temperature for that stress.[89] Crack arrest data is shown in Fig. 9.12 (b) for a low carbon steel plate both of full plate thickness and machined down to reduced thickness. The stress at which a crack is arrested decreases with temperature as the fracture transforms from ductile to cleavage. Thinner plates show a lower temperature and sharper drop in stress because the development of shear as the fracture becomes ductile is more intense. The crack arrest temperature became to be defined by the temperature at which shear lips first appear on the surface.[90] In the 1950s many Robertson crack arrest machines were constructed in the UK and a number of similar crack arrest tests were introduced such as the Esso test in the USA[91] and the double tension test in Japan[92] were introduced.

Irwin and Wells suggested in 1965 that LEFM could be applied to crack arrest.[93] At first it was assumed that the arrest stress intensity factor, K_{Ia}, could be calculated statically without considering the kinetics. This was the view taken by Ed Ripling and his co-workers at the Materials Research Laboratory, Glenwood, Illinois.[94] The opposing view, taken by George Hahn and his colleagues at Battelle, Columbus, Ohio, was that crack arrest was the termination of the propagation phase and that all the kinetic energy had not been recovered at arrest.[95] Jörg Kalthoff at the Fraunhofer-Institut für Werkstoffmechanik, Freiburg, used the method of caustics to measure the dynamic stress intensity directly for a range of transparent brittle polymers and showed that at arrest the dynamic stress intensity factor oscillates about K_{Ia} until the stress waves have dissipated.[96] The ASTM standard for crack arrest toughness measures K_{Ia}.[97]

9.2.4 Welded wide plate tests

The problem with assessing the transition behaviour from small size fracture tests for low carbon steel is that they do not even give the same ranking of different steels. Faced with this dilemma welded wide plate tests were devised in the 1950s to replicate as closely as possible service conditions. The idea was to statically test butt-welded plates that were large enough to develop the residual stresses due to welding that would occur in full size structures. Harry Kennedy was the first to obtain a brittle fracture of a welded joint in the laboratory under static load.[98] Greene built on the work of Kennedy using butt-welded specimens 36 in. (915 mm) wide, which allowed residual stresses similar to those that occur in a full size structures to develop.[99] A saw cut notch was introduced into the weld preparation prior to welding. Using bend tests Greene showed that fracture strength of as welded plates was very low below the transition temperature, but that post weld heat treatment at 650°C, removed the high residual stresses and increased the fracture strength to the yield strength even at temperatures as low as 40°C below the transition. However, it was Alan Wells at the British Welding Research Association, Abington who did most to develop the welded wide plate test to simulate the fracture behaviour of full-size structures.

Fig. 9.13 Wells 6 MN wide plate testing machine (with permission TWI).

Wells made his first wide plate tests on one inch (25 mm) thick butt-welded low carbon steel plates 36 x 36 in. (915 x 915 mm). It required a force of the order of 5 MN to fracture these plates. Testing machines capable of exerting such a force and more were available, but Britain was still recovering from the effects of World War II and research money was hard to find. Fortunately Wells was not only a brilliant researcher but also an inventor of genius.[100] He designed a testing machine, not much larger than the plate itself, using four compact annular diaphragm loading cells with a total capacity of 6 MN, rather than pistons as shown in Fig. 9.13.[101] The largest wide plate testing machine built by Wells, using a similar design, had a capacity of 40 MN and was not much larger than that shown in Fig. 9.13. The 40 MN testing machine designed by Wells is a David compared to the Goliath Balwin-Southwark-Emery machine of less than half the capacity shown in Plate 10. The test plate of Wells, with a longitudinal butt-weld, was itself welded into the end loading assemblies. The transverse stress in the plates was minimised by optimising the position of the loading cells. To simulate the type of welding crack that might occur in a structure, 0.15 mm wide notches, similar to those of Greene, were cut into the double V weld preparation to a depth of 5 mm. These pre-cut notches were not completely buried when the plates were welded together, and when rutile electrodes were used for welding, they frequently produced hot cracks during welding. The plates were stressed in tension along the direction of the butt weld over a range of temperatures until they fractured.

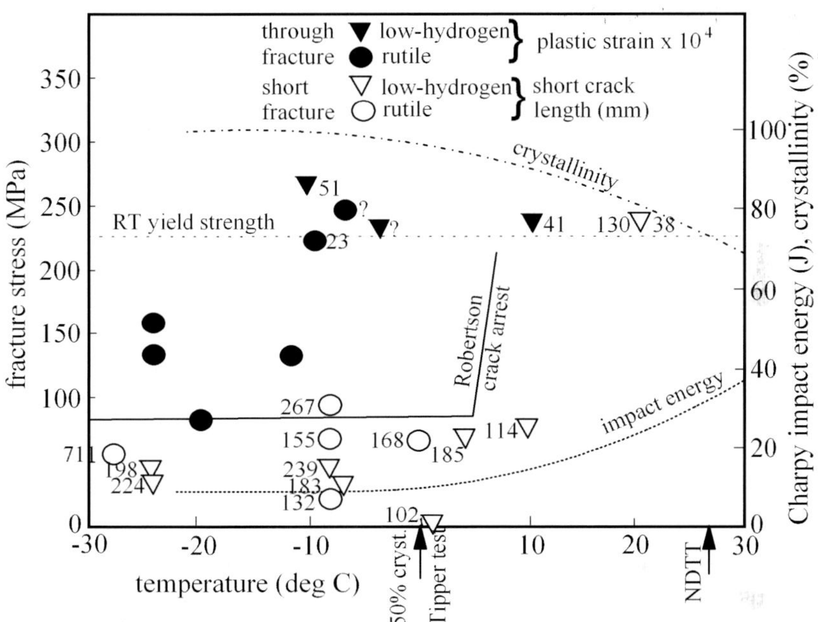

Fig. 9.14 Wells' wide plate test results for a semi-killed mild steel (after Wells 1961b).

The wide plate test results for a typical semi-killed low carbon steel (0.18% C and 0.54% Mn) of the time are shown in Fig. 9.14.[102] At and below 10°C some were fractures initiated at applied stresses well below the yield strength and even on cooling before loading. The stress at the notch was augmented by the longitudinal residual stress of the order of the yield strength due to the butt weld. The butt-weld produced a residual longitudinal tension over a band about 100 mm wide and most fractures were arrested just outside this band. However, below −10°C some low stress fractures propagated right across the specimen. The Robertson crack arrest test predicted the stress level for complete fracture accurately. Charpy V notch results, the Pellini NDTT and the 50% crystallinity transformation temperature obtained from the Tipper test are also indicated on

Fig. 9.14. The correlation of empirical small scale tests with the Wells wide plate test which simulates service conditions is poor. Better correlation with the wide plate test was obtained with the Charpy test when the steel was strained by 4% and aged. Such conditioned steel more nearly matches the base plate near any welding crack before machining.

Post weld heat treatment reduces the residual stresses due to welding and improves the ductility of the strain aged base plate near any welding crack. Wide plate tests showed that post weld heat treatment at temperatures of 450–650°C prevented brittle fracture at low stress at temperatures as low as 100°C below the transition temperature.[103]

Wide plate test rigs similar to those designed by Wells were used by Walter Soete (1912–2002) at the University of Ghent in Belgium, Hiroshi Kihara at the University of Tokyo in Japan and in Czechoslovakia.[104]

9.3 Developments in Steel Making

Improvements in design and steels have led to a greatly reduced incidence of brittle fracture. These improvements can be seen in the decrease in the incidence of brittle fracture in ships.[105] Brittle fractures classed as serious where the vessel was lost or in a dangerous condition occurred once every 10 ship years during the Liberty ship years. Prior to 1949, the Steel Rules of Lloyds Register of Shipping only required that the steel be made by the open hearth process. The Rules were then amended in 1949 and limits to the phosphorus and sulphur contents, a requirement that the manganese percentage be at least 2.5 times that of the carbon, and special approval for the steel of main structural members over 25 mm in thickness added. The amended rules brought about an improvement in steel quality and decreased the rate of serious brittle fractures to once every 100 ship years in the 1950s. Now in the twenty-first-century, further improvements in steel have reduced the rate to about once in 10,000 ship years.

Steel cleanliness is an important factor in steel quality and significant secondary steelmaking developments have reduced the content of impurity elements.[106] Vacuum degassing developed in the 1950s reduced the hydrogen content to below 2 ppm. Gas stirring stations allow the addition of synthetic slags that help to reduce the phosphorus, sulphur and oxides in steel. At the start of this period the usual limiting phosphorous content for low carbon steel was 0.04% now a limit of a tenth of that value can be achieved. Ladle injection was developed in the late 1960s and enables sulphur to be reduced to less than

0.002%. Though manganese sulphide inclusions are not particularly bothersome for low strength steels loaded in the direction of loading they limit the transverse toughness. Ladle reheating can reduce the phosphorus content as well as the nitrogen, oxygen and hydrogen contents.

There are cost benefits in using higher strength steels, but the traditional method of strengthening by going to higher carbon content reduces the toughness and weldability of the steel. In the late 1950s the infamous King-Street Bridge, Melbourne was designed to be constructed in BS 968: 1941, a low-alloy high-strength steel with a maximum carbon content of 0.23% and maximum manganese content of 1.8%. In 1962 fifteen months after it was opened, brittle fracture in a number of its girders caused the bridge to collapse. Two of the main problems were that the need to ensure that the steel's transition temperature was low enough[107] and the potential difficulty involved in welding this steel of higher than usual carbon content[108] were not fully appreciated by the constructors.

The cost benefits of using high strength steel in pipelines built in remote areas are high because of transport costs and in the 1960s a new class of microalloyed steel was developed to meet the huge expansion in pipelines. These steels achieve high strength by alloying very small amounts of strong carbide and nitride forming elements such as niobium, titanium and vanadium which cause precipitation hardening. The carbon content of these steels can be kept to less than 0.1% and they have low transition temperatures and are easily weldable. Such steels can have yield strengths of the order of 500 MPa. However, the introduction new materials frequently bring new problems which happened with the microalloyed steels. These new steels solved the problem of brittle fracture in pipelines, but introduced the problem of ductile fracture. Most pipelines are formed by rolling a plate up and welding it longitudinally. The maximum stress in a pipeline is the hoop tension which, because of the method of manufacture, is transverse to the rolling direction of the plate. In the first microalloyed steels, manganese was used to form inclusions with the sulphur as was the normal practice in steel making. Manganese sulphide is ductile at the rolling temperature and becomes elongated into stringers the direction of rolling. The manganese sulphide inclusions are benign in low-strength steel and the transverse ductile toughness is sufficient to prevent ductile fracture. However, higher transverse toughness is necessary to enable the high strength of the steels to be utilised. The longitudinal manganese sulphide stringers presented notch-like defects that caused the upper shelf toughness of a Charpy specimen to be only about half that in the longitudinal direction. As a consequence ductile fractures occurred in a number of pipelines.[109] There were two remedies to the problem of manganese sulphide. The Japanese steel companies reduced the sulphur content to 0.005% so that there were few manganese sulphide inclusions. Most other steelmakers reduced the sulphur content to around 0.015% and added cerium in the form of mischmetal that has a greater affinity for sulphur than manganese. Cerium sulphide is brittle at the rolling temperature and instead of being rolled into long stringers breaks up into small spherical particles which do not significantly affect the transverse toughness. From the 1980s there has been a development of high strength low alloy (HSLA) steels with yield strengths of up to 700 MPa. These steels have a low carbon content of 0.05%, microalloyed with niobium, titanium and vanadium, and also strengthen by the addition of copper. HSLA steels are now being used extensively in ship construction.

Perhaps the most important development in steel-making during the second half of the twentieth-century was thermo-mechanical processing (TMP), which improved hot rolled steel products by strict control of the rolling and subsequent cooling.[110] In the 1960s TMP was used to improve the mechanical properties by grain refinement and precipitation hardening. Low slab reheating at below the austenite transformation temperature was introduced in the 1970s, which enabled deformation strengthening and superior low temperature toughness to be obtained, because a fine grain size was possible even in thick steel. Accelerated cooling combined with controlled rolling, which became known as the thermo-mechanical controlled process (TMCP), began to be used in the 1980s to utilise transformation toughening. The introduction of the microalloyed steels stimulated TMCP. In the 1990s modelling of the TMCP has led to improvements in the process. Research and development is now enabling the TMCP to produce ultra-fine grain steel.

Steel making has come a long way in the second half of the twentieth-century. In the 1950s Richard Weck (1913–1986), then the Director of the BWRA, described mild steel as a 'natural material' like wood because its production was so simple. That description fitted at the time, but today's steel is a very much more advanced material.

9.4 Elasto-Plastic Fracture Mechanics (EPFM)

Fracture under essentially elastic conditions only occurs in high strength or brittle materials in laboratory sized specimens where the characteristic length, l_{ch}, is of the order of a centimetre or less. In the absence of welding or severe mechanical damage, low-strength steel requires stresses greater than the yield strength to cause fracture even at temperatures below the transition. Thus in the 1950s most of researchers working on the brittle fracture of steel still saw fracture simply in terms of stress. The first attempt at the establishment of a criterion of fracture that could be applied even when a specimen was fully plastic concentrated on a local approach. In analogy with LEFM where fracture depended on the magnitude of the local stress field, it was postulated by Alan Wells that fracture either elastic or plastic would occur when the crack tip opening displacement reached a critical value.[43]

9.4.1 *The crack tip opening displacement (CTOD) concept*

Wells was the first to formulate a theory of fracture at and beyond general yielding which enabled the brittle fracture of low strength steel to be predicted and he considered that this work was his major contribution to fracture research. Although Wells briefly mentioned the CTOD concept at the 1961 Cranfield Conference,[43] his first paper devoted to the CTOD concept was published in 1963.[111] Recognising the direct connection between the critical CTOD, δ_c, and the fracture energy, R, through Eq. (9.15) for stresses well below general yielding, Wells postulated that 'the initiation of brittle fracture [was] uniquely determined over the whole range [elastic and plastic] by a critical value of δ'. Wells saw that the notched three-point-bend test was ideal to measure the CTOD where the notched section was completely yielded because a plastic hinge forms at the notched section and the plastic deformation comes from the rotation of the arms about this hinge. Slip line theory gives the position of the centre of rotation and the CTOD can be found from the plastic rotation of the arms which was originally calculated from the plastic deflection of the load line. The same principle is used today except that a clip gauge is used to measure the crack mouth opening displacement in the notched three-point-bend test and the load line displacement is used in the case of the compact tension test.[112]

In the standard tests to measure the CTOD, deep cracks are used which present a high constraint to plastic flow and provided that the CTOD is very small compared with the specimen thickness and ligament, the critical CTOD is a plane strain value. Shallow cracks and different geometries, such as the centre notch tension specimen provide much less constraint and the critical CTOD is larger than the plane strain value.[113]

To use the CTOD concept in design it is necessary to be able to calculate the CTOD in a structure. For a plate with a central crack of length $2a$ with an applied stress, σ, less that the yield strength, σ_Y, Wells assumed that the CTOD could be calculated from Irwin's estimate of the size of the plastic zone and he obtained an expression for δ given by

$$\delta = \frac{2\pi a \varepsilon_Y}{\left[2\left(\sigma_Y/\sigma\right)^2 - 1\right]}, \quad \sigma < \sigma_Y, \tag{9.23}$$

where ε_Y is the yield strain. For applied stresses greater than the yield he assumed that the CTOD would be proportional to the strain. Choosing the constant of proportionality to agree with Eq. (9.23) for $\sigma = \sigma_Y$, he obtained

$$\delta = 2\pi \varepsilon a, \quad \varepsilon > \varepsilon_Y. \tag{9.24}$$

Wells used these equations to show that the critical CTOD values obtained from notched bend tests could predict the fracture strain in wide plate tests with central slits.[111]

Since Wells original paper the expressions for the CTOD have been improved, but the general concept remains the same. The CTOD design curve, a semi-empirical method for assessing flaws welded structures was developed at the BWRA and its successor TWI from Wells' concept.[114] There is now a British Standard that is based on the R6 method[115] developed by the Central Electricity Generating Board in 1977, which uses the CTOD for the strip yield model given by Eq. (9.14) to combine both fracture and plastic collapse.[116]

9.4.2 *The crack tip opening angle*

The crack tip opening angle (CTOA) concept is a local fracture criterion for stable ductile tearing that is related to the CTOD. It is not clear who first proposed the CTOA as a criterion for ductile crack propagation. Frank McClintock[117] discussed the CTOA as a fracture criterion in 1968 and Henrik Andersson and Fong Shih used it in the 1970s[118], but it is James Newmann[119] who has done the most to promote its use. The CTOA is relatively insensitive to the constraint at a crack tip and providing the crack and remaining ligament lengths are greater than about four times the thickness of the specimen appears to be virtually constant. However, the application of the CTOA concept has been mainly applied to thin metal sheets where the constraint is dominated by through the thickness deformation.

9.4.3 The J-integral and EPFM

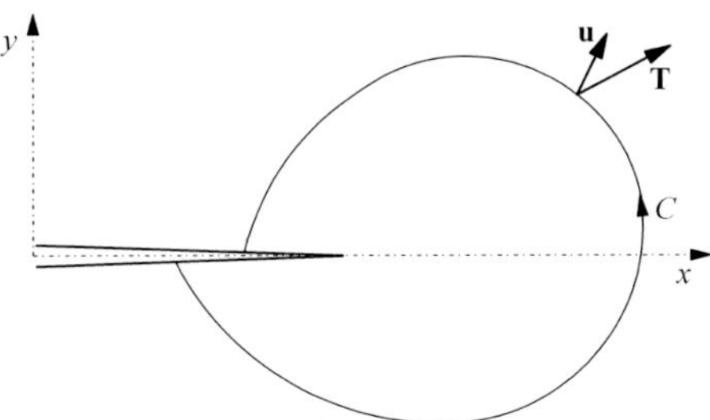

Fig. 9.15 Contour for J-integral.

In the 1960s an elasto-plastic fracture mechanics (EPFM) akin to LEFM was clearly needed. At the first International Conference on Fracture held in Sendai in 1965, Jim Rice presented his ideas of energy balance[120] for material behaviour other than linear elastic which were to culminate in his famous 1968 paper giving the path independent J-integral (see Fig. 9.15) for a non-linear elastic solid[121]

$$J = \int_{C}\left[U dy - \mathbf{T} \cdot \frac{\partial \mathbf{u}}{\partial \mathrm{x}} ds \right],$$
(9.25)

where C is a curve surrounding the crack tip which starts on one surface and continues in a anticlockwise direction and finishes on the other crack surface, s is the distance along the curve; $\mathbf{T} = \boldsymbol{\sigma} \cdot \mathbf{n}$ is the traction vector according to an outward unit vector $\mathbf{n}$ normal to the surface; U is the strain energy density per unit thickness; and x and y are the coordinates along and perpendicular to the crack surface. If the FPZ is infinitesimally small or if there is steady state crack propagation

$$J = -\frac{d\Pi}{da},$$
(9.26)

where Π is the potential energy per unit thickness and a is the crack length. Fractures initiate when $J = R$, the fracture energy. Deformational plasticity, which is a good approximation to the more realistic incremental plasticity, is the same as non-linear elasticity providing there is no unloading. Hence the J-integral is the energy release rate and can be used to characterise crack initiation in an elasto-plastic solid. However, there is a slight difference in interpretation between the fracture energy in LEFM and the fracture energy as measured by the J-integral. In LEFM it is assumed that the K-field characterizes fracture and that

G is the energy release rate that at crack initiation is equal to total work of fracture which is both intimately associated with the separation process as well as the plastic work in the surrounding zone. However, the J-integral is the energy consumed within the FPZ where deformational plasticity breaks down. Thus the fracture energy measured by the J-integral is theoretically smaller than the fracture energy measured by *G*. Since effectively only the plastic part of the J-integral is measured through Eq. (9.26), but the difference is small and is not important. More details of the J-integral criterion may be found in John Hutchinson's review.[122]

The formulation of the J-integral by Rice for elasto-plastic fracture mechanics was a major landmark in the development of fracture mechanics. Similar ideas had been formulated by others at the time. Eshelby had derived conservation integrals in his work on the concept of a "force" on a singularity in 1951,[123] and in 1968 he applied his energy-momentum tensor to fracture mechanics,[124] but he distinguished between his integral and the J-integral. Genady Cherepanov also independently derived the J-integral in a paper published in 1968.[125] However, it was Rice who clearly saw the significance of the J-integral to elasto-plastic fracture mechanics and it is right that it is his name is attached to it.

The measurement of a plane strain initiation J-integral, J_{Ic}, was at first accomplished with multiple specimens using the energy interpretation of *J* given by Eq. (9.26).[126] However, Rice and his collaborators showed that *J* could also be obtained from a single specimen using the energy definition of *J*.[127] For the deeply notched three-point-bend specimen the plastic part of *J* can be found from dimensional arguments

$$J = G_e + \frac{2}{b} \int_0^{\delta_p} P d\delta_p, \tag{9.27}$$

where G_e is the elastic energy release rate, P the load per unit thickness of the specimen, δ_p the plastic component of the load line deflection, and *b* the remaining ligament length. The J-integral for other specimens can be written as

$$J = G_e + \frac{\eta_p}{b} W_p \tag{9.28}$$

where W_p is the plastic work performed up to fracture and the η_p is a function of geometry.

Just as the stresses near a crack tip in LEFM are characterised by the energy release rate, *G*, through its relationship to the stress intensity factor, *K*, so too is the deformational plastic stress distribution characterised by *J*. Two independent papers were published in the same issue of the journal, one by Rice and Rosengren and the other by Hutchinson, giving the relationship for a power hardening solid.[128] The stress field is named the HRR field after all three authors. Fracture initiation will be controlled by J if the FPZ is embedded within the J-field. However, in ductile fracture the FPZ can be large.

The J-integral can be extended somewhat to situations where the FPZ is outside of the dominance of the J-field. Fong Shih and Noel O'Dowd have given a two parameter (J-Q) approximation of the stress field ahead of the crack tip, which can be written as

$$\sigma_{ij} = \left(\sigma_{ij}\right)_{HRR} + Q\sigma_Y \delta_{ij}, \tag{9.29}$$

where $(\sigma_{ij})_{HRR}$ is the HRR stress field, Q is a hydrostatic stress parameter dependent[129] upon the geometry, size and J, σ_Y is the effective yield strength, and δ_{ij} is the Kronecker delta.[130] The hydrostatic stress parameter is negative and large for specimens that offer low constraint to plastic deformation at the crack tip and becomes more positive as the constraint increases. The two-parameter model has been successful in predicting the effects of constraint on cleavage fractures that initiate only after significant plastic deformation[131] and have been used to predict the initiation of ductile tearing.[113]

As is often the case a good concept was pushed beyond its limits of applicability. The Achilles heel of the J-integral is that it is for a deformational plastic material. For fracture initiation there is no problem providing that the FPZ is small, but there is a problem for a propagating crack. The J-integral was used to measure the increasing resistance to fracture with crack growth in the more ductile metals with J_R thought to be the EPFM equivalent of the LEFM G_R. However, with crack growth there is unloading behind the crack tip, and the plastic strain here is not recovered as is required by the non-linear elastic J-integral. It soon became apparent that in many cases the J_R-curves were geometry and size dependent, and the apparent increase in toughness with crack growth was an artefact of the J-integral. Since in a real plastic material, energy left behind the crack tip is not recovered, the J-integral calculated from Eq. (9.27) contains an increasing amount of plastic work. The J-integral can be used for small crack extensions provided the effect of an increased J_R outweighs any unloading effect. Hutchinson and Paris[132] showed that, outside of a core of non-proportional loading or FPZ, the deformation is nearly proportional even during crack propagation provided

$$\frac{b}{J_R}\frac{dJ_R}{da} > N, \tag{9.30}$$

where b is the uncracked ligament and a is the crack length, and N is a number whose magnitude depends upon the constraint. For low constraint N can be as large as 100 and as small as 14 for high constraint.

9.4.4 Plasticity and fracture – work and energy

Energy and work methods have been favoured by many research workers, especially British ones. The classic tension specimen with a central crack considered by Griffith is unstable and quasi-static energy balance can only be used to determine the conditions of fracture initiation. Charles Gurney concentrated on quasi-static crack propagation in specimens where the fracture propagation was stable. Together with his collaborators, Gurney devised analytical and graphical energetic methods to determine the fracture energy[133] from linear and non-linear elastic specimens.[134] Plastic deformation confers extra stability and in many cases fracture specimens that are unstable if elastic, become stable when there is gross plasticity. Tony Atkins, Gurney's student at the University College, Cardiff has developed less well known powerful work methods for plastic fracture.

Atkins' ideas developed from his 1980 paper on cropping.[135] In combined flow and fracture the incremental work done on a specimen is given by

$$Xdu = d\Lambda + d\Gamma + RdA, \tag{9.31}$$

where X is the applied load, u is the load line displacement, Λ is the elastic strain energy, Γ is the plastic work, R is the fracture energy, and A is the crack area. The J-integral method assumes that plasticity can be approximated by non-linear elasticity. Plastic deformation unlike non-linear elastic deformation depends upon the history of loading and not just on the current state of stress. However, as long as there is no significant unloading, the approximation is reasonable. For non-linear elasticity Eq. (9.31) holds if we interpret Γ as the non-linear strain energy. Atkins suggested writing Γ as WV where W is the average non-linear strain density in the non-linear region whose volume is V, Eq. (9.31) can then be rewritten as[136]

$$Xdu = d\Lambda + VdW + WdV + RdA. \tag{9.32}$$

In most situations there is unloading behind the crack tip and in non-linear elastic fracture the strain energy in these regions is recovered but of course if the material is elastoplastic no plastic work can be recovered. If the remaining ligament in a specimen is completely yielded or undergoing non-linear elastic deformation before fracture initiation, dV is negative. Eq. (9.32) is not valid for an elastoplastic material but Atkins has suggested that there is an approximate relationship between the fracture of an elastoplastic material and its non-linear elastic equivalent such that if dV is negative, WdV is the plastic work that is not recovered so that the fracture work is given by

$$Xdu = d\Lambda + VdW + RdA, \qquad (9.33)$$

where Λ and W are obtained from non-linear elasticity.[137,138] In one particular geometry, the DCB specimen, the plastic work that is not recovered can be calculated exactly and the work of fracture is given exactly by Eq. (9.33).[137] In a non-linear elastic specimen the load-displacement point for a specimen whose crack is propagated to a length a would be exactly the same as that for a specimen with an original crack length of a that is loaded up to initiation. However, in a real elasto-plastic DCB specimen the load will be the same in both cases but the displacement of the specimen, which was propagated to a crack length of a, would be greater (see Fig. 9.16).

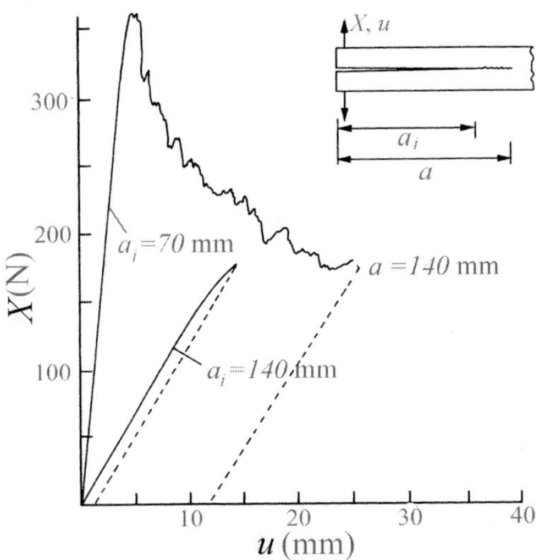

Fig. 9.16 Load-deflection *X-u* curves for two elastoplastic DCB aluminium alloy 6082-TF specimens having different initial crack lengths, but otherwise identical, showing the extra displacement because of non-recoverable plastic deformation (Atkins *et al.* 1998 with permission the Royal Society).

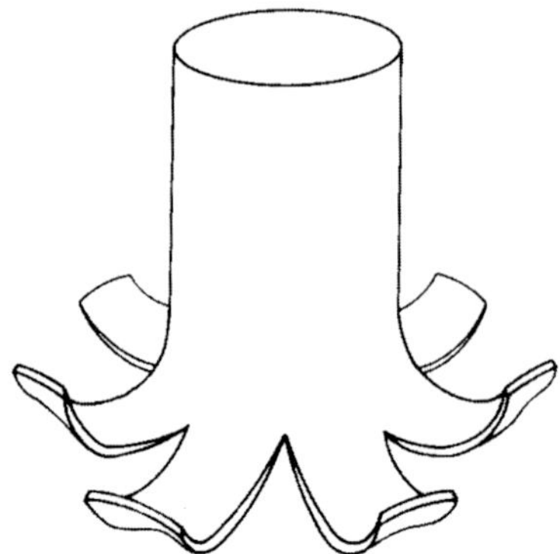

Fig. 9.17 Axial splits in a tube forced over a flared die (Atkins 1987b with permission Elsevier).

In many cases, such as the DCB specimen, where there is gross plasticity, $d\Lambda$ cannot be neglected even if it is small compared with $d\Gamma$ but there are others, especially in cutting and machining[139] where the force is applied directly or near to the crack tip, that it can be. One such non-cutting problem is the axial splitting of ductile metal tubes.[140] Yella Reddy and Stephen Reid performed a series of tests where they axially compressed metal tubes on to a flared die; as the tube flared over the die axial cracks were initiated (see Fig. 9.17).[141] In some experiments no starter cracks were cut into the tube, whereas in other a number of starter cracks were cut. In the series of tests performed on mild steel tubes between 8 and 12 axial cracks form; if the number of starter cracks is smaller than this number then the initiated cracks bifurcate if they are greater some starter cracks do not propagate. In Atkins' analysis[140] the work done, if there is no axial splitting neglecting the elastic strain energy, is

$$Xdu = d\Gamma = WdV + VdW, \tag{9.34}$$

where since material is coming down on the flared end dV is positive and WdV is admissible. During steady state axial splitting since there is no increase in plastic work density ($dW = 0$) only new material is being flared and plastically strained, and

$$XdU = WdV + nRdA, \tag{9.35}$$

where n is the number of axial splits and dA is the increase in the area of each crack. Splitting occurs if

$$WdV + VdW \geq WdV + nRdA, \text{ or}$$

$$n \leq \frac{V}{R}\frac{dW}{dA}. \tag{9.36}$$

Atkins used Eq. (9.36) to predict the number of axial splits as less than 12 which agreed with the experimental results.

9.4.5 The essential work of fracture concept

For a non-linear elastic material the J-integral gives the work per unit area of fracture performed within the FPZ or what has been called the specific essential work of fracture. For a real elasto-plastic material J is the essential work of fracture at initiation and for small crack extensions. In effect the J-integral filters out the plastic work performed outside of the FPZ from the total work of fracture. Apart from the development of shear lips in the early stages of fracture propagation the specific essential work of fracture remains practically constant during slow steady state fracture. Broberg expressed these same ideas using slightly different terms.[142]

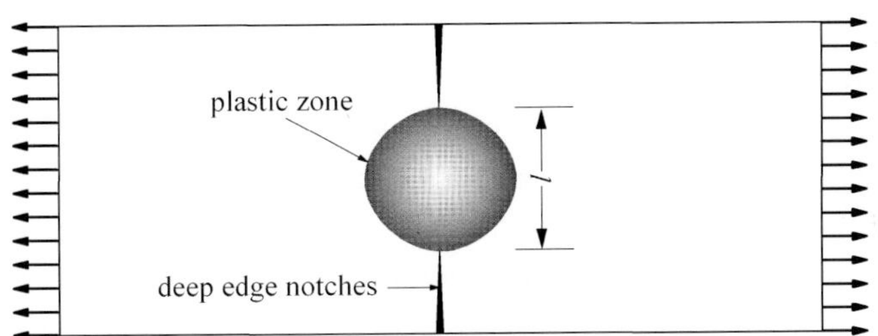

Fig. 9.18 Schematic illustration of the deep edge notch EWF specimen.

If the plastic zone is constrained by an outer elastic field it grows during crack extension and there is no simple way of separating the outer plastic work from the total work of fracture. I conceived a method of separating the essential work of fracture from the total work of fracture in thin ductile metal sheets while John Reddel was a postgraduate student in 1977 working on the ductile fracture of large thin steel sheets.[143] If the whole of an uncracked ligament yields before crack initiation then the plastic zone is limited by the ligament. In specimens such as the deep edge notch tension where yielding does not spread to the outside edges, the plastic deformation during crack propagation remains geometrically similar over the ligament. A schematic illustration of the test is shown in

Fig. 9.18. In ductile metal sheets the plastic zone is practically circular, but it does not matter what the shape is providing it is geometrically similar for different ligament lengths. The area of the plastic zone is proportional to the square of the ligament length, l^2. If there is stable crack growth to complete or nearly complete fracture of the ligament, the total plastic work is proportional to $l^2 h$, where h is the thickness of the specimen. On the other hand the essential work of fracture performed within FPZ is proportional to lh. Hence the total work of fracture, W_f, is given by

$$W_f = lhw_e + l^2 hw_p, \text{ and}$$

$$w_f = \frac{W_f}{lh} = w_e + lw_p, \tag{9.37}$$

where w_f is the specific total fracture work, w_e is the specific essential work of fracture and w_p is a plastic work density. Eq. (9.37) is the basic equation of the essential work of fracture (EWF) method, which in effect is an integrated form Eq. (9.33). By testing a range of ligaments, making sure that, even with the largest ligament, the whole ligament is yielded before fracture initiation, the specific essential work of fracture can be found by plotting the specific work of fracture against the ligament length and extrapolating the resulting straight line to zero ligament length.

The EWF was first applied to the plane stress fracture of thin metal sheets, but has since been applied to other materials, particularly ductile polymers, and to plane strain fracture. The EWF method is not without its difficulties: it needs multiple specimens, the limits for valid ligaments are difficult to set precisely and there is some difficulty in getting consistent results from different laboratories, but it does deliver a single fracture parameter that is representative of crack propagation. A standard for testing thin polymer sheet has recently been proposed.[144]

9.4.6 Modelling the FPZ in elasto-plastic fracture

The only sure way of dealing with crack growth in elasto-plastic fracture is to model the FPZ. Historically modelling of the FPZ was first introduced in 1976 by Arne Hillerborg for concrete which behaves elastically outside of the FPZ, but whose FPZ is large and comparable in size with laboratory specimens making conventional LEFM invalid.[145] In metals it is not necessarily the size of FPZ that makes modelling it important, but the nature of plastic deformation.

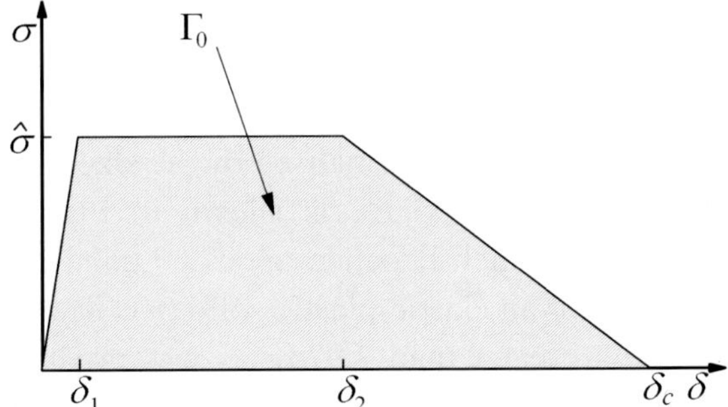

Fig. 9.19 Traction separation relation for fracture process in metals (after Tvergaard & Hutchinson 1992).

The first modelling of the FPZ for metals was by Viggo Tvergaard and John Hutchinson in 1992.[146] Since by definition the stress in a FPZ at the crack tip must be zero, the FPZ is a region of strain softening. In a continuum such a region is unstable and must collapse to a line. Real materials have structure so that a FPZ does not collapse completely, but nevertheless FPZs are narrow and modelling them as a line extension of the crack, called a *fictitious crack* by Hillerborg, is a good approximation. Tvergaard and Hutchinson used such a line model that ultimately derives from Barenblatt[35]. The precise form of the cohesive relationship within a FPZ is not known *a priori*. The most important feature is the cohesive energy, Γ_0, which is the area under the curve. For large FPZs the cohesive strength, σ_m, is important. However the form of the cohesive stress-displacement relationship is comparatively unimportant. The form of the cohesive stress-displacement relationship used by Tvergaard and Hutchinson and is shown in Fig. 9.19. The initial 'elastic' ramp up to the cohesive strength, $\hat{\sigma}$, is introduced in the model to make the incorporation of a FPZ into a finite element programme easy by allowing it to extend completely along the prolongation of the crack, but it is not a real feature and it is essentially for the ramp up to be steep to avoid artefact errors. The loading chosen by Tvergaard and Hutchinson for their analysis was an elastic K-stress field far from a crack tip which is under plane strain conditions. The material was power hardening with a stress-strain curve given by Eq. (1.20). Crack initiation occurs when the stress intensity factor is increased to K_0 which is given by

$$K_0 = \sqrt{\bar{E}\Gamma_0},\qquad(9.38)$$

which conforms to the result obtained from the J-integral. Despite the cohesive energy being constant, there is crack growth resistance caused by plastic deformation outside of the FPZ and the increase in the normalised fracture toughness, K_R/K_0, with crack growth normalised by the size of the plastic zone at crack initiation, D_0 given by Eq. (9.6), is shown in Fig. 9.20. The fracture toughness reaches a plateau which depends upon the cohesive strength and the strain hardening index, n. For an elastic-plastic solid ($n = 0$) fracture cannot occur if the cohesive strength is greater than $2.97\sigma_Y$, as was predicted earlier by Rice from slip line theory.[147] For hardening materials the model predicts that fracture can occur for a moderate cohesive strength of up to about five times the yield strength.

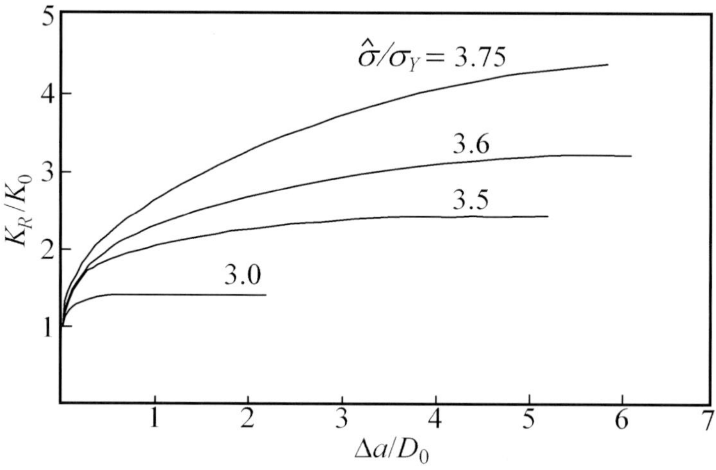

Fig. 9.20 Crack growth resistance curves: $\sigma_Y/E = 0.003$, $n = 0.1$, $\delta_1/\delta_c = 0.15$, $\delta_2/\delta_c = 0.5$ (after Tvergaard and Hutchinson 1992).

Ductile fracture in metals occurs by the nucleation of microvoids at second-phase inclusions which then grow until the plastic flow between the enlarged voids becomes localised and the voids coalesce. Modelling of the mechanism of ductile fracture is usually based on the Gurson model as modified by Tvergaard to improve the accuracy.[148] The heart of the Gurson model is the yield criterion for a material with spherical microvoids

$$\Phi = \left(\frac{\sigma_e}{\bar{\sigma}}\right)^2 + 2q_1 f \cosh\left(\frac{3q_2\sigma_m}{2\bar{\sigma}}\right) - \left[1 + \left(q_1 f\right)^2\right] = 0, \tag{9.39}$$

where σ_e is the macroscopic effective von Mises stress, σ_m is the macroscopic mean stress, $\bar{\sigma}$ is the effective flow stress of the actual microscopic stress state in the matrix, f is the current void volume fraction, and q_1 and q_2 are Tvergaard's adjustment factors. The Gurson model takes care of the effect of constraint through the mean stress σ_m but it says nothing about the nucleation of microvoids or the final coalescence. There is also a problem in mesh sensitivity when the model is incorporated into a finite element analysis. Fong Shih and his collaborators could model the load-displacement and J_R-curves for the pressure vessel steel A533B for geometries of different constraint by ignoring the nucleation of voids and assuming that there was an initial void volume fraction, f_0, and assuming that at some critical void volume fraction, f_E, the material lost all cohesive strength.[149] They dealt with the problem of mesh sensitivity by assuming that the FPZ was confined to single cell width whose size, D, was a material property. The values of f_0 and D which gave the best fit to the experiment J_R-curves obtained from a three-point-bend specimen were used to calculate the load-displacement and J_R-curves for other geometries. These curves agreed reasonably well with the experimental data.

In their modelling of the FPZ using the Gurson model, Lin Xia and Fong Shih calibrated their parameters from a particular geometry and then used these parameters to predict the behaviour of other specimens of different geometries and sizes.[149] This method works well providing the difference in constraint is not too great. However, the Gurson model does not work well over large differences of constraint for two reasons. The first is that the Gurson model uses spherical voids. At low constraint, even if initially spherical, voids quickly deform into more elongated spheroids. The second reason is that void coalescence, the final stage in ductile failure, is not modelled. Jean-Baptise Leblond, and his colleagues have extended the Gurson model to void shape effects[150] and void coalescence by plastic localisation between voids has been addressed by Peter Thomason,[151] Thomas Pardoen and John Hutchinson have extended these modifications to account for strain hardening.[152] For implementation in finite element programmes, Pardoen and Hutchinson have expressed the coalescence model in classic plasticity terms so that there are two yield surfaces: a Gurson-like void growth one and another for coalescence.[153]

Modelling fracture using FPZ can explain how a cleavage fracture can be initiated by prior ductile tearing.[154] A high normal stress is necessary to initiate and propagate a cleavage crack. At low temperatures such a stress can be achieved without ductile tearing and a cleavage crack can initiate and propagate providing the high stress region is sufficiently large. A critical stress over a critical distance criterion was proposed by Richie, Knott and Rice in 1973 for the initiation of a cleavage crack.[155] At the transition temperature the constraint is not high enough to initiate a cleavage fracture. Ductile tearing causes an increase in constraint, especially if the initial constraint is low, and an increase in the volume of material at high stress. Xia and Shih combined their ductile fracture model with Weibull statistics to model the initiation of a cleavage fracture and explained how cleavage is initiated at higher temperatures by prior ductile tearing.[154]

Despite the refinements to the theory of ductile fracture, toughness will continue to be an experimental phenomenological parameter and the models serve to indicate how fracture toughness can be enhanced by changes in the microstructure rather than fully predictive models.[153]

9.5 Fatigue of Metals

Fig. 9.21 Aerial view of Comet Yoke-Uncle in the testing tank at the RAE, Farnborough.

Fatigue has been a problem in aircraft almost from the first flight by Orville Wright in 1903, but it was the 1954 Comet disasters which caused fatigue to become a major consideration in design. In almost premonition of the coming disaster, the Australian aircraft designer turned author Neville Shute (1899–1960) wrote the novel *No Highway*[156] in 1948 where the hero, Mr Honey, a scientist at the RAE Farnborough, predicts fatigue failure of the tail of an airliner in which he is flying after 1,400 flying hours and grounds it by releasing the undercarriage during a stop-over to prevent a catastrophe. The Comet Yoke-Peter crashed after 3,681 flying hours and Yoke-Yoke crashed after 2,740 flying hours, though the crucial issue was the number of pressurisations of the cabin rather than the number of flying hours.[157] The accident investigation of the Comet disasters were carried out at the RAE Farnborough where the salvaged parts of Yoke Peter were reconstructed and the failure traced to a fatigue crack near the starboard corner of the rear ADF window. Comet Yoke-Uncle, obtained from BOAC after flying for 3539h, was subjected to a full-scale fatigue test, pressuring the cabin with water in a water tank while simultaneously applying loads to its wings (see Fig. 9.21). It was concluded from the test that Comet Yoke-Uncle would have failed by fatigue after 9,000 flight hours.

The need for damage tolerance became apparent in the USA during the 1950s.[158] The principal deterrent in the Cold War was the United States Air Force Strategic Command with its fleet of B-47 and B-52 bombers. A series of catastrophic fatigue failures in B-47s during early 1958 after only one to two thousand flying hours, caused critical areas to be strengthened and limitations placed on the performance of the whole fleet. The acceptance of the structural integrity of the B-47 was based on a static test in 1950. No fatigue life was specified for the B-47, but it was expected that they would be kept in service until 1966. In fact the basis for the acceptance of the structural integrity of the B-47 did not really differ from the Wright Brothers' tests made on the *Flyer* in 1903. The disasters caused the implementation of an Aircraft Structural Integrity Program. This programme saw the fatigue testing of aircraft to establish a *safe life* which was the equivalent flight time to failure divided by a safety factor of typically four. The safe life methodology was the basis for the F-111 design in 1962, but did not prevent the crash in 1969 of an F-111 after only 100 flight hours, though this failure was primarily due to a rogue flaw in the wing box rather than fatigue, there were other premature failures. The application of LEFM to fatigue crack growth, the most significant engineering development in fatigue in this period, saw the introduction of the *fail safe* approach, where inspections are used to find cracks before they become dangerous rather than specify a *safe life*.

9.5.1 Low-cycle fatigue

For fully-reversed load cycles the limiting alternating stress, σ_a, is the ultimate strength of the metal and failure occurs in a quarter of a cycle and it is not until the number of cycles is greater than about 10,000 does Basquin's equation (see §7.4.2) hold. Louis Coffin and Sam Manson, working independently in 1954 on thermal fatigue where the number of cycles is low, proposed a similar equation to Basquin's relating the plastic strain amplitude $\Delta\varepsilon_p/2$ to the number of cycles to failure rather than the stress.[159] Their equation is

$$\Delta\varepsilon_p/2 = \varepsilon_f\left(2N\right)^n,\tag{9.40}$$

where ε_f is a fatigue ductility coefficient of the order of the true monotonic fracture strain, and n for most metals is in the range -0.5 to -0.7.

9.5.2 Crack propagation

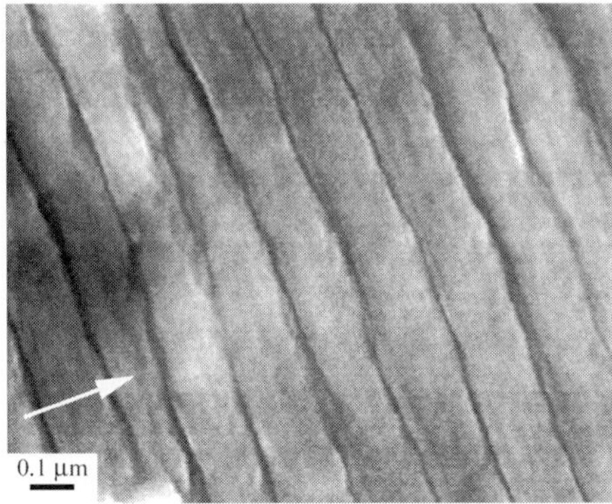

Fig. 9.22 Ductile fatigue striations in aluminium alloy 2024-T3, the arrow shows the direction of fatigue growth (de Castro *et al.* 2007, with permission Elsevier).

In the absence of notches or significant internal defects, fatigue cracks initiate at the surface by the deepening of a slip-band as described in §8.4.1. In 1961 Peter Forsyth described two stages in crack propagation.[160] In Stage I crack growth the crack grows along a slip plane aligned with the maximum shear stress direction and the path can zig-zag. The extent of Stage I growth is usually a grain diameter unless the near-tip plastic zone is small compared with the grain diameter. In stage II growth the fatigue crack turns to propagate normal to the maximum normal stress with slip taking place simultaneously or alternating between the two slip systems. The combined effect of a notch and a high mean stress encourages the immediate onset of Stage II fatigue growth. It is Stage II crack growth that leaves the characteristic fatigue striations on the fracture surface. Ductile fatigue striations on the surface of the aluminium alloy 2024-T3 are shown in Fig. 9.22.

A few crack propagation tests were made prior to 1950,[161] but it was generally thought that the crack propagation phase of fatigue was small. However, in the 1940s it was found that cracks could form after only 10% of the fatigue life at low alternating stress.[162] Alan Head in 1953 was the first to derive an expression for the rate of crack propagation.[163] He assumed that a crack grows through metal whose ductility has been exhausted by cyclic strain hardening. From a simplified model of the material ahead of the crack tip he deduced that

$$\frac{da}{dN} = a^{3/2} d^{-1/2} f(\sigma_a), \tag{9.41}$$

where a is the crack length, d the plastic zone size, and $f(\sigma_a)$ a function of the alternating stress. Head assumed that the size of the plastic zone remained constant and thus decided that the rate of crack growth was proportional $a^{3/2}$, a conclusion supported by the limited experimental data available.[161] However, the data also conformed to a variety of other expressions relating rate of growth and crack length. Norman Frost and Donald Dugdale suggested that the crack growth rate was proportional to the plastic zone size which was proportional to the crack length.[164] Other expressions for the rate of fatigue crack growth were also formulated at this period. The breakthrough came in the late 1950s with the development of LEFM. Paul Paris, who was in close contact with Irwin, spent time with Boeing during 1955 to 1957 and while there he sent a memo to Bill Anderson suggesting that Irwin's stress intensity factor could be used to correlate fatigue crack growth, but verification experiments were postponed because Boeing was fully occupied on component testing for the Boeing 707.[19] A note on fatigue crack propagation was written for Boeing in 1957,[165] but the first paper giving what is now known as the *Paris Law* for fatigue crack growth was finally published in an obscure University of Washington engineering magazine in 1961[166] after being rejected by the AIAA, ASME and *The Philosophical Magazine* because the reviewers believed it impossible to model a fatigue crack growing by plastic deformation with an elastic parameter.[19] The LEFM approach to fatigue crack growth was controversial, but was established by searching experiments on aluminium alloys.[167] The rate of fatigue crack growth is related to the range in the stress intensity factor, $\Delta K = K_{max} = K_{min}$, by the power law

$$\frac{da}{dN} = C\Delta K^m, \tag{9.42}$$

where C is a constant and the index m is typically between two and four for ductile metals.

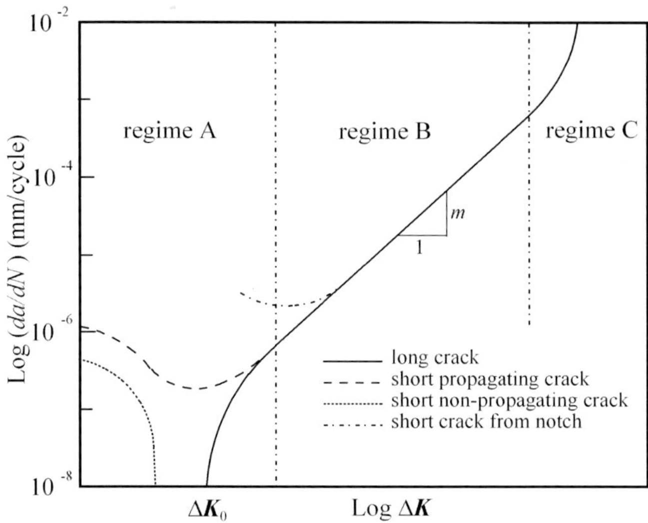

Fig. 9.23 Illustration of the fatigue crack growth of long and short cracks (after Suresh 1998).

The range of applicability of Eq. (9.42) is roughly between a rate of crack propagation of 10^{-6} and 10^{-4} mm/cycle. Cracks that are not microscopically small cannot propagate at stress intensity ranges less than a threshold value ΔK_0[168] and unstable crack propagation occurs when the maximum stress intensity factor reaches the fracture toughness, K_c. Thus for long cracks there are three regimes of fatigue crack growth as shown schematically in Fig. 9.22: a threshold regime A, a Paris regime B, and a high growth rate regime C where K_{max} approaches K_c where the crack growth becomes unstable.[169]

The rate of fatigue crack growth depends upon the stress range, $R = K_{min}/K_{max}$. It was recognized by Rice that, because of the plastic zone in the wake of a propagating crack, a crack would close before the stress intensity factor became negative.[170] Wolf Elber established for a 2024-T3 aluminium alloy that the effective stress intensity range, ΔK_{eff}, defined as $K_{max} - K_{op}$ where K_{op} is the stress intensity at which the crack opens is given empirically by

$$\Delta K_{eff} = \left(0.5 + 0.4R\right)\Delta K, \tag{9.43}$$

and if ΔK_{eff} is used in Eq. (9.42) instead ΔK then the crack growth rate is independent of the stress ratio, R, over the range $-1 < R < 0.7$.[171] There have been many modifications to Eq. (9.43) and it has been realised that ΔK_{eff} is also dependent on specimen geometry. Bernard Budiansky (1925–1999) and John Hutchinson gave an analytical model for crack closure based on the Dugdale model.[172]

Even when the stress remote from the crack tip is low, the material at the crack tip is undergoing plastic strain and is thus a low cycle region. Using the Budiansky-Hutchinson model for the plastic behaviour at a crack tip combined with the Coffin-Mason relationship for low-cycle fatigue it is possible to predict the fatigue crack propagation rate using the Palmgren-Miner law for the accumulation of damage with reasonable accuracy.[173]

The application of fracture mechanics to fatigue has enabled the time between inspections of aircraft to be accurately set so that the probability of detecting a fatigue crack before it becomes long enough to cause unstable fracture is extremely high.

9.5.3 Short fatigue cracks

Most fatigue tests to determine fatigue crack growth rates are performed on specimens that contain cracks millimetres in length. However, the major portion of the life of many engineering components such as turbine discs occurs while the fatigue crack is very small. This situation would not be very worrying for design if the rate of crack propagation of small cracks, for a given stress intensity factor range, was equal to or less than that of long cracks. Unfortunately small cracks grow faster than long ones. Pearson in 1975 was the first to report accelerated growth in short cracks, which can propagate up to a hundred times faster than long cracks at the same stress intensity factor range.[174] Keith Miller (1932–2006), who devoted much of his life to the study of the behaviour of short fatigue cracks, wrote a review of the behaviour of short cracks in 1987.[175] There are two main types of short cracks: *microstructurally small cracks* whose length is comparable to the grain size, *mechanically small cracks* where the plastic zone is comparable to the crack length and LEFM does not apply.[176] For short cracks there is not a one to one relationship between the crack growth rate and the stress intensity factor range. Microscopically short cracks can propagate at stresses less than the fatigue limit, but arrest before they reach the largest microstructural impediment. At the fatigue limit cracks are arrested by the largest microstructural impediment. For stresses greater than the fatigue limit, cracks first grow much faster than long cracks but microstructural features cause the cracks to decelerate. A minimum crack growth rate is reached when the crack size is equal to the largest microstructural impediment after which the crack accelerates and finally merges with the long crack data. The behaviour of different types of short cracks is illustrated schematically in Fig. 9.23.

9.5.4 Multiple site fatigue

During the last 25 years a new form of fatigue failure in aircraft, multiple site fatigue, has been recognised. Most analyses of fatigue damage have been limited to the study of the growth of a single fatigue crack. However, in ageing aircraft fatigue cracks can initiate at a number of similar sites usually at the edge of rivet holes in lap joints in the fuselage. The initiation of these fatigue cracks is often aided by corrosion. Paradoxically, the problem has arisen partly because of the excellent design against fatigue of modern aircraft that has enabled them to keep flying long past their original design life. The accident that first highlighted multiple site fatigue occurred in 1988 to an Aloha Airlines Boeing 737-200 during a flight from the island of Hilo to Honolulu in Hawaii.[177] The aircraft was 19 years old. As many as 60% of US manufactured aircraft are still flying after they are 20 years old and 19 years is not an unusually long life, but the Aloha Airlines operating in Hawaii have very short flights. The B737 had flown an average of 13 flights a day over the 19 years accumulating 35,493 flight hours and more significantly 89,090 pressurisation cycles. After the aircraft levelled out at the cruising altitude of 7,315 m, the fuselage over the first class section tore off; a photograph of the aircraft taken moments after it landed is shown in Plate 1. Miraculously, only a flight attendant in the first class section was sucked out in the resulting decompression and the pilot managed to land the aircraft eleven minutes later without any further fatalities. The initial fracture occurred in the lap joint between two sections of the skin of the fuselage located just above the cabin windows. The lap joint had three rows of rivets and was in addition cold bonded with epoxy impregnated woven cloth. Boeing issued a service bulletin as early as 1972 warning of the debonding, corrosion and fatigue cracking around the fuselage lap joints in the B-737. Boeing had sent a report to Aloha Airlines just two weeks before the accident raising concerns about the general condition of ageing aircraft and the adequacy of the maintenance program to cope with the corrosion problems in the highly corrosive operational environment of Aloha Airlines.

The fail safe design of the fuselage of the B-737 depended on tear straps which were intended to arrest the propagation of a crack in the skin and to enable safe decompression of the cabin through the torn area of the skin. In the Aloha Airlines' B-737 the effectiveness of the tear straps was compromised by debonding. The section of the fuselage where the fracture initiated was lost, but examination of other B-737 lap joints showed multiple site damage in the form of fatigue cracks around the rivet holes. These fatigue cracks linked up to form a critical crack that led to the catastrophic failure of the fuselage skin. It is testimony to fail safe design of the B-737 that it was able it to land without complete loss of the aircraft.

The problem with multiple site fatigue is that the conventional fracture mechanics seems to overestimate the residual strength of a lap joint. It appears that interaction between the plastic zones of the fatigue cracks cause them to link up earlier than might be imagined leading to a large reduction in residual strength. However, multiple site fatigue is still not completely understood.

9.6 Concluding Remarks

Prior to 1950, fracture was not seen as a separate discipline. Textbooks on the strength of materials usually only included a last chapter on the mechanics of materials. Griffiths' theory of fracture was largely not considered by engineers before George Irwin began his work on fracture mechanics in the 1950s. Today fracture mechanics is a recognised discipline.

Fracture mechanics is essentially about scaling. Galileo, some three centuries earlier used stress, though the term and concept had to wait until the nineteenth-century, to scale structures and really nothing fundamentally changed until the twentieth-century. Weighardt and Griffith saw that for brittle materials the scaling of fracture depended upon a length scale as well as stress. As with many scientific and engineering advances, fracture mechanics was not immediately accepted and Irwin had to struggle for its recognition. During Irwin's lifetime fracture mechanics grew to maturity. It has been so successful that it is now often seen as a fully mature discipline that needs no further development. Interest and funding of fracture itself has diminished in the last few years, which is potentially dangerous. All is not known about fracture. My father used to say to me that my answers to engineering questions while I was a student were more complex than the question itself, his observation can certainly be applied to fracture.

Perhaps the most successful application of fracture mechanics has been to fatigue crack growth. Fracture mechanics has also been applied to many other forms of fracture such as stress corrosion and creep fracture, but these important aspects of fracture have been omitted to keep a balance in the size of this chapter. The other main development has been in the physics of fracture, which has led to truly remarkable improvements in steel technology so that the spectre of brittle fracture has almost been banished.

9.7 Notes

[1] Rossmanith (1997).
[2] Smith *et al.* (1951).
[3] Irwin and Kies (1952).
[4] For short plates the energy release rate reached a maximum which could cause a fracture to arrest.
[5] Irwin and Kies (1954). The first trial of this method was made by Lubahn (1959).
[6] Wells (1952).
[7] Taylor and Quinney (1934).
[8] Now the Malaysian Rubber Producers' Research Association.
[9] Rivlin and Thomas (1953).
[10] Called pure shear by Rivlin and Thomas, but this description is a misnomer.
[11] Irwin (1956).
[12] Koehler (1941); Mott and Nabarro (1948).
[13] Forscher (1954).
[14] Bowie (1956).
[15] Wells (1955).
[16] Kies and Smith (1955).
[17] Irwin (1957).
[18] Irwin (1958).
[19] Paris (1997).
[20] Irwin (1960).
[21] Special ASTM committee (1960).
[22] Williams (1957).
[23] Cotterell (1966).
[24] Schabtach *et al.* (1956).
[25] Winne and Wundt (1958).
[26] In 1993 the E24 Committee was merged with E09 Committee on Fatigue to become E08 Committee on Fatigue and Fracture.
[27] AS CA65 – 1972; Leicester (1974).
[28] Irwin *et al.* (1958).
[29] Irwin (1964).

[30] McClintock and Irwin (1965).

[31] E399-70T (1970). Note Eq. (9.8) can also be written as $B > 2.5l_{ch}$.

[32] At first Indian ink was which was drawn into the crack during slow crack growth and left behind during fast unstable crack propagation was used to mark the critical crack length, but later it was realised that corrosion by the ink assisted the slow crack growth and this method was abandoned.

[33] Kraft *et al.* (1961).

[34] See §8.2.3.

[35] Barenblatt (1959, 1962).

[36] See §1.2.2.

[37] Dugdale (1960).

[38] Panasyuk (1960) whose paper was in Ukrainian.

[39] Hahn and Rosenfield (1964).

[40] Goodier and Field (1962).

[41] Burdekin and Stone (1966) unaware of the work of Goodier and Field (1962) also independently derived Eq. (9.14).

[42] Cotterell and Atkins (1996).

[43] Wells (1961a).

[44] Bilby *et al.* (1963).

[45] See §8.2.5.

[46] Benbow and Roesler (1957).

[47] Berry (1963).

[48] Cotterell (1965b).

[49] Gol'dstein and Salanik (1974).

[50] Cotterell and Rice (1980).

[51] Broberg (1987).

[52] Pham *et al.* (2006).

[53] Schardin and Struth (1938).

[54] Barstow and Edgerton (1939).

[55] Wallner, H. (1939).

[56] Fréminville (1907, 1914).

[57] Smekal (1950).

[58] Kerkhof's first paper is a brief note (Kerkhof 1953), but later he gave a review of the method (Kerkhof 1973).

[59] Andrews (1959).

[60] Kies *et al.* (1950).

[61] Kambour (1964).

[62] Cotterell (1965a).

[63] Fineberg and Mander (1999).

[64] Lazar and Hall (1959).

[65] Mott (1948).

[66] The ships sailing to Britain from Australia just after World War II had many Australian academics, who had previously been prevented from travelling, including the famous applied mathematician George Batchelor (1920–2002) who travelled to Cambridge a few months earlier than Yoffe.

[67] Eshelby (1949).

[68] Yoffe (1951).

[69] Broberg (1960).

[70] Eshelby (1970).

[71] Ravi-Chandar (1998).

[72] Murphy *et al.* (2006).

[73] Odette and Lucas (1998).

[74] Hall (1951); Petch (1953).

[75] Orowan (1933b).

[76] Stroh (1957).

[77] Cottrell (1958).

[78] Kelly *et al.* (1967).

[79] Rice and Thomson (1974).

[80] Argon (2001).

[81] Cheung and Yip (1994).

[82] See §8.1.4.1.

[83] Hall and Barton (1963).

[84] Hartbower and Pellini (1951).

[85] Puzak *et al.* (1952).

[86] E208-06 (2006).

[87] Cottrell (1995).

[88] Robertson (1951). A full description of the Robertson crack arrest test is given in Christopher *et al.* (1968).

[89] There was considerable crack tunneling and the arrested crack front had a parabolic form. The arrest point was taken as the approximate focus of the parabolic shape.

[90] Christopher *et al.* (1968).

[91] Feely *et al.* (19554). Developed originally by Standard Oil and called the SOD test, its name changed with the change in the company name.

[92] Yoshiki and Kanazawa (1958).

[93] Irwin and Wells (1965).

[94] Crosley and Ripling (1969, 1980).

[95] Hahn *et al.* (1973); Hoagland *et al.* (1970).

[96] Kalthoff (1985).

[97] E1221-06 (2006).

[98] Kennedy (1945).

[99] Greene (1949).

[100] Perhaps his most noteworthy invention is the *Wells Turbine* where wave power is used to produce compressed air that drives the turbine to generate electric power. The first sizeable turbine capable of generating half a megawatt power was installed on the Island of Islay in Scotland in 2001. A 75 MW generator using 16 Wells Turbines will be completed in Mutriku, Spain in the winter of 2008/2009 and it may be that in the future Wells, like Griffith, will be better known for his work on power generation than fracture.

[101] Wells (1956a).

[102] Wells (1956b, 1961b).

[103] Wells and Burdekin (1963).

[104] See Hall *et al.* (1967).

[105] Sumpter and Kent (2004).

[106] Millman (1999).

[107] The 1941war time specification did not specify the Charpy impact values but, though an additional clause was added, it was not enforced.

[108] The King-street Bridge could have been successfully built in BS 968 if all parties to its design and construction had been competent.

[109] The crack velocity of the ductile fractures in pipelines was only about 60 m/s and did not run far, but that was small consolation.

[110] Ouchi (2001).

[111] Wells (1963).

[112] E1820-06e1 (2006).

[113] Wu *et al.* (1995).

[114] Harrison *et al.* (1968); Burdekin and Dawes (1971).

[115] Harrison *et al.* (1977).

[116] BS 7910:2005

[117] McClintock (1968).

[118] Andersson (1973); Shih *et al.* (1979).

[119] Newmann *et al.* (2003).

[120] Rice (1965).

[121] Rice (1968).

[122] Hutchinson (1983).

[123] Eshelby (1951).

[124] Bilby and Eshelby (1968).

[125] Cherepanov (1968).

[126] Begley and Landes (1972).

[127] Rice *et al.* (1973).

[128] Rice and Rosengren (1968); Hutchinson (1968).

[129] Akin to the T-stress in an elastic field.

[130] O'Dowd and Shih (1991, 1992).

[131] Shih *et al.* (1993).

[132] Hutchinson and Paris (1979).

[133] It was Gurney who introduced the symbol R for fracture energy.

[134] Gurney and Hunt (1967), Gurney and Ngam (1971), Gurney and Mai (1972).

[135] Atkins (1980).

[136] Atkins (1987a).

[137] Atkins *et al.* (1998a).

[138] Atkins *et al.* (2003); Cotterell *et al.* (2000).

[139] See Chapter 11 for examples were this method is used in cutting.

[140] Atkins (1987b).

[141] Reddy and Reid (1986).

[142] Broberg (1975).

[143] Cotterell and Reddel (1977).

[144] Williams and Rink (2007).

[145] Hillerborg (1976).
[146] Tvergaard and Hutchinson (1992).
[147] Rice *et al.* (1980).
[148] Gurson (1977); Tvergaard (1981).
[149] Xia and Fong (1995a, 1995b); Xia *et al.* (1995).
[150] Gologanu *et al.* (1995).
[151] Thomason (1985, 1990).
[152] Pardoen and Hutchinson (2000).
[153] Pardoen and Hutchinson (2003).
[154] Xia and Fong (1996).
[155] Ritchie *et al.* (1973).
[156] Made into the 1951 film *No Highway in the Sky* starring James Stewart.
[157] Withey (1997).
[158] Negaard (1980).
[159] Coffin (1954): Manson (1954).
[160] Forsyth (1962).
[161] Moore (1927); De Forest (1936); Bennett (1946).
[162] Head (1953a).
[163] Head (1953b).
[164] Frost and Dugdale (1958).
[165] Paris (1957).
[166] Paris *et al.* (1961).
[167] Paris and Erodogan (1963).
[168] Paris *et al.* (1972).
[169] Lindley *et al.* (1975).
[170] Rice (1967a).
[171] Eber (1970).
[172] Budiansky and Hutchinson (1978).
[173] Wu *et al.* (1992).
[174] Pearson (1975).
[175] Miller (1987a,b).
[176] Suresh (1998).
[177] Pitt and Jones (1997).

Chapter 10

The Diversity of Materials and Their Fracture Behaviour

The first materials used by man for tools or to construct artefacts were natural ones like wood, skin, bone and stone, but the need for more diverse materials saw the development of man-made materials. In this chapter the fracture of ceramics, cementitious materials, polymers and composites are discussed. The fracture of metals has been covered in the previous chapters.

Pottery, the earliest form of ceramics, has been found in Japan dating to 14,000 BC but, though it is hard, it is brittle. Ceramics are important today because of their hardness and high temperature properties. Transformation toughen ceramics have been developed in the second half of the twentieth century but they have not proved to be the wonder material they were first thought to be.

The Romans developed hydraulic cement that was more durable than the earlier non-hydraulic cements and could harden under water. This new cement enabled the Romans to build bridges and buildings that have lasted to the present time. However, the knowledge of hydraulic concrete died with the Roman Empire and was not rediscovered again until John Smeaton, who rebuilt the Eddystone Lighthouse in 1756, found that a hydraulic lime could be made from limestone containing a considerable amount of clay. Concrete regained its place as a construction material when Joseph Aspdin (1778–1855) patented *Portland cement*, so-called because of its resemblance to Portland stone, the most prestigious building stone in England at the time. Concrete and mortar are strictly composite ceramics, but because of their importance they are usually treated separately as they will be here.

Plastics are now ubiquitous materials, but the first man-made plastic *Parkesine*, the forerunner of celluloid was exhibited by its inventor Alexander Parkes (1813–1890) at the 1862 International Exhibition. At first plastics were seen as cheap substitutes for superior natural materials and did not find wide

289

scale use until after World War II. Tough and strong plastics have now been developed that can take the place of metals in many applications and a world without them is inconceivable to someone born after the Second World War. Most modern adhesives are polymers.

Nearly all biological materials, like wood and bone, are composites and hence composites are the oldest of materials. However composites have only been intensely developed by man in the twentieth century. Although the term composite covers many combinations of materials, it is fibre-reinforced plastics in the aircraft industry that have most captured the imagination. Composites account for 16% of the weight of the Airbus A380 and an incredible 50% of the weight of the Boeing 787 Dreamliner and, since fibre-reinforced plastics are lighter than aluminium, the percentage by volume is even greater.

In the twenty-first century the biggest challenge to civilisation will probably be climate change brought about by a profligate use of energy. Thus materials which require little energy in their production need to be developed. The energy cost to produce a unit volume of some metal, ceramic, plastic, and composite materials are shown in Table 10.1. The approximate energy and financial costs for the same tensile strength relative to that of mild steel are also shown in this table. It is clear that there is a large difference in the relative costs of different materials.[1]

Table 10.1 Approximate energy and financial costs of some metals, ceramics, plastics, and composite materials relative to mild steel.

Material	Energy cost (GJ/m^3)	Energy cost for the same tensile strength as mild steel	Financial cost for the same tensile strength as mild steel
Mild steel	450	1	1
Aluminium	550	2	2
Dense alumina	800	4	13
Macro defect free cement	10	0.1	1
Polyethylene	100	3	0.3
Carbon fibre reinforced plastic	1600	0.8	350

10.1 Ceramics

Stone, a natural ceramic, was one of the first materials used by man. Pottery was the first ceramic manufactured by man and is basically composed of silicates.

Ceramics are usually defined as a combination of one or more metallic element, or semi-metallic element such as silicon, with a non-metallic element, frequently oxygen. They are usually crystalline, but can be glasses. The bonds of ceramics which are compounds of metallic and non-metallic elements such as alumina are primarily ionic whereas the bonds of those that are compounds of two non-metallic elements like silica or are pure elements like diamond or silicon are predominantly covalent. Cement and concrete are also ceramics.

The word *ceramic* was first used to describe pottery in 1850.[2] Ceramics are inert, hard, and resistant to wear which is why they have been a favourite material class since our hominid forebears. Their inertness may be an answer to green power from atomic reactors with safe nuclear waste disposal. Synroc, is a polyphase titanate ceramic designed to immobilise high-level nuclear waste within its crystal structure and was invented in 1978 by a team led by Ted Ringwood (1930–1993) at the Australian National University, Canberra. At the moment, though synroc is very inert, water only dissolves 300 nm/year from its surface, it is not inert enough, but if there is an answer to containing nuclear waste it will be a ceramic.

Ceramics can withstand very high temperatures, which makes them key materials for improving the efficiency of energy generation. The efficiency of gas turbines can be increased by raising the operating temperature. At the moment ceramics are used as coatings for metal blades, but the aim is for un-cooled ceramic blades. However, the problem with ceramic blades is their low fracture toughness. Ceramics are also the key materials for the development of energy efficient solid-state fuel cells with very thin electrodes and electrolytes. Development of high toughness ceramics, especially at high temperature, is the key to the development of these applications and many more. Ceramic composites can be made that are tougher than a homogeneous ceramic material.

Piezoelectric ceramics, where mechanical and electrical strains are coupled, have become increasing used for a wide variety of electronic and mechatronic devices. Fracture in these ceramics is more complex than that in conventional ceramics, because of the non-linear nature of the mechanical and electrical behaviour and the complicated coupling between mechanical and electric fields. A simple treatment of the fracture of piezoelectric ceramics is not possible and they will not be discussed here and the reader is referred to an excellent review by Tong-Yi Zhang and Cun-Fa Gao.[3]

10.1.1 Processing

The processing of ceramics greatly affects their strength and toughness. Crystalline ceramics cannot economically be cast from the molten state because they have a very high melting point and also if cast uncontrolled grain growth causes them to be friable. Glass-ceramics, which were developed from research on photosensitive glasses that crystallised when exposed to UV light or X-rays,[4] are a glass at the working temperature but contain nucleating agents such as titanium dioxide, which cause crystallisation on cooling into tiny crystals, generally smaller than a micron. They are much stronger than glass and can be made with a very low coefficient of thermal expansion. Thus the glass-ceramic, Pyroceram®, developed by Corning in the 1960s is used for cooking ware that can be heated on the top of a stove. Clay products become plastic when water is added and can easily be formed into complex shapes. There is considerable shrinkage during drying to remove water, but pores between clay particles will remain. During firing at between 900 and 1,400°C complex reactions occur. Vitrification or the formation of a glass occurs with some of the constituents in the clay. The liquid glass tends to fill the pores, but some pores remain. The temperature of firing has a large effect on the porosity of the pottery. Bricks fired at 900°C are quite porous whereas highly vitrified translucent porcelain is fired at up to 1,340°C and has almost no pores. Full vitrification cannot be achieved or the body would become soft and slump. High performance ceramics with very few pores are usually solid-state sintered from compacted fine power or hot-pressed at temperatures of 1,500 to 1,800°C. The surface area of fine powders is enormous. A kilogramme of alumina powder with a particle size of one micron has a surface area of about 1,500 m^2, which is roughly the area of four basketball courts. It is the surface energy (1.5kJ for a kilogramme of alumina) associated with this surface area that is the driving force for sintering.

10.1.2 Mechanical properties

The Young's moduli of ceramics are generally larger than those of metals because of the strong ionic and covalent bonds. Yield occurs in crystalline materials by dislocation movement and the yield strength depends upon the theoretical shear strength and the width of influence of dislocations. Rudolf Peierls (1907–1995) and Frank Nabarro (1916–2006) showed in the 1940s that the shear strength is proportional to the shear modulus and increases with decrease the dislocation width.[5] Pure metals have comparatively low shear strengths because dislocations in their close-packed crystal structures are wide and alloying is required to increase their yield strength. Ceramics have both high elastic modulus and narrow dislocations, which combined, give very high yield strengths of the order of 5 GPa. Except at very high temperatures, ceramics in tension fracture before yielding and their yield strength has to be inferred from their hardness. The oxide ceramics like alumina have hardness of 10–15 GPa, carbide ceramics like silicon carbide have hardness over 20 GPa, and diamond has a hardness of 80 GPa. Hardness, the very property that made ceramics desirable in the first place, unfortunately causes them to be brittle because the plastic deformation associated with fracture is tiny. We have already seen in Chapter 9 that glass has a fracture energy of only about 8 J/m^2, diamond has a fracture energy of only about twice this value the toughest high performance ceramics have a fracture energy of the order of 1000 J/m^2. This brittleness is a limiting factor in the use of ceramics.

10.1.3 Fracture

In the absence of plastic deformation the toughness of ceramics is limited and LEFM is valid for all ceramics. The strength and reliability of ceramics can be increased by controlling the size of processing flaws which can be surface cracks, voids, inclusions, or grain boundaries. Control has to start at the powder which must be free of inhomogeneities. Colloid science provides one method of producing homogeneous powders. The powder is processed as slurry and the liquid removed by pressure filtration. During sintering grain growth has to be controlled. The other route to increased strength and reliability is to increase the toughness of the ceramic.

Nothing can be done to increase the inherent toughness of ceramics, but microstructural features can increase the effective toughness. Fracture steps with buried cracks form during transgranular fracture but these do not contribute to significant toughening.[6] Since the fractures in polycrystalline ceramics are usually intergranular there is a small increase in toughness caused by the deflection of the crack and its tortuous path but this too is not very significant.[6] What can produce significant toughening is bridging of the faces of a crack by grains. The bridges hold the surfaces together and impose a negative shielding stress intensity factor on the crack tip. The shielding stress intensity factor develops with crack growth as bridges are left in the wake of the crack to give crack growth resistance causing the fracture energy to increases from its inherent value, R_0, up to a maximum plateau value, R_m. Most bridges eventually pull-out completely from the crack surfaces but some of the bridging grains fracture.[7] In Fig. 10.1 developing bridging grains are illustrated showing both pull-out and transgranular fracture. The pull-out is resisted by frictional forces and the bridging stress, σ_p, decreases with the pull-out, δ, down to zero when the pull-out, δ_c, is of the order of the grain size.[8] For long straight cracks the fracture energy to a steady state value, R_m, given by

$$R_m = R_0 + W_p,\tag{10.1}$$

where W_p is the specific work of complete pull-out given by

$$W_p = \int_0^{\delta_c} \sigma_p\, d\delta.\tag{10.2}$$

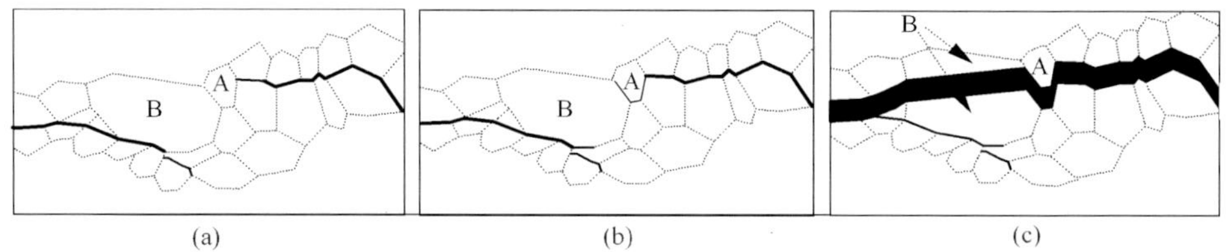

<table>
<tr><td>(a)</td><td>(b)</td><td>(c)</td></tr>
</table>

Fig. 10.1 Sketches of the bridging of crack in a coarse grained alumina at three stages of fracture (a) to (c); the fracture tip is to the right of the sketches; the bridging grain A pulls-out and bridging grain B fractures (after Swanson *et al.* 1987).

Brian Lawn and his colleagues have suggested that the maximum pull-out stress, σ_{pm}, is independent of the grain size, whereas the critical pull-out distance, δ_c, is proportional to the grain size, d.[9] If the shape of the pull-out stress-displacement curve is dependent only on the relative pullout, δ/δ_c, the plateau value, R_m, is a linear function of the grain size. The inherent fracture energy of alumina with a grain size of 16 μm is 20 J/m^2 rising to a maximum plateau value of about 90 J/m^2 after a crack extension of 10 mm.[8] However, the increase in toughness of ceramics caused by a large grain size comes at the expense of a decrease in tensile strength.

The tensile strength of ceramics conforms closely to the Hall–Petch inverse square root grain size relationship[10] as can be seen from Fig. 10.2 where the tensile strength of alumina is shown as a function of the grain size; all the experimental values are very close to the inverse square root relationship except for the smallest grain size of 2.5 μm. The Hall–Petch relationship occurs because the intrinsic flaws in ceramics are proportional to the grain size, but extrinsic flaws are introduced during the manufacture of the tensile specimens which are not directly related to the grain size and these flaws cause a deviation from the Hall–Petch relationship for the smallest grain size. Lawn and his colleagues have used R-curves obtained from strength tests on indented specimens to predict the tensile strength of the alumina specimens, but their results are little different to the Hall–Petch relationship.[9] However, the R-curve does explain the strength of the smallest grain size specimens and they have inferred that the intrinsic flaw size is between 10 and 20 μm. If the intrinsic flaw size is this large, one would expect the results up to a grain size of that order to be controlled by the extrinsic flaw. However, because of the R-curve behaviour, Lawn and his colleagues have shown that the size of the initiating flaw does not have very much effect on the tensile strength for grain sizes greater than about 10 μm.[9] The practical implication of this observation is that the elimination of every small defect when processing ceramics is not necessary.

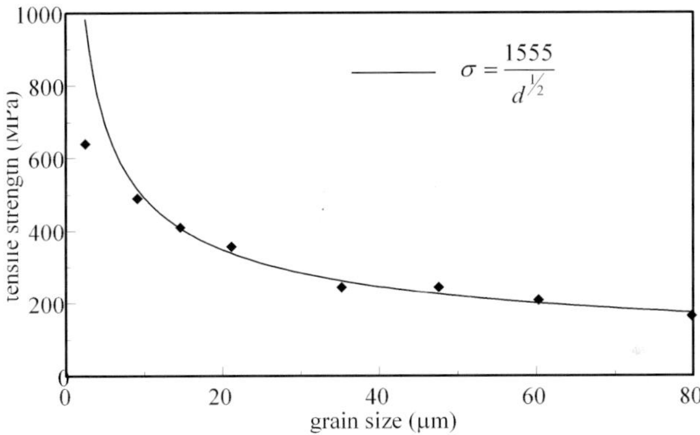

Fig. 10.2 The tensile strength of alumina as a function of the grain size (after Chantikul *et al.* 1990).

10.1.4 *Transformation toughened ceramics*

Transformation toughening of ceramics is based on the martensitic transformation of zirconia from a tetragonal to a monoclinic structure which takes place in unconstrained zirconia with a volumetric dilatational strain of 4 to 5% and a shear strain of 14 to 15%. Zirconium was discovered in the form of the oxide zirconia in 1798 when the German chemist, Martin Klaproth (1743–1817), analysed Sri Lankan zircon which is a cubic crystal of zirconium silicate looking superficially like diamond. Zirconia crystallises from the molten state with a cubic structure which at atmospheric pressure transforms to a tetragonal structure at 2,370°C and to a monoclinic structure at 1,170°C. The early development of zirconia toughened ceramics is due mainly to Ron Garvie, who in 1962 started investigating the high temperature properties of zirconia at the Metallurgical Research Laboratory of the US Bureau of Mines.[11] He used X-ray diffraction to examine the structure of a zirconia powder and found it to be tetragonal. At the time the accepted wisdom was that the tetragonal structure could not be retained at room temperature. Garvie suggested that since the tetragonal structure had a lower surface energy than the monoclinic then it would be possible for the tetragonal structure to be stable at room temperature provided the particle size was smaller than a critical value.[11] From that time onwards, Garvie made the study of zirconia his life's work, changing jobs and countries when necessary to pursue it. Pure zirconia is not itself a suitable structural material because the large volume change when it transforms from a tetragonal to monoclinic structure causes destructive cracking. Garvie worked at methods of partially stabilising the tetragonal phase by alloying the zirconia with calcia (CaO), which lowers the phase transformation temperatures to produce partially stabilised zirconia, a mixture of monoclinic, tetragonal, and cubic structures. A hydrostatic tensile stress can trigger the transformation from the tetragonal to the monoclinc structure. Other metal oxides, such as magnesia and yttria can similarly produce a partially stabilised zirconia. In 1967 Garvie attended a seminar by Earl Parker (1912–1988) one of the inventors of *Transformation Induced Plasticity* (TRIP) steels. TRIP steels undergo a plastic deformation induced martensitic transformation which induces toughness and in a discussion it was suggested that a similar transformation could be occurring in calcia partially stabilised zirconia (Ca-PSZ). After this discussion Garvie produced a simple theory of toughening in these ceramics which he presented to a sceptical audience at a joint meeting of the American and Canadian Ceramic Societies in 1969. One member of the audience, Arthur Heuer, was not sceptical and became another pioneer in transformation toughened ceramics. In 1972 Garvie joined the Division of Materials Science, in the Commonwealth Scientific Industrial Research Organisation (CSIRO), Australia, where the landmark work describing the high strength of bend specimens especially those ground (600 MPa) rather than ground and polished (490 MPa) was performed.[12] X-ray diffraction studies on the surfaces of the specimens showed that the diamond grinding had caused more transformation of the zirconia to the monoclinic structure so the as ground specimens would have had the larger residual compressive surface stress, but most of the strength increase over the value for PSZ with a purely monoclinic structure (250 MPa) came from an increase in the fracture energy of the Ca-PSZ which was 500 J/m^2. Manfred Rühle and Tony Evans gave a review of high toughness ceramics in 1989.[13]

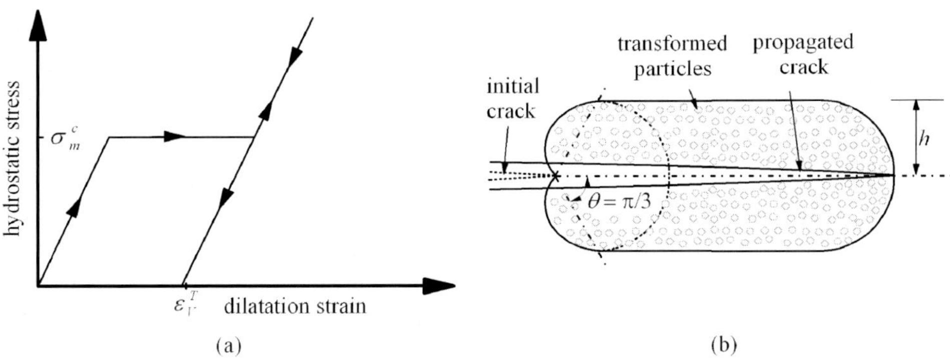

Fig. 10.3 (a) Schematic representation of unconstrained hydrostatic stress versus dilatational strain for tetragonal PSZ. (b) Schematic illustration of the transformed region at the tip of a propagating crack (after McMeeking and Evans 1980).

During crack initiation and propagation those tetragonal particles which experience a hydrostatic stress equal to or greater than some critical value σ_m^c will transform to monoclinic with a free dilatational strain of ε_v^T as shown in Fig. 10.3 (a). The dilatation within the process zone is constrained by the untransformed PSZ outside of it and compressive stresses are developed. At initiation the sectors of the transformed zone where $\theta < \pi/3$ induce a positive stress intensity factor on the crack tip whereas the sectors $\theta > \pi/3$ induce an equal but negative stress intensity factor so the net effective shielding is nil. However, as the crack propagates and a dilatational zone is left in its wake (see Fig. 10.3 (b)) that causes a negative stress intensity factor to be induced at the crack tip. Thus the crack tip is shielded. The shielding stress intensity factor rises to a plateau value after a crack propagation of about five times the half height of the transformed zone, h. Bob McMeeking and Tony Evans calculated that the plateau value of the shielding stress intensity factor, K_T, induced at the crack tip by the transformed zone is

$$K_T = \frac{0.22 E \varepsilon_v^T \mathrm{v}_p \sqrt{h}}{(1-v)}, \qquad (10.3)$$

where v_p is the volume fraction of the tetragonal particles, and the apparent plateau fracture toughness is $(K_0 + K_T)$ where K_0 is the initiation fracture toughness.[14] Comparison of the fracture toughness predicted by Eq. (10.3) with

                    Fracture and Life

experimental values showed that the toughness was under estimated. If the transformation is not activated by the hydrostatic crack tip field, but by shear bands at $\theta = \pi/3$, there are no deleterious transformations in the front sector and the factor 0.22 in Eq. (10.3) is increased to 0.38.[15] The predictions are then much nearer the experimental results. The maximum plateau toughness attainable thorough transformation toughening approaches 20 MPa√m ($R = 2,000$ J/m^2) and is attained after a crack growth of about 500 μm. Ceramics are in demand in high temperature applications, such as gas turbine blades, but unfortunately transformation toughened zirconia cannot be used in these applications. If the difference between the operating temperature and the martensitic start temperature is too large there is no transformation toughening effect. The operating temperature of transformation toughened zirconia is limited to about 300°C which greatly reduces its application.[16]

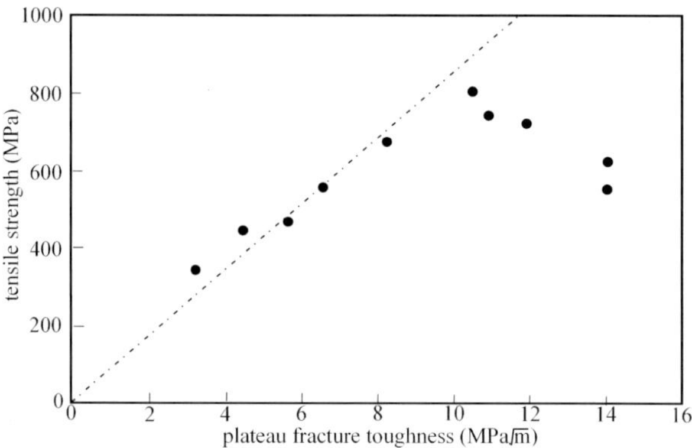

Fig. 10.4 Tensile strength as a function of plateau fracture toughness for Mg-PSZ (after Swain & Rose 1986).

The toughness of PSZ depends on the volume fraction of the tetragonal zirconia particles and one might expect that the maximum toughness would correspond to the maximum strength assuming that the flaw size is reasonably constant. However, it is found that the maximum strength is developed at less than the maximum toughness as can be seen in the results for Mg-PSZ shown in Fig. 10.4. Up to plateau fracture toughness of 9 MPa√m the strength is proportional to the fracture toughness and the slope corresponds to a flaw of about 25 μm.[17] Two reasons for the strength decreasing at higher toughness have been advanced.[15,17] The first is that the crack propagation to attain the plateau value of the fracture toughness increases with toughness and at high toughness values fracture instability occurs before the plateau vale is reached. The second reason is that the high fracture toughness values are reached because the critical transformation stress is low and the process zone width, h, is large. Under these conditions shear bands can precede fracture and cause microcracks as has been observed in steel. Thus when the critical shear stress is small, the onset of yielding causes fracture.

Toughening can also result from microcracking at a crack tip.[18] The microcracking can be initiated either by the dilatational transformation of zirconia from tetragonal to monoclinic, or by a residual tensile stress field formed by thermal expansion mismatch between different particles that in itself is not sufficient to cause microcracking but high enough to cause microcracking in the presence of the additional stress ahead of a crack tip. The materials that have exploited toughening by microcracking are alumina/zirconia, silicon nitride/silicon carbide and silicon carbide/titanium boride. Energy is consumed in the microcracking and contributes to the toughening, but this contribution is minor. The major toughening effect comes from the work performed in expanding the microcracks which is not fully recovered because the microcracks do not fully close. Microcrack toughening is not as powerful as transformation toughening and can only increase the fracture toughness to about 10 MPa√m ($R = 500$ J/m^2).

10.1.5 Cyclic and static fatigue

Metals are susceptible to both cyclic fatigue and time dependent corrosion cracking. Ceramics are susceptible to what is termed by the ceramics community as *static fatigue*, where the strength is dependent on the duration of the load and is akin to stress corrosion, only the crystalline ceramics like alumina, which show an R-curve behaviour, are susceptible to a mechanical effect in cyclic fatigue. The susceptibility of glass to static fatigue was observed first by Louis Grenet (1873–1948) who in 1899 noted a loading rate dependence of strength.[19] The strength of glass in air as a function of loading time obtained in 1939 by Theodore Baker at the Preston Laboratories,[20] Toledo, Ohio, is shown in Fig. 10.5.[21] Moisture in air causes the static fatigue observed here in Fig. 10.5. Terry Michalske and Stephen Freiman were the first give the reason for the susceptibility glass to water in 1982.[22] Moisture attacks the bonds at the tip of a crack in glass. In a water-free environment such as a vacuum, fracture occurs spontaneously when the stress intensity factor reaches the fracture toughness of glass. In a moist environment the crack growth rate increases exponentially with the stress intensity factor, reaching a plateau value when the stress intensity approaches the fracture toughness because the moisture transport cannot keep up with the crack growth. Spontaneous fracture then occurs when the stress intensity factor reaches the fracture toughness.

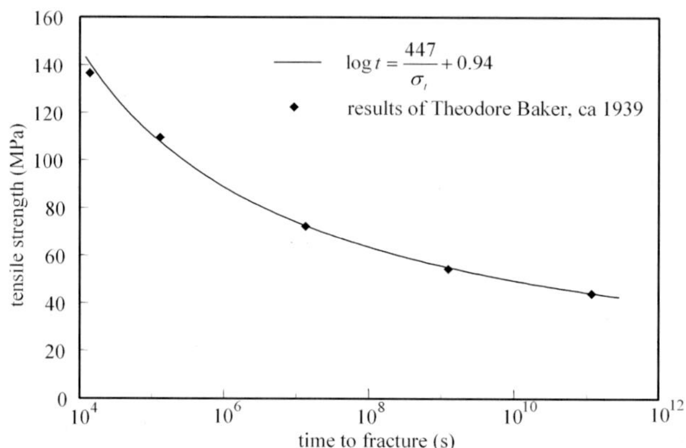

Fig. 10.5 Tensile strength of soda lime glass as a function of time (after Preston 1942).

The molecular structure of glass and the polar character of water enable stress corrosion. Bonds are broken in three stages: (i) water molecules attach themselves to the bridging Si−O−Si bonds; (ii) as the Si−O−Si bond is stretched the water molecule loses an electron to a silicon atom and a proton to the linking oxygen and two new O−H bonds are formed; (iii) the bond ruptures because the polar terminal bonds repel one another. There are some other environments such as ammonia which have a similar structure to water and cause stress corrosion by the same mechanism. Although there are theoretical reasons for the crack growth rate increasing exponentially with the stress intensity factor, the empirical relationship

$$\frac{da}{dt} = AK^n, \tag{10.4}$$

is more usual.[23] Typical values of the power n for stress corrosion of ceramics are in the range 30 to 50, which is much higher than that for metals. Gurney and Pearson demonstrated that glass, while susceptible to static fatigue, is not susceptible to a mechanical effect in cyclic fatigue.[24]

The existence of a mechanical effect in the cyclic fatigue of ceramics was considered unlikely until the late 1970s because of the absence of appreciable crack tip plasticity in most ceramics. The fracture of un-notched specimens subjected to fatigue occurs when a flaw grows to a critical size at which the stress intensity factor at its tip is equal to the fracture toughness. Early analyses of fatigue were made neglecting flaw statistics on the assumption that the flaw which would cause failure in an inert atmosphere is the same as the flaw that leads to failure under stress corrosion. This assumption is only true for uniformly stressed specimens, but in most tests ceramic specimens are subjected to a bending load. In rotational bending the statistical fracture theory always predicts much smaller lifetimes.[25] When data obtained for the lifetimes of alumina specimens, which previously had not shown a definite mechanical effect for cyclic fatigue,[26] were re-analysed using a statistical fracture theory, a definite mechanical effect in cyclic fatigue for the higher loads and frequency was revealed.[25] In polycrystalline alumina the mechanical effect in cyclic fatigue comes from friction wear and disruption of the bridging grains behind the crack tip. Transformation toughened zirconia ceramics also show a mechanical effect in cyclic fatigue, but this mechanism is not fully understood. One problem is that even in transformation toughened ceramics some toughening comes from the bridging of grains behind the crack tip and the degradation of these bridges contributes to cyclic fatigue. Mark Hoffman and his colleagues performed cyclic load tests *in situ* in a SEM on Mg-PSZ and on the basis of these tests suggest that degradation of the bridging grains reduces the critical crack tip stress intensity factor which in turn reduces the width of the transformed zone and hence crack tip shielding.[27]

10.1.6 Refractories and thermal shock

Ceramics are the only materials that have a high enough melting point to be used as refractory linings in furnaces. Because ceramics are brittle they are susceptible to thermal shock that is unavoidable in the operation of a furnace. During sudden cooling of a refractory the surface is put into tension and the centre into compression. The tensile stress at the surface of a slab-like refractory which is suddenly cooled at its surface is given by

$$\sigma = \frac{E\alpha\left(T_a - T_s\right)}{\left(1 - v\right)}, \tag{10.5}$$

                    Fracture and Life

where α is the coefficient of thermal expansion, T_s and T_a are the surface and average temperature of the block.[28] The coefficient of expansion of mullite, an aluminium silicate which is a principal component of fire-clay refractories, has a thermal coefficient of expansion of $5.3.10^{-6}$/deg C and in a dense form a Young's modulus of 70 GPa; thus a temperature difference of about 50°C would cause a stress of 250 MPa, which is above the strength of mullite. Although the analysis of thermal stress fractures in ceramics has a long history, it was the work of David Kingery (1926–2000) in the 1950s that marked the beginning of the modern period. Kingery suggested that there were a number of material resistance factors that could define a materials' resistance to fracture. The first factor is R,[29] which in cases of very large heat transfer coefficients where the surface temperature changes almost instantaneously to that of the environment, is defined as the temperature differential, ΔT_c, that just causes thermal stress fracture[28] or

$$R = \Delta T_c = \frac{\sigma_t (1-\nu)}{E\alpha},\tag{10.6}$$

where σ_t is the tensile strength of the ceramic. Two similar factors R' and R'' have been defined for cases where the heat transfer coefficient is not large and where there is a constant cooling or heating. These three material resistance factors are most suitable for ceramics like glass and porcelain where the design approach is to avoid fracture initiation. Dick Hasselman introduced two more material resistance factors

$$R''' = \frac{E}{\sigma_t^2 (1-\nu)},$$
$$R'''' = \frac{EG_f}{\sigma_t^2 (1-\nu)},\tag{10.7}$$

where G_f is the fracture energy.[30] These two factors are based on fracture mechanics: R''' is the reciprocal of the surface strain energy density at fracture initiation and R'''' compares the fracture energy to the surface strain energy density. These factors assess the ability of cracks to propagate and are more appropriate to granular refractories, R'''' correlates well with thermal shock behaviour as determined by the number of thermal cycles necessary to produce a given weight loss.[31] It should be noted that Hasselman's fourth material parameter is identical to the material's characteristic length, l_{ch}, apart from the factor of $(1-\nu)$.

An alternative assessment of thermal shock damage is to quench a heated refractory and then measure the residual modulus of rupture (MOR). Tapan Gupta performed such tests on small alumina square bars of different grain size and also on a single crystal of sapphire and his results are shown in Fig. 10.6.[32] There is no decrease in MOR until the quenching temperature for the alumina bars is greater than 190°C (for the sapphire bars the critical quench temperature is somewhat higher). The strength of the bars decreases with increase in grain size. For all the grain sizes less than 85 μm there is a sudden drop in MOR at the critical quenching temperature which is more severe the smaller the grain size after which the MOR is constant until finally slowly decreasing with increase in the quenching temperature. The MOR after the drop in value at the critical quenching temperature is greatest for the largest grain size because the length of the cracks formed during quenching is controlled by the fracture toughness which is largest for the large grain size. The behaviour of quenched bars can be qualitatively explained by LEFM.[33]

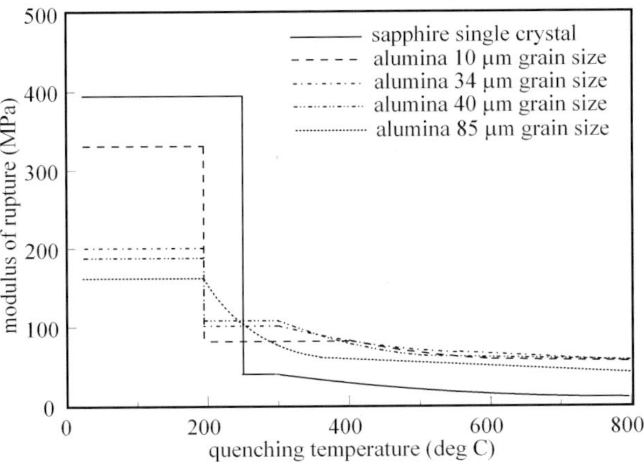

Fig. 10.6 Room temperature moduli of rupture of alumina and sapphire square bars quenched in water at room temperature from various furnace temperatures (after Gupta 1972).

There are two size effects to be considered in the behaviour of quenched square bars of depth W. The first is a thermal size effect which is characterised by the Biot number, β, given by[34]

$$\beta = \frac{Wh}{2k}, \qquad (10.8)$$

where h is the heat transfer coefficient and k is the coefficient of thermal conductivity. If the Biot number is large then heat is extracted from the surface faster than it can be replaced from the interior of the bar by conduction and the surface of the bar very quickly attains that of the quenching medium. For small Biot numbers the surface temperature remains significantly above the quenching temperature for a considerable time. The second size effect is the fracture mechanics one governed by the non-dimensional thickness $\bar{W}$ given by

$$\bar{W} = \frac{W}{l_{ch}}, \tag{10.9}$$

where l_{ch}, is the material's characteristic length. On quenching multiple cracks can initiate from existing pores and flaws in the refractory. Most crack propagation under mechanical loads is unstable, but under thermal stress the crack propagation can be stable and even if at first unstable becomes eventually stable.

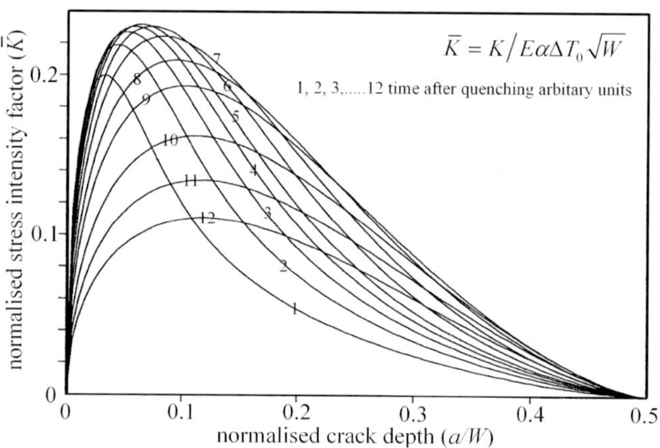

Fig. 10.7 Normalised stress intensity factor as a function of the normalised crack length for increasing time from quenching (after Cotterell *et al.* 1995).

The behaviour of a defect in the form of a single crack gives the qualitative cracking behaviour during quenching. The temperature in a quenched bar, where the Biot number is large, can be found from established solutions at any time after quenching. In Fig. 10.7 the normalised stress intensity factor, $\bar{K} = K/E\alpha\Delta T \sqrt{W}$, where ΔT is the quenching temperature differential, is shown as a function of its normalised crack depth, a/W, for successive times after quenching. The envelope of the curves in Fig. 10.7 determines the degree of cracking under thermal shock. The all time maximum normalised stress intensity factor is 0.23 and occurs for a crack whose normalised depth is 0.065. Hence cracking can only occur if

$$\Delta T > \frac{4.3}{\alpha} \sqrt{\frac{G_f}{WE}}, \tag{10.10}$$

The actual quenching temperature at which cracking occurs depends on the depth of the flaws and can be found from Fig. 10.7. If the depth of the flaws is less than $0.065W$ then the crack growth is initially unstable and a sudden drop in the MOR occurs at the critical quenching temperature. For larger flaws the crack growth is stable the decrease in the MOR at quenching temperatures greater than the critical is gradual. The normalised inherent flaw size, a_0/l_{ch}, can be estimated to be 0.254 which means that whether the MOR drops suddenly or gradually depends only on the normalised bar thickness, $\overline{W}$.[33] If $\overline{W} > 3.9$ the MOR drops suddenly at the critical quenching temperature and if $\overline{W} < 3.9$ the MOR drops gradually.

10.2 Cement and Concrete

Portland cement is made by firing a mixture of chalk and clay to produce three main products $3CaO.Al_2O_3$, $2CaO.SiO_2$, and $3CaO.SiO_2$. With the addition of water Portland cement hydrates to form a hardened cement paste in two reactions. In the first the $3CaO.Al_2O_3$ hydrates to form $3CaO.Al_2O_3.H_6$ in about four hours. The second reaction that hardens the cement is the hydrating of the calcium silicates to form a tobomorite gel ($3CaO.2SiO_2.H_3$) this reaction is slow and takes many days to complete. The structure of hardened cement consists of grains of $3CaO.Al_2O_3.H_6$ with interlocking spines of tobomorite. Concrete is a mixture of sand and stone aggregate bonded together by the cement. Although cement is comparatively cheap, aggregate and sand are even cheaper so a concrete contains as much of these as is consistent with the workability of the concrete.

Concrete is an important material because of its cheapness, ease of construction, and durability. Its fire resistance is also an important aspect in modern building construction. However, it is not a particularly strong material even in compression where its typical strength of 50 MPa is less than that of most rocks and it is weak in tension where its strength is only of the order of 3 MPa. Generally in modern construction steel reinforcement is embedded in the tensile areas and is designed to take the entire tensile load. It was François Hennebique

(1842–1921) who first devised a system of reinforcing concrete. His building system with structural beams of concrete reinforced with longitudinal bars and stirrups to resist the tensile stresses that he patented in 1892 is essentially the system used to day. Only monolithic structures such as gravity dams, which are primarily under a compressive stress, are constructed of un-reinforced concrete. Until comparatively recently concrete has been assumed to have negligible tensile strength for design purposes. However, the trend is to use concrete more efficiently and in un-reinforced beams and pipes, mass structures, concentrated loads in concrete decks, reinforcement bonds and anchorages, tensile fracture can govern the strength. The application of fracture mechanics to concrete structures has thus received considerable attention in the last thirty years especially by the Réunion Internationale des Laboratoires d'Essais et de Recherches sur les Matériaux et les Constructions, known as RILEM.

10.2.1 *Fracture mechanics of cementitious materials*

The application of fracture mechanics to concrete lagged that to metals for a number of reasons. However, the first attempt at applying LEFM to cementitious materials was made by Maurice Kaplan in 1961.[35] He used the compliance method to determine the fracture energy for two sizes of mortar notched bend specimens, but found that the smaller specimens gave a value 38% smaller than the large specimens. He explained the difference in terms of the amount of slow crack growth before instability, but actually the main reason was the large size of the FPZ. In cementitious materials the FPZ is a region where the crack is bridged by sand in mortar and aggregate in concrete. Thus the FPZ size of mortar is about 30 mm and that of concrete is around 500 mm and even bigger for the large aggregate concrete used in the construction of dams where it can be more than 1500 mm. Cementitious materials are essentially elastic but LEFM is not generally applicable to cementitious laboratory specimens, except for hardened cement paste ones, because the FPZ is not small compared with the specimen size. Yiu-Wing Mai and I reviewed the application of fracture mechanics to cementitious materials in 1996.[36]

It was not until Arne Hillerborg at the Lund Institute of Technology proposed his fictitious crack model for the FPZ in 1976 was fracture mechanics successfully applied to cementitious materials.[37] The FPZ in cementitious materials is long and narrow and Hillerborg proposed that it could be replaced by a fictitious crack across whose faces acts a resisting stress that is a function of the opening of the crack. Hillerborg chose a simple linear stress-displacement relationship for strain-softening and with it successfully modelled the fracture of un-notched concrete beams of different sizes using finite element analysis.[38,39] It was assumed that microcracking initiated when the tension side of the beam reached a critical cohesive stress, f_t, and microcracking localised at one point in the beam because of the strain-softening behaviour and grew into a macro crack. The large FPZ effectively makes small specimens behave with some ductility and to be insensitive to shallow cracks or notches so that the initial flaw size does have to be known. The fictitious crack model has since been used by many researchers.

The other physically sound approach to modelling the FPZ in cementitious materials is the crack band model which has been very actively promoted by Zdeněk Bažant at the Northwestern University. Because the FPZ is a region of strain-softening it presents some challenges in FEM where its width can only be represented a single element with the usual definition of strain.[40] A refinement of the crack band model was the introduction of non-local concepts where the strain or stress depends upon not just the state at a point, but upon the average state in the vicinity of the point. Bažant and his co-workers first made use of a non-local strain, but later suggested that a non-local inelastic stress increment led to a simpler finite element solution.[41] The crack band model is more precise than the fictitious crack model but, unless one is specifically interested in very deep notches or the last stages of crack propagation, the extra preciseness is largely illusory because of the large scatter in results from supposedly identical specimens. As Bažant and Cedolin remark: 'the choice of either [the line crack or the crack band model] is basically a question of computational effectiveness'.[40]

Neither the fictitious crack nor the crack band models are easy to implement and equivalent crack models[42] have been proposed that attempt to accommodate the large FPZ by estimating the size of the equivalent crack as Irwin did for metals[43]. However, because of the large FPZs, cementitious materials possess a crack growth resistance which is size dependent except in very large specimens and the critical equivalent stress intensity factor is size dependent.[36]

10.2.2 Size effect

Structures in all materials are subject to a size effect simply because the energy stored varies as the cube of the dimension but the potential fracture area varies only as the square. Hence the nominal strength of a large concrete structure is smaller than that obtained from laboratory sized specimens. Since concrete structures like gravity dams are truly huge one might expect that size effect would be very important. The problem is that 90% of fracture experiments are made on beams that are 500 mm or less in depth. The failure rate of concrete structures is generally low at about one in a million for the normal size concrete structure, but for very large structures it increases to more than one in a thousand.[44] The failure rate for very large structures might be even higher were it not for the dead load safety factor of 1.4 which masks the size effect.[45] Because of the large safety factor failures are due to many causes and have been attributed to causes unrelated to size effect. However, there are proposals for reducing the dead load safety factor which are potentially dangerous unless the size effect is adequately accommodated.

Fig. 10.8 The St Francis Dam disaster of 1928. (Huntington Library, San Marino, California)

The worst civil engineering failure in the USA during the twentieth century was the failure of the St Francis Dam near Santa Clarita in southern California.[46] The St Francis Dam was an arched gravity dam 62 m high and a maximum width of about 185 m. It was completed in 1926 but did not fill completely until 1928. Five days after the water reached the spillway the dam burst and swept all before it to the Pacific Ocean 34 km away killing as many as six hundred people. The scale of the fracture can be judged from the men in the photograph of the fractured dam shown in Fig. 10.8. The failure was attributed to the collapse of the red sandstone conglomerate beneath the western abutment. Water seeping into the foundations of the dam causes an uplift pressure which is the biggest danger to gravity dams.[47] Bažant considers that, though the main cause of the failure was the poor nature of the rock on which it was built, size effect played some part.[48] The failure of the St Francis Dam caused concern over the larger Hoover Dam which had a similar design.

There is little direct evidence of a size effect in the St Francis Dam failure, but more is known about the failure of the New York State Thruway Bridge over the Schoharie Creek in 1987 which killed ten people. The prime cause of this failure was the collapse of the plinth of pier three that had been undermined by scouring. The depth of the plinth was approximately 5 m. The cohesive strength of the concrete was obtained from the Brazilian tests on cylinders cored from the remains of the plinth. The nominal bending strength of the plinth was calculated using the fictitious crack model using the results of the Brazilian test and fracture energies representative of high strength concrete of the aggregate size used in the plinths. These results showed that the modulus of rupture of the plinth was only about 54% of the value deduced from the laboratory tensile strength.[49]

The largest prestressed concrete girder bridge ever constructed, with a main span of 240 m, connected the two main islands Koror and Babelthuap in the Palau Islands which are located between the Philippines and New Guinea.[50] The depth of the cantilever box girders varied from 14.2 m at the main piers to 3.76 m at the centre. The bridge was completed in 1977. By 1990 the centre of the bridge had sagged by 1.2 m and it was repaired by adding eight post tensioned continuous cables under the deck. The central hinge was removed and flat jacks used to give an additional longitudinal compressive force. The flat jacks were then grouted in place to make the bridge continuous. The repair was completed in July 1996 and during good weather in September it collapsed killing two people. The collapse was caused by a large inclined shear–compression fracture emanating from the Main pier where the box girder was 14.2 m. Litigation and out-of-court settlements between the Palauan government and the engineers involved have meant that the exact cause of the collapse has never been officially reported. The conclusions of Chris Burgoyne and Richard Scantlebury were that the failure was probably caused by local damage from over-enthusiastic scabbling[51] of the surface, combined with an insufficiently robust design of the top flange,[50] but it is likely that the large size of the box girder played a part.

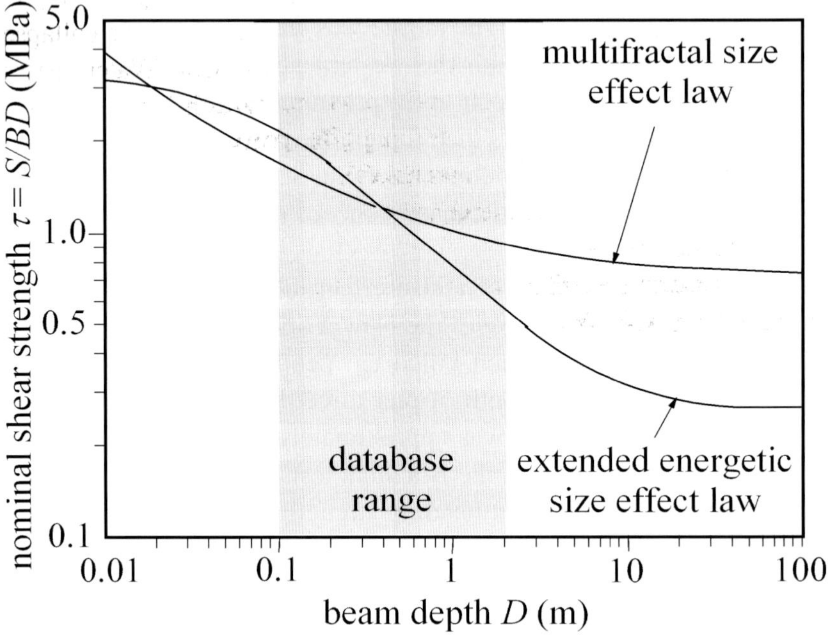

Fig. 10.9 Comparison of Bažant's extended size effect law and Carpinteri's multifractal size effect law, both fitted to shear failures of reinforced concrete beams up to 2 m deep contained in the ACI database, S is the shear force, B is the width of the beam, and D is the depth of the beam (after Bažant and Yavari 2005).

Simple design codes are needed for concrete structures since many thousands large structures are designed each year. In contrast simple design codes are not needed in the aircraft industry such since only a few new large aircraft are designed each decade and sophisticated analyses and large scale tests are affordable. The failures discussed above and others in very large concrete structures demonstrate the necessity of an adequate inclusion of the size effect in the design code. Progress towards such a code has been much delayed because currently there are two fundamentally different theories of size effect structures:

the energetic-statistical theory developed by Zdeněk Bažant in the USA and the multifractal scaling theory developed by Alberto Carpinteri in Italy. Bažant was right in stating that 'the conflict between these two theories is a serious impediment to progress in structural design codes and practice'.[52] In a series of papers Bažant and Carpinteri have been debating the relative merits of the two theories.[53] Bažant's size effect law is based on established energetic principles extended by the inclusion of a non-local form of Weibull's statistical theory, whereas Carpinteri's law is based on fractal geometry a relatively new branch of mathematics not yet fully established in fracture mechanics. The problem is that Carpinteri's theory predicts a smaller size effect for large structures than that of Bažant and it is imperative that the question of which theory most accurately predicts size effect be resolved. Here the best fit of the two size effect laws fitted to results in the American Concrete Institute's database for shear failures of reinforced beams (maximum depth 2 m) without stirrups are compared in Fig. 10.9 without comment. A balanced review doing justice to both of these two theories cannot be undertaken in a book of this nature.

10.2.3 Macro defect free cement

The tensile strength of cement paste is low and it cannot be used to carry load. However, cement paste is not inherently weak, its porous nature makes it so. In the 1980s a macro-defect-free (MDF) cement was developed by James Birchall (1930–1995) and his co-workers which has a flexural strength of 150 MPa, a fracture energy of about 110 J/m^2, and a Young's modulus of 40 GPa.[54] The original MDF cement was based on calcium aluminate cement, but subsequently other MDF cements have been developed based on sulphate-aluminate-ferrite belitic (SAFB) clinkers and or Portland cements. In MDF cements the general porosity is greatly reduced and large macro pores eliminated, by premixing the cement with water-soluble polymers such as polyvinyl alcohol acetate (PVAA) to aid the rheology and using a low water cement ratio so that workable dough is produced from which trapped air can be removed. The MDF cement is typically rolled or pressed into sheets. In the original conception the role of the water-soluble polymer was mainly as a rheological aid, but later work has shown there is a chemical interaction between the polymer and cement.[55] Birchall and his co-workers considered that classic LEFM could be applied to MDF cement and that its strength came solely from the elimination of large flaws. However, it was subsequently shown that MDF cement possesses a FPZ and that the crack resistance increases from 1.4 MPa√m to 2.1 MPa√m over a crack growth of 3 mm.[56] The predominant crack resistance mechanism is due to untorn ligaments between offset cracks; other less significant contributions come from frictional interlocking of adjacent grains on the fracture plane and tearing of polymer fibrils.

Cement is energy efficient and the energy and financial costs of MDF cements are low. It this type of material that should be exploited in the twenty-fist century, but MDF cement has problems that have prevented its exploitation. The prime problem is that under high humidity the PVAA and the interphase regions absorb moisture and the strength degrades. The strength of MDF cement drops to about half after 200 days exposure to 100% humidity.[57] The problem with humidity may be overcome. One method is to cross link the PVAA using an organotitanate crosslinking agent. The initial strength of such modified MDF cement has a lower initial strength of 140–165 MPa, but the degradation in strength under high humidity is much less being still 145–155 MPa after 200 days at 100% humidity.[57]

10.3 Polymers

Polymers are characterised by long carbon chain molecules bonded together by strong covalent bonds. However, they are comparatively weak and more importantly not very stiff because of their structure. When the carbon chains are aligned, as in aramid fibres made from aromatic polyamides, polymers can be both very strong and stiff.

There are a very wide variety of polymers. They can be amorphous or crystalline, a linear thermoplastic, a lightly cross linked elastomer, or a heavily cross linked thermoset, an unfilled single polymer, plasticized, a polymer blend, or filled with fine particles. Their properties depend very much on the temperature and strain rate. In addition, mildly aggressive environments such as moisture and organic solvents can have a large effect on their fracture. From a macroscopic fracture view, polymers can be classified as hard glassy (non-crystalline) and brittle polymers, and ductile semicrystalline polymers.

The toughness of polymers, like metals, comes from energy dissipating deformation at the tip of a crack, but is far less than most metals. There are two modes of non-elastic deformation in polymers: *shear yielding* which takes place with very little change in volume, and *crazing* which takes place with large changes in volume.

10.3.1 Deformation modes

Shear yielding can be diffuse or localized in shear bands. Localization of shear occurs in linear polymers because strain softening occurs when the long polymer chains disentangle and the deformation becomes easier. Often under tension fracture intervenes before shear yielding, but it can be observed under compression load. Shear bands were first observed under compression in polystyrene by Wells Whitney in 1963.[58] The polymer within a shear band is orientated and in transparent polymers they are highly birefringent and are most clearly observed in transmitted light. Shear bands form in a wide range of linear polymers both amorphous and semicrystalline as well as in cross-linked polymers such as epoxy. The shear bands are very thin approximately 100 nm thick but form in clusters of about 1 μm thick.[59] The shear strain within the bands is high and of the order of 1 to 2 in polystyrene.

An alternative mechanism of non-elastic deformation to shear yielding is crazing which was first recognized as a characteristic of glassy linear polymers such as polystyrene but they occur also in semicrystalline polymers. Crazes form under a tensile stress, especially in the presence of moisture or solvents which reduce the stress necessary for crazing. Superficially crazes look like cracks in and they were first showed to be able to sustain a tensile stress by John Sauer and Chih Hsiao at Penn State in 1949.[60] Later Walter Niegisch showed that crazes contained matter which he suggested was in the form of oriented polymer.[61] Roger Kambour showed in 1962 that the refractive index of the craze material is less than that of the polymer, which is the reason crazes look like cracks to the eye.[62] Crazes form normal to the maximum principal stress with a large increase in volume. Kambour wrote a series of classic papers in the 1960s which culminated in his review of 1973 summarizing the knowledge of crazes up to that date.[63] Crazes form by the pulling out of the polymer into long orientated fibrils with voids in between them. The demarcation between the bulk polymer and the craze is very sharp. Short cross-tie fibrils also run between the main fibrils giving the craze some lateral load bearing capacity. The mechanisms, by which the polymer chains become disentangled to form the fibrils, depend upon the temperature. At temperatures about 100°C below the glass transition temperature the mobility of the polymer chains is very low and the chains break rather than slip under high stress. At higher temperatures the polymer chains can disentangle by reptation or snake-like creeping. Near the transition temperature the polymers chains can slip because of the weak van der Waals forces between the chains. Crazing cannot occur in polymers that are significantly cross-linked. Interactions between shear bands and crazes can occur.[59,63] A recent review of crazing has been given by Hans-Henning Kausch and Goerg Michler.[64]

10.3.2 Glassy polymers

The glassy amorphous polymers are transparent and often called organic glasses. The linear polymers polystyrene (PS) and polymethyl methacrylate (PMMA) were first manufactured on a large scale just before World War II in Britain, USA, and Germany. During and after the war PMMA was used for aircraft cockpit canopies. Glassy polymers also include cross-linked thermosets such as epoxy which was first synthesised in Switzerland by Pierre Castan (1895–1985) in 1936 and commercialized by Ciba in 1946. Below their glass transition temperatures, organic glasses are brittle, but not as brittle as inorganic glass. A craze or crazes form at the tip of a notch in linear amorphous polymers and the crack grows within the craze material as is shown schematically in Fig. 10.10. Energy is dissipated in deforming the craze during fracture so that the fracture energies of polystyrene and PMMA are about 300 and 600 J/m^2 respectively whereas soda lime glass has a fracture of energy of only about 10 J/m^2. Polycarbonate (PC) was developed in both the USA and Germany in 1953[65] and is marketed on the basis of being a transparent material with high ductility suitable for demanding glazing applications, but it too is brittle in the presence of a notch especially in thick sections. Cross-linked glassy theromosets are usually more brittle than linear polymers because of their limited plasticity. The fracture energy of epoxy is about 200 J/m^2. Gordon Williams surveyed the fracture of glassy and other polymers in 1984.[66]

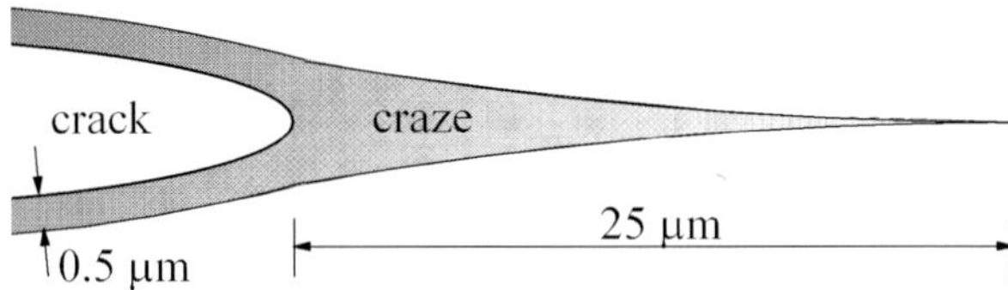

Fig. 10.10 Craze at the tip of a crack, the dimensions are representative of a craze in PMMA (after Kambour 1966).

The measurement of the fracture toughness of glassy polymers presents little real difficulty and some of the first LEFM tests were made on PMMA and cellulose acetate sheets.[67] However, there are complications apart from the strain rate sensitivity, which caused some of the early measurements to be inaccurate. Obtaining a sharp crack needs care. In PMMA a fine saw cut can be extended by pressing a razor blade into its tip to produce a clean sharp crack with a single craze. Even in PMMA a new sharp blade must be used or the force needed to wedge the slot will be excessive and multiple crazes will be produced that cause an artificially high fracture toughness. Polystyrene is notoriously brittle when not toughened, but the early toughness measured by Benbow and Roesler[68] and others in the 1950s and 1960s of around 5000 J/m^2 was artificially high and falsely indicated that PS was tougher than PMMA. In careful experiments Gordon Williams and his co-workers showed that it is impossible to produce a sharp crack in polystyrene without inducing multiple crazing by wedging and fatigue sharpening is necessary. Using fatigue sharpened cracks they found that the room temperature fracture toughness of PS to be 1.05 MPa$\sqrt{m}$ which corresponds to a fracture energy of 340 J/m^2.[69] The Young's modulus of PMMA is very rate sensitive and it is difficult to obtain a comparative fracture energy.[70] Tony Atkins and his co-workers used the Gurney method[71] to obtain the fracture toughness which does not require knowledge of the Young's modulus and representative values of the fracture energy and fracture toughness for PMMA are about 600 J/m^2 and 1.5 MPa$\sqrt{m}$ respectively.[72] Williams and his co-workers have shown that a critical crack tip opening displacement provides an almost unique fracture criterion which is reasonably independent of strain rate for glassy polymers; the critical CTOD for PMMA is about 1.6 µm.[73]

Another problem with some glassy polymers, such as epoxy, is stick-slip fracture. Most fracture toughness testing of polymers is carried out in contoured DCB specimens[74] that exhibit a constant energy release rate independent of the crack length developed by Sheldon Mostovoy and Ed Ripling in 1966.[75] Stable crack growth occurs if the fracture energy increases with crack velocity as it does for PMMA and PS but in epoxy, with a yield strength of less than about 100 MPa, crack tip blunting at low strain rates causes stick-slip behaviour since the crack tip sharpens absorbing less energy as the crack accelerates before arresting again as the fracture energy increases at higher velocities.[76]

Fracture toughness tests are comparatively expensive and time consuming. Industry is interested in simple and cheap methods which accounts for the continuing popularity of the Charpy test. In metals the connection between fracture energy and the Charpy impact energy can only be made with correlations of limited validity. However, for brittle polymers the fracture energy can be formally derived from the impact energy. Williams and his co-workers showed that the impact energy, U_i, is given by

$$U_i = RBW\phi, \tag{10.11}$$

where R is the fracture energy, B is the specimen width, W is the gross specimen thickness, and ϕ is a function the relative depth of the notch and also weakly dependent on the span depth ratio of the specimen.[77] Provided that a sharp crack is used in place of the usual blunt Charpy notch, the fracture energy can be obtained from a series of tests with different crack depths by plotting U_i against $(BW\phi)$ and measuring the slope. Some of the absorbed energy supplies kinetic energy to the specimen and the line has a positive intercept.

10.3.3 Semicrystalline polymers

Symmetrical linear polymers such as polyethylene (PE) are semicrystalline. However the crystalline structure is quite different to that of metals and also there is always a significant portion of amorphous material. The long molecular chains fold back on themselves to produce lamella crystals of the order of 10 nm thick, as shown schematically in Fig. 10.11, which are joined together by regions of amorphous polymer. High density polyethylene is 70–80% crystalline while low density polyethylene is only 40–50% crystalline. Light is scattered by the crystalline structure and polyethylene is translucent but not transparent like the amorphous PMMA. Many polymers that crystallise from the melt or from a solution form spherulites which grow to be spherical with the crystal lamellae radiating outwards. These spherulites look like a Maltese cross under cross-polarised light.

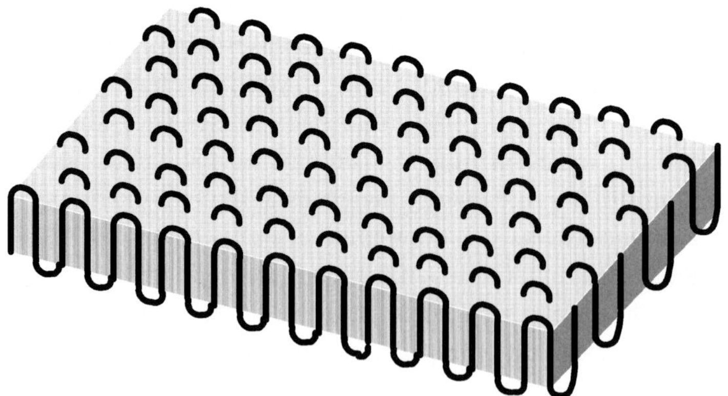

Fig. 10.11 Schematic illustration of the chain folded structure of a polymer crystal.

The glass transition temperatures of the semicrystalline polymers are generally lower than the amorphous polymers and often below room temperature. Hence semicrystalline polymers as a class are more ductile at room temperature than the amorphous glassy polymers. Because the yield strength is generally low LEFM usually cannot be applied to ductile semi-crystalline polymers and EPFM has to be used. The J-integral test methods have been adapted for polymers but are not as well developed as for metals. An ASTM standard[78] exists for obtaining J_R-curves, but there is not one to measure the initiation J; one problem is that crack blunting in polymers can take a number of forms.[79] An alternative method of characterizing ductile polymers that is gaining popularity is the essential work of fracture (EWF) method.[80] One of the problems of measuring the fracture energy of semicrystalline polymers is that there is self crack tip blunting. A recent development has been the proposal to use machining with a sharp tool to measure the minimum toughness.[81]

Polyethylene has become a favoured material for water and gas distribution pipes. The mechanical properties of polyethylene are determined largely by the degree of crystallisation. High density polyethylene (HDPE) with no side branches solidifies with a high degree of crystallinity. It is a relatively stiff and strong semi-crystalline polymer, but is less tough than other polyethylenes with less crystallinity. Gas and water pipe is required to have a very long life. Environmental slow crack growth is the dominant mode of PE pipe failures. Crazes precede fracture in PE as they do in the glassy polymers, but the crazes are large and more open. Solvents can plasticise a craze reducing the craze stress and increasing environmental crack growth rates. One of the most common solvents, a weak solution of detergent, is unfortunately the most aggressive. The detection of crazing in semi-crystalline polymers took longer to uncover than in glassy polymers because a high constraint is necessary for their formation.[82] A craze in a deep circumferentially notched round polyethylene bar is shown in Fig. 10.12. The polyethylene pipe grades are categorized by the hoop stress in MPa that a pipe can withstand at 20°C for 50 years multiplied by ten; thus PE63, the pipe grade HDPE copolymer with some side branches introduced in 1965 can withstand a hoop stress of 6.3 MPa for 50 years. Since 1965 two other pipe grades PE80 and PE100 have been developed. Obviously pipes cannot be tested for 50 years and the results have to be extrapolated from shorter times. There is a transition from ductile to brittle behaviour depending on the applied stress. With improvements in polyethylene blends the transition has been pushed to longer and longer times. At low applied stresses the behaviour is brittle and linear on a log-log plot so that the stress for a life of 50 years can be found by extrapolation. The behaviour of PE pipes has been well reviewed by John Scheirs and his co-workers.[83] Although most pipe failures are by slow crack growth, rapid brittle fractures can propagate in pipes at relatively low stresses if initiated by sudden impact in the earlier grades of PE. However for PE100 at temperatures above -10°C rapid brittle fracture is unlikely.

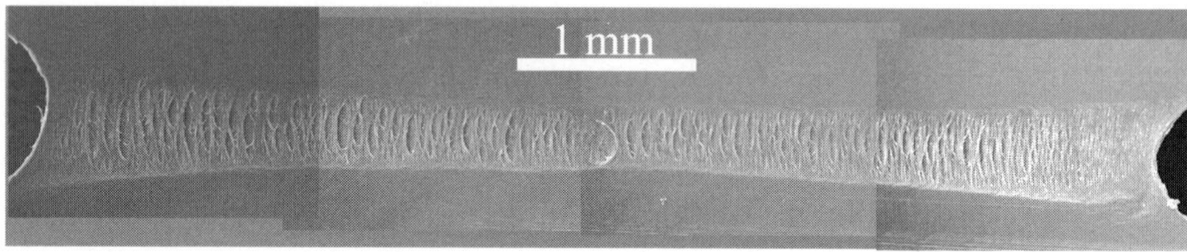

Fig. 10.12 Craze in HDPE deeply circumferentially notched tension specimen (Courtesy Simon Ting).

10.3.4 Toughened polymers

As a class polymers are not particularly tough and much of the expanding use of polymers has been made possible by toughening them. Polymers can be toughened by the introduction of small particles some of which have always been added to reduce costs. It is a question of definition on whether polymers with particles added for toughening are toughened polymers or polymer composites. Here the view is taken that if the toughening comes primarily from energy absorbing mechanisms in the polymer they will be considered toughened polymers, but if they promote toughness by crack bridging, particle pull-out, crack pinning, or crack deflection, they will be considered composites and dealt with in §10.4. A modern trend is to use nano-sized particles, but discussion of such materials is left until Chapter 12.

The most successful method of toughening is the introduction of rubber particles. Such toughening does reduce the Young's modulus and tensile strength, but these losses are usually outweighed by the increase in toughness. The history of rubber toughening dates back to 1927 when Ivan Ostromislensky, working the USA, patented toughened polystyrene made by polymerising a solution of rubber in a styrene monomer, but the polystyrene was cross-linked and could not be moulded. The first commercial rubber toughened polystyrene was developed towards the end of War World II by the Dow Chemical Company and went into production in 1948.[84] Shortage of natural rubber during the war, especially in Germany, led to the development of the artificial rubber styrene-butadiene. High impact polystyrene (HIPS) was produced by dissolving styrene-butadiene rubber in a styrene monomer. The resulting plastic was cross-linked like Ostromislensky's and the key to commercial success lay in the subsequent processing where the rubber network was broken down to produce a mouldable thermoplastic where the rubber was in discrete particles. The modern process enables the rubber to form discrete particles during the polymerisation process. The second commercial rubber-toughened polymer was acrylonitrile-butadiene-styrene (ABS) marketed in 1952 by the US Rubber Company. Polystyrene has an impact toughness of about 0.1 J/mm at room temperature whereas HIPS has an impact toughness of about 0.45 J/mm.[85] The toughness of epoxy can be increased by an order of magnitude by rubber particles.[86]

In both HIPS and ABS the polymer whitens before fracture. Clive Bucknall was the first to recognise that stress whitening was caused by multiple crazing initiated at the rubber particles and that this mechanism produced the high toughness.[87] The rubber particles also act to arrest the crazes so that a very large number of small crazes, which do not become cracks, form instead of a small number of large crazes. Cavitating rubber particles can also form dilatational bands which superficially appear the same as crazes. Often the term craze-like is used to encompass both true crazes and dilatational bands. With rubber toughened epoxies the rubber particles initiate shear yielding. The cavitation of the rubber particles is an important part of the toughening mechanism and occurs before the formation of the crazes.[88] Cavitation of the rubber particles also occurs in other toughened polymers, notably epoxy resins containing carboxyl terminated butadiene acrylonitrile (CTBN) rubber.[89] The rubbers used to toughen polymers are almost incompressible under hydrostatic stress, and have a high bulk modulus despite their low Young's modulus. The stress ahead of a crack is near hydrostatic and without cavitation the stress concentration is not large enough to cause either crazing or shear yielding. The energy released when a rubber particle of diameter, d, cavitates is proportional to d^3 but the surface area of the cavitated particle is proportional d^2, hence cavitation is easier for large rubber particles. Not surprisingly a minimum particle size is needed for toughening.[90]

Low-cost hard particulate fillers, often added to polymers for economy, can toughen glassy polymers. However, when micron-sized particles are added to relatively tough thermoplastics such as nylon 6.6, they can decrease the toughness.[91] In theory debonded hard particles can toughen a polymer by

                                    Fracture and Life

initiating crazes or shear yielding. Once a hard particle debonds near the tip of a crack under essentially hydrostatic stress, it is much like a cavitated rubber particle. The highest toughness is obtained when release agents are used to reduce the bond between the particles and the polymer and lowest when a coupling agent is used to increase the bonding.[91] There is conflicting evidence about the relative importance of matrix energy absorption mechanisms and the crack line toughening mechanisms of crack bridging, particle pull-out, crack pinning, or crack deflection. The efficiency of all the toughening mechanisms depends upon the particle size as well as volume fraction.

10.3.5 Adhesives and adhesion

Until the late 1930s the most adhesives were the natural materials used by man since prehistoric times. Animal and fish glues were used to bond the composite bows developed around the fourth millennium BC in the region east of the Caspian Sea.[92] The Australian aborigine used the gum from trees to haft stone tools.[93] Casein, made from the whey in milk, was used in ancient Egypt for gluing furniture and has been an important adhesive up to modern times.[94] Casein was used during World War II for the thousands of wooden gliders and the very successful wooden fighter plane the *Mosquito*, but it rots in damp conditions.[94] Mark Pryor, a biologist at Cambridge, was sent to the RAE to take charge of the production of the wooden gliders because of the main problem was with the rotting of the casein glue not with its strength or toughness.

Plastics were soon seen to be very suitable adhesives. Nitrocellulose in solution was for many years used as household cement. It was a search for a substitute for shellac to bond mica in electrical apparatus that initially led Leo Baekeland (1863–1914) to develop the phenolic resin *Bakelite*. Modern polymer adhesives began to be developed in the 1930s. Norman de Bruyne (1904–1997),[95] a Fellow of Trinity College, Cambridge, founded Aero Research at nearby Duxford in 1934.[96] De Bruyne had previously used casein as an adhesive to bond plywood for his two light planes *Snark* and *Ladybird* and was well aware of its limitations. In 1937 de Bruyne commissioned Robert Clark to produce a number of urea-formaldehyde resins for Aero Research from which the commercial Aerolite adhesive was developed. Aerolite adhesives are still standard wood glues. However, the most famous adhesive developed by de Bruyne's company was Redux®[97] for use with aluminium.[98] Phenol–formaldhyde (PF) resins adhere well to metals but have two drawbacks: water is evolved during curing and the resins are brittle. The first successful Redux was manufactured as a film of poly(vinyl formal) (PVF) coated on both sides by a PF resole[99]. Redux has gone through many versions some film based, some with a liquid PF resole, and also a powder version. Redux was first used as an aircraft structural adhesive on the de Havilland Hornet in 1944. The most famous use of Redux was on the de Havilland Comet where its use was not a factor in the catastrophic disasters. Most polymer adhesives today are thermosetting plastics such as epoxy. All aspects of adhesives and adhesion have been reviewed by Tony Kinloch.[100]

10.3.5.1 Strength of adhesive joints

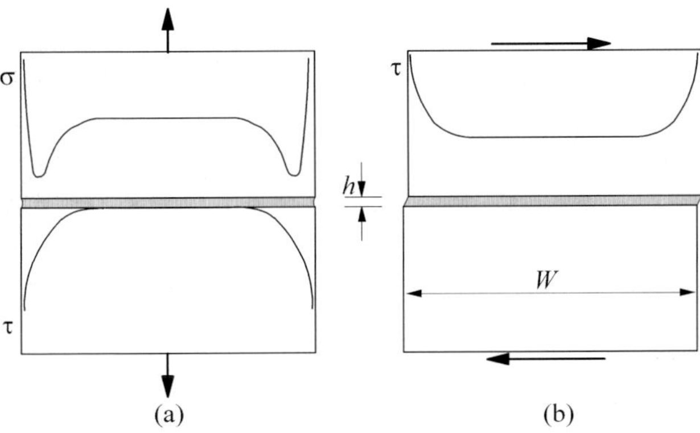

Fig. 10.13 Schematic illustration of the stresses in archetypical joints: (a) Tension. (b) Shear.

Usually the adhesive is much more compliant than the adherends being glued together and the archetypal stresses loaded in tension and shear are shown schematically in Fig. 10.13. The stress is uniform over the central portion of the joint, provided the adhesive thickness/adherend width ratio, h/W, is small, but there are weak singularities at the adhesive/adherend interface at the edges.[101] Cracks start at the edges of the joint and its integrity really depends upon the toughness of the adhesive. The tensile strength of an adhesive joint can only be increased by increasing the width, W, of the joint up to about $10h$, greater joint widths do not increase the strength because the adhesive fails from the edges of the joint. Since the 1920s it has been realised that the strength of tensile joints increase as the bond thickness decreases. Data taken from the 1927 paper by James McBain (1882–1953) and his collaborator W.B. Lee for the strength of shellac bonded aluminium and nickel, obtained from the *poker chip test* where the joint surface is circular, are shown in Fig. 10.14.[102] With ductile adhesives there is less thickness effect on the tensile strength. Joints are much more ductile under shear and usually completely yield before fracture. Even a brittle adhesive with a tensile ductility of only 2% can withstand shear strains in adhesive joints of the order of 100%.[103] Hence ultimate shear strain is a better failure criterion than shear strength. However, the ultimate shear strain decreases with increase in bond thickness except for extremely thin adhesive layers. To understand the strength of adhesive joints it is necessary to understand their fracture toughness.

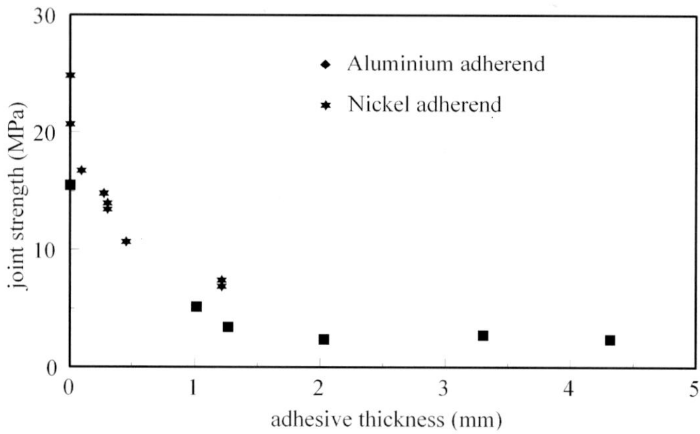

Fig. 10.14 Tensile strength of a shellac adhesive with aluminium and nickel adherends (after McBain and Lee 1927).

10.3.5.2 *Fracture toughness of adhesive joints*

The two archetypical fracture modes for an adhesive joint are mode I and mode II which correspond to the tensile and shear loading on the joints shown in Fig. 10.13. The DCB specimen, especially the contoured DCB where the energy release rate is independent of crack length, is a favoured geometry for mode I fracture toughness testing of adhesives. The mode I toughness of brittle adhesives like unmodified epoxy is independent of the bond thickness and is about 200 J/m² for a piperidine hardened DGEBA epoxy. Rubber toughening increases the toughness of epoxy by more than a magnitude in order and not surprisingly is often used in adhesives. The toughness of rubber toughened epoxy adhesives is strongly dependent on the bond thickness. Willard Bascom (1916–2000) and his colleagues were the first to study the effect of bond thickness on the toughness of such adhesives and their results for a rubber modified piperidine DGEBA epoxy are shown in Fig. 10.15.[104] The toughness of the modified epoxy reached a maximum for a bond thickness of about 0.5 mm before it decreased with further increase in thickness to the bulk value.

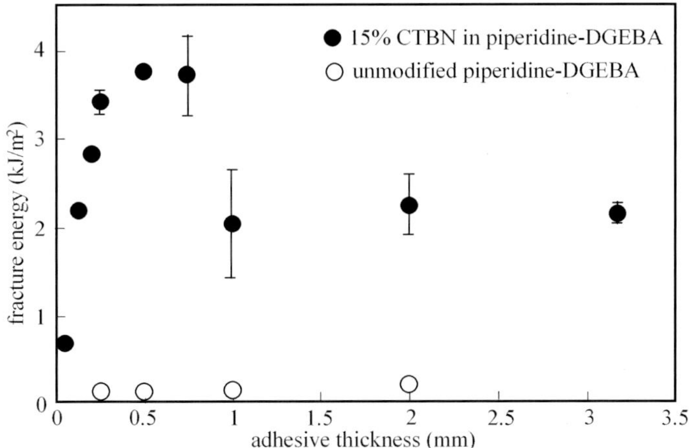

Fig. 10.15 Mode I fracture energy as a function of bond thickness for unmodified piperidine-DGEBA, and for 15% CTBN in piperidine-DGEBA (after Bascom *et al.* 1975).

In ductile polymers, such as rubber toughened epoxy, the plastic zone size is large. The height of the plastic zone is limited by a thin bond thickness and the plastic work performed per unit area of fracture, which is the fracture energy, is roughly proportional to the bond thickness. As the bond thickness increases the plastic zone eventually ceases to occupy the whole thickness as shown in Fig. 10.16.[105] Rather surprisingly the maximum fracture energy in an adhesive bond is greater than that in the bulk adhesive and the maximum plastic zone height is also correspondingly larger. The reason why the maximum plastic zone size and the maximum fracture energy are larger in an adhesive zone than in a bulk polymer lies in the difference in the stress systems. Along the adherend/adhesive interface at the crack tip there is a high shear stress similar to that at the edge of an uncracked joint shown schematically in Fig. 10.13 (a). This shear stress at the adhesive/adherend surface promotes plastic deformation and causes the plastic zone height and the fracture energy to be larger than that in the bulk adhesive over the range in bond thickness indicated in Fig. 10.16. As the bond thickness becomes large compared with the plastic zone for the bulk adhesive, the influence of the interface shear stress becomes negligible and the plastic zone size and fracture energy decrease to the bulk adhesive values.

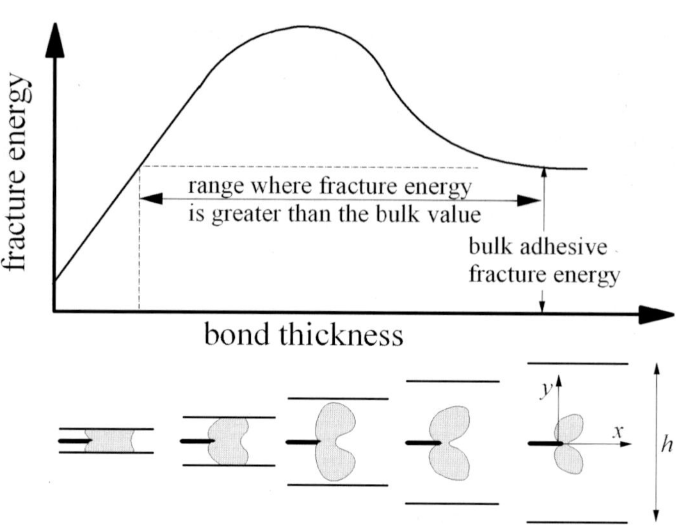

Fig. 10.16 Schematic variation of fracture energy with plastic zone size and adhesive bond thickness (after Crews *et al.* 1988).

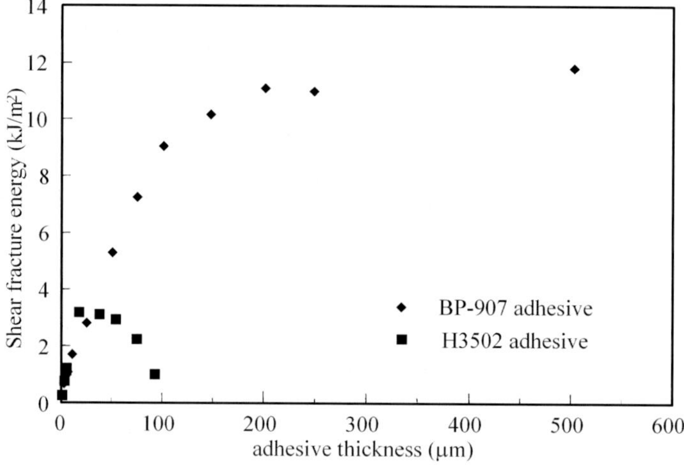

Fig. 10.17 Mode II fracture energy as a function of bond thickness for a brittle epoxy H3502 and a toughened epoxy BP-907 (after Chai 1988).

The mode II fracture energies for a highly cross-linked brittle epoxy H3502 and a toughened epoxy BP-907 are shown as a function of the adhesive thickness in Fig. 10.17.[106] The mode II or shear fracture energy of a brittle adhesive can be up to thirty times larger than the mode I fracture energy whereas for rubber toughened epoxies the mode I and mode II fracture energies are comparable. The mode II fracture energy for the ductile BP-907 adhesive reaches a plateau at a bond thickness of about 200 μm, and the brittle H3502 adhesive reaches a maximum at a bond thickness of about 20 μm and then decreases to be comparable to the bulk mode II fracture energy at a thickness of about 100 μm. The fracture of brittle adhesives under shear, while mode II macroscopically, is mode I microscopically with cracks forming at 45° to the adherends. The reason for the increase in mode II fracture energy with adhesive bond thickness for ductile adhesives is the same as for mode I.

The dependence of strength and ultimate strain on the adhesive thickness can be obtained from knowledge of the effect of thickness on fracture energy. If the adherends are much stiffer than the adhesive and tensile or shear loading is applied to the joint by giving them either a constant tensile displacement, v, or a constant shear displacement, u, failure will occur by a crack propagating from the edge of the specimen[107]. After the crack has propagated a distance, a, roughly equal to the adhesive thickness, h, the energy release rate is independent of the crack length.[108] Jim Rice gave the energy release rate for an elastic-brittle material under a normal load in 1967.[109] In general for an elastic-plastic material the energy release rate is given by

$$G_{I,II} = h \int_0^{(\sigma_c, \tau_c)} (\sigma, \tau) d(\varepsilon, \gamma). \tag{10.12}$$

For mode I loading the fracture energy for an elastic-brittle adhesive joint where the thickness of the adhesive is small compared with the thickness of the adherend is given by

$$G_I = \frac{(1-2v)(1+v)}{2E(1-v)} \sigma^2 h. \tag{10.13}$$

Thus for a brittle adhesive, whose mode I fracture toughness is independent of thickness, the fracture strength is inversely proportional the square root of the thickness. Re-plotting the results of McBain and Lee[102] logarithmically and using linear regression one obtains a slope of −0.607 with a correlation coefficient of 0.932, whereas the simple theory would predict a slope of −0.5. The Mode I fracture energy of a toughened epoxy increases roughly in proportion to the adhesive thickness up to a thickness of the order of about 1 mm. Thus for a toughened epoxy one would expect the tensile strength to be independent of the adhesive thickness up to about 1 mm thick, but then to decrease more rapidly than inversely proportional to the square root as the fracture energy decreases to the bulk adhesive value. Joints under shear usually yield well before fracture even with brittle adhesives and hence stress is not a good measure of their performance. Chai compares the mode II fracture energy obtained from a fracture mechanics test with the inferred value obtained from the shear work performed up to fracture in the napkin test and obtains excellent agreement.[110] Thus it is seen that the adhesive fracture energy controls the fracture strength of joints.

10.4 Composites

Composites are not new but have systematically exploited only from the middle of the twentieth century. The first recorded mention of a composite material occurs in Exodus in the Bible where the use in Egypt of straw in sun dried bricks to prevent cracking during drying is mentioned. Ox hair has long been used to strengthen plaster. Short cellulose fibres in the form of wood-flour were added to Bakelite to produce the first commercial mouldings. Wood-flour and other fillers such as silica-flour, talc, and limestone are frequently added to modern day polymers. The fillers usually increase the stiffness, strength, and toughness of the polymer though one of the main reasons for adding them is to reduce costs. Shortly after the invention of Bakelite, Baekeland experimented on producing composites by impregnating cardboard, pulp board, asbestos and other fibrous and cellular materials with Bakelite.[111] In 1912 George Ellison, an electrical engineer, developed Tufnol®, a tough composite made by impregnating cotton fabric with Bakelite, for use as an insulating material.[112] The invention of Gordon Aerolite[113] by De Bruyne in 1937 marks the real beginning of the development of strong fibre reinforced materials.[95] Gordon Aerolite was made of flax rovings impregnated with Bakelite. The light weight (density 1350 kg/m^3), tensile strength (345 MPa), and stiffness (Young's modulus 34 GPa) of Gordon Aerolite made it a suitable aircraft material. One of the first uses of Gordon Aerolite was for an experimental main spar of the Bristol Blenheim bomber. A specially designed press, purchased from Germany just in time before the Second World War, was needed to fabricate the nine metre long spar in metre length 'bites'. However the spar, a lattice construction the same as the aluminium one and made up of many parts, was not successful.[114] One of Jim Gordon's first jobs at the RAE was to test the de Bruyne spar. A one-off Spitfire fuselage was also built of Gordon Aerolite when there was a fear of aluminium shortage, but was also not successful since it was a replica of the aluminium fuselage and grossly overweight. Jim Gordon realised very early that efficient designs had to be made specifically for the composite rather than simply a replacement of metal parts by composites. Throughout the war cellulose composites were developed with some priority at the RAE. Charles Gurney and Jim Gordon were responsible for one of the most successful applications of cellulose composites: the seat for the Spitfire. The composite seats were die-moulded from manilla paper reinforced with phenolic resins.[114] Apart from the stresses induced in high g turns, the highest stresses came when the pilots scrambled and jumped into their seats. Gurney devised the *Gurney test* where the seat had to withstand him jumping on to it. The culmination of Gordon's 16-year career at the RAE was the production in his laboratory of a series of experimental delta wings, modelled on the Fairey experimental E10/47 aircraft, with a root chord of 3.4 m and a semi-span of 2.5 m, these wings were made from Durestos,[115] an asbestos fibre-reinforced phenolic resin.[114] The wings could sustain loads well in excess of 12g which was required for metallic structures by 1955 but, because of the untried nature of the composite, they were required to sustain 50% higher loads at 18g. This added requirement put an end to the experimental programme.[116] In 1999 the DERA[117]-Cambridge Gordon Laboratory, which focuses on advanced composite materials, was opened commemorating the work of Jim Gordon.

Composites are now tailored for their application; many are functional materials where the mechanical properties are a secondary consideration. Here emphasis is on the fracture and strength of composites, but it must be realised that stiffness is frequently a more important mechanical property. Ceramic particles can increase the stiffness and strength of metal matrix composites, but often reduce the toughness. A huge number of different composites are in use today and in this chapter only fibre reinforced polymer composites are discussed.

10.4.1 Reinforcing fibres

With a growing awareness of the effect of manufacturing on the environment there has been lately an increasing interest in natural fibres and particles for reinforcement. Natural fibres were the first to be used in composites, but they were abandoned in favour of man-made reinforcements.

The first man-made fibre to be used in composites was glass. The process for spinning strong glass fibres from electrically insulating E-glass was developed in the 1920s, but it was not until the Second World War that it was used in composites. Radar needed to be housed in an electrically non-conducting structures and fibreglass was ideal. Jim Gordon gives a delightful story of an Air Marshal inspecting a fibreglass radome intended for a Lancaster bomber. When he was told of what it was made he exploded saying 'Glass! I won't have you putting glass on any of my bloody aeroplanes!'.[94] E-glass has a low alkali content and good resistance to water. A higher strength glass fibre, S-glass, was developed by Owens-Corning in the early 1960s with a strength of more than twice that of E-glass (see Table 10.2).

Table 10.2 Mechanical properties of fibres.

Material	Density (kg/m³)	Tensile strength (MPa)	Failure strain (%)	Young's modulus (GPa)
E-glass	2550	2000	2.5	80
S-glass	2490	4700	5.3	89
Alumina (Saffil)	3280	1950	0.7	300
Carbon (high modulus)	1950	2400	0.6	380
Carbon (high strength)	1750	3400	1.1	230
Kevlar 29	1440	2800	4.5	64
Kevlar 49	1440	3700	2.3	130

Although glass fibres have high strength, their stiffness is low. Carbon fibres have a very high Young's modulus, high strength and low density which make them ideal for composites. Carbon fibres based on rayon were available in the 1950s but their strength and stiffness were low. In 1963 William Watt (1912–1985), Leslie Phillips (1922–1991) and William Johnson, working at that powerhouse of technology the RAE, decided to try to produce carbon fibres whose strength approached that of carbon whiskers. For high strength and stiffness the graphite basal planes need to be aligned parallel to the fibre. Watt based his fibre on a polyacrilonitrile (PAN) precursor. Pyrolytic deposition from pitch has also been used to produce carbon fibres.[118] In the 1960s the potential of boron fibres, formed by depositing boron from the vapour phase onto tungsten wires, for composites was explored under the US Air Force's Project Forecast. Because of their relatively large diameter, boron fibres are not easily handled and did not prove to be as suitable for fibre composites as carbon.

Other fibres that have been developed are Kevlar® and Saffil®. Kevlar is a high modulus organic aramid fibre produced from aromatic polyamide first synthesised in 1964 by Stephanie Kwolek at Du Pont. Saffil (**safe fil**ament) is a short aluminosilicate fibre first produced in the early 1970s as an insulating fibre to replace asbestos and is used in aluminium-based composites.

10.4.2 *Fracture of long fibre composites*

Usually the fibre reinforcement is stronger than the matrix. In composites with long aligned fibres where the fracture strain of the fibre is less than that of the matrix, the fibres only reinforce the matrix if the volume fraction of the fibres, v_f, is greater than a critical value v_{fc} given by[119]

$$v_{fc} = \frac{\sigma_m - \sigma_m'}{\sigma_f + \sigma_m - \sigma_m'},$$
(10.14)

where σ_m is the strength of the matrix, σ_f is the strength of the fibres, and σ_m' is the stress in the matrix at the breaking strain of the fibres. The strength of the composite for $v_f > v_{fc}$ is given by

$$\sigma_c = \sigma_f v_f + \sigma_m' \left(1 - v_f\right).$$
(10.15)

If the fracture strain of the fibres is greater than the matrix fracture strain, then there is always positive reinforcement but the reinforcement is more efficient when the volume fraction of the fibres is greater than a critical value v_{fcc} given by

$$v_{fcc} = \frac{\sigma_m}{\sigma_f + \sigma_f' - \sigma_m},$$
(10.16)

where σ_f' is the stress in the fibres at the fracture strain of the matrix. Depending on the volume fraction and the relative values of the fracture strain, there is either a single fracture or multiple fibre fracture or multiple matrix cracking.

The strength of an aligned fibre composite loaded at an angle to the alignment drops rapidly with the angle of misalignment until it is typically only about 10% of the maximum when loaded at 15° to the direction of alignment. For this reason high performance polymer composites usually consist of layers or laminates stacked together in the manner of plywood. The earliest known use of laminated veneers or plywood was in ancient Egypt but modern plywood, where the veneers are turned from a log in a lathe, dates to the mid-nineteenth century and was an invention of Immanuel Nobel (1801–1872), the father of the more famous Alfred Nobel (1833–1896). In plywood the veneers or lamina are stacked into a cross ply structure with the directions at right angles to each other. That construction is also used in polymer composite laminates, but since composite laminae have little shear strength they are very often stacked into angle plies as well. There are two weaknesses inherent in polymer composite laminates: there is interlaminar shear between layers, especially along edges, and there is little transverse strength or toughness.

                    Fracture and Life

10.4.3 Toughness of fibre composites

One might expect that the toughest fibre composite would the one where the bond between the fibre and matrix was the strongest, but paradoxically that is not so. Quite early in the development of fibre composites it was realized that a comparatively weak bond between the fibre and matrix would give toughness to a composite. Jim Gordon and John Cook pointed out in 1964 that ahead of a crack tip not only is the stress normal to the crack high, but so too is the transverse stress and if the bond between the fibre and the matrix is not too high then the fibres will debond and blunt the crack tip.[120] If the fibres debond the fibres will not necessary break on the same fracture plane and some will pull-out consuming more energy. The bond strength of carbon fibres in an epoxy matrix is relatively high and the fracture surface of the composite is comparatively smooth, but the bond between other fibres and their matrices, such as glass fibres in a polyester matrix, is weak and there is extensive fibre pull-out which causes higher fracture toughness.[118] The toughness of the individual components of fibre reinforced polymer composites is relatively small and they would not be useful were it not for large synergistic effects. The biggest contribution to the toughness of fibre composite materials comes from fibre pull-out. The earliest work on pull-out toughening was done on a model material, copper reinforced by aligned tungsten wires, by Alan Cottrell, Tony Kelly, and Bill Tyson.[121] If the fibres are discontinuous the stress carried by them builds up from their ends through the shear stress at the interface. Metallic matrices yield before fracture and the interfacial shear stress is equal to the shear yield strength k. The stress in a fibre builds up with distance, x, from its ends until the elastic strain in the fibre reaches the strain in the composite; at greater distances from its ends there is no interfacial shear stress and the fibre stress is constant. The stress near the ends of the fibres can be calculated from equilibrium and, neglecting the small matrix stress at the end of the fibre, is given by

$$\sigma = \frac{4kx}{d}, \tag{10.17}$$

where d is the diameter of the fibres. If the length of the fibre, l, is greater than l_c given by

$$l_c = \frac{d\sigma_f}{k}, \tag{10.18}$$

where, σ_f, is the strength of the fibres, the fibres will fracture. If the volume fraction of the fibres is greater than the critical value and there is a single matrix crack, then the strength of the composite is given by

$$\sigma_c = \sigma_f \left(1 - \frac{l_c}{2l}\right) v_f + \sigma_m'' \left(1 - v_f\right), \qquad (10.19)$$

where $\sigma_m'' < \sigma_m'$ is the matrix stress at the average stress in the fibres at fracture. The strength is less than the strength of a composite with continuous fibres given by Eq. (10.15). However, the ends of some fibres will be less than $l_c/2$ from the fracture plane and will pull-out rather than fracture. The contribution of the work of pull-out, R_p, to the fracture energy of the composite is given by

$$R_p = \frac{\sigma_f v_f l_c^3}{12l^2}. \qquad (10.20)$$

Fibres shorter than l_c cannot fracture but only pull-out from the matrix and the composite strength is given by

$$\sigma_c = \sigma_f v_f \frac{l_c}{2l} + \sigma_m'' \left(1 - v_f\right), \qquad (10.21)$$

and the contribution to the fracture energy by fibre pull-out is given by

$$R_p = \frac{\sigma_f v_f l^2}{12 l_c}. \qquad (10.22)$$

For the maximum contribution to the fracture energy the fibres should have a length equal to the critical value, but the strength will be reduced to about half that for continuous fibres. If the strength of a fibre is identical along its length then fibre pull-out could not occur in continuous fibres because the fibre stress must be a maximum at the matrix crack and the contribution to the fracture energy given in Eq. (10.20) tends to zero with increase in the length of the fibres. However if the strength of the fibres has a wide distribution in strength over their length, fibres can fracture at a lower stress away from the matrix crack and the fibres pull-out.[122] Fibre pull-out also occurs with brittle matrices, here the fibres first debond[123] and then pull-out against frictional forces.[124]

The lack of transverse toughness is probably the biggest weakness and a particular concern for carbon fibre reinforced polymers used in aircraft structures which can be damaged simply by dropping a tool on the surface. There are various ways that the interlaminar toughness can be improved. The use of rubber-toughened epoxies is one and another is stitching through the laminate. However, stitching causes damage to the plies and reduces the in-plane strength.

10.5 Concluding Remarks

Fracture is only one aspect of the mechanical properties of materials. Adequate stiffness is often more important than strength or fracture toughness. High strength can only be utilised if the material is stiff. Jim Gordon published a sketch of an aircraft with a strain of 1.6% in its wing spars where the wing tips are bent vertically and remarked that at about 3% strain the wing tips would meet over the pilot's head.[125] Usable strength is limited to the stress that can be developed at strains of no more than about 1%. Of course functional properties can be more important than mechanical properties. It has been realised in this twenty-first century that resources are finite and that our use of energy must be limited if we are to avoid the worst effects of global warming. These considerations put more emphasis on designing materials that have good specific mechanical properties based on mass and energy. The need for light structures not only for aerospace but for most applications was recognised in the last century and was addressed by Gordon in 1964.[125] In Table 10.1 an attempt has been made to quantify the energy cost of a range of materials.

There are three classes of man-made materials: metals, ceramics, and polymers. While there will be improvements in these materials they will not be very large. It is only by combining the classes of materials in composites that major improvements are possible. In this chapter the mechanisms for improving the fracture behaviour of composites have been at the macro- and micro-level. In Chapter 12 the possibility of manipulation of the properties of composites at the nano scale will be addressed.

10.6 Notes

[1]   The relative costs of materials of equal tensile strength are presented because fracture is the topic of this book, but frequently stiffness is more a limitation on the use of a material than strength.

[2]   When it was spelt with a k since it derived from the Greek word κεραμιός meaning burnt stuff.

[3]   Zhang and Gao (2004).

[4]   Shaver (1964).

[5]   Peierls (1940); Nabarro (1947).

[6]   Lawn (1993).

[7]   Swanson *et al.* (1987).

[8]   Mai and Lawn (1987).

[9]   Chantikul *et al.* (1990).

[10] Hall (1951); Petch (1953).

[11] Garvie (1988) contains a detailed history of the early development of transformation toughened ceramics.

[12] Garvie *et al.* (1975).

[13] Rühle and Evans (1989).

[14] McMeeking and Evans (1986).

[15] Evans and Cannon (1986). More recent analytical models have assumed a combined effect of hydrostatic and shear stress on the initiation. These models have predicted a constant in Eq. (10.3) as large as 0.478, see Kelly and Rose (2002).

[16] Hannink and Swain (1986).

[17] Swain and Rose (1986).

[18] Rühle *et al.* (1987).

[19] Grenet (1899). A partial translation of this is given in Grenet (1934).

[20] Founded by Frank Preston (1896–1948) in 1936.

[21] See Preston (1942).

[22] Michalske and Freiman (1982).

[23] Charles (1958) was the first to suggest the power law, but he did not state Eq. (10.4) since he did not use the stress intensity factor, which was only just being formulated, in his paper.

[24] Gurney and Pearson (1948).

[25] Hu *et al.* (1988).

[26] Kohn and Hasselman (1972).

[27] Hoffman *et al.* (1995).

[28] Kingery (1955).

[29] Not to be confused with fracture energy.

[30] Hasselman (1963). G_f has been used instead of R in this section to avoid confusion with R the material resistance factor.

[31] Nakayama and Ishizuka (1966).

[32] Gupta (1972).

[33] Cotterell *et al.* (1995).

[34] Schneider (1955).

[35] Kaplan (1961).

[36] Cotterell and Mai (1996).

[37] Hillerborg *et al.* (1976).

[38] The bilinear stress-displacement introduced by Petersson (1985) gives a closer prediction of the load-displacement curves obtained from notch bend tests.

[39] Although the finite element method is commonly used in analyses using the fictitious model, analyses can more simply be obtained using the superposition of standard LEFM solutions, Cotterell and Mai (1991).

[40] Bažant and Cedolin (1979, 1983).

[41] Bažant and Lin (1988); Bažant (1994).

[42] Jenq and Shah (1985); Nallathambi and Karihaloo (1986).

[43] See §9.1.1.

[44] Melchers (1999).

[45] Bažant and Frangopol (2002).

[46] Jackson and Hundley (2004).

[47] The problem of uplift was fully understood at the time. The French engineer Maurice Lévy (1838–1910) as a result of the failure of the Bouzey dam near Epinal in 1895 proposed what is now known as Lévy's rule that the compressive stress on the upstream face of a dam must not be less than a threshold value near to the water pressure.

[48] Bažant (2002).

[49] Swenson and Ingraffea (1991).

[50] Burgoyne and Scantlebury (2006).

[51] Scabbling is the removal of the top surface of concrete by a machine that pounds the surface with carbide tipped rods. It is performed prior to laying new concrete on old.

[52] Bažant and Yavari (2005).

[53] Carpinteri *et al.* (2005); Bažant and Yavari (2007).

[54] Kendall *et al.* (1983).

[55] Popoola *et al.* (1991).

[56] Mai *et al.* (1990).

[57] Lewis and Boyer (1995).

[58] Whitney (1963).

[59] Bucknall (1977).

[60] Sauer *et al.* (1949).

[61] Niegisch (1961). In the USSR Bessonov and Kuvshinskii (1959) made similar speculations about the nature of crazes.

[62] Kambour (1962).

[63] Kambour (1973).

[64] Kausch and Michler (2005).

[65] The discovery of polycarbonate dates back to the experiments of Einhorn a German chemist in 1898.

[66] Williams 1984.

[67] See §9.1.

[68] Benbow and Roesler (1957).

[69] Marshall *et al.* (1973a).

[70] The difficulty is knowing the appropriate value of Young's modulus for PMMA was the reason why K_c was adopted as a fracture criterion in the first place (see §9.1).

[71] Gurney and Hunt (1967).

[72] Atkins *et al.* (1975).

[73] Williams and Marshall (1975).

[74] Side grooves are used to control the fracture path.

[75] Mostovoy and Ripling (1966).

[76] Kinloch and Williams (1980).

[77] Marshall *et al.* (1973).

[78] D6068-96(2002).

[79] Chung and Williams (1991).

[80] See §9.4.4.

[81] Patel *et al.* (2009).

[82] Chan and Williams (1983).

[83] Scheirs. *et al.* (1996).

[84] Amos (1974).

[85] Since polymers samples often have limited thickness, the impact toughness is quoted in energy absorbed per unit thickness rather than the total energy as they are for metals.

[86] McGarry (1970).

[87] Bucknall and Smith (1965).

[88] Bubeck *et al.* (1991).

[89] Yee and Pearson (1986).

[90] Sultan and McGarry (1977).

[91] Kinloch and Young (1983).

[92] These bows relied for their resilience on the bonding of sinews to the belly of the wooden skeleton and horn on the back of the bow. Armed with such composite bows the Parthian mounted archers were formidable enough to halt the eastwards expansion of Imperial Rome (Hall 1956).

[93] Dickson (1981).

[94] Gordon (1968).

[95] See the biographical memoir Kinloch (2000) for details of his contribution to adhesives and composites.

[96] Aero Research was originally founded in 1931 as the Cambridge Aeroplane Construction Company. Aero Research was taken over by the Swiss company Ciba in 1947.

[97] Coined from **re**search at **Dux**ford.

[98] Bishopp (1997).

[99] The un-cross-linked phenolic resin before curing.

[100] Kinloch (1987).

[101] Williams (1952) has shown that at a clamped-free corner the singularity depends upon the Poisson's ratio. For $v = 0.35$, the stress at a clamped-free corner is given by $\sigma = r^{-0.63}$, where r is the distance from the corner.

[102] McBain and Lee (1927).

[103] Chai (1993).

[104] Bascom *et al.* (1975).

[105] The plastic zones shown in Fig. 10.12 are based on a finite element elasticity solution by Crews *et al.* (1988).

[106] Chai (1988).

[107] In the case of a napkin shear test a crack will grow circumferentially in both directions from some point of weakness.

[108] Finite element analysis has shown that for a tensile loading, the energy release rate becomes constant once $a > 0.8h$ and the condition for shear loading will be similar (Cotterell *et al.* 1996).

[109] Rice (1967b).

[110] Chai (2004).

[111] Baekeland (1909).

[112] Tufnol was a favourite material in the RAE while I was an apprentice during the early 1950s.

[113] The family of Malcolm Gordon (no relation to Jim Gordon), who was de Bruyne's student at Cambridge, was in the linen business and it was Malcolm who suggested using flax.

[114] McMullen (1984).

[115] Patented by the Turner Brothers Asbestos Company in 1947.

[116] Kelly (2005).

[117] Defence Evaluation and Research Agency.

[118] Hull and Clyne (1996).

[119] Aveston *et al.* (1971).

[120] Cook and Gordon (1964).

[121] Cottrell (1964); Kelly and Tyson (1965).

[122] Kelly (1970).

[123] The first fracture mechanics approach to debonding was given by Gurney and Hunt (1967).

[124] The debonding of fibres in a brittle matrix and the various contributions to the fracture toughness are discussed by Gao *et al.* (1988).

[125] Gordon (1964).

Chapter 11

Cutting and Piercing

Until recently, cutting has not been seen as fracture despite the fact that a body is separated into two or more parts. Cutting has been treated differently to fracture because attention has been focussed on metal cutting where the plastic work dominates the work of cutting. However, even if the essential work of fracture is small it can have a significant effect on cutting. Tony Atkins, who has just published a book applying fracture mechanics to all the aspects of cutting,[1] presented the first model of machining that incorporated fracture work in 1974 but that pioneering paper has been largely ignored until recently.[2] Whereas in fracture the applied load is remote from the crack tip, in cutting and piercing the load is applied at or near to the cut or pierced hole. In fracture, energy stored is utilised to produce a fracture, but in cutting and piercing of ductile materials the work of separation comes primarily from the tool work.

Whereas fracture is usually avoided, cutting and piercing can be either desired or avoided. Plants and animals have evolved both efficient means of cutting and avoiding being cut as was discussed in Chapter 3. Man, not having teeth suitable for cutting, circumvented evolution by utilising sharp stone tools. The problem with stone tools was not that they were not sharp, but that they were brittle. In every day life the brittleness of stone tools was not a formidable problem, because either new implements could be made or old ones resharpened, but in battle a broken weapon can be fatal so more durable metal weapons were developed. In war cutting and piercing had to be avoided so armour of different kinds was developed.

Cutting has been an everyday activity since the earliest times. Perhaps since it is such a ubiquitous activity, even the cutting of metals was not studied scientifically before the 1850s and no one saw the need to understand the mechanics of carving the Sunday roast until the last 30 years or so.

11.1 Knives, Microtomes, Guillotines, Scissors, and Punches

The aim of cutting is to separate a body into two parts with as little work as possible, that means doing the least non essential plastic work. How cutting is performed also depends upon whether a thin slice is being taken from the surface of a body, the body is being cut into large chunks, or a thin sheet is being cut perpendicularly to its surface.

The food industry is becoming increasingly interested in the toughness and other mechanical properties of foodstuffs that determine their texture when being eaten rather than rely on the subjective judgement of taste testers. With many foodstuffs it is difficult to perform standard fracture toughness tests and cutting is an attractive alternative.

11.1.1 Cutting thin slices

Thin slices can be cut from soft materials, such as biological tissues and cheese, with little plastic work. Curled slices, such as occur in chips whittled from a piece of wood with a penknife, are evidence of plastic deformation. The strain energy involved in cutting these materials is small and in steady state cutting does not change so that most of the work of cutting goes directly into the work of fracture, though friction work cannot be avoided. If the friction is negligible then the work done by increasing the cut by da is simply $RWda$ where R is the fracture energy and W is the width of the cut. Hence the cutting force per unit width of cut is R, which is independent of the cutting angle. With very floppy materials like cheese there is little contact pressure between the slice and the knife but they can adhere to the cutting surfaces. Friction of soft floppy materials has not been well studied and Atkins assumes it is appropriate to represent the frictional force by an adhesion shear stress, τ_a, and the area of contact so that the adhesive force per unit width of cut is $\tau_a a$, where a is the length of cut.[3] However the notion that the adhesion fails simultaneously over the whole area of contact does not fit well with the underlying notion that fracture takes place sequentially rather than simultaneously. An alternative concept, which has recently been proposed for the quite different situation of the machining of metals and polymers, is that there is an adhesive energy, R_a, which has to be supplied to break the adhesive forces.[4] This alternative concept does not change the form of the expression for the slicing force but simply an identification of R_a with $\tau_a a$. Making this change to Atkins' expression the slicing force per unit width of knife, F, is given by

$$F = R\left[1 + \frac{R_a}{R}\left(1 + \sec\theta\right)\right], \tag{11.1}$$

where θ is the angle of the knife blade. The angle of the knife blade has very little effect. For less floppy materials there can be a pressure between the material and the knife and Coulomb friction may be more appropriate and the slicing force per unit width of cut is given by[5]

$$F = R\frac{\left(1 - \mu^2\right)\tan\theta + 2\mu}{\tan\theta\left(1 - \mu\tan\theta\right)}, \tag{11.2}$$

where μ is the coefficient of friction, though probably the combination of adhesion with Coulomb friction, as has been suggested by Gordon Williams and his colleagues for machining,[4] would be more accurate. If Coulomb friction is assumed the slicing force depends strongly upon the knife blade angle. The slicing forces for the two different friction assumptions are shown in Fig. 11.1. If Coulomb friction applies it is more efficient to use a blade angle of around 20–50° rather than a sharp angle. The typical kitchen knife has a cutting edge angle of 45° or more, though the taper away from the edge is usually less than about 5°. The relatively large edge angle probably results more from a desire to minimise damage than to minimise the cutting force.

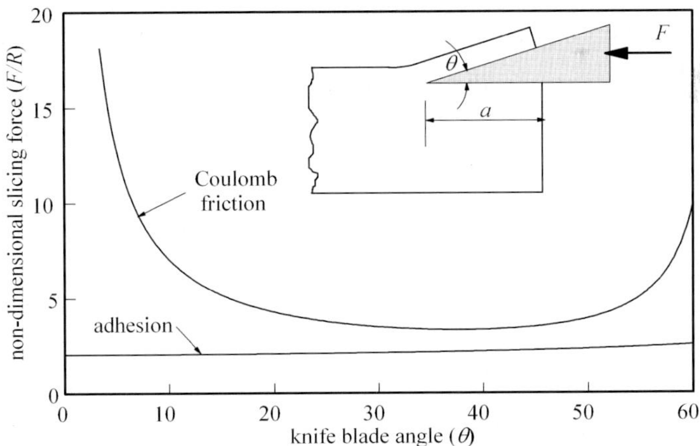

Fig. 11.1 Forces for slicing floppy materials as a function of knife blade angle for: (i) adhesion to the blade, $R_a/R = 0.5$, (ii) Coulomb friction, $\mu = 0.5$.

Provided the material remains elastic, the cutting force is independent of the slice thickness, but with large cutting angles plastic bending occurs which increases the cutting force. Using dimensional arguments similar to those advanced in §3.2.1 for cutting with scissors, it can be shown that the fracture work is a linear function of the slice thickness and the intercept at zero thickness is the specific essential work of fracture. However, for the thinnest slices Williams has shown that the plastic work contribution to the cutting force is proportional to $\sqrt{h}$ and rightly warns about the dangers inherent in extrapolation.[5]

Virtually undamaged thin tissue slices called histological sections can be cut with a microtome that enables the structure of tissues to be observed under the microscope. Histological sections were first prepared during the seventeenth century using sharp razors. Robert Hooke produced the first picture of cells using a section cut 'with a Pen-knife sharpen'd as keen as a Razor'.[6] The first microtome, devised around 1770 by Alexander Cumming for John Hill (1716–1775), could cut sections as thin as 130 μm and is preserved in the Science Museum in London.[7] Charles Chevalier (1804–1859) perfected this device and named it a microtome in about 1839. Precision mechanical devices were developed by the 1890s, which had a metal stage that held the tissue embedded in a paraffin wax or celloidin[8] and either a rotary or rocking mechanism to swing a blade against the specimen's surface. The blade is either steel or glass and can cut sections 2–25 μm thick. In ultramicrotomy, tissues embedded in epoxy resin can be cut with glass or diamond blades into sections as thin as 60–100 nm. The blade tip angle in microtomes is about 40° and has a very sharp edge with a tip radius as small as 10 nm. To avoid damage to the tissue a relief angle of 4–7° is used in microtomy to give clearance between the blade and the specimen. Tony Atkins and Julian Vincent have established that the minimum tissue damage occurs when the microtome is set at an angle that gives the minimum force and have invented a microtome in which the angle of the blade is automatically adjusted to give the minimum cutting force.[9]

Instrumented microtomes have been used to determine the fracture toughness of soft materials.[9,10] Julian Vincent and his student Willis, assuming the plastic work of fracture is proportional to the slice thickness, estimated the specific essential work of fracture for liver and wood to be 37.4 J/m^2 and 102 J/m^2 respectively.[10] Williams reanalysed their results and found that the linear assumption was justified for the liver samples but a square root relationship was more appropriate for the wood samples and estimated that its specific essential work of fracture to be only 28 J/m^2.[5]

11.1.2 *Cutting thick chunks*

When you press down on a soft material such as a steak with a knife, the material first indents before you start to cut it. Not surprisingly the force necessary to start the cut depends upon the sharpness of the knife. As George Orwell wrote in *Down and out in Paris and London*: 'Sharp knives, of course, are the secret of a successful restaurant.' If the knife is really sharp the drop in force as the knife starts to cut is only slight, but with blunt knifes the force can drop sharply.[11] However, when you do cut your steak you probably use a sawing motion because it is easier that way. In a frictionless world a sawing motion would not help in cutting, but Atkins shows that friction at the knife's edge creates a shear stress under the edge in the material being cut which enables transverse work to be done in addition to the normal force.[3] The cutting force per unit width, F, of cut assuming adhesion between the blade and the material being cut, is given in terms of the slice/push ratio, ξ, by

$$\frac{F}{R} = \frac{1 + \dfrac{R_a}{R}\sqrt{\sec^2\theta/2 + \xi^2}}{\sqrt{1+\xi^2}}. \tag{11.3}$$

For Coulomb friction between the blade and the material being cut the cutting force per unit width is given by

$$\frac{F}{R} = \frac{\sqrt{1+\xi^2}\left(\sin\theta/2\sqrt{\sec^2\theta/2} + \mu\right)}{\left(1+\xi^2\right)\left(\sin\theta/2\sqrt{\sec^2\theta/2} + \mu\right) - \mu\left(\sec^2\theta/2 - 1\right)}. \tag{11.4}$$

The cutting force for a knife with a blade angle of 45° is shown in Fig. 11.2 for both adhesion and Coulomb friction for different slice/push ratios. The cutting force decreases with increase in slice/push ratio; the assumption of Coulomb friction predicts the larger decrease in total cutting force with slice/push ratio but in both cases the vertical component of the cutting force decreases quite sharply with the slice/push ratio. Bacon slicers use cutting wheels for efficient slicing for increased cutting efficiency. The transverse motion of the knife would reduce the Coulomb friction so that the assumption of adhesion may be more appropriate than Coulomb friction; Atkins used an instrumented bacon slicer to show that the cutting of cheddar cheese can be predicted by Eq. (11.3).[3] The forces predicted by Eqs. (11.3) and (11.4) are the steady state cutting values. Equally important in starting a cut is the fact that the vertical component of the force is reduced by a transverse motion thus reducing the indentation of a soft solid making the initiation of a cut easier. I shave with a cut-throat razor and so long as I make certain only to drag the razor normal to the blade there is no problem, but the moment there is a slight transverse motion I nick myself.

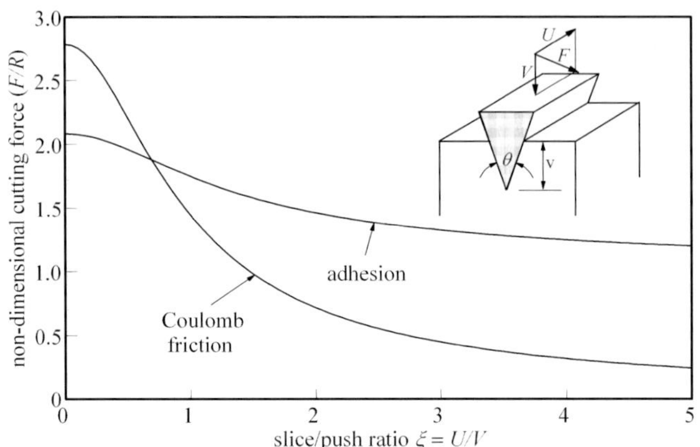

Fig. 11.2 Forces for cutting floppy materials as a function of the slice/push ratio ξ for a knife with a 450 blade angle: (i) adhesion to the blade, $R_d/R = 0.5$, (ii) Coulomb friction, $\mu = 0.5$.

A slice/push ratio in cutting is obtained in oblique cutting where the cutting edge is set at an angle to the movement of the blade. The most notorious use of oblique cutting was the *guillotine* introduced for capital punishment during the French Revolution and named after Joseph-Ignace Guillotin (1738–1814) who proposed its use.[12] Most cutting is performed with an oblique edge; orthogonal cutting is only usual when an oblique edge cannot be used such as in parting off during turning operations.

In cutting cheese with a knife, the cheese adheres to the surface of the knife. For this reason cheese and similar food stuffs are often cut with a wire. The fracture energy of cheese can be conveniently measured from the wire cutting force. Williams and his collaborators have given a simple mechanical model of the wire cutting of cheese.[13] In wire cutting, work is also done to deform the cheese plastically, and against the friction between the wire and the cheese as well as in fracturing the cheese ahead of the wire. As in other orthogonal cutting in the absence of plastic deformation and friction, the cutting force, F, per unit length of wire is equal to the fracture energy, R. The additional force necessary to plastically deform the cheese ahead of the wire and to overcome friction can be found from equilibrium of the forces shown in Fig. 11.3, and the total cutting force is[13]

$$F = R + (1 + \mu)\sigma_Y d, \qquad (11.5)$$

where d is the diameter of the wire. Although Eq. (11.5) obtained from simple equilibrium considerations enables the ranking of the fracture energy of cheese, it is only approximate. One deficiency in the simple model is that cheese behaves viscoelastically. Williams and his co-workers have modified Eq. (11.5) to take into account rate effects and have also performed finite element studies where they modelled the fracture process zone in the cheese.[14]

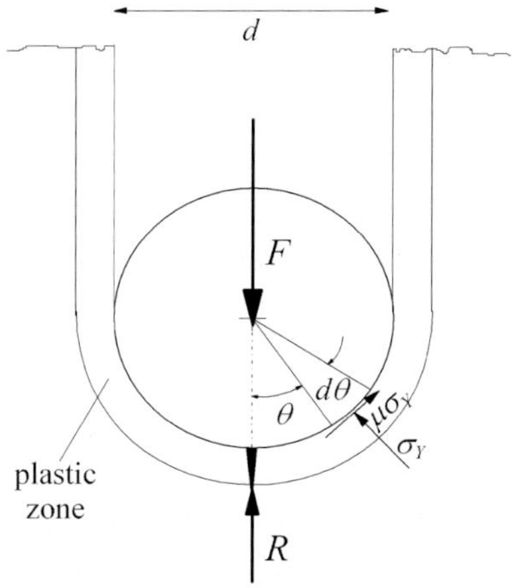

Fig. 11.3 Plastic zones and frictional force in wire cutting (after Kamyab *et al.* 1998).

11.1.3 Wedge indentation

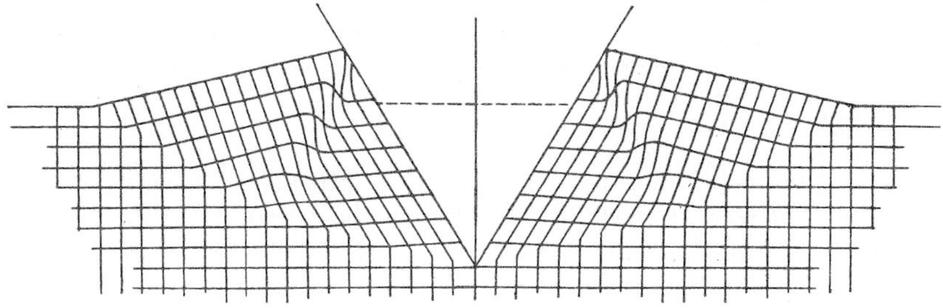

Fig. 11.4 Distortion of a square grid in wedge-indentation from slip-line theory (Hill 1950, with permission Oxford University Press).

There is clearly a cutting action when a wedge is used to indent a metal block under plane strain conditions as can be seen from the distortion of a square grid shown in Fig. 11.4. The classic slip-line theory of wedge indentation of ductile metals of Rodney Hill and his colleagues did not consider fracture.[15] Certainly for large indentations the fracture work is negligible compared with the plastic work, but for small indentations the fracture work can be significant. Expressions for the indentation force derived from slip line theory for a frictionless wedge and the limiting friction case, where either the frictional stress equals the shear yield strength or a dead metal cap forms on the wedge,[15,16] have been used to calculate the maximum indentation depth for the fracture contribution to the indentation force to exceed the contribution from plastic deformation and friction. The results are shown in Fig. 11.5 as a function of the wedge angle for a hypothetical material where the shear yield strength is 100 MPa and the fracture energy is 10 kJ/m^2. For small wedge angles fracture is important for indentations of the order of 50 μm and should be considered in nano and micro hardness tests on metals using wedges. However, fracture has a negligible effect on the usual metal hardness tests which use a pointed indenter such as a Vickers or Berkovich indenter, where the fracture surface area is negligible.

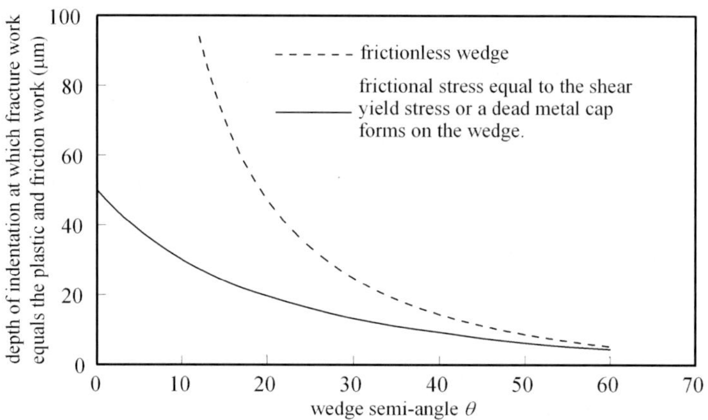

Fig. 11.5 The depth of indentation at which the plastic and friction work equals fracture work in wedge indentation (shear yield strength 100 MPa, fracture energy 10 kJ/m^2).

11.1.4 Cutting thin sheets and plates

Scissors[17] and guillotines are used to cut thin, flexible sheet materials. The blades of scissors are sprung together to ensure a clean cut and the blade of a paper guillotine is similarly sprung causing considerable friction which is an order of magnitude greater than any friction between the blade and the material being cut.

In the cutting of thin flexible sheet materials such as paper or biological tissue, all the work performed, apart from the frictional work, goes into fracture. The friction work can be found by operating the scissors or guillotine without any material. Scissors and guillotines cut obliquely; scissors have straight edges so that the cutting angle decreases during cutting, but paper guillotines very often have curved blades so that the cutting angle remains constant. Oblique cutting reduces the cutting force as discussed in the previous section but it does not change the fracture work. Tony Atkins and Yiu-Wing Mai discussed the guillotining of thin sheets of various materials in 1979 and derived the cutting force as a function of the cutting angle and sheet thickness that, though expressed differently, is a frictionless version of Eq. (11.3).[18] For guillotining thin flexible sheet materials, as well as cutting with scissors, the fracture energy is approximately a linear function of the thickness with the specific essential work being the zero thickness intercept.[19] In cutting metal shim, the knife of the guillotine can be forced away from base plate and a burr formed on the cutting edge, which increases the fracture energy.[18]

The *Titanic* hit an iceberg in 1912 which tore a long cut along its hull with catastrophic consequences. In 1989 the *Exxon Valdez* grounded on Prince William Sound's Bligh Reef and ripped a concertina-like tear in its hull with disastrous environmental consequences. The fractures in both accidents form a class of plate cutting where a plate parallel to the direction of travel hits a wedge. Bo Cerup Simonsen and Tomasz Wierzbicki describe three types of cutting that can occur in these circumstances:[20]

(i) Stable or clean curling cut where the plate, cut obliquely to its plane by a sharp wedge, folds to the same side during the entire process.[21]

(ii) Braided cut where the plate, separated at a sharp narrow wedge with little rake angle, deforms fold back-and-forth to give a braided appearance.

(iii) Concertina tearing where the plate folds back-and-forth in front of a blunt wedge and the tear in the plate diverges as discussed in §1.5.3.3.[22]

11.1.5 Cropping bars

In 1950 Herbert Swift (1894–1960) together with his student Chang published a detailed paper where the cropping of metal bars with orthogonal lines scribed on the sides was interrupted to allow the development of deformation and fracture to be seen.[23] The results for shearing 12.7 mm thick mild steel bars with no clearance between the punch and the die are shown in Fig. 11.6. A crack has initiated at close to the maximum load in Fig. 11.6 (a) and by 25% penetration of the thickness Fig. 11.6 (b) shows that the cracks have penetrated a considerable distance from both the punch and the die. The two cracks do not meet and a tongue of metal is left which is compressed, as shown in Fig. 11.6 (c, d), causing the two halves of the bar to move apart and the tongue to become burnished.

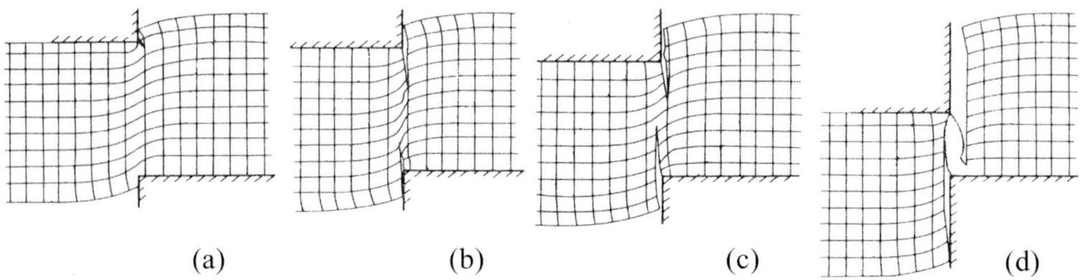

(a)   (b)   (c)   (d)

Fig. 11.6 Progressive fracture in shearing mild steel with no clearance: (a) at maximum load. (b) 25% penetration. (c) 32% penetration. (d) 60% penetration (Chang and Swift 1950 with permission Institute of Materials, Minerals and Mining).

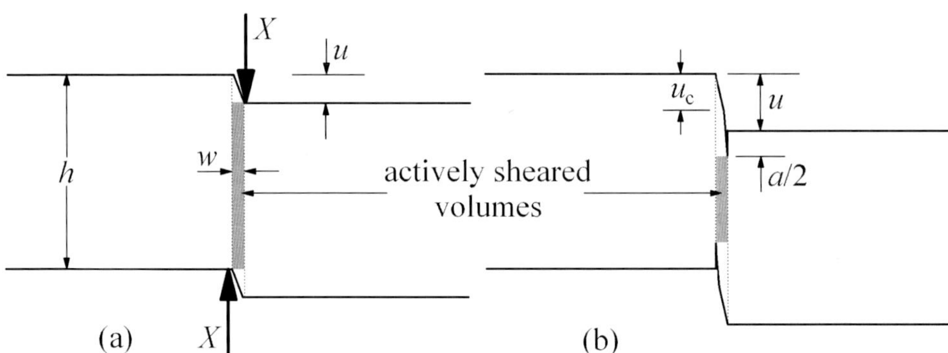

Fig. 11.7 Idealized plate punching (a) plastic shearing without cracking. (b) plastic shearing with cracking, (after Atkins 1980).

Atkins has analysed these results ignoring any elastic strain energy using his method described in §9.4.4.[24] The idealised punching before and after cracks have started to propagate is shown in Fig. 11.7. Atkins assumed that the shear strain was constant over a small band of width w. Before crack initiation the plastic shear strain is given by $\gamma = u/w$. Assuming that the shear stress is given by

$$\tau = \tau_0 \gamma^n, \tag{11.6}$$

The accumulated work done per unit width of the punch as a function of the punch displacement, u, if there is no cracking is given by

$$\int_0^u Xdu = \Gamma(u) = h\int_0^u \tau\left(1 - \frac{u}{h}\right)du = \tau_0 h^2 \left(\frac{u}{h}\right)\left(\frac{u}{w}\right)^n \left[\frac{1}{(1+n)} - \frac{1}{(1+2n)}\left(\frac{u}{h}\right)\right], \quad (11.7)$$

where Γ is the plastic work, friction between the punch and die is neglected. By a suitable choice of parameters Eq. (11.7) can be made to fit the experimental curves of Chang and Swift[23] up to crack initiation at a punch displacement u_c and such a curve is shown schematically in Fig. 11.8. The fracture energy, R, can be estimated from the critical punch displacement, u_c, because up to crack initiation the material can be treated as if it were nonlinear elastic. Hence[25]

$$R = \frac{\tau_0 u_c}{1+n}\left(\frac{u_c}{w}\right)^n. \quad (11.8)$$

After crack initiation the height of the actively shearing volume is $w(h-u-a)$ where a is the crack length relative to the punch and the crack length is $(a+u-u_c)$. Thus the accumulated work done per unit length is given by

$$\int_0^u Xdu = \Gamma(u_c) + h\int_{u_c}^u \tau\left(1 - \frac{u}{h} - \frac{a}{h}\right)du + R(a+u-u_c). \quad (11.9)$$

Atkins assumed that the crack length, a, is proportional to $(u-u_c)$ and calculated the constant of proportionality from the punch displacement at which the shearing force becomes zero. With this assumption the accumulated work done after the initiation of cracks, shown schematically in Fig. 11.8, can be calculated.

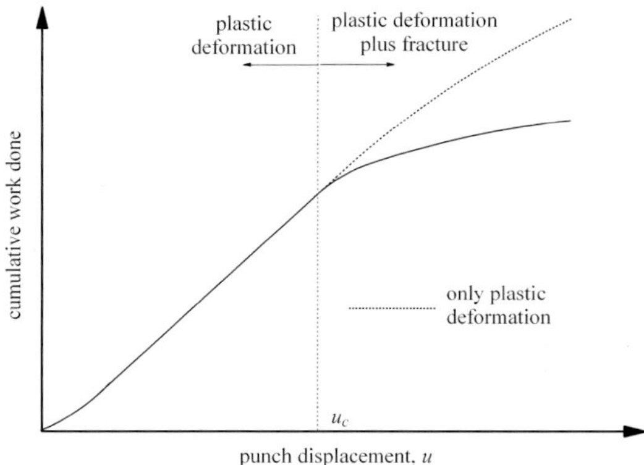

Fig. 11.8 Schematic work done in punching as a function of punch displacement.

11.2 Machining of Metals

The engraving of softer metals occurred very early. Metal cutting files, developed from woodworking bronze rasps in Egypt, date to the twentieth dynasty (1190–1077 BC); Assyrian iron rasps have been found dating to the 7th century. The Greeks developed complicated instruments that used gear wheels and screws, such as the Antikythera instrument which was discovered as part of the cargo of a vessel that sunk off the island of Antikythera during the second century BC.[26] The Antikythera instrument showed the movement of the planets through a large number of brass gear wheels that were accurately cut. However, while much has been written about such instruments, there is virtually no mention of how they were made.

Vertical boring machines for cannon were developed in the fifteenth century. Wood cutting lathes are very ancient and a bronze turning tool dating to the beginning of the second millennium BC was found in the South Osetin region of Georgia.[27] However metal cutting lathes were not used until the seventeenth century. The construction of James Watt's steam engine was made possible by the development in 1774 of an accurate horizontal boring machine by John Wilkinson (1728–1808) that could bore a cylinder true to 'within the thickness of a worn shilling'. Interchangeable parts for mass production require a much higher accuracy. In France the parts for the mechanism of the musket lock were interchangeable by 1785 when Thomas Jefferson (1745–1826), who was then the United States Minister in France, assembled a musket lock from randomly selected parts. By the mid nineteenth century all the traditional metal cutting machines had been developed and studies of the science of metal cutting began. Hans Ernst (1892–1978)[28] and Iain Finnie[29] have written reviews of the historical development of metal-cutting science.

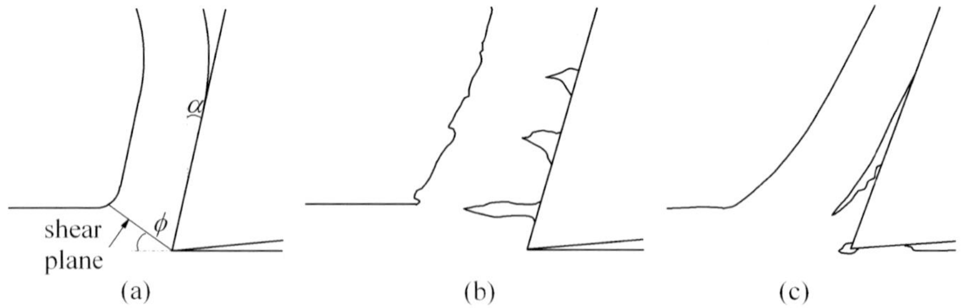

Fig. 11.9 (a) Continuous chip, (b) Discontinuous chip. (c) Continuous chip with a built-up edge. (after Ernst 1938).

The first scientific study of metal cutting according to Finnie was that of drilling by Cocquilhat in 1851.[30] Chip formation in machining has been studied since Time's 1870 monograph on the cutting of metal and wood.[31] In 1881 Arnulph Mallock (1851–1933) made precise drawings of chips formed in a number of metals showing two of the main chip types: continuous and discontinuous illustrated in Figs. 11.9 (a, b) using a microscope attached to the tool holder.[32] Ernst also identified continuous chips with a built-up edge shown in Fig. 11.9 (c).[33] Mallock was the first to show that the metal cutting process was one of shear and occurred along a sharply defined plane.

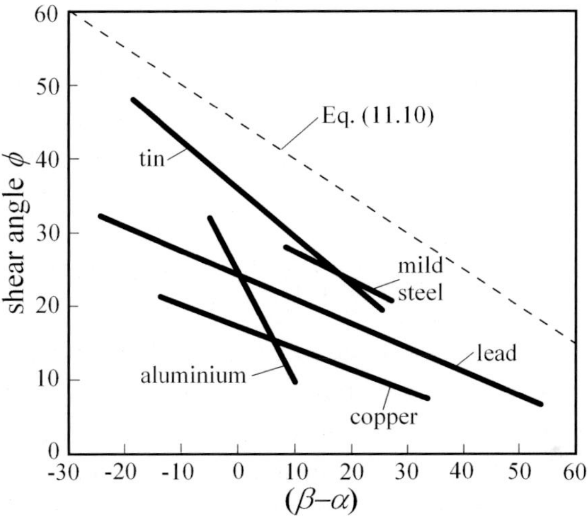

Fig. 11.10 Comparison of Merchant's expression, Eq. (11.10) with experimental data (after Astakhov 2005).

Much of the mechanics of orthogonal machining has been concerned with single plane shear models and the prediction of the angle, ϕ, that the shear plane makes with the direction of cutting; the experimentally measured shear angles for a number of metals are shown in Fig. 11.9 (a). In 1941 Eugene Merchant (1913–2006) and Hans Ernst derived the often quoted expression

$$\phi = \pi/4 + \alpha/2 - \beta/2, \qquad (11.10)$$

where α is the rake angle and $\beta = \tan^{-1}\mu$ is the friction angle on the rake face from two considerations: first that the shear on the shear plane is a maximum[34] and secondly from the shear angle that makes the work done by the cutting force a minimum in 1945[35]. Fracture was not considered in deriving this expression.

Although Eq. (11.10) is usually attributed to Merchant and Ernst, it was already known in Europe. It is not certain who first derived Eq. (11.10),[29] Konstantin Zvorykin at the Khar'kov Technological Institute published it in 1893 but stated that it had already been proposed at an earlier date.[36] Since 1893 Eq. (11.10) was published at least three more times before Merchant and Ernst rediscovered it.[29] The problem is that the shear plane angle is not well predicted for metals by Eq. (11.10) as can be seen from Fig. 11.10.

Since the work of Merchant and Ernst there have been a number of other relationships developed for the shear plane angle which have the general form

$$\phi = C_1 + C_2 (\beta - \alpha),$$

(11.11)

where C_1 and C_2 are constants[37] but all of these, like the Merchant-Ernst expression, are only dependent on the rake and friction angle and are size independent. Other analyses of machining with deformation over an extended zone that do not consider the work of fracture are similarly only dependent on the plastic behaviour and friction characteristics.

11.2.1 *The role of fracture in machining*

Merchant did consider the work required to create new surfaces in metal cutting and estimated its magnitude in comparison with the plastic work.[38] However Merchant was working before the Irwin-Orowan concept[39] of the fracture energy of metals was known, and used the surface energy which he knew was of the order of 1 J/m^2, instead of the fracture energy which is least 10^4 J/m^2, and so estimated that the work of separation would only be important for cuts as thin as 100 nm.[40] In retrospect if Merchant had used a more realistic value for the fracture work he would have found that a more realistic thickness of 1 mm at which fracture work becomes significant. The famous German engineer Franz Reuleaux (1829–1905) reported a crack about 200 μm long running ahead of the cutting tool in 1900 and concluded that the action of a cutting tool was comparable to an axe splitting wood.[41] It is quite possible that Reuleaux was mistaken, as Kick suggested in 1901, and there was no crack ahead of the tool.[42] In splitting wood a crack can run ahead of the tool because the elastic strain energy is not negligible compared with the plastic work and the fracture work can come from its release, but the tougher the material the shorter the crack ahead of the tool, and as the toughness increases, the crack tip eventually coincides with the tip of the tool. As Milton Shaw and his colleagues wrote in

1954[43]: 'When a ductile metal is cut to produce… a continuous ribbon there is usually no evidence of fracture or crack formation. Yet, new surface is generated and this must involve fracture.' Certainly cracking along the shear plane is evident with discontinuous chip formation as described by Shaw and his colleagues. Yet the mention of cracking and fracture during machining seems to have been a heresy to many researchers. Ernst wrote[28]: 'the fallacies of Reuleaux have persisted in some cases, to the present day' and Finnie[29] refers to Reuleaux's 'backward step' despite being earlier a co-author of Shaw's paper.[43]

With machining models based only on the plastic flow there is no size effect in continuous chip machining. However with increase in the depth of cut, continuous chip formation becomes discontinuous.[37,43] Even glass can be machined providing the depth of cut is small enough. In the 1960s Busch filmed the machining of glass showing ribbons emerging from the rake face.[44] Keith Puttick and his colleagues showed that continuous chip formation could be obtained in spectrosil, an optical quality fused quartz, when the depth of cut was 250 nm or less.[45] The size effect in metal-cutting arises because the plastic work per unit width of cut is proportional to the depth of cut, whereas the fracture work is independent of the chip thickness.

The major difference between the mechanics of metal forming, which only needs to consider deformation and machining where new surfaces are produced has been pointed out by Viktor Astakov.[46] This difference has largely been ignored by the machining community, but finite element simulations of machining cannot be run without a *separation criterion* at the tip of the tool as Atkins has rightly emphasized.[47] A wide variety of separation criteria have been used;[48] most of these criteria do not include the fracture energy explicitly but Atkins has inferred that the fracture energy is about 10 kJ/m^2.[47]

11.2.2 Mechanics of machining

Gordon Williams and his colleagues have published an exhaustive paper analysing cutting and machining using fracture mechanics concepts that covers the complete range of possible deformation mechanisms.[49] Relatively small rake angles are used in the machining of metals and polymers, but bending rather than shearing solutions are possible if the rake angle is larger than about 60°. Some plastic bending must occur in machining even with small rake angles to account for the curling of the chip. The debate of the first few decades of the twentieth century as to whether or not there was a crack at the tip of the cutting tool is resolved. When the cutting tool touches the crack tip energy can be directly supplied by the tool for both deformation and fracture. If the tool does not touch the crack tip, energy for fracture can only come from the elastic strain energy and the crack tip can only leave the tip of the tool if the rake angle is large. In the traditional analyses of cutting it has been assumed that Coulomb friction exists along the rake face of the cutting tool. Williams and his colleagues have carefully examined the existing machining data and shown that the results are not consistent with just Coulomb friction. Adhesion on the rake face does occur and is in part responsible for the formation of built-up edge. Adhesion generally occurs on the rake face in addition to Coulomb friction thus if R_a is the adhesion energy the force per unit width of cut, the force, S, acting along the rake face is given by

$$S = R_a + \mu N \tag{11.12}$$

where N is the force normal to the rake face and μ is the coefficient of friction. The adhesion energy, R_a, is similar in magnitude to the fracture energy, R, of the material. Here discussion will be limited to machining with small rake angles where the tool touches the crack tip and shearing is the dominant deformation mechanism, this condition and the other possibilities are given by Williams and his colleagues.[49] The forces per unit width of cut acting on the tip of the tool, as shown in Fig. 11.11, are the same as those of Ernst and Merchant[34] except for the fracture force on the crack, R, which they neglected. The equilibrium of the forces on the cutting tool together with the relationship given by Eq. (11.12) gives the transverse force

$$F_t = \left(F_c - R \right) \tan \left(\beta - \alpha \right) + \frac{R \cos \beta}{\cos \left(\beta - \alpha \right)}, \tag{11.13}$$

where F_c and F_t are the cutting and transverse forces respectively and, $\beta = \tan^{-1} \mu$ is the friction angle. Although the shear plane is assumed to be infinitesimally thin, it is shown in Fig. 11.11 as having a finite width so that the forces acting on each side of the shear plane can be shown. Thus the shear force, F_s, per unit width of cut acting on the shear plane is

$$F_s = \left(F_c - R \right) \cos \phi - F_t \sin \phi, \tag{11.14}$$

and the cutting force can be expressed as

$$F_c = \frac{1}{1 - Z \tan \phi} \left\{ kh \left[\frac{1}{\tan \phi} + \tan \phi \right] + \frac{R_a}{\cos \alpha + \mu \sin \alpha} \tan \phi \right\} + R, \tag{11.15}$$

where k is the shear yield strength and $Z = (\mu + \tan\alpha)/(1 + \mu\tan\alpha)$. Applying the principle of minimum work, as did Merchant and Ernst, the shear plane angle, ϕ, is given by

$$\cot\phi = \sqrt{1 + Z^2 + \frac{(R\sin\beta + R_a\cos\alpha\cos\beta)}{kh\cos\alpha\cos(\beta - \alpha)}}, \qquad (11.16)$$

and the cutting force becomes

$$F_c = 2kh\cot\phi + R. \qquad (11.17)$$

This solution is similar to that of Atkins.[47] Williams and his colleagues have extended this solution by including bending as well as shear.[49] Existing experimental machining data on the cutting force fits this new theory of machining. This theory also explains the finite cutting force when the data is extrapolated to a zero cut that was ignored in previous analyses that did not consider fracture. However, the shear angle predicted, though good for polymers, is not accurate for all metals.

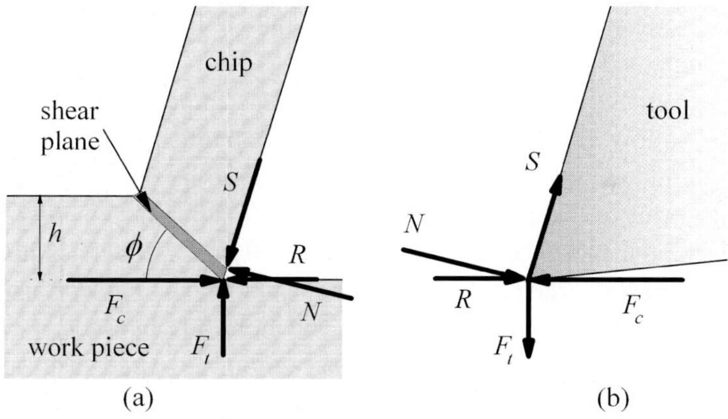

Fig. 11.11 Free body diagram of machine cutting showing the forces acting on the tool where the tip of the tool touches the machining crack and plastic shearing is the deformation mechanism. (a) Forces acting on the work piece and chip. (b) Forces acting on the tool.

11.3 Piercing

Although piercing has its peaceful uses, such as getting jab from the doctor against the flu, it is also important in war. Arrows have great penetrating power. At the siege of Abergavenny in 1182 Welsh arrows shot from elm longbows are said to have penetrated an oak door 4 inches (100 mm) thick.[50] Heavy English medieval arrows with bodkin points could pierce plate armour.[51] Composite bows were especially powerful; Francis Bacon (1561–1626) wrote that an arrow fired from a Turkish composite bow could pierce a steel or brass target two inches (50 mm) thick.[52]

Experiments were carried out by John Greaves (1602–1652) at Woolwich on the penetration power of cannon balls using a target of three butts made of 19 inches (480 mm) thick oak and elm, the second being 14 yards (13 m) behind the first and the third 8 yards (7 m) behind the second.[53] A 32 lb (14.5 kg) shot from an iron demi-cannon fired with 10 lb (4.5 kg) of powder perforated the first two butts and hit the third. Although shells were first used in the fourteenth century and became widely used in land mortars, their use at sea was considered too dangerous because of the danger of premature bursting. The first use of shells at sea was in 1788 when Samuel Bentham (1757–1831), in the naval service of the Tsar of Russia, fitted a flotilla of longboats with brass cannon and gained a notable victory against a large Turkish squadron. Henri-Joseph Paixhans (1783–1854) invented a delaying mechanism for shells in 1823 which enabled them to be fired with safety. However, until the middle of the nineteenth century the cannon ball remained supreme at sea.

Splinters from the timbers of a hull hit by a cannon ball caused the most casualties to sailors. In 1813 William Moore[54] published an essay on naval gunnery as an addendum to a book on rockets where he discusses the technique of gunnery to achieve the maximum splintering.[55] Rather surprisingly the cannon ball that causes the most damage by producing flying splinters is not one fired with the maximum charge. A high velocity cannon ball passes through a hull knocking out a clean frustum of timber. The hole left by such a cannon ball partially closed and was easy to repair. A cannon ball fired at a velocity just sufficient to pierce the hull, now known as the *ballistic limit*, caused the maximum splintering and damage. No reason is given for this statement by Moore which was repeated by Howard Douglas (1776–1861) in 1820.[56] The answer to this apparent paradox is that a cannon ball hitting a timber hull at high velocity is similar to punching a hole in a hull supported by a die because, though the velocity is low compared with the velocity of longitudinal waves, which are about 5,000 m/s for oak, it is high enough to prevent much bending occurring. A cannon ball fired at 600 yards (550 m) with one sixth the maximum charge,[55] which is only just able to penetrate the hull, caused bending and much splintering.

11.3.1 Deep penetration of soft solids

Predators must have strong enough jaws to enable their teeth to penetrate their prey. Deer and bovids need a sufficiently tough skin so that fatalities from antlers and horns are kept to a minimum during ritualised combat for females. Tyres must resist penetration from nails. Conversely, hypodermic needles must easily pierce the skin. To understand the resistance to penetration it is necessary to study the mechanics.[57,58] Here the analysis of Oliver Shergold and Norman Fleck is given for rubber-like incompressible materials such as skin, which can be described by the one term Ogden strain energy density function,[59] U, given by

$$U = \frac{2\mu}{\alpha^2}\left(\lambda_1^\alpha + \lambda_2^\alpha + \lambda_3^\alpha\right),\tag{11.18}$$

where μ is the infinitesimal shear modulus, λ_i are the principal stretch ratios, and α[60] is a strain hardening exponent. The mechanisms of essentially elastic penetration by blunt and sharp punches is quite different: a blunt punch creates a cylindrical crack, as shown in Fig. 11.12 (a), whereas a planar crack forms with a sharp punch, as shown in Fig. 11.12 (b).

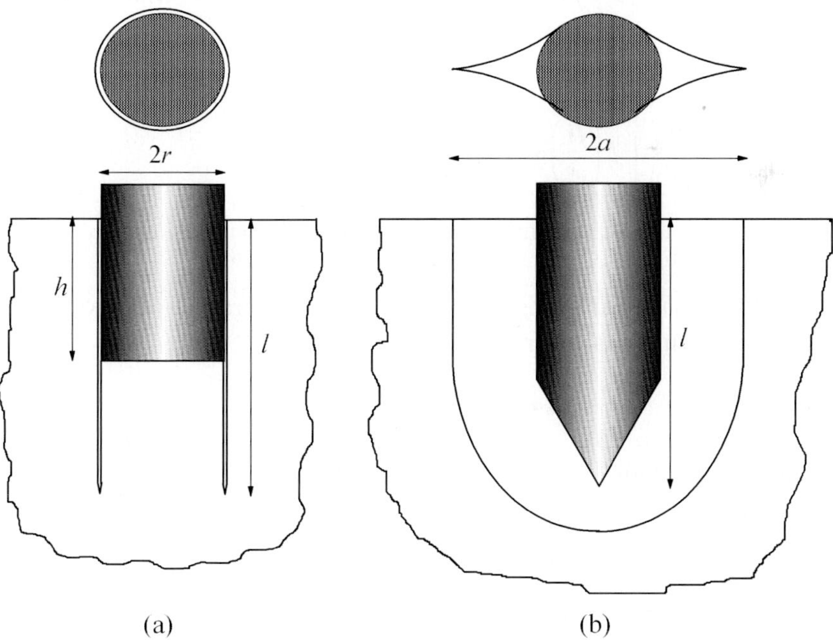

Fig. 11.12 Steady state penetration of a soft solid (a) with a blunt punch. (b) with a sharp punch (after Shergold and Fleck 2004).

Since we are concerned with highly extensible materials, the cylindrical crack, which forms at the tip of a blunt punch of radius r, has a significantly smaller radius, b, in the undeformed state. The penetration force compresses the column of length, l, and radius, b, into a column of length $(l - h)$ and radius r, while the material outside of the ring crack is expanded from a radius of b to r. Since the solid is incompressible

$$\pi b^2 l = \pi r^2 \left(l - h\right).$$ (11.19)

The penetration of a frictionless punch under a force, P, is steady state. The work done in increasing the penetration by dh is given by

$$Pdh = \frac{\partial \Lambda}{\partial l} dl + 2\pi b R dl,$$ (11.20)

where Λ is the strain energies stored and R is the fracture energy. On dimensional grounds the rate of change in the strain energy, Λ, is given by

$$\frac{\partial \Lambda}{\partial l} = \pi \mu r^2 f \left(\frac{b}{r}\right)$$ (11.21)

where the function $f\left(b/r\right)$ can be found from the Ogden strain energy density function. Using Eqs. (11.19) and (11.20) the penetration pressure, p, can be written as

$$\frac{p}{\mu} = \left[f\left(\frac{b}{r}\right) + \frac{2b}{r}\left(\frac{R}{\mu r}\right) \right] \Big/ \left[1 - \left(\frac{b}{r}\right)^2 \right].$$ (11.22)

The value of b/r is that which minimises the penetration pressure from minimum work principles.

The work done in increasing the steady state penetration of a soft solid by a sharp frictionless punch is given by

$$Pdl = \frac{\partial \Lambda}{\partial l} dl + 2aR dl.$$ (11.23)

On dimensional grounds

$$\frac{\partial \Lambda}{\partial l} = \mu \pi r^2 g \left(\frac{a}{r}\right),$$ (11.24)

and Shergold and Fleck have calculated the function $g\left(a/R\right)$ using finite elements. The penetration pressure is given by Eqs. (11.23) and (11.24) as

$$\frac{p}{\mu} = \frac{1}{\pi} g\left(\frac{a}{r}\right) + \frac{2}{\pi}\left(\frac{R}{\mu r}\right)\left(\frac{a}{r}\right),$$

(11.25)

and the ratio a/r is that which gives the minimum penetration pressure.

Not surprisingly a sharp frictionless punch penetrates a soft solid more easily than a blunt frictionless punch as can be seen from the comparison of the penetration pressures for blunt and sharp punches for an Ogden material with a strain hardening exponent of $\alpha = 9$, which is typical of skin, shown in Fig. 11.13.

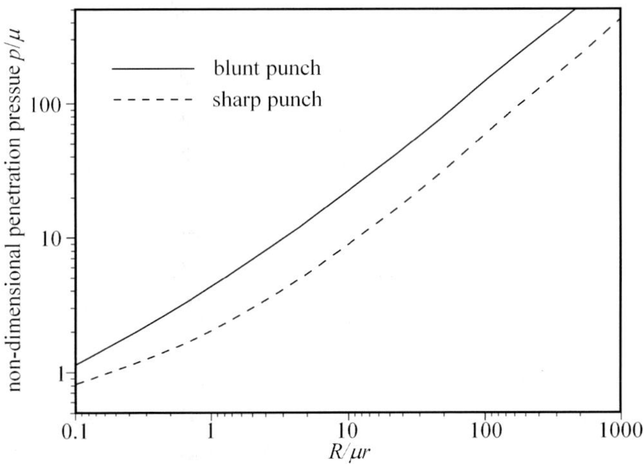

Fig. 11.13 Comparison of the indentation pressure for blunt and sharp punches for a soft solid with strain hardening coefficient typical of skin (after Shergold and Fleck 2004).

11.3.2 Deep penetration of stiff solids

The deep penetration of unconstrained stiff solids is mainly one of hole expansion.[61] In the deep penetration of ductile metal targets by sharp cylindrical punches, fracture takes place but is in the form of initiating a tiny hole that is then expanded by plastic flow;[62] a cone of dead metal forms on a blunt punch that then acts like sharp one. Hence the fracture work is small and the penetration can usually be modelled by plasticity alone. Deep penetration of less ductile polymers can cause fractures because of the circumferential tension induced during the expansion of the initial hole. Fleck and his co-workers have modelled the deep penetration of polycarbonate where a hackle zone forms around the punch.[63] Since this hackle zone does not consist of well defined cracks, the penetration can only be modelled semi-empirically by assuming that a hackle zone forms when the strain reaches some critical value.

11.3.3 Piercing of sheets and plates

The deformation of sheets pierced normal to their plane with conical indentors depends upon the depth to which the indentor has penetrated and the conical semi-angle. In 1948 Geoffrey Taylor experimented on the piercing of lead sheet with a very sharp cone (semi-angle 1.5°) and the results are shown in Figs. 11.14 (a, b).[64] If the hole pierced and enlarged was less than 7 to 10 times the sheet thickness the lead was deformed symmetrically as in Fig. 11.14 (a), but if the hole was enlarged to a greater diameter an unsymmetrical deformation resulted as shown in Fig. 11.14 (b) where the tube could be formed on either side of the sheet. In both of these cases there was no fracture other than the initial piercing of the hole. A more usual form of piercing of an aluminium alloy with a cone of semi-angle of 20° is shown in Fig. 11.14 (c) where five cracks have formed to give a petal shaped deformation. Tony Atkins and his colleagues have shown that the number of petals, which decrease with cone angle, can be predicted by the application of the principle minimum work.[65]

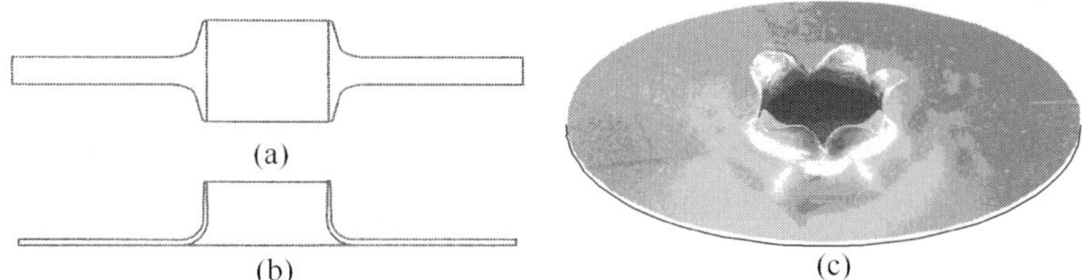

Fig. 11.14 Sheet piercing (a) piercing of lead by a sharp conical punch to a diameter of less than 7–10 times the sheet thickness. (b) piercing of lead by a very sharp conical punch with a diameter of greater than 7–10 times the sheet thickness. (c) piercing of an aluminium sheet by a cone of semi-angle 20° (a,b after Taylor 1948, c after Atkins *et al.* 1998b).

11.4 Armour and Piercing Impact

Animals and man have developed armour against piercing impact. Animals evolved two methods to defeat impact penetration: the first was the evolution of hard materials, like chitin for cuticles or shells, where the aim is to defeat penetration by preventing the indentation that precedes penetration; the second method relied on evolving high toughness materials. Man too has used these two basic techniques for armour. Until the seventeenth century warriors relied mainly on chain mail and plate armour, though this was did not always offer good protection. The Spanish conquistadors thought that their chain mail would be adequate protection against Indian stone-tipped arrows. In Florida the Spaniards tested the effectiveness of native archery by offering a captive Indian his freedom if he could shoot an arrow through a coat of mail. To their surprise the arrow did not just pierce one coat of mail but two.[66] After experiences such as these the Spaniards abandoned heavy mail for more effective padded cloth armour. The Aztec nobles used armour made of feathers sewn to a backing fabric.[67] Jim Gordon writing in 1978 commented that the fracture mechanism of feathers is something of a mystery, it still is.[68] Perhaps the modern equivalent of padded cloth and feather armour is aramid fibre body armour though this is often used in conjunction with ceramic plates.

11.4.1 Perforation mechanisms in metal plates

There are many mechanisms of perforating a metal plate by impact depending upon the relative hardness of the plate and penetrator, the penetrator diameter to plate thickness ratio, sharpness of the penetrator, the velocity of impact, and toughness, which have been identified by Marvin Backman and Werner Goldsmith.[69] Some of the mechanisms such as petalling, when the penetrator diameter is small compared with the plate thickness, or plugging that occurs at impact velocities significantly higher than the ballistic limit are the same as occur during quasi-static piercing, or punching, but other such as *scabbing* are unique to high velocity impact.

(a)   (b)

Fig. 11.15 (a) Scab torn from a 19 mm thick mild steel plate when gun-cotton is detonated in contact with one side. (b) Cracking caused when a similar charge of gun-cotton is detonated in contact with a 32 mm thick mild steel plate (after Hopkinson 1912).

   Bertram Hopkinson coined the term *scabbing* in 1912 during experiments on the effects of detonating gun-cotton near to and in contact with mild steel plates.[70] Hopkinson found that if a cylinder of gun-cotton weighing between 28 and 56 g was placed in contact with mild steel plate 12 mm or less thick and detonated a hole was punched through the plate of about the same diameter as the gun-cotton. However, if the plate was 19 mm thick 'the curious result … [is that a] scab of metal is torn off, and projected away with a velocity sufficient to enable it to penetrate a thick wooden plank', as shown in Fig. 11.15 (a). To investigate the mechanism of scabbing, Hopkinson detonated a 56 g cylinder of gun-cotton in contact with a 32 mm thick mild steel plate; no separation of metal was visible, but when the plate was sectioned an internal crack was found, as shown in Fig. 11.15 (b) which was the beginning of the separation of a scab. The detonation of the gun-cotton sent a compression wave propagating through the plate. At the back surface the compression wave is reflected as a tension wave. The stress at successive time intervals is shown schematically in Fig. 11.16 where the three dimensional spreading and diminution of the wave is ignored. After time interval 3 the wave is reflected, after time interval 4 tension is created which becomes a maximum at time interval 6 at a position P some way inside the plate. If the tension is sufficient to form a scab, the velocity of the particles becomes trapped within the scab and it flies off with high velocity. In order to produce a scab the compression wave must have duration shorter than about the thickness of the plate divided by the velocity of longitudinal waves. For the 19 mm plate this duration would be less than about 3 μs. Since scabbing involves the formation of cracks, the fracture toughness of the metal plate must be important.

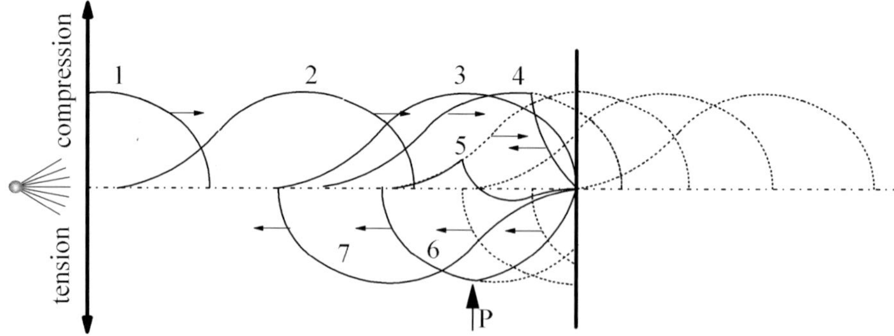

Fig. 11.16 Schematic illustration of the propagation of a compression wave into a plate and its reflection as a tension wave at successive time intervals.

Soft steel capped ammunition was introduced around 1905 for piercing hard armour plate, and is still being used. A M59 37 mm soft steel capped shell from World War II is illustrated in Fig. 11.17. Hopkinson explained in 1912 that the soft steel cap deformed on impact forming a supportive ring around the shell proper which greatly increased the impact force that could be withstood without yielding.[70] In experiments in penetrating on Hadfield Era hard faced armour plate with 105 mm standard British uncapped and Helcon capped armour piercing uncharged shells, both using Hadfield cast steel, the uncapped shell fragmented without piercing the plate whereas the capped shells completely pierced the plate and, apart from the soft steel cap, were virtually undamaged.[70]

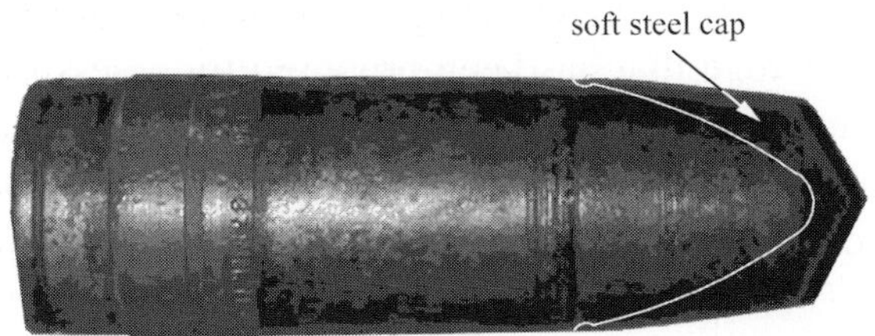

Fig. 11.7 A World War II M59 soft steel capped shell.

There are so many variables involved in piercing by impact that little can be said in general. High hardness of the impacted plate is seen as the best defence against piercing but there is no simple correlation between hardness and resistance to penetration.[71] Toughness must play a part in some forms of piercing, but a reasonably recent review of the impact of projectiles does not once mention toughness.[72]

11.4.2 Helmet development

Our first real knowledge of helmets comes from the ancient Greeks. The Egyptian soldier did not wear a helmet before the New Kingdom (1550 BC) or perhaps as late as the twentieth dynasty (1200 BC). The most effective helmets of the Greeks were the Corinthian ones made from bronze. Corinthian helmets date from the 7th century BC and the early ones were made of bronze with an average thickness of 1.2 mm and hardness of 90 DPH and weighed around 1.4 kg. One of the problems with all metal armour is their weight. By the last quarter of the 6th century BC the weight of the Corinthian helmet had been reduced to about 0.7 kg by decreasing the thickness to about 0.7 mm. The hardness of these new helmets was increased to 130–80 DPH by hammering and also the padded distance between the helmet and skull was increased by about 10 mm under the cap, and by about 20 mm in the centre of the forehead[73]. A late Corinthian helmet dating to about 460 BC which was captured by the Argives in battle is shown in Plate 10. Increasing the hardness of a metal by cold working makes indentation more difficult, but it also decreases the toughness and the piercing by a sharp pointed weapon would not have been that difficult. The strategy adopted may have been to limit the penetration of the weapon to prevent wounding. Obviously it is not possible to perform destructive tests on the Corinthian helmets so that Henry Blyth and Tony Atkins undertook to measure the force required to penetrate a series of metal sheets of different toughness and hardness with a triangular knife with a single sloping edge that has been proposed for assessing the performance of modern body armour against stabbing attacks.[74] They also compared their experimental results with the theory developed by Tomasz Wierzbicki and his co-workers for wedge cutting though metal sheets.[20,75] This theory enabled the penetration force, F_{new}, of the new style Corinthian helmet to be compared with the penetration force, F_{old}, for the old one to give

$$\frac{F_{new}}{F_{old}} = \left(\frac{\text{VPH}_{new}}{\text{VPH}_{old}}\right)\left(\frac{h_{new}}{h_{old}}\right)^{1.6}\left(\frac{\overline{\delta}_{cnew}}{\overline{\delta}_{cold}}\right)^{0.2},$$

$$= (1.7 \sim 2)\left(\frac{0.7}{1.2}\right)^{1.6}\left(\frac{0.77}{1.53}\right)^{0.2} = 0.63 \sim 0.74,$$

(11.26)

where h_{new}, h_{old}, are the thickness of the old and new style helmets and $\overline{\delta}_{cnew}$, $\overline{\delta}_{cold}$ are the normalised critical crack tip opening displacements respectively. So, the new helmets were easier to penetrate than the old ones. The increased distance between the helmet and the skull may have been an attempt to compensate by increasing the work of penetration. Probably the change in helmet design was related to a change in tactics to a more mobile warfare which needed lighter armour.

At the start of World War I soldiers wore no head protection. In 1915 the French equipped their soldiers with a steel cap liner, the casque Adrian, named after Quartermaster Louis Adrian (1859–1933) who, so the story goes, noted that a soldier had escaped death from a bullet owing to his habit of wearing his metal food bowl under his cap. The British first issued the Mark I steel helmet designed by John Brodie in 1916. The original Type A helmet was formed from mild steel, which would not have offered much protection. After a few weeks Type B formed from a modified Hadfield Manganese Steel and weighing 1.5 kg was introduced. Robert Hadfield (1858–1940) patented his austenitic manganese steel containing 13% manganese and 1.2% carbon in 1883. This steel has remarkable strain hardening capabilities caused by stress induced transformation of austenite to martensite, which enabled it to achieve strength of about 1.6 GPa after a strain of 0.5%, and it had high toughness. The Brodie helmet was 0.9 mm thick and primarily designed to give protection from shrapnel, especially from above which was the reason for the brim; the ballistic limit was 183 m/s when impacted by a 15 g, 0.45 calibre bullet. The advance on the Corinthian helmet was to use a tough as well as a hard metal.

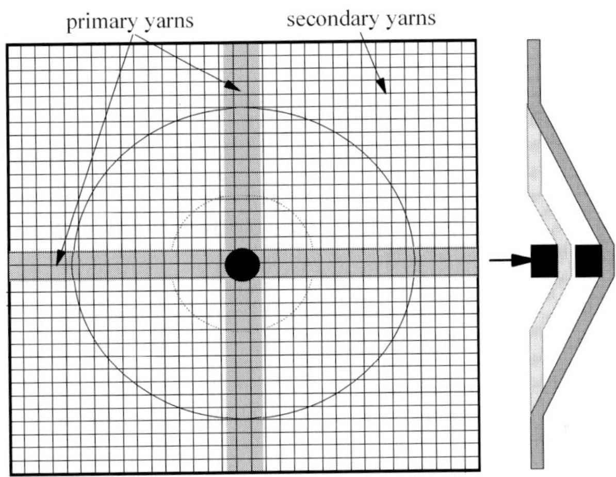

Fig. 11.8 Deformation cone formed when a composite is impacted by a bullet.

Modern helmets, like the American Advanced Combat Helmet or the Spectra Helmet used by French and other forces, rely on polymer reinforced composites using aramid, ultra high molecular weight polyethylene, or other high toughness fibres. The modern helmet is able to be comparatively light and resist penetration because of the very high resilience of the aramid fibres. Kevlar 29 has a capacity to absorb 37 kJ/kg of energy before breaking. When hit by a projectile, transverse waves propagate out from the area of impact and the composite deforms into a cone that expands with time maintaining the same conical angle as shown schematically in Fig. 11.18. The yarns to fail first are the primary yarns which are the ones that are continuous across the impact area. The major energy absorption comes from the deformation and fracture of the primary yarns, elastic deformation of the secondary yarns, and the kinetic energy of the deformation cone, with the latter two being the dominant ones.[76] With this simple model of the penetration mechanics the ballistic limit can be estimated. The American Advanced Combat Helmet uses Kevlar fibres and is designed to give protection from shrapnel fragments and has a ballistic limit to 9 mm submachine bullets of 430–440 m/s; though the helmet at 1.36–1.63 kg depending upon size is lighter than the previous American PASGT helmet, but is still not lighter on average than the steel helmets worn in the two World Wars.

11.4.3 Development of battleship armour

Until the end of the Napoleonic wars the wooden hulls of ships of the line, which were some 600 mm thick, were shot-proof, except at close range, to the cannon of the day. However proposals were made in the early nineteenth century to clad the wooden hulls of ships with wrought iron. The three French floating batteries: *Dévastation*, *Lave*, and *Tonnante* took part in the bombardment of Fort Kinburn in 1855 during the Crimean War. The *Dévastation* was hit '67 times without any other effect on stout iron plates other than to dint them at most one and a half inches'.[77] Bill Johnson, who has long experience in impact mechanics, wrote a much needed review of the development in warship armour and expressed the opinion that, in view of the practical and security difficulties in obtaining information from modern full scale tests, much can be learnt from a study of the historical records. His review is widely used in this section.[78]

The success of the French muzzle-loading guns in the Crimean War led to the adoption of armour plate. Although the French initially led the way with ship armour, and the wooden frigate *Gloire* in 1859 was the first naval ship to be protected by wrought iron plates (120 mm thick), the repayment of the costs of the 1848 revolution halted further experimentation. The answer by the British to the *Gloire* was to build the *Warrior*, the first all iron warship, with a 114 mm thick armour belt and powered by both steam and sail. The *Warrior* launched in 1860 was finally armed with ten Armstrong 110 pounder and four 70-pounder rifled breech loading guns as well as twenty-six 68-pounder muzzle loading smooth bore guns, but none were ever fired in anger.[79]

William Armstrong realised that a breech mechanism of the day could not be made to withstand the explosive power required to send a shell through thick iron. By 1859 innovations in rifling techniques made heavy muzzle loading guns possible that could pierce the armour of ships like the *Gloire* and the *Warrior*. The *Bellerophon* launched in 1865 had armour 150 mm thick. However improvements in ordnance quickly made that obsolete. To indent a flat plate an indentor without a soft steel cap needs to have hardness at least about 2.5 times that of the plate because the metal under the indentor is highly constrained. Hence the effectiveness of ordnance can be improved by increasing its hardness. In 1867 William Palliser (1830–1882) used chilled cast iron to produce an ogive-pointed shot with a hardened nose that became the standard for the British navy until 1909. The limit to the practical thickness of wrought iron armour was reached with the *Inflexible* of 1881 that had an armour belt 600 mm thick made up of two layers. Hardened steel plate with about 0.45% carbon was tried as armour but did not have sufficient toughness and fragmented. Compound armour made by casting steel onto wrought iron plates was more successful and was used by the British navy up to the early 1890s. The compound armour relied on the hard steel surface to break up the shot. Forged steel amour-piercing shot introduced in 1886 pierced the hardened steel face. In 1889 nickel was used to alloy the steel to produce both hardness and toughness and marks the transition to the use of alloyed steels. During the Battle of Jutland in 1916, the Germans pierced 230 mm thick plates from a distance of about 1.34 km, while the British pierced 280 mm thick plates at 1.59 km.

11.5 Concluding Remarks

Cutting and piercing are wide fields covering all materials and many different mechanisms, but the common ground is material separation or fracture. Surprisingly though fracture is now a widely known mature discipline there are still areas where it has seen little application. Two of these are machine cutting and impact piercing. Those using finite element methods in these areas have been forced to use some criteria for separation but these have not always been well based on fracture concepts. Fortunately established fracture mechanists have started to take an interest in these topics and a greater understanding of machining cutting and impact piercing must be the result.

11.6 Notes

[1]   Atkins (2009).
[2]   Atkins (1974).
[3]   Atkins *et al.* (2004).
[4]   Patel *et al.* (2009).
[5]   Williams (1998).

[6] Hooke (1665).

[7] Bracegirdle (1978).

[8] A sulphated and nitrated cellulose which in solution can infiltrate tissues.

[9] Atkins and Vincent (1984).

[10] Willis and Vincent (1995).

[11] McCarthy *et al.* (2007).

[12] The guillotine was used in many parts of Europe during the Middle Ages under various names.

[13] Kamyab *et al.* (1998).

[14] Goh *et al.* (2005).

[15] Hill *et al.* (1947).

[16] Grunzweig *et al.* (1954). For wedge semi-angles 40° or less the frictional limitation is the shear strength and for larger angles it is the formation of a cap of dead metal. Shear yielding along the wedge is caused by coefficients of friction of 0.389 or less and a dead metal zone forms when the coefficient of friction is 0.242 or less.

[17] Discussed in §3.1.

[18] Atkins and Mai (1979).

[19] See §3.1 and Eq. (3.8).

[20] Simonsen and Wierzbicki (1996).

[21] Zheng and Wierzbicki (1996).

[22] Wierzbicki *et al.* (1998).

[23] Chang and Swift (1950).

[24] Atkins (1980).

[25] Atkins (1988).

[26] De Solla (1959).

[27] Zagorskii (1982).

[28] Ernst (1951).

[29] Finnie (1956).

[30] Cocquilhat (1851).

[31] Time (1870).

[32] Mallock (1881).

[33] Ernst (1938).

[34] Ernst and Merchant (1941).

[35] Merchant (1945b).

[36] Zvorykin (1893).

[37] Armarego and Brown (1969).

[38] Merchant (1945a).

[39] See §8.2.6.

[40] Surprisingly Merchant did speculate that local plastic deformation at the cutting edge would be necessary in addition to the true surface energy (see §8.2.6).

[41] Reuleaux (1900).

[42] Kick (1901).

[43] Cook *et al.* (1954).

[44] Busch (1968).

[45] Puttick *et al.* (1989).

[46] Astakhov (1999).

[47] Atkins (2003).

[48] Vaz *et al.* (2007).

[49] Williams *et al.* (2009).

[50] Featherstone (1974).

[51] Hardy (1976).

[52] Bacon (1622).

[53] Johnson (1986).

[54] The dates of Moore's birth and death are unknown and nothing is known of him except between the years of 1806 and 1823 when he was on the staff of the Royal Military Academy at Woolwich.

[55] Moore (1813), see Johnson (1995).

[56] The 4th edition of 1855 was reprinted in 1982 (Douglas 1855). Douglas was also on the staff of the Military College at Woolwich.

[57] Stevenson and Abmalek (1994).

[58] Shergold and Fleck (2004).

[59] Ogden (1972).

[60] For skin $\alpha \approx 9$, and for silicone rubber $\alpha \approx 2$–3.

[61] Bishop *et al.* (1945).

[62] Metal hardness tests with sharp pointed indenters depend only on the plastic deformation unlike wedge indenters where fracture can be important for low loads and shallow indentations.

[63] Wright *et al.* (1992).

[64] Taylor (1948).

[65] Atkins *et al.* (1998b).

[66] Pope (1909).

[67] Hassig (1988).

[68] Gordon (1978).

[69] Backman and Goldsmith (1978).

[70] Hopkinson (1912b).

[71] Sangoy *et al.* (1988).

[72] Corbett *et al.* (1996).

[73] Blyth (1993).

[74] Blyth and Atkins (2002).

[75] Wierzbicki and Thomas (1993).

[76] Morye *et al.* (2000), Naik *et al.* (2006).

[77] Dahlgren (1856).

[78] Johnson (1988).

[79] The *Warrior* was restored in the 1980s and can be seen in the Portsmouth Historical Dockyard.

Chapter 12

Recent Developments and the Twenty-First Century

At the beginning of the twenty-first-century fracture is a mature discipline. The recent developments in the study of fracture are largely due to a tremendous growth in computing power and in material characterisation techniques such as atomic force microscopy (AFM) invented by Gerd Binnig, Calvin Quate and Christoph Gerber in 1985.[1] Fracture behaviour at the atomic scale obviously ultimately controls engineering fracture behaviour and early in the development of fracture theory the linking of the two scales was seen as the holy grail for the understanding of the fracture behaviour of materials. Modern techniques are making this quest seem possible. The exploration of the properties of nanocrystalline materials and nanocomposites needs an understanding of the fracture of materials at the atomic scale. Nature, of course, has learnt through evolution how to build strong and tough materials from the bottom up and natural materials are being studied to try to learn her secrets and produce biomimetics.

In the mid-twentieth-century the impetus for the development of fracture mechanics was catastrophic structural failures, a new driving force has been the phenomenal growth in microelectronics and the need for mechanical reliability at a very small scale. Continuum mechanics has been successful in assessing the integrity of thin films and multilayers down to the order of a micron. While future nano-devices will need modelling at a smaller scale, it seems appropriate to start this chapter with a short review of the success of conventional fracture mechanics in assessing the integrity of micro-devices.

12.1 Integrity of Thin Films and Multilayers

Technology from the late-twentieth-century onwards has been dependent on microelectronic devices whose mechanical, as well as electrical, integrity is essential. Consequently fracture mechanics has been applied to the thin films and multilayers contained in these devices. While this application may not be as heroic as the prevention of the engineering catastrophes described in earlier chapters, it is essential to modern technology and may be crucial for the maintenance of our environment in the future. Thin films are also important in other engineering applications such as films in solid state fuel cells, organic light emitting diodes (OLED), wear-resistant coatings to metal-cutting tools, hard transparent coatings on optical polymers, and ceramic thermal barrier coatings. Because of the square-cube relationship between fracture energy and strain energy these thin films behave differently to macroscale components, but continuum mechanics are still applicable. The reviews by Bill Nix, John Hutchinson, Zhigang Suo, and Tony Evans give excellent summaries of the issues involved.[2,3]

The stresses in traditional engineering components are usually due to external loads, but in micro-devices the stresses in thin films and microlayers are intrinsic. Micro-devices are typically made by growing or depositing thin films onto a substrate, which is often a silicon wafer. These processes usually occur at high temperature and because the films usually have a different coefficient of thermal expansion to the substrate, high thermal stresses are induced during cooling and the films are under a biaxial residual stress. Since the substrates are usually thick compared with the films, the stresses in the substrates are low. Devices containing thin films and multilayers obviously contain many interfaces and delamination along the interface between dissimilar materials is a common failure mechanism. The study of interfacial cracks between dissimilar materials has a long history; it is included here because of its importance to modern technology.

12.1.1 Interfacial toughness

The theoretical elastic stress system at the tip of an interfacial crack between two different isotropic elastic materials generally has a characteristic oscillating stress first determined by Max Williams.[4] In the 1960s solutions were given to a number of interfacial problems.[5] For many years afterwards little attention was paid to interfacial cracks, but from the late 1980s there was resurgence in interest driven by the emerging new technologies such as microelectronics. John Dundurs showed in 1969 that for a wide class of plane strain problems of elasticity for isotropic bimaterials that the stresses depend only on two elastic parameters.[6] The two Dundurs' parameters are α, which is a measure of the plane strain elastic modulus mismatch, and β which is a measure of the mismatch in bulk modulus. The expressions for α and β are:

$$\alpha = \frac{\overline{E}_1 - \overline{E}_2}{\overline{E}_1 + \overline{E}_2},$$

$$\beta = \frac{1}{2} \frac{\overline{E}_1(1-\nu_1)(1-2\nu_2) - \overline{E}_2(1-\nu_2)(1-2\nu_1)}{\overline{E}_1(1-\nu_1)(1-\nu_2) + \overline{E}_2(1-\nu_2)(1-\nu_1)},$$

(12.1)

where $\overline{E} = E/(1-\nu^2)$ is the plane strain elastic modulus and the subscripts 1 and 2 refer to the material on either side of the interface. Because of the physical limitations on elastic constants, α and β are restricted within the limits of $\alpha = \pm 1$, and $\beta = (\alpha \pm 1)/4$. For many bimaterials, β is small and Hutchinson and Suo have argued that the effect of a non-zero value of β is usually of secondary consequence and that assuming $\beta = 0$ makes little practical difference.[3] If β is non-zero the stress field very close to the crack tip oscillates violently from tension to compression and there is a problem with apparent interpenetration of the crack surfaces. However, since there is a fracture process zone at the tip of a crack, these theoretical difficulties are largely illusory. Interfacial cracks are generally mixed mode. For $\beta = 0$ the two modes are separable and the mode-mixity angle ψ can be defined by

$$\psi = \tan^{-1}\left(K_I / K_{II}\right),$$

(12.2)

where K_I and K_{II} are the mode I and mode II stress intensity factors. The fracture energy for mixed mode fracture is dependent on the mode-mixity angle. Alex Volinsky and his colleagues have reviewed the methods of measuring the interfacial toughness of thin films.[7] The measurement of the interfacial toughness as a function of mode-mixity is difficult for thin films and the scatter in the results increases with the mode-mixity angle. The increase in interfacial toughness with mode-mixity is more marked with metallic films than ceramic ones. There is also significant crack growth resistance with metallic films if the cohesive strength is more than about three times the yield strength and the steady state interfacial toughness increases with film thickness. For a copper/tantalum nitride/silica/silicon system[8] the TaN/SiO$_2$ interfacial toughness energy increased from its intrinsic value of 5 J/m^2 when the copper film was 300 nm or less thick to 15 J/m^2 when the copper film was 1 μm thick and 80 J/m^2 for the thickest copper film tested, which was 16.4 μm thick.[9]

12.1.2 *Film cracking and delamination*

Thin films can crack normal to the film or delaminate. The stress in the film must be tensile for it to crack without delamination, but delamination with or without film cracking can occur by buckling under compression. Classic fracture mechanics has been applied very successfully to the geometries of the thin films and multilayers. In engineering applications fracture is most usually unstable but with thin films and multilayers stable steady-state crack propagation is usual. One of the difficulties with classic fracture mechanics is that the flaw or crack size has to be known in order to predict the fracture strength. Steady state crack propagation is more difficult than initiation and Tony Evans and his colleagues have suggested a robust fail safe integrity criterion based on preventing crack propagation that is independent of flaw size.[10]

12.1.2.1 *Delamination and cracking under tensile residual stress*

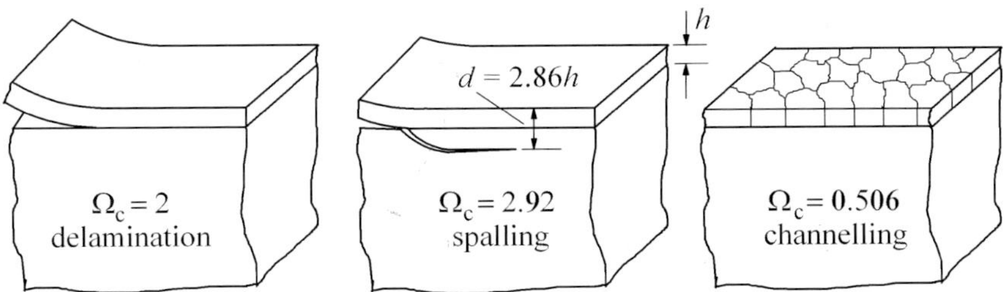

Fig. 12.1 Delamination and cracking of films under tensile residual stress (after Hutchinson 1996).

The robust fail-safe criterion for elastic fracture can be expressed in terms of a dimensionless cracking number, Ω, given by

$$\Omega = \frac{h\sigma^2}{\overline{E}_f R},$$
(12.3)

where h is the film thickness, σ, is the residual stress in the film, $\overline{E}_f$ is the film plane strain Young's modulus, and R is the fracture energy of the film or the delamination toughness.[11] If the cracking number is less than a critical value, Ω_c, which depends on the fracture mode, cracking will not occur. Three modes of fracture and their critical cracking number, Ω_c, are shown in Fig. 12.1, the cracking number for film delamination is independent of Dundurs' parameters, but for the other two cases it is dependent on α and β and is shown for the case where the film and substrate are very similar ($\alpha \approx 0$, $\beta \approx 0$). The cracking number for film delamination is very easily obtained since the strain energy per unit width, Λ, released by a unit increase in the length of the delamination is $\Lambda = \sigma^2 h / 2\overline{E}_f$, hence $\Omega_c = 2$. If an interfacial crack kinks into the substrate to deform and forms a spall its depth, d, is determined by the depth at which the crack becomes mode I. For films stiffer than the substrate ($\alpha > 0$) the relative depth of the spall will be greater than 2.86 and cracking index will be greater than 2.92.[12]

Channelling, also called crazing, is a common form of cracking in many systems: pottery glazes, if they have a lower coefficient of thermal expansion than the pottery, often craze with time in moist atmospheres, clay soils craze during the drying of a surface layer which creates residual tension and, on a geological scale, the Giant's Causeway, in Northern Ireland (see Plate 11) where the giant prismatic forms were created by the rapid cooling of lava flows by the sea.. For films stiffer than the substrate ($\alpha > 0$) the critical cracking number, Ω_c, is reduced.[13] If the film is subjected to a uniaxial tension rather than biaxial residual tension, the channelling cracks are straight and parallel and Michael Thouless has estimated the minimum distance between the cracks.[14]

Since the cracking number is proportional to the film thickness, thin films are stronger than thick ones. Thus there is an advantage in using the thinnest possible film. Sol-gel coatings are widely used as a protective film on polymeric substrates in optical lenses, automobiles, safety windows, and flexible display panels and their scratch resistance increases with decrease in film thickness.[15]

12.1.2.2 Delamination by buckling with or without film cracking

Films deposited at high temperature, such as metal lines on polymeric substrates in electronic packages, can be under substantial residual compressive stresses. If the interfacial toughness is low such films can delaminate by buckling. Away from an edge, a film will initially delaminate under biaxial residual compressive stress by forming an axisymmetric buckle. However, the front of an axisymmetric buckle is not stable and subsequent growth develops into a fascinating variety of forms. The *telephone cord*, shown in Fig. 12.2, is the most ubiquitous buckled pattern but a wide range of other folded patterns exist.[16] Thouless observed that amorphous silicon could delaminate from a glass substrate either to form a telephone cord buckled form or the silicon could crack in which case the buckled delamination is straight-sided.[17] Similar straight-sided cracked buckles form in indium-tin oxide (ITO) coatings on polyethylene terephthalate (PET), which are used in flexible organic light emitting diodes (OLED), when the PET film is bent so that the ITO coating is on the compression side.[18] Both the telephone cord and the straight-sided blister propagate at one end with a constant width of the blister.

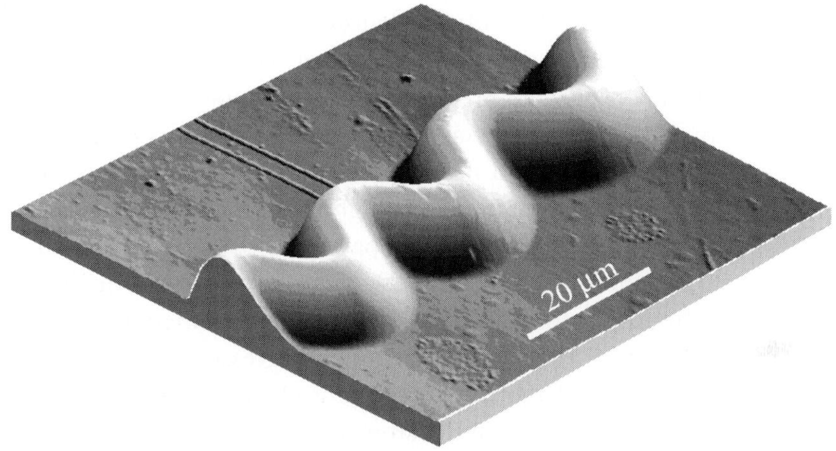

Fig. 12.2 AFM picture of a telephone cord buckle in a 125 nm tungsten film on a silicon substrate (Courtesy Yiu-Wing Mai).

If the substrate is not substantially stiffer than the film ($\alpha < 0.5$), the energy release rate, G, at the edge of a straight sided buckle of half width, b, can be analysed as if the substrate is rigid and are[3,19]

$$\frac{G}{G_0} = \left[1 - \left(\frac{b_0}{b}\right)^2\right]\left[1 + 3\left(\frac{b_0}{b}\right)^2\right],$$

(12.4)

where

$$G_0 = \frac{\sigma^2 h}{2\bar{E}_f},$$

(12.5)

and b_0 is the half width at the onset of buckling at a residual stress of σ given by

$$b_0 = \pi h \sqrt{\frac{\bar{E}_f}{12\sigma}}.$$

(12.6)

Under steady-state conditions the average energy released at the head of a straight sided blister is given by

$$\frac{G}{G_0} = \left[1 - \left(\frac{b_0}{b} \right)^2 \right]^2 . \tag{12.7}$$

The energy release rate at the side and head of the blister are shown in Fig. 12.3. The energy release rate at the sides of the blister reaches a maximum at $b/b_0 = 1.73$ which limits the sideways expansion of the blister, but the energy release rate at the head is always less than that at the sides and it seems at first strange that it should propagate at the head. The explanation of this apparent paradox, given by Hutchinson and Suo, is that the mode-mixity angle at the sides of the blister is greater than that at its head with the mixity at the sides tending to mode II at $b/b_0 = 2.75$, since the mode II toughness is much greater than mode I, it is the mode-mixity that controls the propagation.[3] If the film is substantially stiffer than the substrate ($\alpha > 0.5$), as in the case of ceramic films on polymer substrates, the substrate enables rotation to take place at the edge of the blister and the energy release rates are substantially greater than that given by Eqs. (12.4) and (12.7).[18,20] If the film is brittle it can crack behind the tip of the buckle and the analysis is somewhat different.[17,18]

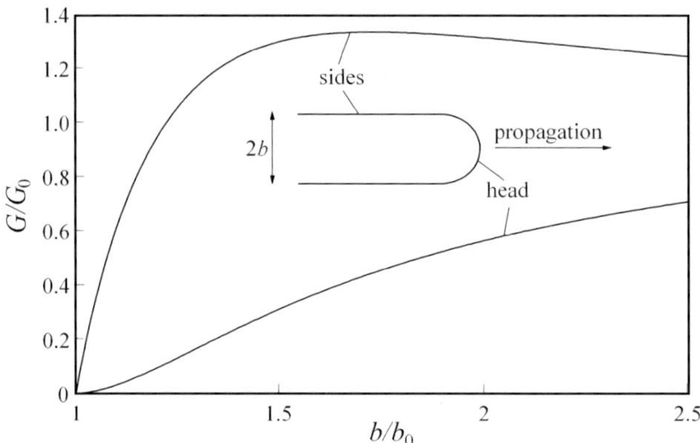

Fig. 12.3 Energy release rate at the head and sides of a tunnelling blister (after Hutchinson 2001).

12.2 Multiscale Modelling

Fracture involves all scales from the atomic to macroscale and the knowledge of the interdependence of the scales in-between is necessary. There are two main approaches to the multiscale modelling of fracture: top-down and bottom-up; which is best depends upon what is being modelled. Until the last decades of the twentieth-century the top-down phenomenological continuum mechanics approach based on experimentally determined material properties was dominant. This approach was refined during the development of fracture mechanics and is supreme for the purpose for which it was developed, the prediction of fracture, and can be used down for the films of a micron or even less in microelectronic devices as has been argued by John Hutchinson and Tony Evans.[21] The top-down continuum mechanics approach has also been successful in the design of the composite materials developed during the twentieth-century. As such the top-down approach will always have a major place in fracture mechanics. However, the development of new materials and composites, and the prediction of the mechanical integrity of nanoelectronic devices with physical features of the order of 10 nm or less during the twenty-first-century may require a bottom-up approach. A bottom-up approach needs a seamless coupling of the scales.

12.2.1 *Continuum mechanics*

Continuum mechanics is extraordinarily successful down to a very small scale. Proponents of the bottom-up approach often use quite misleading graphs to show the areas where the different mechanics apply in a space-time field where, starting from atomistic modelling at extremely small size and time area, there is a logarithmic increase in size and time up to continuum mechanics. Conventional continuum mechanics, where the stress depends upon the strain at a point and there is no length scale, can be used down to the order of a micron and times as short as a nanosecond if the strain gradient is small. The yield strength measured in a tension specimen is reasonably constant down to 10 μm thick wires but in torsion, where there is a strain gradient, the yield strength increases by a factor of three as the wire diameter decreases from 170 to 10 μm.[22] There is always a strain gradient near the tip of a crack, but in ductile materials the gradient is limited by a relatively large fracture process zone. Thus Hutchinson and Evans have shown that the steady state fracture energy can be accurately predicted by conventional continuum mechanics, provided the fracture process zone is modelled, even if the film thickness is of the order of a micron.[21] At the mesoscale, where conventional continuum mechanics becomes inaccurate because of high-strain gradients or small-scale, size matters. At the other end of the spectrum to continuum mechanics is quantum mechanics which describes the behaviour of materials at the atomic scale.

12.2.2 Mesomechanics

Since fracture mechanics has been developed using conventional mechanics the simplest means of exploring the mesoscale is through non-local continuum plasticity. However, down at the micron scale, plasticity in crystalline solids is discrete and occurs by the movement of dislocations. The interaction of dislocations with cracks is important for the fundamental understanding of fracture. An accumulation of a high density of dislocations is the cause of fatigue initiation. Dislocation dynamics covering the scale from a fraction of a micron to tens of microns is providing answers to some fundamental questions in fracture and fatigue.

12.2.2.1 Strain gradient plasticity

Non-local continuum mechanics began with the two French brothers; Eugène (1866–1931) and François (1852–1914) Cosserat, who formulated a fully consistent continuum theory where each point in a body has the freedom of displacement and rotation.[23] In the 1960s couple-stress or micropolar elasticity was developed where, as well as stress at a point, there is also a couple-stress with the units of force/length.[24] Strain gradient plasticity grew out of couple-stress theory. The original Fleck-Hutchinson phenomenological theory of strain gradient plasticity was based only on rotation gradients and had a single material length parameter, l.[25] Although this theory accounted for the increase in the torsional strength of thin wires, it did not show a significant increase in the stress level near a crack tip above that given by conventional plastic theory. However in fracture experiments on the interface between single crystals of niobium and sapphire the crack tip was atomistically sharp even though niobium is ductile and there were a large number of dislocations present.[26] A stress of around ten times the yield strength is required to cause atomic decohesion. In an earlier analysis of ductile fracture using a fracture process zone and conventional plastic theory Tvergaard and Hutchinson showed that the maximum stress was limited to about five times the yield strength.[27] Conventional plasticity is perfectly adequate to model ductile fracture, which was the purpose of the Tvergaard and Hutchinson paper, where the fracture process zone is comparatively large and the stress gradient is relatively small. However, conventional plasticity is inadequate to model atomic decohesion where the fracture process is very small. Fleck and Hutchinson improved their phenomenological strain gradient plasticity theory in 1997 by introducing stretch as well as rotation gradients and introduced an additional material length parameter.[28] John Hutchinson and Yuegang Wei repeated the earlier 1992 analysis of elasto-plastic fracture with a fracture process zone at the tip of the crack using the 1997 version of the Fleck-Hutchinson strain gradient plasticity theory.[29] The steady-state work of fracture Γ_{ss} normalised by the cohesive energy, Γ_0, is shown as a function of the normalised cohesive strength, $\hat{\sigma}/\sigma_Y$, in Fig. 12.4 for different normalised values of the material length parameter, l/D_0, where D_0 is the plane strain plastic zone size at crack initiation given by Eq. (9.6). When $l/D_0 = 0$, the steady state work of fracture is the same as that obtained by Tvergaard and Hutchinson using conventional plasticity.[27] For metals the material length parameter, l, is of the order of a micron (4 μm for copper)[22] and D_0 is the range or 0.1–1 μm for atomic decohesion thus the ratio l/D_0 can be expected to be at least unity. Hence fracture can occur if the cohesive strength is ten or more times the yield strength. For ductile fracture by void growth and coalescence D_0 is of the order of a millimetre and l/D_0 is tiny and strain gradient effects are negligible.

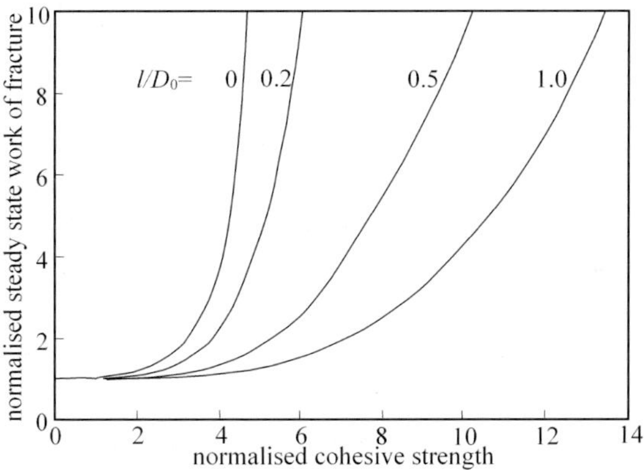

Fig. 12.4 Normalised steady-state work of fracture as a function of the normalised cohesive strength for various values of l/D_0 (after Wei and Hutchinson 1997).

The Fleck-Hutchinson theory was based on the macroscopic uniaxial stress-strain curve with micromechanical tests used to determine the material length scale. In 1998 Bill Nix and Huanjian Gao used the concept of geometrically necessary dislocations to model the hardness, H, of metals as a function of indentation depth, h, and obtained the relationship

$$\frac{H}{H_0} = \sqrt{1 + \frac{h^*}{H}}, \tag{12.8}$$

where H_0 is the hardness for macroscopic hardness and h^* is a characteristic length dependent upon the material and the shape of the indenter.[30] From this analysis they found that the material length parameter, l, for strain gradient plasticity to be

$$l \approx \left(\frac{\mu}{\sigma_Y} \right)^2 b, \qquad (12.9)$$

where μ is the shear modulus and b is the Burgers vector. This work was then used as the basis of a mechanism based strain gradient plasticity (MSG) theory.[31]

The Fleck-Hutchinson theory is a higher order theory that requires additional field and boundary conditions whereas the Nix-Gao theory is a lower order theory involving only classic second order stress and strain tensors. The advantage of the lower order theory is that it can be easily accommodated in standard finite element programmes. Tony Evans and John Hutchinson have recently written a searching review of the two approaches.[32]

12.2.2.2 Dislocation dynamics

The dislocations are modelled as line singularities in an elastic continuum, which is reasonably accurate down to about ten atoms from the core. Although the theory of dislocations matured by 1950, calculations involving large numbers of interacting dislocations were not able to be made until the late 1980s and even then focused on macroscopically uniform stress states with simple boundary conditions. In 1995 Erik Van der Giessen and Alan Needleman formulated a superposition method where the singular elastic dislocation fields for an infinite planar solid are represented analytically and a finite-element solution is used for the non-singular field with the superimposed boundary conditions.[33] Dislocation dynamics can be chaotic with the specimen geometry and stress distribution determining whether chaos emerges.[34] In plane strain uniaxial tension with two slip systems the stress-strain curve is little affected by small initial perturbations in the position of the dislocations, but the crack growth resistance curve is significantly affected by changes in the position of the initial dislocations by as little as $10^{-3}b$. The presence of gradients in the crack problem necessitate the presence of geometrically necessary dislocations, whereas in the uniaxial tension case there are only statistical dislocations present and may be the two types of dislocation play different roles in the development of chaos.

A problem with dislocation dynamics studies of fracture is that, unlike strain gradient plasticity where the input parameters are relatively easy to determine from experiment, the behaviour depends upon the density of dislocation sources and dislocation obstacles which cannot be easily determined. Two dimensional dislocation dynamics does throws some light on the short fatigue crack problem.[35] Vikram Deshpande, Alan Needleman and Erik Van der Giessen have used dislocation dynamics to analyse the fatigue growth of mode I edge cracks in specimens whose material is representative of aluminium; the material is assumed to be initially dislocation free but with randomly distributed Frank-Read sources with a density of 64 μm^{-2} that nucleate edge dislocations.[36] The cohesive energy and cohesive strength of the material were taken as 1 J/m^2 and 750 MPa respectively. The threshold stress intensity range is independent of crack length for $a > 300$ μm, but decreases to about half that value for $a = 10$ μm. The large reduction in the threshold range is due to an increase in internal stress associated with the dislocation structure.

Three-dimensional dislocation dynamics analyses were pioneered by Ladislas Kubin and his colleagues in the early 1990s.[37] The description of the individual dislocations and the treatment of the boundary value problem are the main issues in the numerical implementation of discrete dislocation dynamics in three dimensions. Erik Van der Giessen and Alan Needleman and their colleagues' description of dislocations is based on the mixed-character straight segment discretization of Robert Kukta[38] and their own decomposition[33] to solve the boundary value problem.[39]

12.2.3 Atomistic mechanics

Quantum mechanics is the foundation stone of the bottom-up approach to mechanical behaviour. All the methods discussed above are an approximation to the quantum mechanical description of material behaviour. Quantum mechanics has long been established and in 1929 the great theoretical physicist Paul Dirac (1902–1984) could write: 'The general theory of quantum mechanics is now almost complete... [apart from] the exact fitting in of the theory with relativity ideas...The difficulty is only that the exact application of these laws leads to equations much too complicated to be soluble. It therefore becomes desirable that approximate practical methods of applying quantum mechanics should be developed.'[40] Computing power is now catching up with theory, but *ab initio* calculations are still limited to a hundred or so atoms over a time interval of a few picoseconds. Molecular dynamics, where the degrees of freedom of electrons are removed, enables larger volumes and times to be analysed.

12.2.3.1 Quantum mechanics

Erwin Schrödinger (1887–1961) formulated the equation for the motion of the electrons and the nuclei that govern the observable properties of materials in 1926.[41] However, even with modern computers the many-body Schrödinger equation cannot be solved even approximately. Density functional theory, first formulated by Pierre Hohenberg, Walter Kohn, and Lu-Jen Sham in the 1960s, has been a very powerful quantum-mechanical method where the many-bodied electron wave function is replaced by the electron density and the problem of interacting electrons becomes a problem of non-interacting electrons moving in an effective potential.[42] The density functional theory has been used to extract the properties of the core of dislocations using a semi-discrete generalisation of the Peierls-Nabarro model.[43]

An important use of density functional theory is to calculate the effective interatomic potentials, which can then be used in molecular dynamics. However, density functional theory does not provide the desired accuracy for many materials and more accurate means of calculating the interatomic potentials have been developed such as the tight binding method.[44]

12.2.3.2 Molecular dynamics

Molecular dynamics, where the interaction between atoms is represented by potential functions, enables both the length and time scales to be extended. Some 10^{12} atoms can be analysed making possible the modelling of cubes up to 1 μm with current computers. The number of atoms that can be modelled increases with computer power according to Moore's law so that in fifteen or so years cubes of up to 10 μm should be able to be modelled.

The basic idea of molecular dynamics is to assume that the electrons are fixed to the nuclei of the atoms and the interaction between atoms is determined by potential functions. Newtonian mechanics can then be used to determine the dynamic evolution of the position of all the atoms. The first molecular dynamics simulation of this type was performed on liquid argon in 1964 by Aneesur Rahman using 864 atoms.[45] Farid Abraham and his group have been at the forefront of using molecular dynamics to simulate fracture.[46] One of earliest molecular dynamics studies by this group was the examination of dynamic mode

I fracture in 1994 using a 2-D solid rare gas model material, with the same intent as Griffith when he used glass in his classic experiments namely to have a material not governed by complexities.[47] The specimen was a wafer with 1,000 atoms on each side large enough to simulate the first stages of dynamic fracture. The wafer was a perfect 2-D crystal with no defects and a fracture was initiated from a notch midway along one side of the wafer. Three views of the propagating fracture at different times are shown in Fig. 12.5. In Fig. 12.5 (a) the crack has smoothly accelerated normal to the applied strain to a velocity of less than one-third of the speed of Rayleigh waves forming the equivalent to the mirror surface in glass fracture. At velocities around one-third of the velocity of Rayleigh waves, the crack surface begins to roughen slightly forming a mist-like surface, see Fig. 12.5 (b). At six-tenths of the speed of Rayleigh waves the crack path starts to be unstable and the crack surface becomes rough and a hackle surface develops as shown in Fig. 12.5 (c).

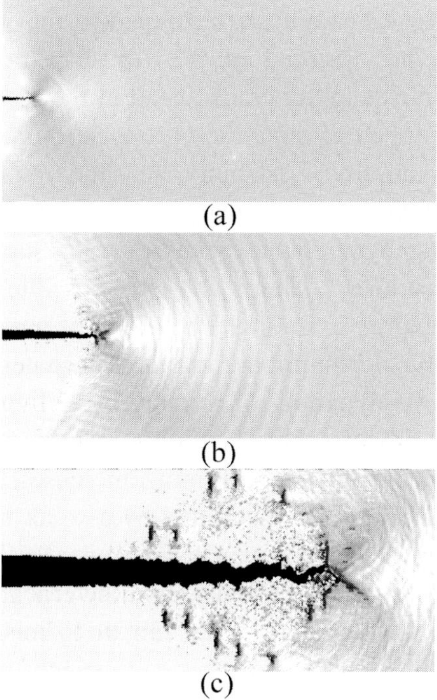

(a)

(b)

(c)

Fig. 12.5 Three snapshots in the development of a 2-D propagating fracture in a rare-gas model material: (a) Crack velocity less than 1/3 one-third of the velocity of Raleigh waves. (b) Crack velocity about one-third of the velocity of Rayleigh waves. (c) crack velocity about six-tenths of the velocity of Rayleigh waves (Abraham *et al.* 1998 with permission the Institute of Physics).

The limiting crack velocity in mode I fracture is the velocity of Rayleigh waves at which there is no energy available for fracture. In practice the velocity of naturally propagating cracks falls far short of the velocity of Rayleigh waves because of the path instability. Mode II crack growth is not natural and a brittle crack will turn to propagate under mode I, but a crack can be made to run in mode II if two plates are weakly bonded together. Ares Rosakis and his colleagues bonded together two plates of Homalite-100, a transparent birefringent polymer, with a notch along the bond line and under asymmetrical impact a mode II a crack propagated along the interface at velocities between the shear and longitudinal waves.[48] Bertram Broberg has shown using continuum mechanics that, though mode I cracks cannot propagate faster than the velocity of Raleigh waves, mode II cracks can propagate between the velocities of shear waves and longitudinal waves, but that velocities between those of Rayleigh and shear are forbidden.[49] Abraham and his colleagues simulated the same experiment using molecular dynamics.[47] In their experiments the crack reached the velocity of Rayleigh waves and then jumped to the velocity of longitudinal waves with a Mach cone forming at the tip; the transition occurs by the nucleation of an intersonic daughter crack ahead of the mother crack.

Although the dynamics of elastic crack propagation are usually described by continuum mechanics using linear elasticity, near the tip of a crack the strains are hyperelastic and non-linear. It has been suggested independently by Huajian Gao and Farid Abraham that hyperelastic behaviour can have an important effect on the dynamics of fracture.[50] There are two possible types of hyperelastic behaviour either stiffening or softening. The effect of hyperelastic behaviour has been studied using artificial biharmonic interatomic potentials.[51] Hyperelasticity has a marked affect on the dynamics of fracture if the hyperelastic region, which in effect is the fracture process zone, is comparable in size to the characteristic length of the material. Hyperelastic stiffening increases the energy flow to the crack tip and if the crack path is constrained by a weak layer the crack velocity can exceed the velocity of Raleigh waves. The energy flow to the crack tip is reduced in a hyperelastic softening material and the crack path becomes unstable at lower velocities and the transition from mirror to mist and hackle occurs at lower crack velocities.[52]

Hyperelasticity also can also affect the conditions for cleavage. Abraham and his colleagues have and made a molecular dynamics study of a 3-D model of a solid fcc rare gas and containing 10^8 atoms, which shows some unexpected results.[53] fcc crystals are typically ductile, irrespective of whether they are a rare gas solid or a metal. The solid rare gas is used as a model material; Abraham and his colleagues quote Tony Kelly and Norman Macmillan's description of krypton at liquid hydrogen temperature as having 'the hardness of butter on a cold day'.[54] In their simulation of fracture, cleavage occurs on a (110) plane. If cleavage occurs at all it usually is on the crystal plane with the lowest surface energy. However, ductile fracture occurs on a (111) which has a lower surface energy.

12.3 Nanocrystalline Materials and Polymer Nanocomposites

The use of nanoparticles to strengthen metals is not new and was the basis of the age-hardening aluminium-copper-magnesium alloy, discovered by Wilm in 1909, where precipitated particles typically 10 nm thick and 100 nm in diameter impede the motion of dislocations.[55] Piano wire is an even older example of the use of ultra fine-grained microstructures formed in very heavily drawn pearlitic steel to obtain high strength. The 1960s saw the beginning of the development of ultrafine grain-size metals to exploit the Hall-Petch relationship.[56] However, the development of nanostructured materials as a class in materials science really began in the 1980s. Nanocrystalline materials are usually defined as having a grain size of less than about 100 nm, whereas those with a grain size from 100–1,000 nm materials are described as having an ultra-fine grain-size.

Much effort has been given to polymer nanocomposites where the nanoparticles have one dimension of the order of ten nanometres or less. The nanoparticles can be spherical, such as silica or calcium carbonate, plate like clay particles, or carbon nanotubes. While much of the driving force for these nanocomposites has been functional properties, such as fire and moisture permeability resistance and some specific mechanical properties such as creep and wear resistance, fracture resistance is important.

The processing of nanocrystalline materials and nanocomposites is difficult and much effort has been necessary to develop the techniques, but here only the mechanical properties and the main processing problems are discussed.

12.3.1 Nanocrystalline materials

A perfect crystal has the lowest free energy of a solid. Nanocrystalline materials are much more disordered than other materials such as glasses or other imperfect crystalline solids. With very small grain sizes the volume fraction of intercrystalline disordered material becomes very significant. At a grain size less of 100 nm, the volume fraction of intercrystalline material is about 3%, which increases to about 30% for a 10 nm grain size and the density of defects so high that about half the atoms are in the core of the defects. The highly disordered state of nanocrystalline materials results in many enhanced physical properties such as higher self-diffusion, specific heat, thermal expansion, increased optical absorption of thin films, superior magnetic properties and the importance of nanocrystalline materials is largely because of these functional properties.

There have been two recent reviews of mechanical properties of nanocrystalline materials.[57,58] The Young's modulus of a metal is usually the mechanical property least affected by processing, but in nanocrystalline metals the average atomic spacing can be significantly greater than usual causing the Young's modulus to be significantly smaller. Decrease in grain size increases the yield and fracture strength of crystalline metals according to the Hall–Petch relationship,[59] but there appears to be a possible inverse Hall–Petch relationship for some metals for grain sizes less than 10 nm.[60] For grain sizes greater than about 10 nm the yields strength increases as the grain size decreases because the number dislocations piled up at grain boundaries decreases with grain size and a larger stress is necessary to initiate dislocations in adjacent grains. Very small grains cannot support dislocation pile-ups and the Hall–Petch relationship breaks down. There have been a number of proposed models for the yield strength behaviour for very small grain sizes and a consensus opinion on what is the most likely mechanism has not yet been formed. Some metals, like nickel, do not seem to have an inverse Hall–Petch relationship even at very small grain size. This exception for nickel agrees with one model for the inverse Hall–Petch relationship based on the relative probability of a dislocation being absorbed in the grain boundary rather than propagated across the boundary because the high melting temperature and consequent high activation energy halts the grain-boundary absorption of dislocations.[60]

There is little information on the fracture toughness of nanocrystalline materials because the processing methods, such as electro-deposition or e-beam deposition, which are used to produce fully dense materials typically yield only thin foils. As a consequence, the plane stress fracture toughness is measured, which is dependent on the thickness of the specimen, making comparison with other data difficult. Fracture toughness tests on compact tension specimens of nickel between 0.22 and 0.35 mm thick formed by electro-deposition with grain sizes 19 to 25 nm are very sensitive to the annealing temperature.[61] For annealing temperatures of 100°C or less the specimens showed crack growth resistance with the fracture toughness increasing from 60–70 MPa√m at initiation to plateau out at about 120 MPa√m after a crack extension of about 5 mm. The specimen annealed at 200°C showed no crack growth resistance and had a fracture toughness of about 34 MPa√m. A commercial pure polycrystalline nickel with a grain size of 21 μm also showed no crack growth resistance and had a fracture toughness of about 60 MPa√m. There is some evidence from experiments on a recrystallised amorphous $(Fe_{0.99}Mo_{0.01})_{78}Si_9B_{13}$ ribbon that the fracture toughness decreases with grain size for grains less than 35 nm.[62]

The fatigue limit of nanocrystalline metals is higher than that for microcrystalline metals in keeping with their higher strength, but the fatigue crack growth rate is higher because their fracture surface is smoother.[57]

12.3.2 Nanocomposites

Nanocomposites are often more important for their functional properties than their mechanical properties. Both properties are dependent on obtaining good dispersion of the nanoparticles which requires complicated processing techniques. These techniques and the mechanical properties of nanocomposites have been reviewed by Sie Chin Tjong.[63] The functional and mechanical properties of carbon nanotube composites can be improved by alignment as well as good dispersion and Yiu-Wing Mai's group have reviewed the necessary processing techniques.[64]

The general trend is that nanoparticles toughen the glassy polymers, but do not always toughen the more ductile polymers and in many cases embrittle them. A slight embrittlement from an already adequate toughness does not necessarily detract from the usefulness of functional nanocomposites. However, what is disturbing is the belief by some in the almost magical properties of nanoparticles just because of their very tiny size. One of the problems is that since the processing of nanocomposites depends upon polymer chemists, much of the development of nanocomposites has been led by people with very little mechanical background. Of course by concentrating on the mechanical properties and neglecting the processing details, as is done here, is open to a similar criticism, but more thought on the mechanical behaviour is presently required to redress the balance.

12.3.2.1 Nanoparticles

Nano-sized particulate fillers such as wood flour, carbon black, silica flour, talc, limestone, and clay were added to polymers long before the development of nanocomposites. Very often the fillers were used to simply reduce the cost, but they can also improve the mechanical properties. In these uses the particles are in agglomerates. What is new is that processing techniques have been developed to break-up the agglomerates into nanoparticles. There are three types of nanoparticles: compact particles of low-aspect ratio, silicate clay plate-like particles, and nanotubes.

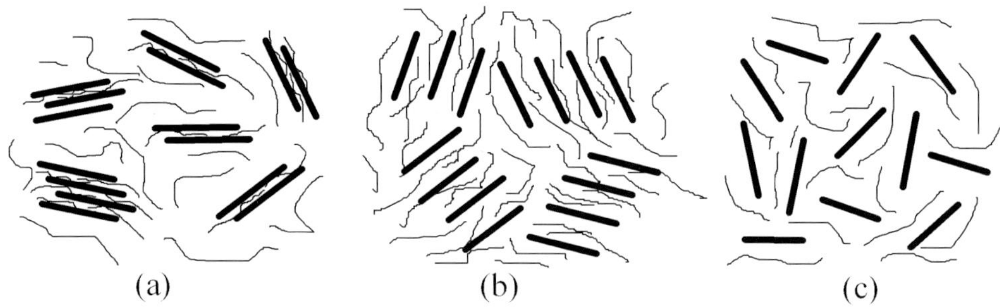

Fig. 12.6 Schematic illustration of (a) intercalated clay, (b) ordered exfoliated, (b) disordered exfoliated clay, polymer nanocomposites.

Clays, montmorillonite is a favourite because it is environmentally friendly, readily available, and low in cost, are layered aluminium silicates. Montmorillonite, named after Montmorillon in France, is available as micron-sized tactoids or agglomerates made up of platelets about 1 nm thick and 100–1,000 nm wide. Composites formed of unmodified clay are not really nanocomposites at all. To make the platelets more compatible with polymer matrices, clays are generally organically modified resulting in a larger interlayer spacing. For use in nanocomposites the clay particles are intercalated or exfoliated as shown in Fig. 12.6. In an intercalated nano-composite, polymer chains enter the galleries between the clay platelets forcing them further apart and the particles are broken up into stacks a few platelets high. However, the tactoids still need breaking up or they act like micron-sized particles. There are a wide variety of means of breaking up the tactoids which have various degrees of success. In a fully exfoliated structure the platelets are completely separated, but full exfoliation is never achieved. Often the structure of a clay nanocomposite looks well exfoliated in a transmission electron microscopy where the field of view is of the order of a micron but optical microscopy, with a field of view of a 500 μm, shows that the composite consists of clusters of nano particles of the order of 20 μm or more.[65,66]

The first transmission electron microscopy evidence of carbon nanotubes was found by two Russian physicists Leonid Radushkevich and Lukyanovich in 1952,[67] subsequently carbon nanotubes were rediscovered by a number of researchers but none have had the impact made by the 1991 paper by Sumio Iijima[68]. Carbon nanotubes (CNTs) can be classified into single-walled nanotubes (SWNTs) with a diameter of 1–2 nm, and multi-walled nanotubes (MWNTs) with a diameter of 3–10 nm. The aspect ratio of carbon nanotubes is of the order of 1,000. Measurement and interpretation of the Young's modulus and strength of CTNs is difficult, but they are of the order of 1000 GPa and 10 GPa respectively. At present the biggest problem is one of cost: currently MWNTs cost about US$1,000/kg and SWNTs $75,000/kg. For this reason most CNT polymer composites use MWNTs. The use of carbon nanotubes polymer composites is more for their functional properties, especially electrical conductivity, than their mechanical properties. There is also the health risk associated with fibres or nanotubes whose diameter is less than about 10 μm which can penetrate the pulmonary regions of the lung. Asbestos was a wonder-material until its carcinogenic properties became known. A recent study on mice suggests a potential link between inhalation exposure to long CNTs and mesothelioma, but it is not known whether there would be sufficient exposure to such particles in the workplace to reach a threshold dose in the mesothelium.[69]

As far as mechanical properties are concerned, a natural nanotube has more potential benefits than carbon nanotubes.[70] Halloyosite is an aluminosilicate clay mineral with the empirical formula $Al_2Si_2O_5(OH)_4$ consisting of numerous numbers of nanotubes of diameter around 50 nm and length 500–2,000 nm. The mineral was first described in 1826 and named after Belgian geologist Omalius d'Halloy (1783–1875). Halloysite historically has been used to make fine china and ceramics but its structure has only recently been studied. One of the first suggested uses of halloysite nanotubes (HNT) was in drug-delivery systems. HNT does not have the high strength of CNT but it is comparatively cheap. The health risks with halloysite nanotubes have not been established, but because of their fine diameter they should be treated with caution. Not surprisingly, the small number of companies that have halloysite nanotube development projects claim that there is little risk, but then companies manufacturing asbestos products were long in denial.

Nanoparticles usually require surface modification to either increase or reduce the adhesion to the polymer and to improve the dispersion of the particles. Nanoparticles can also have a significant effect on the structure of the adjacent polymer.

12.3.2.2 Toughening mechanisms

The extremely large surface area of nanoparticles and their associated surface energy is the prime reason for the belief that nanoparticles must toughen polymers, provided they are well dispersed. The total surface area of the particles in a one cubic centimetre nanocomposite comprising of a 5% volume fraction of spherical nanoparticles 20 nm in diameter is a staggering 15 m^2, the area of a large room. However, the contribution to the work of fracture from the debonding of the nanoparticles is not dependent on the size of the particles but is only dependent on their volume fraction.[71] Also many of the toughening mechanisms used in polymer composites are not available if the particles are nano-sized.

Two related toughening mechanism in composites are crack pinning and crack deflection, which were originally proposed for ceramic materials. A crack bows out from particles, which pin it until it breaks away from the particles similar to the mechanism of dispersion hardening of metals where a dislocation line pinned by particles bows out and needs an increased shear stress to move it. To account for the resistance of a dislocation to bowing, Nevill Mott and Frank Nabarro introduced the concept of the tension in a dislocation line.[72] This concept was borrowed by Fred Lange to account for the increased toughness caused by crack pinning.[73] The analogy is not perfect but does enable the crack pinning effect to be visualised. If the crack line tension is T the effective fracture energy, R_e, is given by

$$R_e = 2\left(\gamma + \frac{T}{d}\right), \tag{12.10}$$

where γ is the surface energy and d is the particle spacing. The line tension of a dislocation is constant, but the line tension of a crack is not.[74] However, qualitative conclusion can be drawn from Eq. (12.10), assuming that the crack front between two particles is circular in form, the smallest radius of curvature when the crack breaks away from its pinning is a half the distance between particles. Thus Eq. (12.10) suggests that, since the distance between particles for a given volume fraction is proportional to the particle size, the crack pinning mechanism should be a powerful toughening mechanism. However, just as in the case of a pinned dislocation which can cut through its pinning particles if they are too small, so too can a crack cut through small particles as Tony Kinloch and his co-workers have suggested.[75] The crack pinning mechanism can only function if the particle size is larger than the crack tip opening displacement. Even for brittle plastics such as epoxy the crack tip opening displacement is of the order of one micron, hence Kinloch and his co-workers suggest that this mechanism is not available for polymer nanocomposites where the particle size is much less than a micron.[75] There was no evidence of crack pinning in fracture surface of epoxy specimens with 20 nm silica particles tested by either the Kinloch[75] or Klaus Friedrich's group[76]. In contrast crack bowing lines are clearly evident in epoxy specimens with 50 μm glass spherical particles.[77]

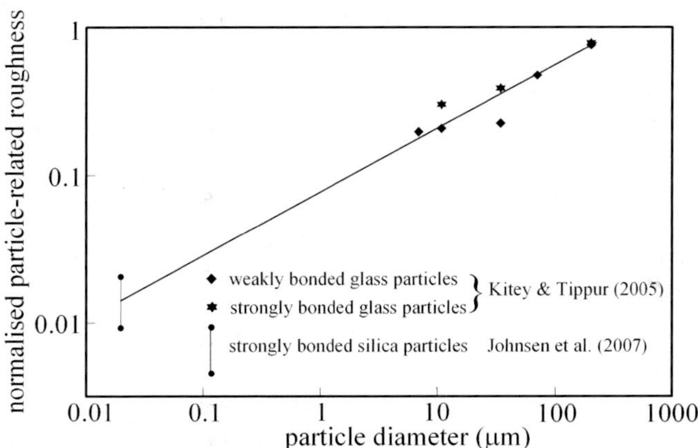

Fig. 12.7 The normalised particle-induced surface roughness, Ra_c/Ra, as a function of the particle diameter for epoxy reinforced by spherical glass and silica particles.

Particles can cause a crack to deflect out of its plane, or twist, absorbing more energy because the fracture surface is larger as well as because the fracture is mixed-mode.[78] Both crack pinning and crack deflection cause roughening of the surface. Rajesh Kitey and Hareesh Tippur have suggested the interesting concept of separating the surface roughness, Ra,[79] into two components: Ra_p the surface roughness due to the deviation of a fracture plane around the particles and Ra_f the fracture-induced surface roughness.[80] The particle-induced surface roughness Ra_p is given approximately by

$$\mathrm{Ra}_p \approx \frac{3\pi}{16} \frac{\mathrm{v}_p d}{\left(1 - \mathrm{v}_p\right)} \tag{12.11}$$

where d is the diameter of the particle, v_p is the volume fraction of the particles. The ratio of the particle induced surface roughness to the total surface roughness, $\mathrm{Ra}_p/\mathrm{Ra}$, as a function of the particle size, d, is shown in a logarithmically in Fig. 12.7 for the data of Kitey and Tippur[80] and Kinloch's group[75] from experiments on epoxy reinforced with spherical glass and silica particles. Both groups of experimenters used notched three-point-bend specimens, but Kitey and Tippur used impact loading whereas Kinloch's group used quasi-static loading. The volume fraction of the glass spherical particles, whose diameter ranged from 7–203 µm, was 10% and the volume fraction of the 20 nm diameter silica particles ranged from 2.5–13.4%. Despite the differences in the experiments the ratio, the results of both sets of experiments are given quite well by the equation

$$\frac{\mathrm{Ra}_p}{\mathrm{Ra}} = 0.076 d^{0.43}, \tag{12.12}$$

where d is in microns. The particle-induced roughness, Ra_p, is a very small portion of the total fracture surface roughness for nano-sized particles. Thus the fracture-induced roughness becomes the dominant roughness when the particles are nano-sized, but the total fracture surface roughness increases with both the volume fraction of the particles and the particle size. Extra fracture work must be expended to create this increase in fracture surface area.

Bridging of a crack by fibres or particles behind the crack tip is a powerful crack growth toughening mechanism in polymer composites, but low aspect ratio nanoparticles are too small to provide much toughening by this mechanism. However, long carbon and halloysite nanotubes are of the order of 10 µm in length and may provide some toughening by bridging.

The main toughening of nanocomposites comes from inducing energy absorbing deformation in the matrix. These mechanisms are matrix specific and are discussed in the following sections.

12.3.2.3 Glassy matrices

One of the most important of the glassy matrices is the thermoset epoxy which is widely used for adhesives and the matrix of fibre reinforced composites. The toughness of epoxy can be increased by the addition of a second phase of a dispersed rubber or a thermoplastic polymer, but a rubbery phase increases the viscosity of the monomer hindering the impregnation of fibres and reduces the stiffness of the cured polymer. Nanosilica particles are an attractive alternative toughening method, since they do not increase the viscosity significantly and stiffen the cured polymer. Nanosilica particles can be obtained commercially as a colloidal sol in an epoxy matrix which gives good dispersion when mixed with epoxy.

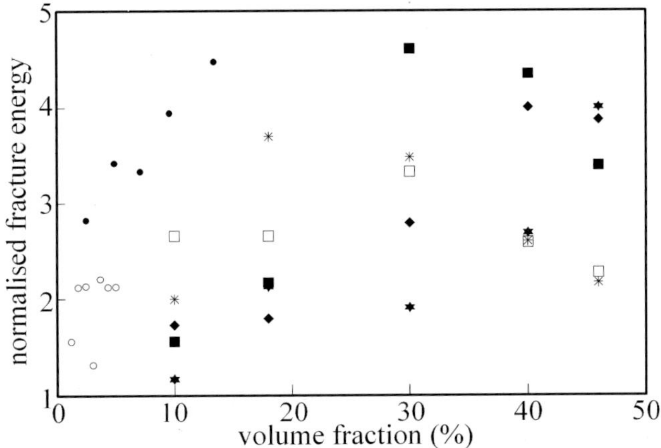

Fig. 12.8 Normalised fracture energy of epoxy reinforced with silica and glass particles as a function of volume fraction for different sized particles.

The toughening of epoxy with nanosilica particles is one of the success stories for the mechanical properties of nanocomposites and in this instance they appear to do better than larger particles. The fracture energy of a range of epoxy toughened by silica or glass particles normalised by the fracture energy of the pure epoxy are plotted against the volume fraction of the particles, varying from 20 nm to 62 μm in diameter in Fig. 12.8. The results of the Kinloch[75] and Friedrich[81] groups, both using the same well dispersed 20 nm silica particles, but different epoxies, are very consistent and clearly give a greater fracture energy than the results of Robert Young's group[82] for glass particles greater or equal to 4.5 μm in diameter for the same volume fraction. The particle size distributions for these latter particles are much wider than those for the 20 nm particles and the dispersion was not as good. Stick slip fracture was also observed by Young's group, which accounts for the far less consistency in the results. The Kinloch and Friedrich groups disagree on the probable cause of the toughening with 20 nm particles. Kinloch's group consider there to be little toughening by crack pinning or deflection and that the major toughening came from shear yielding around delaminated particles, whereas Friedrich's group believe that the major toughening comes from crack pinning and deflection. As yet the toughening of epoxy by nano-sized stiff particles has not been successfully modelled. Obviously more work is needed on the modelling of the toughening of epoxy by nanoparticles.

Micron sized dispersed rubber particles have long been used to toughen epoxy. The results of Tony Kinloch's group indicate that if nano silica particles are added to a rubber toughened epoxy the fracture energy is increased substantially more than would be expected from a simple addition of the two fracture energies.[83] More recent results from Yiu-Wing Mai's group show no synergistic effect and the fracture energy for epoxy containing both nanosilica and nano-sized rubber particles is simply the sum of the two fracture energies of epoxy/nanosilica and epoxy/rubber nanocomposites.[84]

The first research on the use of nanoclay composites was in nylon at the Toyota Central Research Laboratory in the early 1990s. Research on epoxy-clay nanocomposites started in 1995.[85] Epoxy reinforced with either intercalated or exfoliated clay does not increase the toughness as well as nanosilica. The maximum fracture energy of intercalated or exfoliated clay epoxies seems to be 2–2.5 times that of the matrix and occurs at a weight fraction of 1–5%; at higher volume fractions the fracture energy decreases presumably because of lack of tactoid break-up or dispersion problems.[65,66,86,87] However, Rolf Mülhaupt's group report a fracture energy of about four times that of the matrix at a volume fraction of about 5%, but there seems to be little break-up of the tactoids.[88] Hence intercalated or exfoliated clay is not very successful at toughening epoxy, but there are other functional properties that are increased such as moisture resistance. Albert Yee's group[65] observed a considerable number of microcracks between clay layers around the fracture plane which had opened up considerably, and the dilatation produced by this opening could produce a significant shielding of the crack tip such as occurs with ceramics[89]. However, since the break-up of tactoids is far from complete there is really still little evidence as to the efficiency of well dispersed intercalated or exfoliated clay.

The interest in carbon nanotubes polymer composites is for their functional not mechanical properties. Carbon black has long been used as a filler for polymers primarily because of its cheapness and the fracture properties are comparable with those of extremely expensive carbon nanotubes. A weight fraction of 0.1% increases the fracture energy of epoxy by about 90%, whereas the same weight fraction of an amino-functionalised form of double walled carbon nanotubes only increases the fracture energy by about 40%.[90] However the fracture toughness of 0.1% weight fraction carbon black and carbon nanotubes reinforcements are both increased by about 40% because carbon black is not as efficient at increasing the Young's modulus as carbon nanotubes. The big problem with carbon nanotubes, as with nanoclay, is the difficulty of dispersing the particles.

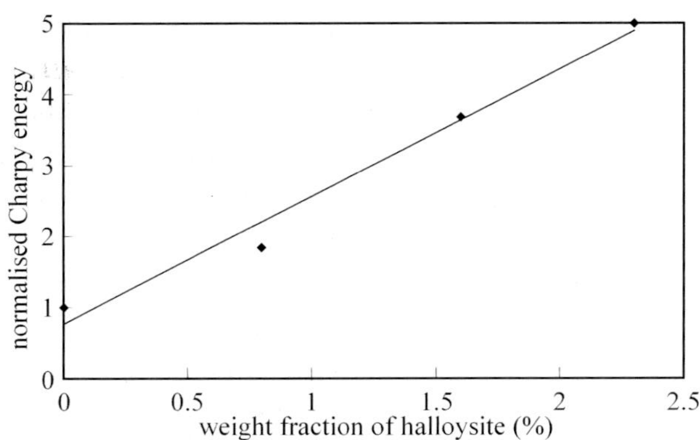

Fig. 12.9 Normalised Charpy energy as a function of the weight fraction of halloysite in epoxy (after Ye *et al.* 2007).

Halloysite nanotubes (HNT) can be comparatively well dispersed in acetone and simply added to epoxy.[70] The fracture energy, as measured by the Charpy test, increases linearly with weight fraction and the energy is increased by a factor of 5 at a relatively modest 2.3% weight fraction (see Fig. 12.9).[70] There is massive microcracking near the main fracture, while the HNTs are not long enough to provide substantial bridging of the main crack, they do arrest the microcracks. It is the energy absorbed by the microcracking activity that is thought to provide the main toughening action.

12.3.2.4 Semicrystalline matrices

The major toughening in semicrystalline polymers reinforced with spherical particles comes from the energy absorbed in massive crazes or craze-like regions and shear yielding near to the crack tip.[91] Nanoparticles also provide nucleation sites for the crystallisation of semicrystalline polymers. Isotactic polypropylene (iPP) crystallizes with a spherulitic structure with a diameter of the spherulites of the order of 100 μm. The addition of nano-sized calcium carbonate particles causes the spherulites to be much reduced in size and distorted. The phase is also changed. Isotactic polypropylene has three different forms: α-phase, the most common, β-phase, a metastable phase, and γ-phase. Particles form ready nucleation sites for the β-phase. Additives such as stearic acid promote the β-phase. The β-phase can also be induced by allowing crystallisation to take place at elevated temperature. The β-phase has significantly lower yield strength than the α-phase and consequently the β-phase spherulites are more highly strained and make crazing easier.[92,93] Treating calcium carbonate particles with a non-ionic modifier, polyoxyethylene nonphenol (PN) greatly improves their dispersion and also increases the β-phase, crazing, and toughness.[94] Optical micrographs of the multiple craze like bands and the relative impact energies for iPP and 44 nm diameter $CaCO_3$ particles with and without PN modifier are shown in Fig. 12.10. The extent of crazing and the fracture energy increase with the addition of the nanoparticles and the fraction of the β-phase in the polypropylene.

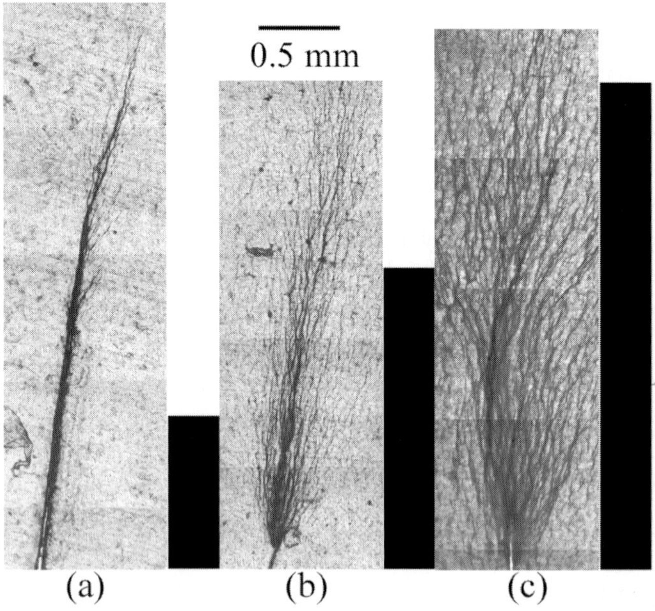

Fig. 12.10 Optical micrographs of multiple craze-like bands in (a) iPP, 0% β. (b) iPP + 15 wt% $CaCO_3$, 12.4% β. (c) iPP + 15 wt% $CaCO_3$ + 1.5 wt% PN, 15.8% β; the bars show the relative fracture energy (courtesy Qing-Xiu Zhang).

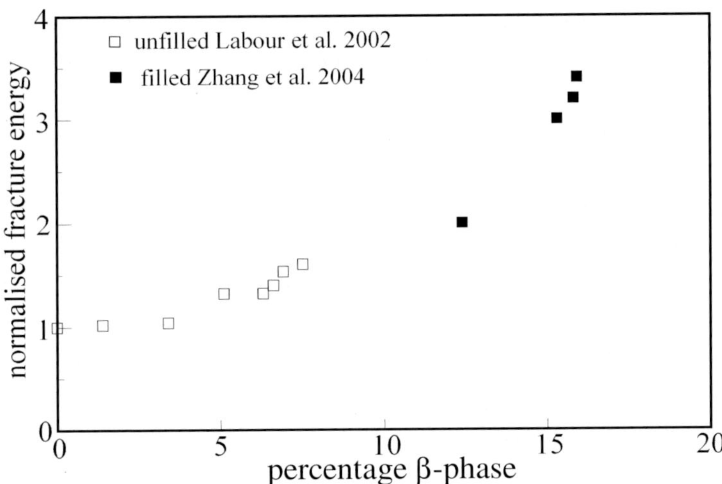

Fig. 12.11 The normalised fracture energy for filled and unfilled iPP.

Clearly the β-phase has a large effect on the fracture energy but it is unclear how much of the toughening is due to the mechanical effect of the particle. The Materials, Engineering and Sciences (MATEIS) Group of INSA de Lyon have obtained J_R curves for iPP with up to 7.5% β-phase obtained allowing crystallisation at high temperature.[92] The fracture energy for unfilled iPP with various percentages of β-phase normalised by the value for α-phase iPP, by arbitrarily taking the J_R value at 1 mm crack growth as an indicative value, has been plotted against the percentage of the β-phase in Fig. 12.11 and the normalised impact fracture energy for 15 weight percent of 44 nm diameter calcium carbonate obtained from the results of Yiu-Wing Mai's group[94] have been added to this plot. While it must be allowed that there is some arbitrariness about the combining of these two sets of data, the two sets of data do seem to lie on the same curve. The MATEIS group also tested iPP filled with 10% by weight of stearic acid treated 100 nm $CaCO_3$ particles.[92] Up to 29% of the β-phase was obtained in these filled specimens, but they all gave lower J_R-curves than the unfilled α-phase specimens. Although the MATEIS group consider that stearic acid treatment would give good dispersion it is possible that during the 20 minutes the specimens were held at 190°C that significant agglomeration of the particles occurred. Certainly the results of the MATEIS group for the filled iPP are at variance with the results of other research groups.

The normalised impact energy for iPP reinforced by calcium carbonate particles for a range of volume fractions and different particle sizes is shown in Fig. 12.12. Maintaining a good dispersion of the particles is more difficult with increased volume fraction especially for the very small sizes and only those results where the dispersion was thought to be good have been included in Fig. 12.12.[95] The trend lines shown in Fig. 12.12 are remarkably consistent and indicate a clear nano-effect for well dispersed particles with the largest 25 μm particles embrittling the iPP. However, this nano-effect cannot be fully exploited because with very small particles good dispersion cannot be obtained except at small volume fractions and for 44 nm particles 9% is about the limit at which good dispersion can be obtained. The best toughness appears to be obtained with particles of about 700 nm where good dispersion is possible at volume fractions of the order of 30%.

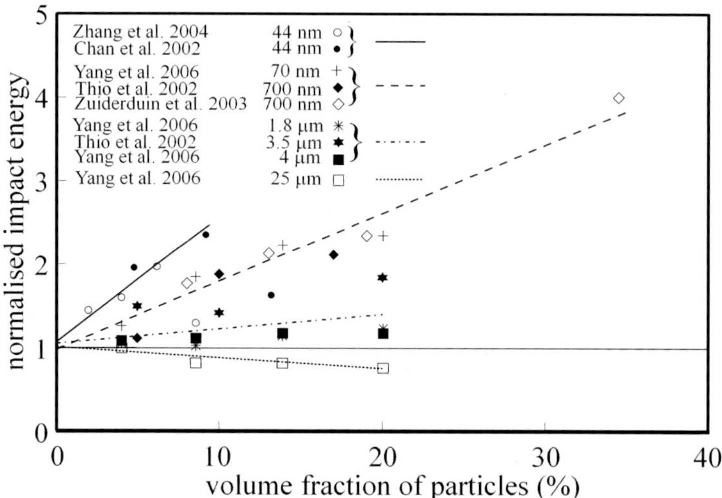

Fig. 12.12 Normalised impact energy as a function of the volume fraction of well dispersed calcium carbonate particles in isotactic polypropylene, the following treatments of the $CaCO_3$ particles were used: no stated treatment (Chan *et al.* 2002; Zang *et al.* 2004), stearic acid (Zuiderduin *et al.* 2003), calcium stearate (Thio *et al.* 2002), silane (Yang *et al.* 2006).

The β-phase nucleates at the surface of the particles and the percentage of that phase must be related to the surface area of the particles per unit volume of the nanocomposite which is proportional to v_p/d. The normalised fracture energy, for the results shown in Fig. 12.12, has been replotted against v_p/d in Fig. 12.13. While there is considerable scatter it does seem that the impact energy is a function of the surface area of the particles per unit volume. Combining the observations made from Figs. 12.11 and 12.13 it seems probable that the main role of nanoparticles in toughening polypropylene is promoting the nucleation of the β-phase in the iPP.

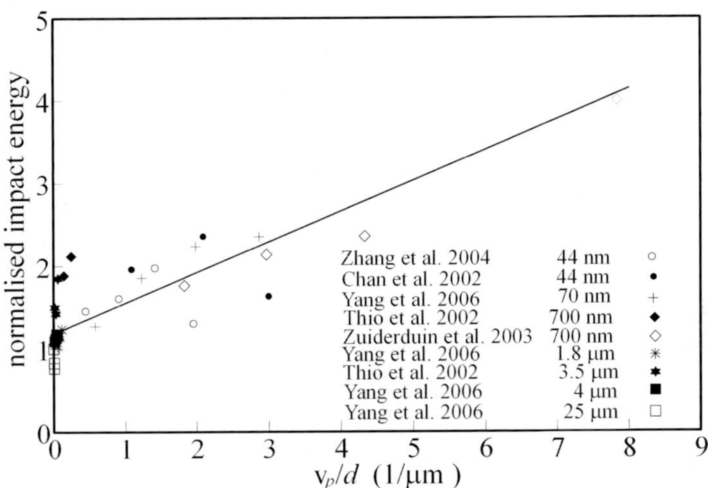

Fig. 12.13 Normalised impact energy as a function v_p/d for the data in Fig. 12.12.

The nano-effect in semicrystalline nanocomposites was ascribed to the existence of a critical inter-particle spacing which was originally proposed by Souheng Wu[96] for rubber toughened nylon in 1985. This concept was extended by Ali Argon's group to rigid particles in 1999.[97] The average inter-particle distance, l_p, is given by[96]

$$l_p = d\left[\left(\pi/6v_p\right)^{1/3} - 1\right].$$
(12.13)

The critical inter-particle spacing concept has been linked to transcrystallinity where the crystal lamellae orientate normal to a particle surface for a distance of the order of 100 nm.[98] However, crystal lamellae can also orientate in the normal direction of flow during injection moulding.[99] The complete picture of the formation of transcrystalline regions in nanocomposites has been given by Yiu-Wing Mai's group.[100] In the core region of injection moulded nanocomposites crystalline lamellae form normal to the nanoparticles, but in the near surface regions where there is shear induced flow the lamellae are aligned normal to the flow. Since fracture specimens are usually made with injection moulded bars and fracture involves deformation in all regions, Mai's group suggest that preferential alignment of the lamellae may not control the toughness of semicrystalline polymer nanocomposites.[100] Hence the critical inter-particle distance concept does not have a good physical foundation. The results for the impact energy for iPP nanocomposites shown in 12.12 have been replotted in Fig. 12.14 against the inter-particle spacing, l_p, and it does not seem there is a relationship for these results.

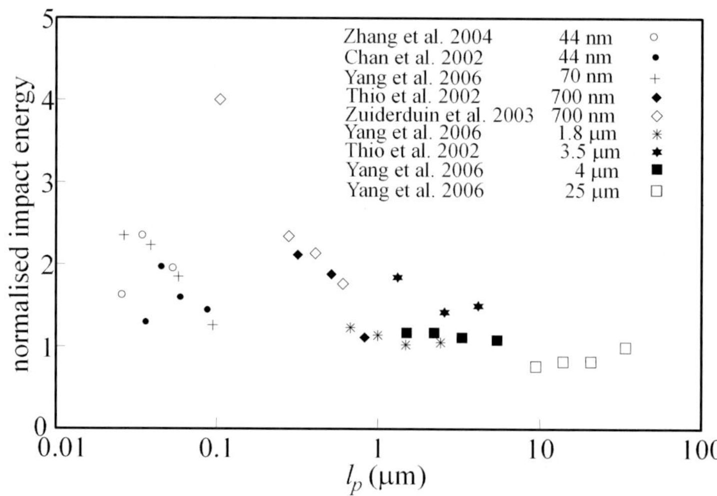

Fig. 12.14 Normalised impact energy as a function the inter-particle distance, l_p, for the data in Fig. 12.12.

Dispersive mixing of clay needs an anhydride-modified polypropylene and even then the exfoliation of clay is poor compared with nylon.[101] Hence there has been more research on nylon, especially nylon 6, than polypropylene matrices. Nylon 6 has two major crystalline phases, monoclinic α-phase and monoclinic or pseudo hexagonal γ-phase.[102] The α-phase is the most stable crystalline phase, and is obtained by slowly cooling from the melt. The γ-phase is less stable and is promoted by clay particles but the overall crystallinity is reduced.[103,104] The γ-phase has a higher ductility.[102] However clay nano-particles, even when well exfoliated or intercalated, embrittle nylon 6 rather than toughen.[103,104] The formation of craze-like bands and shear yielding are the mechanisms by which the nylon absorbs energy. Transmission optical micrographs of fractured double notched four point bend specimens of nylon 6 with and without exfoliated organoclay shown in Fig. 12.15 show crazes or craze-like bands at the tip of the non-fractured notched section. The relative fracture energy of the four specimens is indicated by the bars. Clearly the addition of exfoliated clay has suppressed crazing and very greatly embrittled the nylon 6, but transmission electron microscopy shows that craze nucleation occurred near the interface of the clay platelets with the nylon matrix.[104] No birefringence was detected using crossed polarisers suggesting that there was negligible shear yielding. The reason why spherical particles can induce multiple crazing yet intercalated or exfoliated clay particles suppress it is not clear. Because of the embrittling effect of nanoclay particles on nylon 6, researchers have turned their attention to ternary nanocomposites with elastomeric particles to increase toughness. However as Mai's group remark, though 'in the case of a ternary composite with micro-sized fillers, the relationship between fracture properties and deformation behaviour is reasonably well-established and understood. When the reinforcement is reduced to nano-scale in a ternary composite, however, basic knowledge of the deformation and fracture mechanisms is still in its infancy.'[105]

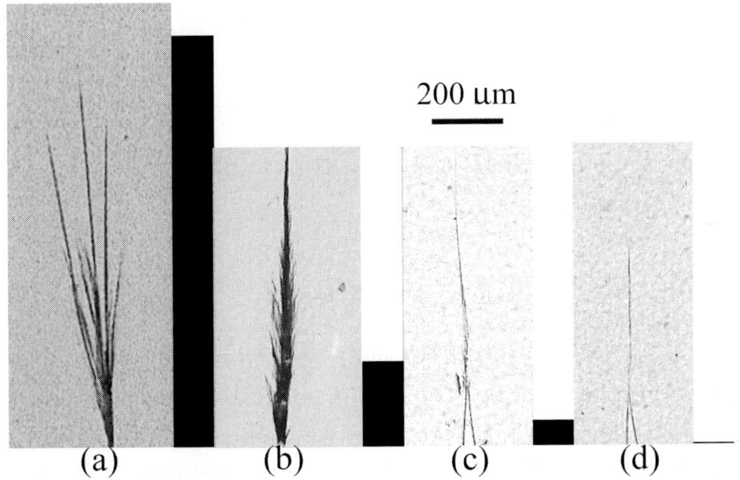

Fig. 12.15 Transmitted optical micrographs of craze-like bands in nylon 6 clay nanocomposites (a) unfilled, (b) 2.5 wt% clay, (c) 5 wt% clay, (d) 10 wt% clay; the bars show the relative fracture energy (He *et al.* 2008 with permission American Chemical Society).

In isotactic polypropylene, MWCNTs and HNTs unlike calcium carbonate particles, do not nucleate the β-phase and the only phase is α.[106,107] The fracture behaviour of polypropylene reinforced with MWCNTs or HNTs is very similar as shown in Fig. 12.16. The room temperature impact energy of polypropylene is modestly increased by about a quarter for weight fractions of MWCNTs of 1–3% and long CNTs are slightly tougher than short ones.[108] Dispersion problems cause the toughness to peak in the 1–3% weight fraction range. At higher temperatures the MWCNTs do toughen polypropylene somewhat more substantially. The enhancement in impact energy by HNTs is also modest and reaches a maximum at low weight fractions of about 5% because of dispersion problems.[107,109]

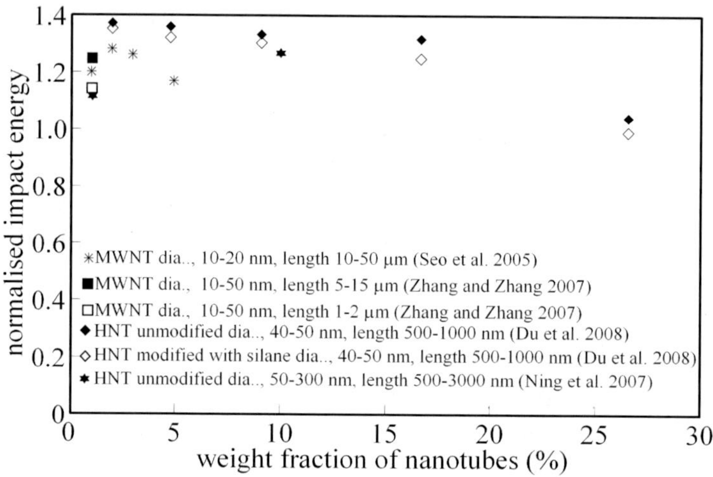

Fig. 12.16 Normalised room temperature impact energy as a function of the weight fraction of MWCNTs and HNTs in polypropylene.

12.4 Biomimetics, Strength, and Toughness

There are a number of fundamental fracture problems that we humans have not been able to solve that restricts our use of materials. In many applications we require stiff and strong materials yet when we look at traditional stiff and strong materials like ceramics or high strength metals they are usually also brittle which greatly affects the way they can be used. Nature has found ways to get round the nexus between stiffness, strength, and toughness. The common minerals of biological tissue have stiffness but not toughness, but bone, nacre, and antler have a toughness of up to three orders of magnitude higher (see Fig. 3.2). Reliability is also an engineering problem. Components either have flaws when made, or develop fatigue cracks during use, which reduce their strength. Yet we can break a rib and it will heal and be as strong again without any support other than that afforded by our own body. Also nature is energy efficient whereas man-made materials consume vast quantities of energy. She paints with a small palette and evolves new materials from those that already exist, whereas man tends to design more from scratch using a vast palette of materials. The word *biomimetics* was coined by the American biophysicist Otto Schmitt (1913–1998) in 1969 to describe the transfer of technology from nature to engineering.

We have seen in the previous section that methods developed to make tough composites with macro- and micro-size particles do not simply scale down to the nano-size particles. Natural materials have hierarchical structures from the atomic to the macro scale created by evolution to be fit for the purpose. Engineering materials are much simpler and it is easier to understand the mechanisms that create strength and toughness. Stiffness only depends weakly on structure being mainly determined by the stiffness of the reinforcing particles and their volume fraction. To get a stiff material one simply has to pack in as large a volume fraction of hard particles as is possible without reducing the other properties, it is strength and toughness requirements that determine how the hard particles are packed, how many can be tolerated and what is their size and shape.

Understanding how six or seven levels of hierarchy in materials, such as bone, work together to produce their material properties is an extremely difficult task. One development, because of this difficulty, is to use a systems approach. A successful engineering problem solving system TRIZ, the acronym of *Teorija Reshenija Izobretatel'skih Zadach* or the 'Theory of Inventive Problem Solving', whose development was initiated by Genrich Altshuller (1926–1998) and Rafik Shapiro in 1946, is being used for biomimetics by Julian Vincent and the Centre for Biomimetics and Natural Technologies, at the University of Bath.[110] As yet this ambitious programme has not been used to design materials with high stiffness, strength, and toughness and only time will tell how successful this approach is. What is clear is that the next-generation materials based on biomimetics will require a complete multidisciplinary approach with team members who have a good general knowledge of all the disciplines as well as expert knowledge of their own field. It is a lack of general knowledge of other disciplines that in many cases has held back the development of nanocomposites

There can be no general theory of biomimetics since the biomaterials have evolved along many different paths. Here a few examples are given of the application of biomimetics to materials where resistance to fracture is an important property. As yet, none of these has led to commercial exploitation, in fact biomimetics as a whole has produced few commercial products. Julian Vincent gives three examples: the Anglepoise lamp designed from the human arm by George Carwardine (1887–1948) and first manufactured in 1934, Velcro, the now ubiquitous fastener for clothes invented by George de Mestral (1907–1990) who in 1955 observed the hair on his dog pulling hooked seeds from burdock, and self-cleaning paints with the trade name Lotusan®[111], based on the work of Wilhelm Barthlott and Christoph Neinhaus on very high water repellency of some species of leaves.[112] One hopes that the rate of successful commercial application of biomimetics in the twenty-first-century is much higher.

12.4.1 Composites modelled on wood tracheids

An early attempt at using biomimetics in designing composites with a high toughness was made by Jim Gordon and George Jeronimidis in 1983.[113] Based on the observation that much of the toughness of wood comes from the splitting of the microfibrils in the S2 layer which are spirally wound along the cell,[114] Gordon and Jeronimidis suggested that macrofibres could be constructed with fibres wound at an angle of 15° of a polymer tube that could be used as efficient reinforcement for polymer composites. They constructed hollow macrofibres formed of epoxy tubes 1.5 mm in diameter wound with glass fibres and used these macrofibres to construct a high toughness composite. The biomimetic design was successful, but the macrofibres could not be produced cheaply and the product was not commercialised.

12.4.2 Artificial Nacres

Not all levels of hierarchy in biological materials are necessarily essential to the development a limited range of mechanical properties some are simply needed as a means to obtain the higher levels in the hierarchy. Nacre is very stiff because the volume fraction of the aragonite mineral phase is 95%. It has some six levels of hierarchy but the toughness is obtained primarily on the micron level where aragonite platelets 0.5 µm thick with sides of 5–10 µm are cemented together with a thin (~20 nm) organic layer in a brick-like manner (see Fig. 3.9). Hence the mechanisms by which nacre obtains its stiffness and toughness are reasonably well understood[115] and the problem is mainly one of how to fabricate the brick-like structure. There have been two quite different approaches to fabricating artificial nacres: a group at Oklahoma State University[116] have made nacre-like thin films by depositing alternate layers of montmorillonite clay platelets and polyelectrolytes by layer-by-layer assembly, another group at the Lawrence Berkeley National Laboratory, California[117] have constructed an alumina scaffold by freezing a solution containing alumina particles which they sintered and then infiltrated with an aluminium silicon alloy.

The artificial nacre made by the Oklahoma State University group most nearly matched the dimensions of nacre. Only tensile tests were made on the films, which for 200 layers was about 5 μm thick and showed rainbow diffraction colours similar to the pearl tint of sea shells. Although the Young's modulus of the artificial nacre was very much smaller than that of nacre, their strength was comparable and the fracture strain considerably greater.

The Lawrence Berkeley National Laboratory group's artificial nacre had a much smaller ceramic content of 45% than nacre. The specimens were thick enough to perform standard fracture toughness tests, obtaining a fracture toughness of 5.5 MPa$\sqrt{m}$ which is similar to nacre in the dry state, as well as tension tests, obtaining a strength of 400 MPa which is about four times that of nacre. Aluminium alloys do not wet alumina, but small additions of titanium reduce the surface tension and improve the wetting. When 0.5% of titanium was added to the aluminium alloy the fracture toughness was increased to about 10 MPa$\sqrt{m}$ and the strength to 600 MPa. Presumably reducing the surface tension enabled the aluminium alloy to be infiltrated more completely. In nacre the fracture toughness increases when the nacre is hydrated and the protein layers between the aragonite platelets are plasticized so that there is more pull-out. However, in the artificial nacre most of the fracture work probably comes from plastic deformation in the aluminium which forms 55% of the volume of the composite where the protein layer is only 5% of the volume of nacre. Clearly though this artificial nacre is superficially similar to nacre, its toughness is obtained differently.

12.4.3 Self healing polymers

One of the remarkable properties of bone is its ability to heal itself and it continually regenerates itself to adapt to stress changes. A polymer with a self-healing capability has been recently developed by Scott White and his group at the University of Illinois.[118,119] There have been several previous methods of healing materials including glass, concrete, asphalt and a range of polymers proposed,[120] but in none, before the work of White and his group, was the healing self-induced. The self-healing method for epoxy relies on microcapsules of urea-formaldehyde containing dicyclopentadiene (DCPD) fracturing during crack propagation and being cured by Grubb's catalyst that is also embedded in the epoxy. The curing of the DCPD is complete in about 24 hours. The microcapsules need to be reasonably large for them to be fractured by cracking and release sufficient quantities of the DCPD and a diameter of about 180 μm is optimum. The microcapsules toughen the virgin epoxy from 0.55 MPa√m to about 1.2 MPa√m at about 15% volume fraction of 180 μm microcapsules. Tapered double-cantilever beam were used to test the specimens where the energy release rate is independent of crack length and though the fractures were brittle they were not complete and after testing the specimens were unloaded allowing the crack faces to come together and heal. Provided the volume fraction of the microcapsules was greater than 5% the specimens regained more than 97% of the toughness of the neat epoxy; at 20% volume fraction the healed toughness was 0.7 MPa√m.

Naturally the self healing system of White and his group is not suitable for practical applications where fracture occurs under essential static loading because in most cases the fracture would be unstable and fracture complete before healing could start. However, it can control high cycle fatigue at relatively low frequencies of loading. In fatigue experiments on the same system White's group showed that providing the maximum stress and stress range were small enough that crack propagation could be prevented.[121] The DCPD was drawn to the crack tip by surface tension and the meniscus form effectively blunted the crack tip.

12.5 Concluding Remarks

At the beginning of the twenty-first-century the fundamental concepts in fracture mechanics have been established and applied to many fields. Most of the fundamental concepts have been briefly discussed here but there are some, such as the application of fractals, which have not. In the case of fractals they have not been discussed for two reasons: one they are difficult to discuss simply and secondly because they do not really help in understanding the mechanics of fracture. The surface roughness of cracks as well as the distributions of microcracks can be described, over limited ranges by fractal concepts. However, there is no universal relationship between fractal dimension and fracture properties and as Zdeněk Bažant and Arash Yavari have said 'simply knowing the fractal dimension of a fracture surface or a distribution of microcracks does not help in understanding the mechanics of failure'.[122]

The main recent developments have been in the application of fracture mechanics to smaller scales. At present the development of the fracture properties of nanocomposites is slow because mechanisms developed for the microscale are being used which are not all appropriate to the nanoscale and new concepts of fracture at the nanoscale have not been developed. One of the challenges is still to understand the fracture mechanisms across the scales from angstroms to millimetres. Such understanding is necessary before all nature's hierarchical design secrets can be unlocked and biomimetics become more fruitful.

One of the strengths of the early pioneers of the understanding of fracture was their broad knowledge of many disciplines. Today the breadth of science and engineering has become extremely wide. No one can be an expert in all disciplines, but a broad education should not be neglected at the expense of specialisation. The days are of course gone when someone like Geoffrey Taylor could work largely by himself, now large research teams are needed because of the many different expertises that are necessary for modern research. However, it is essential the team members have a general appreciation of all the disciplines involved. Vast funds are now necessary for research leading to the situation where obtaining funds becomes more important than the research itself. Governments set priority areas and often research on those areas is undertaken not because the researcher has developed ideas that are worth following but just because funds are available. Most of the important breakthroughs in any research have occurred because of the ideas of individual researchers not because effort has been directed. Alan Cottrell has made many major contributions to fracture research yet he told me that he did not believe that he had ever received specific funding for any of that research. Bright people will always produce the best research if left to work on what they see as important. Research is not just another commodity to be managed.

In what direction will fracture research lead in the twenty-first-century? The answer to that question lies with those entering the field. Research by its nature is unpredictable.

12.6 Notes

[1] Binnig *et al.* (1986).
[2] Nix (1989); Evans and Hutchinson (1995).
[3] Hutchinson and Suo (1992).
[4] Williams (1959).

[5] Cheropanov (1962); England (1965); Erdogan (1965); Rice and Sih (1965).

[6] Dundurs (1969).

[7] Volinsky *et al.* (2002).

[8] Tantalum nitride is used as a very thin barrier ($\approx$25 nm) layer to prevent copper diffusion. The failure was in the TaN/SiO$_2$ interface for all the thickness of copper.

[9] Lane *et al.* (2000).

[10] Evans *et al.* (1988).

[11] A non-dimensional cracking number $Z = 1/Q$ was first introduced by Evans *et al.* (1988). Hutchinson (1996) used the inverse number which is presented here.

[12] Suo and Hutchinson (1989).

[13] Beuth (1992).

[14] Thouless (1990).

[15] Chen and Wu (2008).

[16] Gioia and Ortiz (1997).

[17] Thouless (1993).

[18] Cotterell and Chen (2000).

[19] Hutchinson (2001).

[20] Yu and Hutchinson (2002).

[21] Hutchinson and Evans (2000).

[22] Fleck *et al.* (1994).

[23] Cosserat and Cosserat (1909).

[24] Mindlin and Tiersten (1962).

[25] Fleck and Hutchinson (1993).

[26] Elssner *et al.* (1994).

[27] Tvergaard and Hutchinson (1992); see also §9.4.6.

[28] Fleck and Hutchinson (1997).

[29] Wei and Hutchinson (1997).

[30] Nix and Gao (1998).

[31] Gao *et al.* (1999).

[32] Evans and Hutchinson (2009).

[33] Giessen and Needleman (1995).

[34] Deshpande *et al.* (2001).

[35] See §9.5.3.

[36] Deshpande *et al.* (2003).

[37] Kubin *et al.* (1992).

[38] Kukta (1998).

[39] Weygand *et al.* (2002).

[40] Dirac (1929).

[41] Schrödinger (1926).

[42] Hohenberg and Kohn (1964); Kohn and Sham (1965).

[43] See §10.1.2; Lu *et al.* (2000).

[44] Cohen *et al.* (1994).

[45] Rahman (1964).

[46] Abraham (2002).

[47] Abraham *et al.* (1994).

[48] Rosakis *et al.* (1999).
[49] Broberg (1989).
[50] Gao (1996); Abraham *et al.* (1997).
[51] Buehler *et al.* (2003).
[52] Buehler and Gao (2006).
[53] Abraham *et al.* (1998).
[54] Kelly and Macmillan (1986).
[55] Street (1962).
[56] Armstrong *et al.* (1966).
[57] Kumar *et al.* (2003).
[58] Meyer (2006).
[59] See §9.2.
[60] Carlton and Ferreira (2007).
[61] Mirshams *et al.* (2001).
[62] Gan and Zhou (2001).
[63] Tjong (2006).
[64] Xie *et al.* (2005).
[65] Wang *et al.* (2005).
[66] Qi *et al.* (2006).
[67] Radushkevich and Lukyanovich (1952).
[68] Iijima (1991).
[69] Poland *et al.* (2008).
[70] Ye *et al.* (2007).
[71] Cotterell *et al.* (2007).
[72] Mott and Nabarro (1948).
[73] Lange (1970).
[74] Evans (1972).
[75] Johnsen *et al.* (2007).
[76] Zhang *et al.* (2006).
[77] Kinloch *et al.* (1985).
[78] Faber and Evans (1983).
[79] Mean deviation from the surface centre line.
[80] Kitey and Tippur (2005).
[81] Deng *et al.* (2007).
[82] Spanoudakis and Young (1984).
[83] Kinloch *et al.* (2005).
[84] Private communication.
[85] Lan *et al.* (1995).
[86] Kinloch and Taylor (2006).
[87] Dean *et al.* (2007).
[88] Kornmann *et al.* (2002).
[89] See §10.1.3.
[90] Gojny *et al.* (2004).
[91] Zebarjad *et al.* (2004); Weon *et al.* (2006).
[92] Labour *et al.* (2002).

[93] Aboulfaraj *et al.* (1995).

[94] Zhang *et al.* (2004).

[95] Nevertheless in both sets of data for the smallest 44 nm particles there is a decrease in impact energy for the largest volume fraction, this decrease is assumed to be because of agglomeration and these values have been ignored in fitting the line of best fit.

[96] Wu (1985).

[97] Bartczak *et al.* (1999).

[98] Muratoğlu *et al.* (1995a,b).

[99] Corté *et al.* (2005).

[100] Dasari *et al.* (2007).

[101] Chen *et al.* (2003).

[102] Ito *et al.* (1998).

[103] Liu *et al.* (1999).

[104] He . (2008).

[105] Lim *et al.* (2007).

[106] Assouline *et al.* (2003); Leelapornisit *et al.* (2005).

[107] Ning *et al.* (2007).

[108] Seo *et al.* (2005); Zhang and Zhang (2007).

[109] Du *et al.* (2008).

[110] Vincent *et al.* (2006).

[111] The first experiments were done with Tropaeolum a family of South American climbers, by Wilhelm Barthlott in 1977 and he and Christoph Neinhaus developed sustainable industrial products based on the lotus leaf in 1997.

[112] Vincent (2003).

[113] Gordon and Jeronimidis (1983).

[114] See §3.2.

[115] See §3.3.2.1.

[116] Tang *et al.* (2003).

[117] Deville (2006).

[118] White *et al.* (2001).

[119] Brown *et al.* (2004).

[120] A review of crack healing in polymeric materials is given by Wu *et al.* (2008).

[121] Brown *et al.* (2005).

[122] Bažant and Yavari (2005).

Appendix

Glossary of Symbols and Abbreviations

Symbols

a	Interatomic distance. Crack length for cracks with one tip, half crack length for cracks with two tips.
a_0	Equilibrium interatomic distance. Undeformed crack length.
b	Width. Slot length. Burgers vector. Half width of buckle.
b_0	Half width of initial buckle, defined Eq. (12.6).
c_2	Velocity of shear waves.
c_R	Velocity of Rayleigh waves.
d	Diameter. Grain size.
d_0	Size of a plastic particle formed under plane stress, see Eqs. (9.5) and (9.12).
e	True strain, defined Eq. (1.22).
g	Acceleration due to gravity.
h	Thickness. Heat transfer coefficient.
k	Coefficient of thermal conductivity. Shear yield strength.
k_f	Fatigue stress concentration factor.
k_t	Theoretical stress concentration factor.
l	Length. Material length parameter.
l_c	Critical fibre length, defined Eq. (10.18).
l_{ch}	Characteristic length, defined Eq. (1.31).
l_p	Interparticle distance, see Eq. (12.13).
p	Pressure.
q	Electronic charge. Notch sensitivity index defined Eq. (8.14).
m	Weibull index, defined Eq. (8.11).
n	Number. Index.
n_i	Number of fatigue cycles at stress σ_i.
u, v	Displacements in the x and y directions.

u_c	Critical displacement.
v_p	Volume fraction of particles.
v_f	Velocity of the fracture front. Volume fraction of fibres.
w	Weakness, the inverse of strength.
w_e	Specific essential work of fracture.
w_f	Specific work of fracture.
w_p	Specific plastic or non-essential work of fracture.
x, y, z	Cartesian coordinates.
A	Area.
B	Plate thickness. Specimen width.
C	A constant.
D_0	Size of a plastic particle formed under plane strain, see Eq. (9.6).
E	Energy. Young's modulus.
$\bar{E}$	Plane strain Young's modulus, defined Eq. (1.14).
E_c	Young's modulus of a composite.
$\bar{E}_f$	Plane strain Young's modulus of film.
E_m	Young's modulus of a matrix.
E_p	Young's modulus of a particle.
E_s	Energy radiated by an earthquake.
F	Force.
F_c	Cutting force.
F_t	Transverse force in cutting.
G	Energy release rate or crack extension force.
G_I, G_{II}	Mode I, II energy release rates.
G_f	Fracture energy, an alternative to R.
G_{Ic}	Plane strain fracture energy.
G_0	Reference energy release rate, defined Eq. (12.5).
H	Horizontal force.
I	Second moment of area of a section.
J	J-integral, defined Eq. (9.24).
J_R	The J crack growth resistance.
K	Stress intensity factor.
K_I, K_{II}, K_{III}	Mode I, II, III stress intensity factors.
$\bar{K}$	Non-dimension stress intensity factor.
K_c	Fracture toughness.

K_{Ia}	Arrest fracture toughness.
K_{Ic}	Plane strain fracture toughness.
K_R	The K crack growth resistance.
K_T	Shielding stress intensity factor induced in transformation toughening.
K_0	Initiation fracture toughness.
ΔK	Stress intensity range.
L	Length.
M	Bending moment. Mass. Magnitude of an earthquake on the Richter scale.
N	Force acting normal to a cutting tool.
N_i	Fatigue life in cycles at a stress of σ_i.
P	Probability.
Q	Hydrostatic stress parameter, see Eq. (9.28).
R	Fracture energy. Radius. Stress ratio.
$R, R', etc.$	Material resistance factors of Kingery and Hasselman, see Eqs. (10.6) and (10.7).
R_0	Fracture energy at initiation.
R_a	Surface roughness.
R_a	Adhesive energy.
R_m	Maximum fracture energy.
R_p	Contribution to fracture energy by fibre pull-out.
S	Force acting along the face of a cutting tool.
S_0	Inherent shear strength of the rock, see Eq. (2.4).
T	Stress vector. Absolute temperature. T-stress, the second term in the series expansion for the elastic stress at a mode I crack. Crack line tension, see Eq. (12.10).
ΔT	Quenching temperature differential.
U	Strain energy density.
U_i	Impact energy.
V	Volume.
X	Force.
W	Plastic work density. Depth of a beam.
W_e	Essential work of fracture.
W_f	Work of fracture.
W_p	Plastic work. Specific work of fibre pull-out.
$\overline{W}$	Non-dimensional depth of a beam, defined Eq. (10.9).

Y, Y'	Form factors defined Eqs. (1.26) and (1.29).
Z	Section modulus.
α	Rake angle of cutting tool. Coefficient of thermal expansion. Strain hardening exponent in Ogden strain energy density function, defined Eq. (11.7).
α, β	Dundurs parameters, defined Eq. (12.1).
β	Plastic work factor. Biot number, defined Eq. (10.8). Friction angle, defined §11.2.
δ	Deformation or deflection. Crack opening displacement (CTOD).
δ_c	Critical CTOD.
γ	Intrinsic surface energy. Shear strain.
γ_p	Extrinsic surface energy.
ε	Normal strain.
ε_f	Fatigue ductility coefficient, see Eq. (9.39).
ε_V^T	Free dilatational transformation strain.
$\Delta\varepsilon_p$	Plastic strain range.
η	Viscosity, see Eq. (2.1). Strain hardening index.
η	Eta factor, see Eq. (9.27).
λ	Stretch ratio, defined Eq. (1.12).
$\lambda_1, \lambda_2, \lambda_3$	Principal stress ratios.
$\lambda_\delta, \lambda_P$	Constants.
ξ	Slice/push ratio in cutting.
μ	Shear modulus. Coefficient of friction.
ν	Poisson's ratio.
$\bar{\nu}$	Plane strain Poisson's ratio, defined Eq. (1.14).
ρ	Radius of curvature. Density.
σ	Stress.
σ_n	Normal stress.
σ_a	Alternating stress.
σ_c	Critical stress.
σ_{coh}	Cohesive stress.
σ_e	Von Mises equivalent stress, defined Eq. (1.24).
σ_f	Fibre strength.
σ_m	Mean or hydrostatic stress. Strength of matrix.
σ'_f	Fibre stress at fracture strain of the matrix.
σ'_m	Stress in matrix at fibre fracture strain.

σ_m''	Stress in matrix at the average stress in the fibres when they break.
σ_p	Fibre pull-out stress.
σ_{pm}	Maximum fibre pull-out stress.
σ_t	Ideal theoretical tensile strength, see Eq. (1.25).
σ_{UTS}	Ultimate tensile stress.
σ_Y	Yield strength.
σ_0	Reference stress.
θ	Angle.
τ	Shear stress.
τ_0	Reference shear stress, defined Eq. (11.6).
ϕ	Parameter, mainly a function of relative crack depth, used to describe impact energy, see Eq. (10.11). Angle of shear plane in cutting.
Γ	Plastic work.
Γ_0	Cohesive energy.
Λ	Strain energy.
Π	Potential energy.
Σ	Strength.
$\Delta\Sigma$	Limiting fatigue stress range.
$\Delta\Sigma_a$	Limiting stress range for fully reversed stress cycles.
Ω	Dimensionless cracking number, defined Eq. (12.3).

Abbreviations

ABS	Acrylonitrile butadiene styrene.
ADF	Automatic direction finder.
AFM	Atomic force microscope.
bcc	Body centred cubic.
BWRA	British Welding Research Association.
Ca-PSZ	Calcia partially stabilized zirconia.
CNT	Carbon nanotube.
CTBN	Carboxyl terminated butadiene acrylonitrile.
CTOA	Crack tip opening angle.
CTOD	Crack tip opening displacement.
DCB	Double cantilever beam.
DCPD	Dicyclopentadiene.
DGEBA	Diglycidyl ether of bisphenol A.
DPH	Diamond pyramid number.
EPFM	Elasto-plastic fracture mechanics.
EWF	Essential work of fracture.
fcc	Face centred cubic.
FEM	Finite element method.
FPZ	Fracture process zone.
HDPE	High density polyethylene.
HIPS	High impact polystyrene.
HNT	Halloysite nanotubes.
HRR	Hutchinson, Rice and Rosengren.
HSLA	High strength low alloy [steels].
iPP	Isotactic polypropylene.
ITO	Indium-tin oxide.
LEFM	Linear elastic fracture mechanics.
MATEIS	Materials, Engineering and Sciences Group of INSA de Lyon.
MDF	Macro defect free cement.
Mg-PSZ	Magnesia partially stabilized zirconia.
MSG	Mechanism based strain gradient plasticity.
MWNT	Multi walled nano tube.
NDTT	Nil ductility transition temperature.
NRL	Naval Research Laboratory.
OLED	Organic light emitting diode.

PAN	Polyacrilonitrile.
PASGT	Personal armour system for ground troops.
PC	Polycarbonate.
PE	Polyethylene.
PET	Polyethylene terephthalate.
PF	Phenol formaldehyde.
PMMA	Polymethyl methacrylate.
PN	Polyoxyethylene nonphenol.
PP	Polypropylene.
PS	Polystyrene.
PSB	Persistent slip band.
PVAA	Polyvinyl alcohol acetate.
PVF	Poly(vinyl formal).
RAE	Royal Aircraft Establishment, Farnborough, England.
RILEM	Réunion Internationale des Laboratoires d'Essais et de Recherches sur les Matériaux et les Constructions.
SAFB	Sulphate aluminate ferrite belitic clinkers.
SF	Safety factor.
SWNT	Single walled nanotubes.
TMCP	Thermo-mechanical controlled process.
TMP	Thermo-mechanical processing.
TRIP	Transformation induced plasticity.
TRIZ	*Teorija Reshenija Izobretatel'skih Zadach* or the "Theory of Inventive Problem Solving".
TWI	The Welding Institute.
UTS	Ultimate tensile strength.

Bibliography

Aboulfaraj, M., G'Sell, C., Ulrich, B., and Dahoun, A. (1995) In situ observation of the plastic deformation of polypropylene spherulites under uniaxial tension and simple shear in the scanning electron microscope. *Polymer*, 36:731–42.

Abraham, F.F. (2002) Very large scale simulations of materials failure, *Phil. Trans. Math. Phys. Engng. Sci.*, 360:367–382.

Abraham, F.F., Brodbeck, D., Rafey, R.A., and Rudge, W.E. (1994) Instability of fracture: a computer simulation investigation, *Phys. Rev. Lett.*, 73:272–275.

Abraham, F.F., Brodbeck, D., and Rudge, W.E. (1997) A molecular-dynamics investigation of rapid fracture mechanics, *J. Mech. Phys. Solids*, 45:1595–1619.

Abraham, F.F., Brodbeck, D., Rudgey, W.E. Broughton, J.Q., Schneider, D., Land, B., Lifka, D., Gerner, J., Rosenkrantz, M., Skovira, J. and Gao, h. (1998) *Ab initio* dynamics of rapid fracture, *Modelling Simul. Mater. Sci. Eng.*, 6:639–670.

Adam, J.-P. (1994) *Roman Building: Materials and Techniques*, Trans. A. Mathews, Indiana University Press.

Adams, F.D., and Nicolson, J.T. (1900) An experimental investigation into the flow of marble, *Phil. Trans Roy. Soc.*, A195:363–401.

Agassiz, L. (1840) *Étude sur les glaciers*, Zenf et Gassman, Nicolet.

Albert, W.A.J. (1838) Über Treibseile am Harz, *Archive für Mineralogie, Geognosie, Bergbau und Hüttenkunde*, 10:215–234.

Amicus (1859) Letters to *The Times*, on the 1st November and the 6th December 1859.

Amos, J.L. (1974) The Development of Impact Polystyrene–A Review, *Polym. Engng. Sci.*, 14:1–14.

Anderrson, H. (1973) A finite-element representation of stable crack-growth, *J. Mech. Phys. Solids*, 21:337–356.

Andrews, E.H. (1959) Stress waves and fracture surfaces, *J. Appl. Phys.*, 30:740–743.

Anon. (1894) Boiler Explosions and Boiler Laws, *Manufacturer and Builder*, 26, p. 232.

Anon. (1946) Edwin Gilbert Izod Obituary, *Journal of the Institute of Metals*, 72:718–719.

Argon, A.S. (2001) Mechanics and physics of brittle to ductile transitions in fracture, *J. Eng. Mater. Tech.*, 123:1–11.

Aristotle, School of, (1936) Mechanical Problems. *In Aristotle's minor works*, trans. W. S. Hett, Heinemann, London.

Armarego, E.J.A., and Brown, R.H. (1969) *The Machining of Metals*, Prentice-Hall, Englewood Cliff, New Jersey.

Armstrong, R.W., Chou, Y.T., Fisher, R.M., and Louat, N. (1966) Limiting grain size dependence of strength of a polycrystalline aggregate, *Phil. Mag.*, 14:943–951.

Arnold, D. (1991) *Building in Egypt: Pharaonic Stone Masonry*, Oxford University Press, Oxford.

AS CA65-1972, *SAA Timber Engineering Code*, Standards Association of Australia, Sydney, 125pp.

Astakhov, V.P. (1999) *Metal cutting mechanics*, CRC Press, Boca Raton.

Astakhov, V.P. (2005) On the inadequacy of the single-shear plane model of chip formation, *Int. J. Mech. Sci.*, 47:1649–1672.

Atkins, A.G. (1974) Fracture toughness and cutting, *Int. J. Prod. Res.*, 12:263–274.

Atkins, A.G. (1980) On cropping and related processes, *Int. J. Mech. Sci.*, 22:215–231.

Atkins, A.G. (1987a) Plastic fracture, *EGF Newsletter*, Nr. 4

Atkins, A.G. (1987b) On the number of cracks in the axial splitting of ductile metal tubes, *Int. J. Mech. Sci.*, 29:115–121.

Atkins, A.G. (1988) Mechanics of the transition from plastic flow to ductile fracture in guillotining, *Proc. ECF-7*, Budapest, Ed. Czoboly, E., Vol. 1, pp. 81–87.

Atkins, A.G. (1995) Opposite paths in tearing of sheet materials, *Endeavour*, 19:2–10.

Atkins, A.G. (2003) Modelling metal cutting using modern ductile fracture mechanics: quantitative explanations for some longstanding problems, *Int. J. Mech. Sci.*, 45:373–396.

Atkins, A.G. (2009) *The Science and Engineering of Cutting*, Butterworth Heinemann, Oxford.

Atkins, A.G., Chen, Z., and Cotterell, B. (1998a) The essential work of fracture and J_R curves for the double cantilever beam specimen: an examination of elastoplastic crack propagation, *Proc. Roy. Soc. Lond.*, A454:815–833.

Atkins, A.G., Chen, Z., and Cotterell, B. (2003) Prediction of the energy dissipation rate in ductile crack propagation, *Fatigue Fract. Engng. Mater. Struct.*, 26:67–77.

Atkins, A.G., Khan, M.A., and Liu, J.H. (1998b) Necking and radial cracking around perforations in thin sheets at normal incidence, *Int. J. Impact Engng.*, 21:521–539.

Atkins, A.G., Lee, C.S., and Caddell, R.M. (1975) Time-temperature dependent fracture toughness of PMMA, Part 1, *J. Mater. Sci.*, 1975:381–1393.

Atkins, A.G., and Mai, Y.W. (1979) On the guillotining of materials, *J. Mater. Sci.*, 14:2747–2754.

Atkins, A.G., and Mai, Y.W. (1985) *Elastic and Plastic Fracture*, Ellis Horwood, Chichester.

Atkins, A.G., and Vincent, J.F.V. (1984) An instrumented microtome for improved histological sections and the measurement of fracture toughness, *J. Mater. Sci.* Letters, 3:310–312.

Atkins, A.G., Xu, X., and Jeronimidis, G. (2004) Cutting, by 'pressing and slicing,' of thin floppy slices of materials illustrated by experiments on cheddar cheese and salami, *J. Mater. Sci.*, 39:2761–2766.

Auerbach, F. (1891) Measurement of hardness, *Ann. Phys. Chem.*, 43:61.

Aveston, G., Cooper, G.A., and Kelly, A. (1971) Single and multiple fracture, *The Properties of Fibre Composites*, Conference Proceedings National Physical Laboratory, pp. 15–26.

Bacon, F. (1622) *Historia Naturalis et Experimentalis*, M. Lownes and G. Barret, London.

Backman, M.E., Goldsmith, W. (1978) The mechanics of penetration or projectiles into targets, *Int. J. Engng. Sci.*, 16:1–99.

Baekeland, L.H. (1909) The synthesis constitution and uses of Bakelite, *Ind. Eng. Chem.*, 1:149–161.

Barenblatt, G.I. (1959) On the equilibrium cracks due to brittle fracture (in Russian), *Dokaldy AN SSSR*, 127:47–50.

Barenblatt, G.I. (1962) The mathematical theory of equilibrium cracks in brittle materials, *Advances in Applied Mechanics*, 7:55–129.

Barlow, P. (1817) *An Essay on the Strength of Timbers*, J. Taylor London.

Barnaby, N. (1879) The use of steel in navy construction' *J. Iron Steel Inst.*, 15:45–54.

Barstow, F.E., and Edgerton, H.E. (1939) Glass fracture velocity, *J. Am. Ceram. Soc.*, 22:302–307.

Bartczak, Z., Argon, A.S., Cohen, R.E., and Weinberg, M. (1999) Toughness mechanism in semi-crystalline polymer blends: II. High-density polyethylene toughened with calcium carbonate filler particles, *Polymer*, 40:2347–2365.

Barthelat, F., and Espinosa, H.D. (2007) An experimental investigation of deformation and fracture of nacre–mother of pearl, *Exp. Mech.*, 47:311–324.

Bascom, W.D., Cottington, R.L., Jones, R.L., and Peyser, P. (1975) The fracture of epoxy- and elastomer-modified epoxy polymers in bulk and as adhesives, *J. Appl. Poly. Sci.*, 19:2545–2562.

Basquin, O.H. (1910) The exponential law of endurance tests, *Proc. Annual Meeting, American Society for Testing Materials,* 10:625–630.

Bathe G., and Bathe, D. (1935) *Oliver Evans: A Chronicle of Early American Engineering*, Historical Society of Pennsylvania, Philadelphia.

Bažant, Z..P. (1994) Nonlocal Damage theory based on micromechanics of crack interactions, *J. Eng. Mech. Div. ASCE*, 120:593–617.

Bažant, Z..P. (2002) Concrete fracture models: testing and practice, *Eng. Fract. Mech.*, 69:165–205.

Bažant, Z.P., and Cedolin, L. (1979) Blunt crack band propagation in finite element analysis, *J. Eng. Mech. Div. ASCE*, 105:297–315.

Bažant, Z.P., and Cedolin, L. (1983) Finite-element analysis of crack band propagation, *J. Eng. Mech. Div. ASCE*, 109:93–108.

Bažant, Z.P., and Frangopol, D.M. (2002) Size effect hidden in excessive dead load factor, *J. Struct. Eng. ASCE*, 128:80–86.

Bažant, Z.P., and Lin, F.B. (1988) Nonlocal smeared cracking model for concrete fracture, *J. Struct. Eng. ASCE*, 114:2493–2510.

Bažant, Z.P., and Yavari, A. (2005) Is the cause of size effect on structural strength fractal or energetic–statistical? *Eng. Fract. Mech.*, 72:1–31.

Bažant, Z.P., and Yavari, A. (2007) Response to A. Carpinteri, B. Chiaia, P. Cornetti and S. Puzzi's Comments on ''Is the cause of size effect on structural strength fractal or energetic-statistical?'' *Eng. Fract. Mech.*, 74:2897–2910.

Begley, J.A., and Landes, J.D. (1972) The J integral as a fracture criterion, In *Fracture Toughness, Proceedings of the 1971 National Symposium on Fracture Mechanics, Part* II, ASTM STP 514, American Society for Testing and Materials, pp. 1–20.

Benbow, J.J., and Roesler, F.C. (1957) Experiments on controlled fracture, *Proc. Phys. Soc.* B, 70:201–210.

Bennett, J.A. (1946) A study of the damaging effect of fatigue stressing on X4130 steel. *Proc. Amer. Soc. Test. Mater.*, 46:693–714.

Bernays, E.A. (1991) Evolution of insect morphology in relation to plants, *Phil. Trans. Roy. Soc. Lond.*, B333:257–264.

Bernays, E.A., and Janzen, D.H. (1988) Saturniiid and Sphngied caterpillars: Two ways to eat leaves, *Ecology*, 69:1153–1160.

Berry, J.P. (1963) Determination of fracture surface energies by the cleavage technique, *J Appl. Phys.*, 34:62–68.

Bessemer H. (1905) *Sir Henry Bessemer, F.R.S.: An Autobiography*, Offices of "Engineering", London.

Bessonov, M.I., and E. V. Kuvshinskii E.V. (1959) Cracks in transparent plastics, their growth and structure, *Fizika Tuerdogo Telu*, 1:1441–1452.

Beuth, J.L. (1992) Cracking of thin bonded films in residual tension, *Int. J .Solids Struct.*, 29:1657–1675.

Bilby, B.A., Cottrell, A.H., and Swinden, K.H. (1963) The spread of plastic yield from a notch, *Proc. Roy. Soc. Lond.*, A272:304–314.

Bilby, B.A., and Eshelby, J.D. (1962) Dislocations and the theory of fracture, In *Fracture, An Advanced Treatise*, Ed. Liebowitz, H., Academic Press, New York, Vol. 1, pp. 99–182.

Biggs, W.D. (1960) *The Brittle Fracture of Steel*, MacDonald and Evans Ltd., London.

Binnig, G., Quate, C..F., and Gerber, Ch. (1986) Atomic force microscope, *Phys. Rev. Lett.*, 56:930–933.

Biringuccio, V. (1540) *Dela pirotechnia*, Venetia.

Bishop, R.F., Hill. R., and Mott, N.F. (1945) The theory of indentation and hardness tests, *Proc. Phys. Soc.*, 57:147–159.

Bishopp, J.A. (1997) The history of Redux® and the Redux bonding process, *Int. J. Adhesion and Adhesives* 17:287–301.

Blanchard, J., Bussel, M., Marsden, A., and Lewis, D. (1997) Appraisal of existing ferrous structures, *Stahlbau*, 66:333–345.

Blyth, P.H. (1993) Metallurgy of Two Fragmentary Archaic Greek Helmets. *J. Hist. Met. Soc.*, 27:25–36.

Blyth, P.H., and Atkins, A.G. (2002) Stabbing of metal sheets by a triangular knife. An archaeological investigation, *Int. J. Impact Engng.*, 27:459–473.

Böker, R. (1915) Die Mechanik der bleibenden Formanderung in kristallinisch aufgebauten Körpern, *Verhandl. Deut. Ingr. Mitt. Forsch.*, 175:1–51.

Bonaparte, J.F., and Coria, R.A. (1993) Un nuevo y gigantesco saurópodo titanosaurio de la Formación Río Limay (Albiano-Cenomaniano) de la Provincia del Neuquén, Argentina, *Ameghiniana*, 30:271–282.

Bond, P.E., and Langhorne, P.J. (1997) Fatigue behaviour of cantilever beams of saline ice, *J. Cold Reg. Engrg.*, 11:99–112.

Boodberg, A., Davies, H.E., Parker, E.R., and Troxell, G.E. (1948) Causes of cleavage fracture in ship plates – tests of wide notched plates, *Weld. J. Res. Suppl.*, 27:186s–199s.

Bowen, P.B. and Raha, S. (1970) Microshear bands in polystyrene and polymethyl methacrylate, *Phil. Mag.*, 22:455–491.

Bowie, O.L. (1956) Analysis of an infinite plate containing radial cracks originating at the boundary of an internal circular hole, *J. Math. Phys.*, 35:60–71.

Boyd, G.M. (1970) *Brittle Fracture in Steel Structures*, Butterworth & Co., London.

Bracegirdle, B. (1978) *A History of Micro-technique*, Cornell University Press, Ithaca, N.Y.

Braithwaite, F. (1854) On the fatigue and consequent fracture of metals, *Min. Proc. Inst. Civil Engrs. Session 1853–1854*, 13:463–473.

Broberg, K.B. (1960) The propagation of a brittle crack, *Arkiv för Fysik*, 18:159–192.

Broberg, K.B. (1975) On stable crack growth, *J. Mech. Phys. Solids*, 23:215–237.

Broberg, K.B. (1987) On crack paths, *Eng. Fract. Mech.*, 28:663–679.

Broberg, K.B. (1989) The near-tip field at high crack velocities, *Int. J. Fract.*, 39:1–13.

Brodsley, L., Frank, C., and Steeds, J.W. (1986) Prince Rupert's drops, *Notes Rec. R. Soc. Lond.*, 41:1–26.

Brown, E.N., White, S.R., and Sottos, N.R. (2004) Microcapsule induced toughening in a self-healing polymer composite, *J. Mater. Sci.*, 39:1703–1710.

Brown, E.N., White, S.R., and Sottos, N.R. (2005) Retardation and repair of fatigue cracks in a microcapsule toughened epoxy composite—Part II: In situ self-healing, *Comp. Sci Technol.*, 65:2474–2480.

Bryant, R.W., and Dukes, W.A. (1966) The effect of joint design and dimensions on adhesive strength, *Appl. Poly. Symp.*, 3:81–98.

BS 7910:2005 *Guide to methods for assessing the acceptability of flaws in metallic structures*, British Standards, London.

Bubeck, R.A., Buckley, D.J., Kramer, E.J., and Brown H.R. (1991) Modes of deformation in rubber-modified thermoplastics during tensile impact, *J. Mater. Sci.*, 26:6249–6259.

Bucknall, C.B. (1977) *Toughened Plastics*, Applied Science Publishers, London.

Bucknall, C.B., and Smith R.R. (1965) Stress-whitening in high-impact polystyrenes, *Polymer*, 6:437–446.

Budiansky, B., and Hutchinson, J.W. (1978) Analysis of closure in fatigue crack growth, *J. Appl. Mech.*, 45:267–276.

Buehler, M.J., Abraham, F.F., and Gao, H. (2003) Hyperelasticity governs dynamic fracture at a critical length scale, *Nature*, 426:141–146.

Buehler, M.J., and Gao, H. (2006) Dynamical fracture instabilities due to local hyperelasticity at crack tips, *Nature*, 439:307–310.

Buffon, G. (1775) *Histoire Naturelle*, Paris.

Burdekin, F.M, and Dawes, M.G. (1971) Practical use of linear elastic and yielding fracture mechanics with particular reference to pressure vessels, In *Proc. Inst. Mech. Conf.*, London, pp. 28–37.

Burdekin, F.M., and Stone, D.F.W. (1966) The crack opening displacement approach to fracture mechanics in yielding materials, *J. Strain Analysis*, 1:145–153.

Burgan, M. (2007) *Robert Hooke: Natural Philosopher and Scientific Explorer*, Compass Point Books, Minneapolis.

Burgoyne, C., and Scantlebury, R. (2006) Why did the Palau Bridge collapse, *Struct. Eng.*, 84:30–37.

Burke, J.G. (1966) Bursting boilers and the Federal power, *Technology and Culture*, 7:1–23.

Busch, D.M. (1968) *Dissertation*, Technical University of Hanover.

Campbell, G.S. (1981) A note on fatal aircraft accidents involving metal fatigue, *Int. J. Fat.*, 3:181–185.

Carlton, C.E., and Ferreira, P.J. (2007) What is behind the inverse Hall–Petch effect in nanocrystalline materials? *Acta Mater.*, 55:3749–3756.

Carpinteri, A., Chiaia, B., Cornetti, P., and Puzzi, s. (2007) Comments on "Is the cause of size effect on structural strength fractal or energetic-statistical?" by Bažant & Yavari [Engng Fract Mech 2005;72:1–31], *Eng. Fract. Mech.*, 74:2892–2896.

Cauchy, A. (1823) Recherches sur l'équilibre et le mouvement intérieur des corps solides ou fluides, élastiques ou non élastiques, *Bulletin Société Philomatique*, pp. 9–13.

Cauchy, A. (1829) *Exercices de mathématique*, Vol. IV, De Bure, Paris.

Célarié, F., Prades, S., Bonamy, D., Ferrero, L., Bouchaud, E., Guillot, C., and Marlière, C. (2003) Glass breaks like metal, but at the nanometer scale, *Phys. Rev. Letters*, 90:075504.

Cerling, T.E., Harris, J.M., MacFadden, B.J., Leakey, M.G., Quadek, J., Eisenmann, V., and Ehleringer, J.R. (1997) Global vegetation change through the Miocene/Pliocene boundary, *Nature*, 389:153–158.

Chadwick, P. (2002) Ian Naismith Sneddon, O.B.E. 8 December 1919–4 November 2000, *Biog. Mem. Fellows Roy. Soc.*, 48:417–437.

Chai, H. (1988) Shear fracture, *Int. J. Fract.*, 37:137–159.

Chai, H. (2004) The effects of bond thickness, rate and temperature on the deformation and fracture of structural adhesives under shear loading, *Int. J. Fract.*, 130:497–515.

Chan, C.-M., Wu, J., Li, J.-X., and Cheng, Y.-K. (2002) Polypropylene/calcium carbonate nanocomposites, *Polymer*, 43:2981–2992.

Chan M.K.V., and Williams J.G. (1983). Slow stable crack growth in high density polyethylene. *Polymer* 24:234–244.

Chang T.M., and Swift, H.W. (1950) Shearing of metal bars, *J. Inst. Metals*, 78:119–146.

Chantikul, P., Bennison, S.J., and Lawn, B.R. (1990) Role of grain size in the strength and R-Curve properties of alumina, *J. Am. Ceram. Soc.*, 73:2419–2427.

Charles, R.J. (1958) Static fatigue of glass II, *J. Appl. Phys.*, 29:1554–1560.

Charpy, G. (1901) Esaay on the metals impact bend test of notched bars, (translated from the original French by E Lucon), reprinted in: *The Pendulum Impact Testing: A Century of Progress*, 2000 (Eds: T.A. Siewert and S. Manahan), ASTM STP 1380, West Conshohoken, Pennsylvania.

Charpy, G. (1906) Sur l'influence de la température sur la fragilité des métaux, *Proceedings of the International Association for Testing Materials*, [A].

Chen, L., Wong, S.-C., and Pisharath, S. (2003) Fracture properties of nanoclay-filled polypropylene. *J. Appl. Poly. Sci.*, 88:3298–3305.

Chen, Z., and Wu, L.Y.L. (2008) Scratch resistance of protective sol-gel coatings on polymeric substrates, In: *Tribology of Polymeric Nanocomposites: Friction and Wear of Bulk Materials and Coatings*, Eds. Friedrich, K. and Schlarb A.K., Elsevier, pp. 325–353.

Cheropanov, G.P. (1962) The stress state in a heterogeneous plate with slits (in Russian), *Izvestia AN SSSR, OTN, Mekhan. I Mashin*, 1:131–137.

Cherepanov, G.P. (1968) Crack propagation in continuous media (in Russian), *Prikladnaia Matematica I Mekhanica*, 31:476–488. English translation, (1998) In *Fracture a topical*

Encyclopedia of Current Knowledge, Ed. Cherepanov, G.P., Krieger Publishing Company, Malabar, Florida, pp. 41–53.

Cheung K.S., and Yip, S. (1994) A molecular-dynamics simulation of crack-tip extension: the brittle-to-ductile transition, *Modelling Simul. Mater. Sci. Eng.*, 2:865–892.

Chiu, W.C., Thouless, M.D., and Endres, W.J. (1998) An analysis of chipping in brittle materials, *Int. J. Fract.*, 90:287–298.

Christopher, P.R., Baker, R., Burton, D., Cotton, H.C., Craik, R., Fearnehough, G., Lessells, J., Lismer, R., Wakins, B., and Wortley, L. (1986) The Roberston crack arrest test, *Brit. Weld. J.*, 15:387–394.

Chung, W.N., and Williams, J.G. (1991) Determination of J_{Ic} for polymers using the single specimen method, *Elastic-Plastic Fracture Test Methods: The User's Experience*, Vol 2, ASTM STP 1114, Ed. J. A. Joyce, American Society for Testing and Materials, Philadelphia, pp. 320–339.

Clark, G. (1969) *World Prehistory: A New Outline*, Cambridge University Press, Cambridge.

Clark, G., and Piggott, S. (1970) *Prehistoric Societies*, Penguin, Harmondsworth,.

Clarke, S., and Engelbach, R. (1930) *Ancient Egyptian Construction and Architecture*, Dover Publications Reprint [1990], Mineiola, NY.

Clayton, M. (2006) Leonardo da Vinci: Experience, experiment and design, *Apollo–The International Magazine of Art and Antiques*, 164, p 82.

Cocquilhat, M. (1851) Expériences sur la résistance utile produites dans le forage, *Annales des Travaux Publics en Belgique*, 10:199–215.

Coffin, L.F. (1954) A study of the effects of cyclic thermal stresses on a ductile metal, *Trans. Amer. Soc. Mech. Engng.*, 76:931–950.

Cohen, R.E., Mehl, M..J., and Papaconstantopoulos, D..A., Tight-binding total-energy method for transition and noble metals, *Phys. Rev.*, B 50:14 694–14697.

Cook, J., and Gordon, J.E. (1964) A mechanism for the control of crack propagation in all-brittle systems, *Proc. Roy. Soc. Lond.*, A282:113–123.

Cook, N., Finnie, I., and Shaw, M.C. (1954) Discontinuous chip formation, *Trans. ASME*, 76:153–162.

Corbett, G.G., Reid, S.R., and Johnson, W. (1996) Impact loading of plates and shells by free-flying projectiles: A review, *Int. J. Impact Engng.* 18:141–230.

Cornell, J.W., Foley, F.B., Stewart, H.G., Jackson, W.F., and LePage, W.L. (1963) Report of the committee for the preservation of the Liberty Bell, *J. Franklin Inst.*, 275:161–191.

Corté, L., Beaume, F., and Leiblera, L. (2005) Crystalline organization and toughening: example of polyamide-12, *Polymer*, 46:2748–2757.

Coserat, E., and Cosserat, F. (1909) *Théorie des Corps Déformables*, Herman, Paris.

Cotterell, B. (1965a) Velocity effects in fracture propagation, *Applied Materials Research*, 4:227–232.

Cotterell, B. (1965b) On brittle fracture paths, *Int. J. Fract. Mech.*, 1:96–103.

Cotterell, B. (1966) Notes on the paths and stability of cracks, *Int. J. Fract. Mech.*, 2:526–533.

Cotterell, B. (1969) The paradox between the theories for tensile and compressive fracture, *Int. J. Fract. Mech.*, 5:R251–R252.

Cotterell, B. (1972) Brittle fracture in compression, *Int. J. Fract.*, 8:195–208.

Cotterell, B. (1997) A.A. Griffith and the foundations of fracture mechanics, In *Fracture Research in Retrospect*, ed.Rossmanith, H.P. pp. 105–122.

Cotterell, B., and Atkins, A.G. (1996) A review of the J and I integrals and their implications for crack growth resistance and toughness in ductile fracture, *Int. J. Fract.*, 81:357–372.

Cotterell, B., Atkins, A.G., and Chen, Z. (2000) A discussion of the extension of the J_R concept to significant crack growth, In *Multi-Scale Deformation and Fracture in Materials and Structures, The James R. Rice 60th Anniversary Volume*, Eds. Chung, T.J., and. Rudnicki, J.W., Kluwer Academic, Dordrecht, The Netherlands, pp. 223–236.

Cotterell, B and Chen, Z. (2000) The buckling and cracking of thin films on compliant substrates, *Int. J. Fract.*, 104:169–179.

Cotterell, B., Chia, J.Y.H., and Hbaieb, K. (2007) Fracture mechanisms and fracture toughness in semicrystalline polymer nanocomposites, *Eng. Fract. Mech.*, 74:1054–1078.

Cotterell, B., Kamminga, J., and Dickson, F.P. (1985) The essential mechanics of conchoidal flaking, *Int. J. Fract.*, 29:205–221.

Cotterell, B., and Kamminga, J. (1986) Finials on stone flakes, *J. Arch. Sci.*, 13:451–461.

Cotterell, B., and Kamminga, J. (1987) The formation of flakes, *Amer. Antiquity*, 52:675–708.

Cotterell, B., and Kamminga, J. (1990) *Mechanics of Pre-Industrial Technology*, Cambridge University Press, Cambridge.

Cotterell B,, and Mai, Y.W. (1991) Size effect on the fracture of concrete. In *Fracture of Engineering. Materials and Structures*, Eds. Teoh, S.H., and Lee, K.H., pp. 348–354, London: Elsevier Applied Science, 1991.

Cotterell, B., and Mai, Y.W. (1996) *Fracture Mechanics of Cementitious Materials*, Blackie Academic & Professional, Glasgow.

Cotterell, B., Ong, S.W., and Qin, C. (1995) Thermal shock and size effects in castable refractories, *J. Am. Ceram. Soc.*, 78:2056–2064.

Cotterell, B., and Reddel, J.K. (1977) The essential work of plane stress ductile fracture, *Int. J. Fract.*, 13:267–277.

Cotterell, B., and Rice, J.R. (1980) Slightly curved or kinked cracks, *Int. J. Fract.*, 16:155–169.

Cotterell, B., Sim, M.C., Armrutharaj, G., and Teoh, S.H. (1996) A fracture test method for mode I fracture of thin metal materials, *J. Test. Eval.*, 24:316–319.

Cottrell, A.H. (1958) Theory of brittle fracture in steel and similar metals, *Trans AIME*, 212:192–203.

Cottrell, A.H. (1964) Strong solids, *Proc. Roy. Soc. Lond.*, A282::2–9.

Cottrell, A.H. (1996) Fifty years on the shelf, *Int. J. Pres. Ves. & Piping*, 64:171–174.

Cottrell, A.H., and Hull, D. (1957) Extrusion and intrusion by cyclic slip in copper, *Proc. Roy. Soc. Lond.*, A242:211–213.

Coulomb, C.A. (1776) Sur une application des réglés de maximums et minimes à quelques problèmes de statique relatifs à l'architecture, *Mém. Acad. Sci. Savants Étrangers*, Vol. VII, Paris.

Coulomb, C.A. (1785) Premier-[Troisième] *Mémoire sur l'Electricité et le Magnétisme*, Académie Royale des Sciences, Paris.

Crews, J.H., Shivakumar, K.N., and Raju, I.S. (1988) Factors influencing elastic stresses in double cantilever beam specimens, In *Adhesively Bonded Joints: Testing, Analysis, and Design*, Ed.

Johnson, W.S.,. ASTM STP 981, American Society for Testing and Materials, Philadelphia, pp. 119–132.

Crosley, P. B., and Ripling, E. J. (1969) Dynamic fracture toughness of A533 steel, *J. Basic Engn., Trans. AMSE*, 91:525–534.

Currey, J.D. (1979) Mechanical properties of bone tissues with greatly differing properties, *J. Biomech.*, 12:313–319.

Currey, J.D. (1980) In Symposium XXXIV *The Mechanical Properties of Biological Materials* (Symposia of the Society for Experimental Biology, Cambridge University Press, Cambridge) pp. 75–97.

Currey, J.D. (2002) *Bones: Structure and Mechanics*, Princeton University Press, Princeton.

D6068-96(2002) *Standard Test Method for Determining J-R Curves of Plastic Materials*, ASTM Standard.

Dahlgren, J.A. (1856) *Shells and Shell-Guns*, King and Baird, Philadelphia.

Daniels, H.E. (1945) A statistical theory of the strength of bundles, *Proc. Roy. Soc. Lond.*, A183:405–435.

Dasari, A., Yu, Z.-Z., and Mai, Y.-W. (2007) Transcrystalline regions in the vicinity of nanofillers in polyamide-6, *Macromol.*, 40:123–130.

Davidenkov, N.N. (1936) *Dinamicheskaya ispytanya metallov*, Moscow.

de Castro, P.M.S.T., de Matos, P.F.P., Moreira, P.M.G.P., and da Silva L.F.M. (2007) An overview on fatigue analysis of aeronautical structural details: Open hole, single rivet lap-joint, and lap-joint panel, *Mater. Sci. Engng. A*, 468–470:144–157.

De Forest, A.V. (1936) The rate of growth of fatigue cracks. *Trans. ASME, J. Appl. Mech.*, 58:A23–A25.

de Fréminville, Ch. (1907) Caractères des vibrations accompagnant le choc déduits de l'examen des cassures, *Rev. de Metallurgie*, 4:833–884.

de Fréminville, Ch. (1914) Recherches sur la Fragilité — L' éclatement, *Révue de Metallurgie*, 11:971–1056.

De Morgan, F. (1894) *Fouilles à Dahchow*, Holzhausen, Vienna.

De Solla, D.J. (1959) An ancient Greek computer, *Scientific American*, 200:60–67.

Dean, K., Krstina, J., Tian, W., and Varley, R.J. (2007) Effect of ultrasonic dispersion methods on thermal and mechanical properties of organoclay epoxy nanocomposites, *Macromol. Mater. Eng.*, 292:415–427.

Deng, S., Ye, L., and Friedrich, K. (2007) Fracture behaviours of epoxy nanocomposites with nano-silica at low and elevated temperatures, *J. Mater. Sci.*, 42:2766–2774.

Deshpande, V.S., Needleman, A., and Glessen, Van der, E. (2001) Dislocation dynamics is chaotic, *Scripta Mater.* 45:1047–1053.

Deshpande, V.S., Needleman, A., and Glessen, Van der, E. (2003) Discrete dislocation plasticity modeling of short cracks in single crystals, *Acta Mater.*, 51:1–15.

Deville, S., Saiz, E., Nalla, R.K., and Tomsia, A.P. (2006) Freezing as a path to build complex composites, *Science*, 311:515–518.

Dewar, J., and Hadfield, R.A. (1904) The effect of liquid air temperatures on the mechanical and other properties of iron and its alloys, *Proc. Roy. Soc. Lond.*, 74:326–336.

Dickson F.P. (1981) *Australian Stone Hatchets*, Academic Press, Sydney.

Dieter, G.E. (1961) *Mechanical Metallurgy*, McGraw-Hill, New York.

Dinsmoor, W.B. (1922) Structural iron in Greek Architecture, *Amer. J. Arch.*, 26:148–158.

Dirac, P.A.M. (1929) Quantum mechanics of many-electron systems, *Proc. Roy. Soc. Lond.*, A123:714–733.

Dobb, M.G., Fraser, R.D.B., and MacCrae, T.P. (1967) The fine structure of silk fibroin, *J. Cell Biol.*, 32:289–295.

Dobzhansky, T.G. (1964) Biology, molecular and organismic, *Amer. Zool.*, 4:443–52.

Docherty, J.G. (1932) Bending tests on geometrically similar notched bar specimens, *Engineering*, 133:645–647.

Docherty, J.G. (1935) Slow bending tests on large notched bars, *Engineering*, 133:211–213.

Dodwell, E. (1834) *Views and descriptions of Cyclopian, or Pelasgic, remains in Greece and Italy, with constructions of a later period, from drawings by the late Edward Dodwell.* Intended as a supplement to his classical and topographical tour in Greece during the years 1801, 1805, and 1806, A. Richter, London.

Domanski, M., and Webb, J.A. (1992) Effect of heat treatment on siliceous rocks used in prehistoric lithic technology, *J. Arch. Sci.*, 19:601–614.

Doolittle, T. (1693) *Earthquakes Explained and Improved*, Printed for John Salisbury, London.

Douglas, H. (1855) *A Treatise on Naval Gunnery*, Conway Maritime Press, 4th Edition, reprinted 1982.

Du, M., Guo, B., Cai, X., Jia, Z., Liu, M., and Jia, D. (2008) Morphology and properties of halloysite nanotubes reinforced polypropylene nanocomposites, *e-Polymers*, No. 130.

Dugdale, D.S. (1960) Yielding of sheets containing slits, *J. Mech. & Phys. Solids*, 8:100–104.

Dundurs, J. (1969) Edge-bonded dissimilar orthogonal wedges, *J. Appl. Mech.*, 36:650–652.

Dunnell, R.C., McCutcheon, P.T., Ikeya, M., aned Toyada, S. (1994) Heat treatment of Mill Creek and Dover cherts on Maldon Plains, southeast Missouri, *J.Arch. Sci.*, 21:79–89.

E208-06 (2006) *Standard Test Method for Conducting Drop-Weight Test to Determine Nil-Ductility Transition Temperature of Ferritic Steels*, ASTM Standard.

E399-70T (1970) *Test Method for Plane Strain Fracture Toughness of Metallic Materials*, ASTM Standard.

E1802-06e1 (2006) *Standard Test Method for Measurement of Fracture Toughness*, ASTM Standard.

E1221-06 (2006) *Standard test method for determining plane-strain crack-arrest fracture toughness, K_{Ia}, of ferritic steels*, ASTM Standard.

Editorial Article (1861) Effects of percussion and frost upon iron, *The Engineer*, 18th June.

Editorial Article (1862a) The hydrostatic test for boilers, *The Engineer,* 11th July.

Editorial Article (1862b) Notes of the week, *The Engineer*, 4th July.

Elber, W. (1970) Fatigue crack closure under cyclic tension, Engng. Fract. Mech., 2:37–45.

Elices, M., Pérez-Rigueiro, J., Plaza, G.R., Guinea, G.V. (2005) Finding inspiration in Argiope trifasciata spider silk fibers, *J. Mater.*, 57:60–65.

Ellsberg, E. (1925) Structural damage sustained by the SS Majestic, *Marine Engineering and Shipping News*, 30:430–431.

Elssner, G., Korn, D., and Rühle, M. (1994). The influence of interface impurities on fracture energy of UHV diffusion-bonded metal-ceramic bicrystals, *Scripta Metall. Mater.*, 31:1037–1042.

Emerson, W. (1754) *Mechanics or the Doctrine of Motion*, Knight and Lacey, London.

Engelbach, R. (1922) *The Aswan Obelisk*, L'Institut Français d' Archéology Orientale, Cairo.

England, A.H. (1965) A crack between dissimilar media, *J. Appl. Mech.*, 32:400–402.

Ernsberger, F.M. (1960) Detection of strength-impairing flaws in glass, *Proc. Roy. Soc. Lond.*, A257:213–223.

Erdogan, F.G. (1962) On the stress distribution in plates with collinear cuts under arbitrary loads, *Proceedings of the Fourth US National Congress of Applied Mechanics*, p. 547.

Erdogan, F.G. (1965) Stress distribution in bonded dissimilar materials with cracks, *J. Appl. Mech.*, 32:403–410.

Ernst, H. (1938) *Physics of Metal Cutting: Machining of metals*, American Society for Metals, Cleveland, Ohio.

Ernst, H. (1951) Fundamental aspects of metal cutting and cutting fluid action, *Annals New York Acad. Sci.*, 53:936–961.

Ernst, H., and Merchant, M.E. (1941) Chip formation, friction and high quality machined surfaces, *Surface Treatment of Metals*, ASM, Cleveland, Ohio, pp. 299–378.

Eshelby, J.D. (1949) Uniformly moving dislocations, *Proc. Phys. Soc. Lond. Section* A, 62:307–314.

Eshelby, J.D. (1951) The force on an elastic singularity, *Phil. Trans. Roy. Soc. Lond.*, A244:87–112.

Eshelby, J.D. (1970) Energy relations and the energy momentum tensor in continuum mechanics, In *Inelastic Behaviour of Solids*, eds. Kanninen, M.F., Alder, W.F., Rosenfield, A.R., and Jaffee, R.I., pp. 77–115, McGraw Hill, New York.

Evans, A.G., and Cannon, R.M. (1986) Toughening of brittle solids by martensitic transformations, *Acta Metall.*, 34:761–800.

Evans, A.G., Drory, M.D., and Hu, M.S. 1988 The cracking and decohesion of thin films, *J. Mater. Res.*, 3:1043–1049.

Evans, A.G., and Hutchinson, J.W. (1995) The thermomechanical integrity of thin films and multilayers, *Acta. Metall. Mater.*, 2507–2530.

Evans, A.G., and Hutchinson, J.W. (2009) A critical assessment of theories of strain gradient plasticity, *Acta Mater.*, 57:1675–1688.

Evans, O. (1805) *Abortion of the Young Steam Engineer's Guide*, Philadelphia.

Ewing, J.A. and Humfrey, J.C.W. (1903) The fracture of metals under repeated alternations of stress, *Phil. Trans. Roy. Soc. Lond.*, 200A:241–250.

Ewing, J.A., and Rosenhain, W. (1900) The crystalline structure of metals, *Phil. Trans. Roy. Soc. Lond.*, 193A:353–375.

Faber, K.T., and Evans, A.G. (1983) Crack deflection processes – I Theory, *Acta Metall.*, 3:565–576.

Fairbairn, W. (1850) An experimental inquiry into the strength of wrought-iron plates and their riveted joints as applied to ship-building and vessels exposed to severe strains, *Phil. Trans. Roy. Soc. Lond.*, 140:677–725.

Fairbairn, W. (1864) Experiments to determine the effect of impact, vibratory action, and long continued changes of load on wrought-iron girders, *Phil. Trans. Roy. Soc. Lond.*, 154:311–325.

Farren, L., Shayler, S., and Ennos, A.R. (2004) The fracture properties and mechanical design of human fingernails, *J. Exp. Biology*, 207:735–741.

Featherstone, D. (1974) *Bowmen of England*, New England Library.

Feely, F.J., Hrklo, D., Kleppe, S.R., and Northrup, M.S. (1954) Report on brittle fracture studies, *Weld. J. Res. Sup.*, 33:99s.

Fineberg, J., and Marder, M. (1999) Instability in dynamic fracture, *Physics Reports*, 313:1–108.

Finnie, I. (1956) Review of the metal-cutting analyses of the past hundred years, *Mech. Engng.* 78:715–721.

Finnie, I., and Mayville, R.A. (1990) Historical aspects in our understanding of the brittle transition in steels, *Journal of Engineering Materials and Technology, Trans. ASME,* 112:56–60.

Fleck, N.A., and Hutchinson, J.W. (1993) A phenomenological theory for strain gradient effects in plasticity, *J. Mech. Phys Solids*, 41:1825–1857.

Fleck, N.A., and Hutchinson, J.W. (1997) Strain gradient plasticity. In J.W. Hutchinson and T.Y. Wu (Eds.) *Advances in Applied Mechanics*, Vol. 33. Academic Press, New York pp. 295–361.

Fleck, N.A. Muller, G.M., Ashby, M.F. and Hutchinson, J.W. (1994) Strain gradient plasticity: Theory and experiment, *Acta Metal. Mater.*, 42:475–487.

Foley, R., and Lahr, M.M. (2003) On stony ground: lithic technology, human evolution, and the emergence of culture, *Evol. Anthro.*, 12:109–122.

Forscher, F. (1954) Crack Propagation, *Weld. J. Res. Suppl.*, 33:579s–584s.

Forsyth, P.J.E. (1962) A two stage process of fatigue crack growth, *Proceedings of the Crack Propagation Symposium*, Cranfield, pp. 76–94.

Frank, F.C., and Lawn, B.R. (1967) On the theory of Hertzian fracture, *Proc. Roy. Soc Lond.*, A299:291–306.

Freudenthal, A.M. A random fatigue testing procedure and machine, *Amer. Soc. Test. Mater.*, 53 pp. 896–910.

Frost, N.E., and Dugdale, D.S. (1958) The propagation of fatigue cracks in sheet specimens, *J. Mech. Phys. Solids.*, 6:92–110.

Fujimoto, K., and Shioya, T. (2005) On the wavy crack propagation behaviour in internally pressurized brittle tube, *JSME International Journal*, Series A 48:178–182.

Galilei, G. [1638] (1933) *Dialogues Concerning the Two New Sciences*, Trans. H. Crew and A. de Salvico, Northwestern University Press, Evanston, Illinois.

Gan, Y., and Zhou, B. (2001) Effect of grain size on the fracture toughness of nanocrystalline FeMoSiB, *Scripta Mater.*, 45:625–630.

Gao, H. (1996) A theory of local limiting speed in dynamic fracture, *J. Mech. Phys. Solids*, 44:1453–1474.

Gao, H., Huang, Y., Nix, W.D., and Hutchinson, J.W. (1999) Mechanism based strain gradient plasticity: I Theory, *J. Mech. Phys. Solids*, 47:1239–1263.

Gao, Y.C, Mai, Y.W., and Cotterell, B. (1988) Fracture of fiber-reinforced materials, *Zeit Ange Math. Phys.*, 39:550–572.

Garvie, R.C. (1988) A personal history of the development of transformation toughened PSZ ceramics, *Mat. Sci. Forum*, 34–36:65–77.

Garvie, R.C., Hannik, R.H., and Pascoe, R.T. (1975) Ceramic steel? *Nature*, 258:703–704.

Gassner, E. (1939) Festigkeitsversuche mit wiederholter Beanspruchung im Flugzeugbau, *Deutsche Luftwacht, Ausg. Luftwissen* 6:61–64.

Gassner, E. (1941) *Auswirkung betriebsähnlicher Belastungsfolgen auf die Festigkeit von Flugzeugbauteilen.* Kurzfassung der Dissertation gleichen Titels, Jahrbuch 1941 der deutschen Luftfahrtforschung, S. pp. 1472–1483.

Geber, H. (1874) Bestimmung der zulässigen Spannungen in Eisen-konstructionen, *Zeitschrift des Bayerischen Architeckten und Ingenieur-Vereins*, 6:101–110.

Geller, R.J., Jackson, D.D., Kagan, Y.Y., and Mulargia, F. (1997) Earthquakes cannot be predicted, *Science*, 275:1616–1619.

Gillespie, (1877) On flint cores as tools, *J. Anthro. Inst. Gt. Britain Ireland*, 6:260–263.

Gioia, G., and Ortiz, M. (1997) Delamination of compressed thin films, *Adv. Appl. Mech.*, 33:119–192.

Girard, P.S. (1798) *Traité Analytique de la Résistance des Solides*, Paris.

Giraud, J.L. (1780) *Histoire naturelle de la France méridionale*, Chez C. Belle, Paris.

Glynn, J. (1844) On the causes of fracture of the axles of railway carriages, *Min. Proc. Inst. Civil Engrs. For 1844*, 3:201–203.

Goh, S.M., Charalambides, M.N., and Williams, J.G. (2005) On the mechanics of wire cutting cheese, *Eng. Fract. Mech.*, 72:931–946.

Gojny, F.H., Wichmann, M.H.G., Köpe, U., Fiedlera, B., and Schulte, K. (2004) Carbon nanotube-reinforced epoxy-composites: enhanced stiffness and fracture toughness at low nanotube content, *Comp. Sci. Tech.*, 64:2363–2371.

Gol'dstein, R.V., and Salganik, R.L. (1974) Brittle fracture of solids with arbitrary cracks. *Int J Fract.*, 10:507–523.

Gologanu, M., Leblond, J.-B., Perrin, G., and Devaux, J., (1995) Recent extensions of Gurson's model for porous ductile metals. In *Continuum Micromechanics*, Ed. Suquet, P., Springer-Verlag, pp. 61–130.

Goodier, J.N., and Field, F.A. (1962) Plastic energy dissipation in crack propagation, In *Fracture of Solids*, eds Drucker, D.C. and Gilman, J.J., Interscience Publishers, pp. 103–117.

Goodman, J. (1899) *Mechanics Applied to Engineering*, Longman and Green, London.

Gordon, J.E. (1964) Some considerations in the design of engineering materials based on brittle solids, *Proc. Roy. Soc. Lond.*, A:282:16–23.

Gordon, J.E. (1968) *The New Science of Strong Materials, or Why You Don't Fall Through the Floor*, Penguin Books Ltd., Harmondsworth.

Gordon, J.E. (1978) *Structures, or Why Things Don't Fall Down*, Penguin Books Ltd., Harmondsworth.

Gordon, J.E., and Jeronimides, G. (1974) Work of fracture of natural cellulose, *Nature*, 252:116.

Gordon, J.E., and Jeronimides, G. (1980) Composites with high work of fracture, *Phil. Trans. Roy. Soc. Lond.*, A294:545–550.

Gordon, R.B. (1988) Strength and structure of wrought iron, *Archeomaterials*, 2:109–137.

Gotti, E., Oleson, J.P., Bottalico, L., Brandon, C., Cucitore, R., and Hohlfeder, R.L. (2008) A comparison of the chemical and engineering characteristics of ancient Roman hydraulic concrete with a modern reproduction of Vitruvian hydraulic concrete, *Archaeometry*, 50:576–590.

Gurson, A.L., 1977. Continuum theory of ductile rupture by void nucleation and growth, Part I, *J. Engng. Mater. Tech. – Trans. ASME*, 99:2–15.

Gough, H.J. (1924) *The Fatigue of Metals*, Ernest Benn Ltd, London.

Gough, H.J. (1933) Crystalline structure in relation to failure of metals especially by fatigue, *Proc. Amer. Soc. Test. Mat.*, 33, Part II, 3–114.

Gough, H.J., and Sopwith, D.G. (1932) Atmospheric action as a factor in fatigue of metals, *Journal of the Institute of Metals*, 49:93–122.

Gramberg, J. (1965) Axial cleavage, a significant process in mining and geology, *Engng. Geol.*, 1:31–72.

Greaves, R.H. and Jones, J.A. (1925) The effect of temperature on the behaviour of iron and steel in the notched-bar impact t test, *J. Iron & Steel Inst.*, 62:123–165.

Greene, T.W. (1949) Evaluation of the effect of residual stresses, *Weld. J. Res. Suppl.*, 28:193s–211s.

Greenwell, W. (1865) Notices of the examination of ancient grave-hills in the North Riding of Yorkshire, *Arch. J.*, 22:95–105.

Grenet, L. (1899) The mechanical strength of glass (in French), *Bull. Soc. d'Encouragement de l'Industrie, Nat.*, Paris (Ser. 5), 4:838–848.

Grenet, L. (1934) The mechanical strength of glass, Trans. Preston F.W., *Glass Industry*, 15:277–280.

Griffith, A.A. (1920) The phenomena of rupture and flow in solids, *Phi. Trans. Roy. Soc. Lond.*, 221:163–198.

Griffith, A.A. (1924) The theory of rupture, In *Proc. First Int. Congr. Appl. Mech.*, Technische Boekhandel en Drukkerij, Delft, eds. Biezono, C.B., and Burgers, J.M., pp. 55–623,

Griffith, A.A., and Taylor, G.I. (1917) The use of soap films in solving torsion problems, *Proc. Inst. Mech. Engrs.*, 93:755–789.

Grunzweig, J., Longman, I.M., and Petch, N.J. (1954) Calculations and measurement on wedge-indentation, *J. Mech. Phys. Solids*, 2L:82–84.

Guin, J.-P., and Wiederhorn, S.M. (2005) Crack-tip structure in soda–lime–silicate glass, *J. Am. Ceram. Soc.*, 88:652–659.

Guin, J.-P., and Wiederhorn, S.M. (2006) Surfaces formed by subcritical crack growth in silicate glasses, *Int. J. Fract.*, 140:15–26.

Gupta, T.K. (1972) Strength degradation and crack propagation in thermally shocked $Al2O_3$, *J. Am. Ceram. Soc.*, 55:249–53.

Gurney, C. (1994) Continuum mechanics and thermodynamics in the theory of cracking, *Phil. Mag. A*, 69:33–43.

Gurney, C., and Hunt, J. (1967) Quasi static crack propagation, *Proc. Roy. Soc. Lond.*, A299:508–524.

Gurney, C., and Mai, Y.W. (1972) Stability of cracking, *Eng. Fract. Mech.*, 4:853–863.

Gurney, G., and Ngam, K.M. (1971) Quasi-static crack propagation in non-linear structures, *Proc. Roy. Soc. Lond.*, A325, 207–222.

Gurney, C, and Pearson, S. (1948) Fatigue of mineral glass under static and cyclic loading, *Proc. Roy. Soc. Lond.*, A192:537–544.

Gurson, A.L. (1977) Continuum theory of ductile rupture by void nucleation and growth. 1. Yield criteria and flow rules for porous ductile media, *J. Eng. Mater. Techn.*, 99:2–15.

Hahn, G.T., and Rosenfield, A.R. (1964) Local yielding and extension of a crack under plane stress, *Ship Structure Committee Report*, SSC–165.

Hahn, G.T., Hoagland, R.G., Kanninen, M.F., and Rosenfield, A.R. (1973) A preliminary study of fast fracture and arrest in the DCB test specimen, In *Proc. Int. Conf. Dyn. Crack Prop.*, ed Sih, G.C., pp. 649–662.

Haigh, B.P. (1917) Experiments on the fatigue of brasses, *J. Inst. Metals*, 18:55–77.

Haimson, B. (2006) True triaxial stresses and the brittle fracture of rock, *Pure Appl. Geophys.*, 163:1101–1130.

Hall, A.R. (1956) Military technology, In *A History of Technology*, Eds. Singer, C., Holmyard, E.J., Hall, A.R., and Williams, T.J., Vol. II pp. 695–730.

Hall, E.O. (1951) The deformation and ageing of steel: III Discussion of results, *Proc. Phys. Soc.*, B64:747–753.

Hall, W.J., and Barton, E.W. (1963) Summary of some studies of brittle-fracture propagation, *Ship Structure Committee*, SSC–149.

Hall, W.J., Kihara, H., Soete, W., and Wells, A.A. (1967) *Brittle Fracture of Welded Plate*, Prentice-Hall, Englewood Cliffs, N.J.

Hallbauer, D.K., Wagner, H., and Cook, N.G.W. (1973) Some observations concerning the microscopic and mechanical behaviour of quartzite in stiff triaxial compression tests, *Int. J. Rock Mech. Min. Sci., Geomech., Abstr.*, 10:713–726.

Hannink, R.H.J., and Swain, M.V. (1986) Particle toughening in partially stabilized zirconia: influence of thermal history, In *Tailoring of Multiphase and Composite Ceramics* York, *Conference*, Eds. Messing, G.L., Pantano, C.G., and Newman, R.E., Plenum Press, New York, p. 259–265.

Hardy, B.L., and Raff, R.A. (1997) Recovery of mammalian DVA from Middle Paleolithic stone tools, *J. Arch. Sci.*, 24:601–611.

Hardy, R. (1976) *Long bow: A Social and Military History*, Patrick Stephens, Cambridge.

Harrison, J.D., Burdekin, F.M., and Young, J.G. (1968) *Second Conference on the Significance of Defects in Welded Structures*, Weld. Inst. London.

Harrison, R.P., Loosemore, K., and Milne, I. (1977) CEGB Report no. R/H/RG-Rev. 1.

Hartbower CE, and Pellini WS, (1951) Explosion bulge test studies of deformation of weldments, *Weld. J. Res. Suppl.*, 30:307s–318s.

Hasselman D.P.H. (1963) Elastic energy at fracture and surface energy as design criteria for thermal shock, *J. Am. Ceram. Soc.*, 46:535–540.

Hassig, R. (1988) *Aztec Warfare: Imperial Expansion and Political Control*, University of Oklahoma Press, Norman, Oklahoma.

Head, A.K. (1953a) The mechanism of fatigue in metals, *J. Mech. Phys. Solids*, 1:134 to 141.

Head, A.K. (1953b) The growth of fatigue cracks, *Phil. Mag.*, 44:925–938.

Head, A.K., and Hooke, F.H. (1956) Random noise fatigue testing, *Proceedings of the International Conference on Fatigue of Metals*, Institution of Mechanical Engineers, London, pp. 301–303.

Herodotus, *The Histories*, Trans. De Selincourt, 1977, Penguin Books, Harmondsworth.

Hess H.H. (1962) History of ocean basins. In eds. A.E.J. Engel, H.L. James, and B.F. Leonard, *Petrologic studies: A Volume to Honor A.F. Buddington*, New York: Geological Society of America, pp. 599–620.

Heyman, J. (1966) The stone skeleton, *Int. J. Solids Struct.*, 2:249–279.

Heyman, J, (1967) On shell solutions for masonry domes, *Int. J. Solids Struct.*, 3:227–241.

Heyman, J. (1969) The safety of masonry arches, *Int. J. Mech. Sci.*, 11:365–385.

Heyman J. (1972) 'Gothic' construction in ancient Greece, *J. Soc. Arch. Hist.*, 31:3–9.

Heyman, J. (1988) Poleni's problem, *Proc. Instn, Civ. Engrs., Part 1*, 84:737–759.

Heyman, J. (1998) *A Structural Analysis: A Historical Approach*, Cambridge University Press, Cambridge.

Hertz, H. (1881) *J. Reine Angew. Math.* 92:156.

Hertz, H. (1882) *Verhandlungen des Verins zur Bedföderung des Gewerbe Fleisses*, 61:449. These two papers by Hertz are reprinted in English In *Hertz Miscellaneous Papers*, Macmillan, London 1896 Chs. 5,6.

Hill, R. (1950) *The Mathematical Theory of Plasticity*, Clarendon Press, Oxford.

Hill, R., Lee, E.H., and Tupper, S.J. (1947) The theory of wedge indentation of ductile materials, *Proc. Roy. Soc. Lond.*, A188:273–289.

Hillerborg, A., Modéer, M., and Petersson, P.E. (1976) Analysis of crack formation in concrete by means of fracture mechanics and finite elements, *Cement Concr. Res.*, 6:773–781.

Hillerton, J.E., Reynolds, S., and Vincent, J.F.V. (1982) On the indentation hardness of insect cuticle, *J. Exp. Biology*, 96:45–52.

Hoagland, R.G., Rosenfield, A.R., Gehlen, P.C., and Hahn, G.T. (1977) A crack measuring technique for K_{Im}, K_{ID}, and K_{Ia} properties, *ASTM STP* 627, pp. 177–202.

Hodgkinson, E. (1838) On the relative strength and other properties of cast iron obtained by hot and cold blast, *Report of the Seventh (or Liverpool) Meeting of the British Association*.

Hodgson, J., and Boyd, G.M. (1958) Brittle fracture in ships; an empirical approach from recent experience, *Trans. Inst. Naval Arch.*, 100:141–180.

Hoek, E., and Brown, E.T. (1980) *Underground excavations in rock, Inst. Min. Metall.*, London.

Hoek, E., and Brown, E.T. (1997) Practical estimates of rock mass strength, *Int. J. Rock Mech. Min. Sci.*, 34:1165–1186.

Hoffman, M.J., Mai, Y.W., Wakayama, S., Kawahara, M., and Kishi, T. (1995) Crack-tip degradation processes observed during *in situ* cyclic fatigue of partially stabilized zirconia, *J. Am. Ceram. Soc.*, 78:2801–2810.

Hohenberg, P., and Kohn, W. (1964) Inhomogeneous electron gas, *Phys. Rev.*, 136:B864–B871.

Holloman, J.H., and Zener, C. (1946) Problems in the fracture of metals, *J. Appl. Phys.*, 17:82–90.

Holloway, D.G. (1959) The strength of glass fibres, *Phil. Mag. Ser.* 8, 4:1101–1106.

Holloway, D.G., and Hastilow, P.A.P. (1961) High strength glass, *Nature*, 189:385–386.

Hood, C. (1842) On some peculiar changes in the internal structure of iron, independent of, and subsequent to, several processes of its manufacture, *Min. Proc. Inst. Civil Engrs.*, 2:180–181.

Hooke, R. 1665 *Micrographia or Some Physiological Descriptions of Minute Bodies made by Magnifying Glasses with Observations and Inquiries thereupon*, J. Martin and J. Allestry, London.

Hooke, R. (1678) *Lectures de Potentia Resititutiva; or, Of Spring, Explaining the Power of Springing Bodies: to Which are Added some Collections*, John Martyn, London.

Hopkinson, B. (1912a) A high-speed fatigue-tester, and the endurance of metals under alternating stresses of high frequency, *Proc. Roy. Soc. Lond.*, A86:131–149.

Hopkinson, B. (1912b) The pressure of a blow, In *The Scientific Papers of Bertram Hopkinson*, eds. Ewing, J.A., and Larmor, J., Cambridge University Press, 1921, pp. 423–437.

Hu, X.Z., Mai, Y.W., and Cotterell, B. (1988) A statistical theory of time-dependent fracture for brittle materials, *Phil. Mag.*, 58:299–324.

Hull, D., and Clyne, T.W. (1996) *An Introduction to Composite Materials*, Cambridge University Press, Cambridge, 2nd edition.

Humphrey, J.W., Oleson, J.P., and Sherwood, A.N. (1998) *Greek and Roman Technology: A Source Book*, Routledge, London.

Hunt R.A., and McCartney, L.N. (1979) A new approach to Weibull's statistical theory of brittle fracture, *Int. J. Fract.*, 15:165–375.

Hutchinson, J.W. (1968) Singular behaviour at the end of a tensile crack in a hardening material, *J. Mech and Phys. Solids*, 16:13–64.

Hutchinson, J.W. (1983) Fundamentals of the phenomenological theory of nonlinear fracture mechanics, *J. Appl. Mech.*, 50:1042–1051.

Hutchinson, J.W. (1996) *Stresses and Failure Modes in Thin Films and Multilayers: Notes for a DCAMM Course*, Technical University of Denmark.

Hutchinson, J.W. (2001) Delamination of compressed films on curved substrates, *J. Mech. Phys. Solids*, 49:1847–1864.

Hutchinson, J.W., and Evans, A.G. (2000) Mechanics of materials: top-down approaches to fracture, *Acta Mater.* 48:125–135.

Hutchinson, J.W., and Paris, P.C. (1979) Stability analysis of J-controlled crack growth, In *Elasto-Plastic Fracture*, Ed. Landes, J.D., Begley, J.A., and Clarke, G.A., ASTM STP 668, American Society for Testing and Materials, pp. 37–64.

Hutchinson, J.W., and Suo, Z. (1992) Mixed mode cracking in layered materials, *Adv. Appl. Mech.*, 29:63–189.

Iijima, S. (1991) Helical microtubules of graphite carbon, *Nature*, 354:56–58.

Inglis, C.E. (1913) Stresses in a plate due to the presence of cracks and sharp corners, *Trans. Inst. Naval Arch.*, 55:291–230.

Irwin, G.R. (1948) Fracture dynamics, In *Fracturing of Metals*, Symposium of the American Society of Metals, Cleveland, Ohio, 1947. pp. 147–166.

Irwin, G.R. (1956) Onset of fast crack propagation in high strength steel and aluminum alloys, *Sagamore Research Conference Proceedings*, Vol 2, pp. 289–305.

Irwin, G.R. (1957) Analysis of stresses and strains near the end of a crack traversing a plate, *Trans. ASME, J. Appl. Mech.*, 24:361–364.

Irwin, G.R. (1958) Fracture, In *Encylopedia of Physics*, Vol V1 – *Elasticity and Plasticity*, Ed. S. Flügge, Springer-Verlag, Berlin, pp. 551–590.

Irwin, G.R. (1960) Fracture mechanics, In *Structural Mechanics, First Symposium on Naval Structural Mechanics*, August 1958, Stanford University, Pergamon Press, pp. 557–594.

Irwin, G.R. (1964) Dimensional and geometric aspects of fracture, In *Fracture of Engineering Materials*, Papers presented at a Conference held at Rensselaer Polytechnic Institute, August 1959. pp. 211–230.

Irwin, G.R., and Kies, J.A. (1952) Fracturing and fracture dynamics, *Weld. J. Res. Suppl.*, 31:95s–100s.

Irwin, G.R., and Kies, J.A. (1954) Critical energy release rate analysis of fracture strength, *Weld. J. Res. Suppl.*, 33:193s–198s.

Irwin, G.R., Kies, J.A., and Smith, H.L. (1958) Fracture strengths relative to onset and arrest of crack propagation, *Proc. ASTM,* 58:640–660.

Irwin, G.R., and Srawley, J.E. (1962) Progress in the development of crack toughness tests, *Materialprüfung*, 4.

Irwin, G.R., and Wells, A.A. (1965) A continuum mechanics view of crack propagation, *Metall. Rev.*, 10:223–270.

Ito, H., (1979) Rheology of the crust based on long-term creep tests, *Tectonophysics*, 52:624–641.

Ito, M., Mizuochi, K., and Kanamoto, T. (1998) Effects of crystalline forms on the deformation of nylon 6, *Polymer* 39:4593–4598.

Izod, E.G. 1903) Testing brittleness of steel, *Engineering*, 76:431–432.

Jackson, A.P., Vincent, J.F.V., and Turner, R.M. (1988) The mechanical design of nacre, *Proc. Roy. Soc. Lond.*, B234:415–440.

Jackson, D.C., and Hundley, N. (2004) Privilege and responsibility: William Mulholland and the St Francis Dam disaster, *California History*, 82:8–47.

Jaeger, J.C., and Cook, N.G.W. (1979) *Fundamentals of Rock Mechanics*, 3rd Edition, Chapman and Hall, London.

Jenkin, C.F. (1925) High-frequency fatigue tests, *Proc. Roy. Soc. Lond.*, 109:119–143.

Jenkin, C.F., and Lehmann, G.D. (1929) High frequency fatigue tests, *Proc. Roy. Soc. Lond.*, 125:83–119.

Jiang, H., Huang, Y., Zhuang, Z., and Hwang, K.C. (2001) Fracture in mechanism-based strain gradient plasticity, *J. Mech Phys. Solids*, 49:979–993.

Joffé, A., Kirpitschewa, M.W., and Lewitzky, M.A. (1924) Deformation und Festigkeit der Kirstalle, *Zeitschrift fur Physik.* 22:286–302.

Johnsen, B.B., Kinloch, A.J., Mohammed, R.D., Taylor, A.C., and Sprenger, S. (2007) Toughening mechanisms of nanoparticle-modified epoxy polymers, *Polymer*, 48:530–541.

Johnson, W. (1986) Mostly on oak targets and 19th century naval gunnery, *Int. J. Impact Engng.*, 4:175–183.

Johnson, W. (1988) Some conspicuous aspects of the century of rapid changes in battleship armours, ca 1845–1945, *Int. J. Impact Engng.*, 7:261–284.

Johnson, W. (1995) Contents and commentary on William Moore's A Treatise on the Motion of Rockets and an Essay on Naval Gunnery, *Int. J. Impact Engng.*, 16:499–521.

Johnson, W.R. (1832) Remarks on the strength of cylindrical steam boilers, *Journal of the Franklin Institute*, 14:149–154.

Kahn, N.A., and Imbemko, E.A. (1948) A method of evaluating the transition from shear to cleavage-type failure in ship plate. *Weld. Journal Res. Suppl.,*. 27:169s–182s.

Kalthoff, J.F. (1985) On the measurement of dynamic fracture toughnesses – a review of recent work, *Int. J. Fract.* 27:277–298.

Kambour, R.P. (1962) Optical properties and structure of crazes in transparent glassy polymers, *Nature*, 195:1299–1300.

Kambour, R.P. (1964) Refractive index and composition of poly(methyl methacrylate) fracture surface layers, *J. Poly. Sci.*, A2:4165–4168.

Kambour, R.P. (1966) Mechanism of fracture in glassy polymers. III. Direct observation of the craze ahead of the propagating crack in poly(methyl methacrylate) and polystyrene, *J. Poly. Sci.*, A–2, 4:349–358.

Kambour, R.P. (1973) A review of crazing and fracture in thermoplastics, *J. Poly. Sci, Macromo. Rev.*, 7:1–154.

Kamminga, J. (1985) The pirri graver, *Aust. Aboriginal Studies*, II:2–25.

Kamminga, J. (1982) *Over the Edge. Functional Analysis of Australian Stone Tools*, Occasional; papers in Anthropology, No. 12, Anthropology Museum, Queensland University.

Kamyab, I., Chakrabarti, S., and Williams, J.G. (1998) Cutting cheese with wire, *J. Mater. Sci.*, 33:2763–2770.

Kaplan, M.F. (1961) Crack propagation and the fracture of concrete, *J. Am. Concrete Inst. Proc.*, 58:591–603.

Kaspi, M.A., and Gosline, J.M. (1997) Design complexity and fracture control in the equine hoof wall, *J. Exp. Biology*, 200:1639–1659.

Kausch, H.-H. and Michler G.H. (2005) The effect of time on crazing and fracture, *Adv. Poly. Sci.*, 187:1–33.

Keckes, J., Burgert, I., Frühmann, K., Müller, M., Kölln, K., Hamilton, M., Burghammer, M., Roth, S.V., Stanzl-Tschegg, S., and Fratzl, P. (2003) Cell-wall recovery after irreversible deformation of wood, *Nature Materials*, 2:810–814.

Kelly, A. (1970) Interface effects and the work of fracture of a fibrous composite, *Proc. Roy. Soc. Lond.*, A319:95–116.

Kelly, A. (2005) Very stiff fibres woven into history – very personal recollections of some of the British scene, *Comp. Sci. Techn.* 65:2285–2294.

Kelly, A., and Macmillan, N.H. (1986) *Strong Solids*, Clarendon Press, Oxford.

Kelly, A., and Tyson, W.R. (1965) Tensile properties of fibre-reinforced metals: copper/tungsten and copper/molybdenum, *J. Mech. Phys. Solids*, 13:329–350.

Kelly, A., Tyson, W.R., and Cottrell, A.H. (1967) Ductile and brittle crystals, *Phil. Mag.*, 15:567–586.

Kelly, P.M., and Rose, L.R.F. (2002) The martensitic transformation in ceramics and its role in transformation toughening, *Prog. Mater. Sci.*, 47:463–557.

Kendall, K., Howard, A.J., and Birchall, J.D. (1983) The relationship between porosity microstructure and strength, and the approach to advanced cement-based materials, *Phil. Trans. Roy. Soc. Lond.*, A310:139–153.

Kennedy, H.E. (1945) Some causes of brittle failure in welded mild steel structures, *Weld. J. Res. Suppl.*, 24:588s–597s.

Kerkhof, F. (1953) Analyse des sproden Zugbruches von Glasern Ultraschall, *Naturweis.*, 40, p. 478.

Kerkhof, F. (1973) Wave fractographic investigations of brittle fracture dynamics, In *Dynamic Crack Propagation*, ed. Sih, G.C., Noordof, Leyden, pp. 3–35.

Kick, F. (1901) Zur Frage der Wirkungsweise des Taylor-White und des Bohler-Rapid-Stahles, *Baumaterialenkunde*, 6:227–229.

Kies, J.A., Smith, H.L. (1955) Toughness testing of hot stretched acrylics, *Proceedings of the Aircraft Industries Association and Air Research and Development Command Joint Conference*, Dayton.

Kies, J.A., Sullivan, A.M., and Irwin, G.R. (1950) Interpretation of fracture markings, *J. Appl. Phys.*, 21:716–720.

Kingery, W.D. (1955) Factors affecting thermal stress resistance of ceramic materials, *J. Amer. Ceram. Soc.*, 38:3–15.

Kinloch, A.J. (1987) *Adhesion and Adhesives*, Chapman and Hall, London.

Kinloch, A.J. (2000) Norman Adrian De Bruyne, *Biog. Mem. Fellows Roy. Soc.*, 46:127–143.

Kinloch, A.J., Maxwell, D.L., and Young, R.J. (1985) The fracture of hybrid-particulate composites, *J. Mater. Sci.*, 20:4169–4184.

Kinloch, A.J., Mohammed, R.D., Taylor, A.C., Eger, C., Sprenger, S., and Egan, D. (2005) The effect of silica nano particles and rubber particles on the toughness of multiphase thermosetting epoxy polymers, *J. Mater. Sci.*, 40:5083–5086.

Kinloch , A.J., and Taylor, A.C. (2006) The mechanical properties and fracture behaviour of epoxy-inorganic micro- and nano-composites, *J. Mater. Sci.*, 4:3271–3297.

Kinloch, A.J., and Williams, J.G. (1980) Crack blunting mechanisms in polymers, *J. Mat. Sci.*, 15:987–996.

Kinloch, A.J., and Young, R.J. (1983) *Fracture Behaviour of Polymers*, Applied Science Publishers, London.

Kirkaldy D. (1862) *Results of an Experimental Inquiry into the Tensile Strength and other Properties of various kinds or Wrought-Iron and Steel*, Published by the author, Glasgow.

Kirsch, G. (1898) Die Theorie der Elastzität die Bedürfuisse der Festigheitslehre, *Zeitschrift des Vereines deutscher Ingenieure*, 42:797–807.

Kitchener, A. (1987) Fracture toughness of horns and a reinterpretation of the horning behaviour of bovids, *J. Zoology, London*, 213:621–639.

Kitey, E.R., and Tippur, H.V. (2005) Role of particle size and filler–matrix adhesion on dynamic fracture of glass-filled epoxy. II. Linkage between macro- and micro-measurements, *Acta Mater.*, 53:1167–1178.

Klemm, D.D., and Klemm, R. (2001) The building stones of ancient Egypt – a gift of its geology, *African Earth Sci.*, 33:631–642.

Koehler, J.S. (1941) On the dislocation theory of plastic deformation, *Phys. Rev.*, 60:397–410.

Kohn, W., and Sham, L.-J. (1965) Self-consistent equations including exchange and correlation effects, *Phys. Rev.*, 140:A1133–A1138.

Kornmann, X., Thomann, R., Mülhaupt, R., Finter, J., and Berglund, L.A. (2002) High performance epoxy-layered silicate nanocomposites, *Poly. Engng. Sci.*, 42:1815–1826.

Kraft, J.M., Sullivan, A.M., and Boyle, R.W. (1961) Effect of dimensions on fast fracture instability of notched sheets, *Proceedings of the Crack Propagation Symposium*, Cranfield, pp. 8–28.

Krohn, D.A., and Hasselman, D.P.H. (1972) Static and cyclic fatigue behaviour of a polycrystalline alumina, *J. Am. Ceram. Soc.*, 55:208–211.

Kubin, L.P., Canova, G., Conat, M., Devincre, B., Pontikis, V., and Brechet, Y., Dislocation microstructures and plastic flow: a 3D simulation, *Solid State Phen.*, 1992, 23–24:455–472.

Kukta R V (1998) *PhD Thesis*, Brown University, Providence.

Kumar, K.S., Swygenhoven, H. Van, and Suresh, S. (2003) Mechanical behavior of nanocrystalline metals and alloys, *Acta Mater.*, 51:5743–5774.

Kuntze, W. (1928) Der Bruch gerberten Zugpoben, *Archiv Eisenhüntenwesen*, 2:109–111.

Kuntze, W. (1932) Kohäsionfestigkeit, *Mitt. Mat. Prüf. Anstalten*, 20:1–61.

Labour, T., Vigier, G., Séguéla, R., Gauthier, C., Orange, G., and Bomal, Y. (2002) Influence of the β-crystalline phase on the mechanical properties of unfilled and calcium carbonate-filled polypropylene: ductile cracking and impact behaviour, *J. Poly. Sci. Part B: Poly. Phys.* 40:31–42.

Lamé, G., and Clapeyron, B.P (1831) .*Mémoire sur l'équilibre intérieur des corps solides homogènes*, 7.

Lake, G.J. (1970) Application of fracture mechanics to failure in rubber articles, *Proc. Conf. On Yield, Deformation and Fracture of Polymers*, Institute of Physics and the Physical Society.

Lake, G.J., and Yeoh, O.H. (1978) Measurement of rubber cutting resistance in the absence of friction, *Int. J. Fract.* 14:509–526.

Lan, T., Kaviratna, P.D., and Pinnavaia, T.J. (1995) Mechanism of clay tactoid exfoliation in epoxy-clay nanocomposites, *Chem. Mater.*, 7:2144–2156.

Lane, M., Dauskardt, R.H., Vainchtein, A., and Gao, H.J. (2000) Plastic contributions to interface adhesion in thin-film interconnect structures, *J. Mater. Res.*, 15:2758–2769.

Lange, F.F. (1970) Interaction of a crack front with a second phase dispersion, *Phil. Mag.*, 22:983–992.

Lauterwasser, B.D., and Kramer, E. (1979) Microscopic mechanisms and mechanics of craze growth and fracture, *Phil. Mag.*, 39:469–495.

Lawn, B.R. (1993) *Fracture of Brittle Solids*, Cambridge University Press, Cambridge.

Lawn, B.R., and Wilshaw, T.R. (1975) Indentation fracture: principles and applications, *J. Mater. Sci.*, 10:1049–1081.

Layton, E. (1971) Mirror-image twins: The communities of science and technology in 19th-century America, *Technology and Culture*, 12:562–580.

Lazar, R., and Hall, W.J. (1959) Studies of the brittle fracture propagation in six-foot wide structural steel plates, *Ship Structure Committee Report*, SSC–112.

Le Chatelier, A. (1892) *On the Fragility after Immersion in a Cold Fluid*, French Testing Commission, Vol. 3.

Leicester, R.H. (1974) Fracture strength of wood, *First Australian Conf. Engng. Mater.*, University of NSW, pp. 729–742.

Lévy, M. (1871) Extrait du mémoire sur les équations générales des mouvements intérieurs des corps solides ductiles au-delà des limites où l'élasticité pourrait les ramener à leur premier état, *Journal de Mathématiques*. 16:368–372.

Lewis, J.A., and Boyer, M.A. (1995) Effects of an organotitanate cross-linking additive on the processing and properties of macro-defect-free cement, *Advn. Cem. Bas. Mat.*, 2:2–7.

Lim, S.-H., Dasari, A., Yu, Z.-Z., Mai, Y.-W., Liu, S., and Yong, M.S. (2007) Fracture toughness of nylon 6 organoclay/elastomer nanocomposites, *Comp. Sci. Techn.*, 67:2914–2923.

Lindley, P.B. (1972) Energy for growth in model rubber components, *J. Strain Anal.*, 7:132–140.

Lindley, T.C., Richards, C.E., and Ritchie, R.O. (1975) The mechanics and mechanisms of fatigue crack growth in metals, In *The Mechanics and Physics of Fracture*, Metals Society/Institute of Physics Joint Meeting, Cambridge.

Liu, L., Qi, Z., and Zhu, X. (1999) Studies on nylon 6/clay nanocomposites by the melt-intercalation process. *J. Appl. Poly. Sci.*, 71:1133–1138.

Love, A.E.H. (1906) *A Treatise on the Mathematical Theory of Elasticity*, 2nd edition, Cambridge University Press, Cambridge.

Lu, G., Kioussis, N., Bulatov, V.V., and Kaxiras, E. (2000) Generalized-stacking-fault energy surface and dislocation properties of aluminum, *Phys. Rev.*, B62:3099–3108.

Lubahn, J.D. (1959) Experimental determination of energy release rate for notch bending and notch tension, *Proc. ASTM*, 59:885–915.

Lucas , A., and Harris, JR. (1962) *Ancient Egyptian Materials and Industries*, 4th ed., Edward Arnold, London Lucas, P.W. (2004) *Dental Functional Morphology: How Teeth Work*, Cambridge University, Cambridge.

Lucas, P.W., Darvell, B.W. Lee, P.K.D., and Yuen, T.D.B. (1995), The toughness of plant cell walls, *Phil Trans. Roy. Soc. Lond.*, B348:363–372.

Lucas, P.W., and Pereira, B. (1990) Estimation of the fracture toughness of leaves, *Functional Ecology*, 4:819–822.

Lucas, P.W., Turner, I.M., Dominy, N.J., and Yamashita, N. (2000) Mechanical Defences to Herbivory, *Ann. Bot.* 86:913–920.

Ludwik, P. (1909) *Elemente der Technologischen Mechanik*, Springer, Berlin.

Ludwik, P. (1927) Die Bedeutung des Gleit- und Reisswiderstandes für die Werkstoffprüfung, *Zeitscrift des Vereines deutscher Ingenieure*, 71:1532–1538.

Ludwik, P., and Scheu, R. (1923) Kerbwirkugen bei Flusseisen, *Stahl und Eisen*, 43:999–1001.

Lyell, C. (1830–33) *Principles of Geology*, Selected text published by Penguin, 1997, Harmondsworth.

MacAyeal, D.R., Okal, E.A., Aster, R.C., Bassis, J.N., Brunt, K.M., Cathless, L.M., Drucker, R., Fricker, H.A., Kim, Y.-J., Martin, S., Okal, M.H., Sergienko, OV., Sponsler, M.P., and Thom, J.E. (2006) Transoceanic wave propagation links iceberg calving margins of Antarctica with storms in the tropics and Northern Hemisphere, *Geophys. Res. Letters*, 33:L17502.

McBain, J.W., and Lee, W.B. (1927) Adhesives and adhesion: gums, resins, and waxes between polished metal surfaces, *J. Phys. Chem.* 31:1674–1680.

McCarthy, C.T., Hussey, M., and Gilchrist, M.D. (2007) On the sharpness of straight edge blades in cutting soft solids: Part I – indentation experiments, *Eng. Fract. Mech.*, 74:2205–2224.

McClintock, F.A. (1968) Local criteria for ductile fracture, *Int. J. Fract. Mech.*, 4:102–130.

McClintock, F.A., and Irwin, G.R. (1965) Plasticity aspects of fracture mechanics, In *Fracture Toughness Testing and Its Applications*, ASTM STP 381, pp. 84–110.

McClintock F.A., Walsh J.B. (1962) Friction on Griffith cracks in rocks under pressure, *Proc. 4th US Nat. Congr. Appl. Mech.*, 1015–21.

McConnell, J.E. (1849) On railway axles, *Min. Proc. Inst. Mech. Engrs (1847–1849)*, pp. 13–27.

MacDonald, W.L. (1982) *The Architecture of the Roman Empire: An Introductory Study*, Yale University Press, New Haven.

McGarry, F.J. (1970) Building design with fibre reinforced materials, *Proc. Roy. Soc. Lond.*, A319:59–68.

McMeeking, R.M., and Evans A.G. (1986) Mechanics of transformation-toughening in brittle materials, *J. Am. Ceram. Soc.*, 65:242–246.

McMullen, P. (1984) Fibre/resin composites for aircraft primary structures: a short history, 1936–1984, *Composites*, 15:222–230.

Mai, Y.W., Barakat, B., Cotterell, B., and Swain, M.V. (1990) R-curve behaviour in a macro-defect-free cement paste, *Phil. Mag.*, 62:347–361.

Mai, Y.W., and Lawn, B.R. (1987) Crack-interface grain bridging as a fracture resistance mechanism in ceramics: II Theoretical fracture mechanics model, *J. Am. Ceram. Soc.*, 70:289–294.

Mallock, A. (1881) The action of cutting tools, *Proc. Roy. Soc. Lond.*, 33:127–139.

Manson, S.S. (1954) Behaviour of materials under conditions of thermal stress, *NACA Rpt.* 1170, Lewis Flight Propulsion Laboratory, Cleveland.

Mark, R. and Hutchinson, P. (1986) On the structure of the Roman Pantheon, *The Art Bulletin*, 68, 5, pp. 24–34.

Mariotte, E (1686) *Traité du Mouvement des Eaux et des Autres Corps Fluides*, Paris.

Marshall, G.P., Culver, L.E., and Williams, G.P. (1973a) Fracture Phenomenon in polystyrene, *Int. J. Fract.*, 9:295–309.

Marshall, G.P., Williams, J.G., and Turner, C.E. (1973) Fracture toughness and absorbed energy measurements in impact tests on brittle materials, *J. Mat. Sci.*, 8:949–956.

Marshall, G.W., Balooch, M., Gallagher, R.R., Gansky, S.A., Marshall, S.J. (2001) Mechanical properties of the dentinoenamel junction: AFM studies of nanohardness, elastic modulus, and fracture, *J. Biomed. Mater. Res.*, 54:87–95.

Mazzetta, G.V., Christiansen, P., and Farin, R.A. (2004) Giants and bizarres: body size of some southern South American cretaceous dinosaurs, *Hist. Biol.*, 16:71–83.

Melchers, R.E. (1999) *Structural Reliability: Analysis and Prediction*, 2nd edition, Wiley, New York.

Merchant, M.E. (1945a) Mechanics of the metal cutting process. I Orthogonal cutting and a type 2 chip, *J. Appl. Phys.* 16:267–275.

Merchant, M.E. (1945b) Mechanics of the metal cutting process. II Plasticity conditions in orthogonal cutting *J. Appl. Phys.*, 16:318–324.

Mesnager, A. (1906) Essais sur barreaux entailles fait ay laboratoire de l'école des ponts et chausses, *Association Internationale pour l'Essai des Matériaux, Congrès de Bruxelles*, Rapport Non Officiel No A6f pp. 1–16.

Meyers, M.A. (2006) Mechanical properties of nanocrystalline materials, *Prog. Mater. Sci.*, 51:427–556.

Meyers, M.A., Lin, A.Y.M., Seki, Y., Chen, P.Y., Kad, B.K., and Bodde, S. (2006) Structural biological composites: An overview, *J. Mater.*, 58:35–41.

Michalske, T.A., and Freiman, S.W. (1982) A molecular interpretation of stress corrosion in silica, *Nature*, 295:511–512.

Miller, K.J. (1987a) The behaviour of short fatigue cracks and their initiation. Part I A review of two recent books, *Fatigue Fract. Engng Mater. Struct.*, 10:75–91.

Miller, K.J. (1987b) The behaviour of short fatigue cracks and their initiation. Part II A general summary, *Fatigue Fract. Engng Mater. Struct.*, 10:93–113.

Millman, M.S. (1999) Secondary steelmaking developments in British Steel, *Ironmaking and Steelmaking*, 26:169–175.

Mindlin, R.D., and Tiersten, H.F. (1962) Effects of couple stresses in linear elasticity, *Arch. Ration. Mech. Anal.*, 11:415–448.

Miner, M.A. (1945) Cumulative damage in fatigue. *Trans. ASME J. Appl. Mech.*, 12:A159–A164.

Mirshams, R.A., Xiao, C.H., Whang, S.H., and Yin, W.M. (2001) RCurve characterization of the fracture toughness of nanocrystalline nickel thin sheets, *Mater. Sci. Engng.*, A315:21–27.

Mogi, K. (1967) Effect of the intermediate principal stress on rock failure, *J. Geophys. Res.*, 72:5117–5131.

Mogi, K. (1971) Fracture and flow of rocks under high triaxial compression, *J. Geophys. Res.*, 76:1255–1269.

Mohr, O. (1882) Uber die Darstellung des Spannungszustandes unddes Deformationszustandes eines Korpere lementes und uber die Anwendung derselben in der Festiakeitslehve, *Zivilingenieur*, 28:112–155.

Mohr, O. (1990) Welche Umstäbedingen die Elastizitätsgrenze und den Bruch eines Materials, *Zeit Ver. Deut. Ing*, 44:1524–1530 and 1572–1577.

Moore, H.F. (1927) *Univ. of Illinois Eng. Exp. Sta. Bull.*, No 165.

Moore, H.F. and Kommers, J.B. (1927) *Fatigue of Metals*, McGraw Hill, New York.

Moore, W. (1813) *Treatise on the Motion of Rockets to which is added an Essay on Naval Gunnery*, G. and S. Robinson, Paternoster-Row.

Morgan, T.D., Baker, P., Kramer, K.J., Basibuyuk, H.H., and Quicke, D.L.J. (2003) Metals in mandibles of stored product insects: do zinc and manganese enhance the ability of larvae to infest seeds? *J. Stored Prod. Res.*, 39:65–75.

Mostovoy, S., and Ripling, E.J. (1966) Fracture toughness of an epoxy system, *J. Appl. Poly. Sci.*, 10:1351–1371.

Mott, N.F. (1948) Brittle fracture in mild steel plates — II, *Engineering*, 165:16–18.

Mott, N.F., and Nabarro, F.R.N. (1948) *Report of a Conference on the Strength of Solids*, The Physical Society, London, pp. 1–19.

Morye, S.S., Hine, P.J., Duckett, R.A., Carr, D.J., and Ward, I.M. (2000) Modelling of the energy absorption by polymer composites upon ballistic impact, *Comp. Sci. Techn.*, 60:2631–2642.

Muhly, J.D. (1980) The bronze age setting, In: *The Coming of the Age of Iron*, Eds. Wertime, T.A., and Muhly, J.D., Yale Univ. Press, New Haven, pp. 25–68.

Mulmule, S.V., and Dempsey, J.P. (2000) LEFM size requirements for fracture testing of sea ice, *Int. J. Fract.*, 102:85–98.

Mulvaney, J., and Kamminga, J. (1999) *Prehistory of Australia*, Allen and Unwin, St Leonards, NSW.

Muratoğlu, O.K., Argon, A.S., Cohen, R.E., and Weinberg, M. (1995a) Toughening mechanism of rubber-modified polyamides, *Polymer*, 36:921–930.

Muratoğlu, O.K., Argon, A.S., and Cohen, R.E. (1995b) Crystalline morphology of polyamide-6 near planar surfaces, *Polymer*, 36:2134–52.

Murgatroyd, J.B. (1942) Significance of surface markings on fractured glass, *J. Soc. Glass Tech.*, 26:155–171.

Murphy, M., Ali, M., and Ivankovic, A. (2006) Dynamic crack bifurcation in PMMA, *Eng. Fract. Mech.*, 73:2569–2587.

Murray, J., and Segall, P. (2002) Testing time predictable earthquake recurrence by direct measurement of strain accumulation, *Nature*, 419:287–291.

Muscat-Fenech, C.M., and Atkins, A.G. (1994a) Elastoplastic convergent and divergent crack paths in tear testing of sheet materials, *Fat. Fract. Engng. Struct.*, 17:133–143.

Musschenbroek, P. (1729) *Physicae Experimentales et Geometricae*, Samuelem Luchtmans.

Nabarro, F.R.N. (1947) Dislocations in a simple cubic lattice, *Proc. Phys. Soc.*, 59:256–272.

Nakayama, J., and Ishizuka, M. (1966) Experimental evidence for thermal shock resistance, *Bull. Am. Ceram. Soc.*, 45:666–669.

Naik, N.K., Shrirao, P., and Reddy, B.C.K. (2006) Ballistic impact behaviour of woven fabric composites: Formulation, *Int. J. Impact Engng.*, 32:1521–1552.

Nallathambi P., and Karihaloo, B.L. (1986) Determination of specimen-size independent fracture toughness of plain concrete, *Mag. Conc. Res.*, 25:315–321.

Navier, C.L.M.H. (1926) *Résumé des Leçons Données à l'École des Ponts et Chaussées, sur l'Application de la Mécanique à l'Établissement des constructions et des machines*, Firmin-Didot, Paris.

Needham, J., and Wang, L. (1959) *Science and Civilisation in China*, Vol. 3, *Mathematics and the Sciences of the Heavens and the Earth*, Cambridge University Press, Cambridge.

Needham, J. Wang, L., and Lu, G.-W. (1971) *Science and Civilsation in China. Vol IV. Physics and Physical Technology.* Part III. *Civil Engineering and Nautics*, Cambridge University Press, Cambridge.

Negaard, G.R. (1980) The history of the aircraft structural integrity program, *ASIAC Report* No. 680.18.

Newmann, J.C., James, M.A., and Zerbst, U. (2003) A review of the CTOA/CTOD fracture criterion, *Eng. Fract. Mech.*, 70:371–385.

Niegisch, W.D., The morphology of crazing in some amorphous thermoplastics, *Fourth Meeting Polymer and Fiber Microscopy*, Norwalk, Conn..

Ning, N.-Y., Yin, Q.-J., Luo, F., Zhang, Q., Du, R., and Fu, Q. (2007) Crystallization behavior and mechanical properties of polypropylene/halloysite composites, *Polymer*, 48:7374–7384.

Nix, W.D. (1989) Mechanical properties of thin films, *Metal. Trans.*, 20A:2217–2245.

Nix, W.D., and Gao, H. (1998) Indentation size effects in crystalline materials: a law for strain gradient plasticity, *J. Mech. Phys. Solids*, 46:411–425.

Noël, P. (1965) *Technologie de la Pierre de Taille*, SDRBTP, Paris.

Obreimoff, J.W. (1930) The splitting strength of mica, *Proc. Roy. Soc. Lond.*, A127:290–297.

Oda, S., and Keally, C.T. (1973) Edge-ground stone tools from Japanese preceramic culture, *Mater. Cult.*, No. 22 Society for the Study of Material Cultures, Tokyo.

442   *Fracture and Life*

Odette, G.R., and Lucas, G.E. (1998) Recent Progress in Understanding Reactor Pressure Vessel Embrittlement, *Rad. Effects and Defects in Solids*, 144:189–231.

O'Dowd, N.P., and Shih, C.F. (1991) Family of crack-tip fields characterized by a triaxiality parameter — I. Structure of fields. *J. Mech. Phys. Solids*, 39:989–1015.

O'Dowd, N..P., and Shih, C.F. (1992) Family of crack-tip fields characterized by a triaxiality parameter — II. Fracture applications. *J. Mech. Phys. Solids*, 40:939–963.

Ogden, R.W. (1972) Large deformation isotropic elasticity: on correlation of theory and experiments for incompressible rubber-like solids, *Proc. Roy. Soc. Lond.*, A326:565–584.

O'Keefe, R. (1994) Modeling the tearing of paper, *Amer. J. Phys.*, 62:299–305.

Orowan, E. (1933a) Die Zugfestigkeit von Glimmer und das Problem der technischen Festigkeit, *Zeitschrift für Physik*, 82:235–266.

Orowan, E. (1933b) Die erhöhte Festigkeit dünner Fäden, der Joffé-Effekt und verwandte Erscheinungen von Standpunkt der Griffithschen Bruchthorie. *Zeitschrift für Physik*, 86:195–213.

Orowan, E. (1934) Die mechanischen Festigkeitseigenschaften und die Realstruktur der Kristalle, *Zeitschrift für Kistallographie*, 89:327–343.

Orowan, E. (1945) Notch brittleness and the strength of metals, *Trans. Inst. Engrs and Shipbuilders* Scotland, 89:165–215.

Orowan, E. (1949) Fracture and strength of solids, *Reports on Progress in Physics*, 12:185–232.

Otto, W.H. (1955) Relationship of tensile strength of glass fibers to diameter, *J. Am. Ceram. Soc.*, 38:122–124.

Ouchi, C. (2001) Development of steel plates by intensive use of TMCP and direct quenching processes, *ISIJ Int.*, 41:542–553.

Palmgren, A (1924) Die Lebensdauer von Kugellagern, *Zeitschrift des Vereines deutscher Ingenieure*, 68:339–341.

Panasyuk, V.V. (1960) To the theory of crack propagation in brittle solids under deformation (in Ukrainian), *Doklady Akademii Nauk Ukrainy*, 9:1185–1188.

Pardoen, T., Hachez F., Marchioni, B., Blyth, P.H., Atkins, A.G. (2004) Mode I fracture of sheet metal, *J. Mech. Phys. Solids*, 52:423–52.

Pardoen, T., and Hutchinson, J.W. (2000) An extended model for void growth and coalescence, *J. Mech. Phys. Solids*, 48:2467–2512.

Pardoen, T., and Hutchinson, J.W. (2003) Micromechanics-based model for trends in toughness of ductile metals. *Acta Mater.*, 51:133–148.

Paris, P.C. (1957) *A Note on the Variables Affecting the Rate of Crack Growth Due to Cyclic Loading*, D–17867, The Boeing Company Addendum.

Paris, P.C (1997) George Irwin, Boeing professor of mechanics at Lehigh University: The history of a fruitful research period, , In *Fracture Research in Retrospect: An anniversary volume in Honour of G.R. Irwin's 90th Birthday*, (Ed. Rossmanith, H.P.) A.A. Balkema Rotterdam, pp. 227–236.

Paris, P.C, Bucci, R.J., Wessel, E.T., Clark, W.G., and Mager, T.R. (1972) Extensive study of low fatigue crack growth rates in A533 and A508 Steels, *Stress Analysis and Growth of Cracks, Proceedings of the 1971 National Symposium on Fracture Mechanics*, Part I, ASTM STP 513, American Society for Testing and Materials, pp. 141–176.

Paris, P.C., and Erdogan, F. (1963) A critical analysis of crack propagation laws, *J. Basic Engng. – Trans ASME*, 85:528–534.

Paris, P.C., Gomez, M.P., and Anderson, W.P. (1961) A rational analytic theory of fatigue, *The Trend in Engineering*, 13:9–14.

Parsons, W. B. (1968) *Engineers and Engineering in the Renaissance*, Massachusetts Institute of Technology, Cambridge, Mass. Originally published in 1939 by The Williams and Wilkins Co..

Patel, Y., Blackman, B.R.K., and Williams, J.G. (2009) Measuring fracture toughness from machining tests, *Proc. Inst. Mech. Eng. Part C: J. Mech. Eng. Sci.*, Vol. 223, in press.

Pearson, G. (1795) Experiments and observations to investigate the nature of a kind of steel, manufactured at Bombay, and there called wootz: with remarks on the properties and composition of the different states of iron, *Phil. Trans. Roy. Soc. Lond.*, 85:322–346.

Pearson, S. (1975) Initiation of fatigue cracks in commercial aluminium and the subsequent propagation of very short cracks, *Eng. Fract. Mech.*, 7:235–247.

Peierls, R. (1940) The size of a dislocation, *Proc. Phys. Soc.*, 52:34–37.

Perrot, G., and Chipiez, C. (1882) *Historie de l'Art dans l'Antiquité*, Hachette, Paris.

Petch, N.J. (1953) The cleavage strength of polycrystals, *J. Iron and Steel Inst.*, 174:25–28.

Peterson, R.E. (1933) Stress concentration phenomena in fatigue of metals, *Trans. AMSE, J. Appl. Mech.*, 55:157–171.

Peterson, R.E., and Wahl, A.M. (1936) Two and three dimensional cases of stress concentrations and comparison with fatigue tests, *Trans. AMSE, J. Appl. Mech.*, 58:A15–A22.

Petersson, P.-E. (1985) *Crack growth development of fracture zones in plain concrete and similar materials*, Rpt. TVBM-1006 Div. Bldg. Mats., Lund Institute of Technology.

Petrie, W.M.F. (1883) *The Pyramids and Temples of Gizeh*, Field and Tuer, London.

Petrie, W.M.F. (1884) On the mechanical methods of the ancient Egyptians, *J. Anth. Inst.*, 13:88–109.

Petrie, W.M.F. (1917) *Tools and Weapons*, British School of Archaeology in Egypt, London.

Petrovic, J.J. (2003) Review, mechanical properties of ice and snow, *J. Mater. Sci.*, 38:1– 6.

Pham, V.B., Bahr, H.-A., Bahr, U., and Fett, (2006) Crack paths and the problem of global directional stability, *Int. J. Fract.*, 141:513–534.

Pitt, S., and Jones, R. (1997) Multiple-site and widespread fatigue damage in ageing aircraft, *Eng. Fail. Anal.*, 4:237–257.

Plato, Gorgias. In *The Collected Dialogues of Plato*, Eds. E. Hamillton and H. Cairns, Bollinger Series 71 (1961) Pantheon Books, New York.

Plutarch, Lives Marcellus. In *Agesilaus and Pompey, Pelopidas and Marcellus*, Vol. 5, Trans. B. Perrin, 1917, Heinemann, London.

Poland, C.A., Duffin, R., Kinloch, I., Maynard, A., Wallace, W.A.H., Seaton, A., Stones, V., Brown, S., MacNee, W., and Donaldson, K. (2008) Carbon nanotubes introduced into the abdominal cavity of mice show asbestos like pathogenicity in a pilot study, *Nature Nanotech.*, 3:423–428.

Polanyi, M. (1921) Die Natur vom zerreißenden Verfahren, *Zeitschrift für Physik*, 7:323–327.

Poleni, G. (1748) *Memoire istoriche della gran Cupola del Tempo Vaticano*, Padova.

Poncelet, J.V. (1839) *Introduction à Mécanique, Industrielle, Physique Our Expérimentale,* Deuxième édition, de Gauthier-Villars, Paris.

Pope, S. (1909) *A Study of Bows and Arrows,* University of California Press, Berkely.

Popoola, O., Kriven, W.M., and Young, J.F. (1991) Microstructural and microchemical characterization of a calcium aluminate-polymer composite (MDF cement), *J. Amer. Ceram. Soc.,* 74:1928–1933.

Prandtl, L. (1903) Zur Torsion von prismatischen Staben, *Physikalische Zeitschrift,* 4:758–770.

Prandtl, L. (1923) Anwendugsbeispiele zu Einem Henkyschen Satz uber das Plastische Gleichgewicht, *Zeitschrift fur Angewande Mathematik und Mechanik,* 3:401–406.

Prange, H.D., Anderson, J.F., and Rahn, H. (1979) Scaling of skeletal mass to body mass in birds and mammals, *American Naturalist,* 113:103–122.

Preston, F.W. (1926) A study of the rupture of glass, *J. Soc. Glass Tech.,* 10:234–243.

Preston, F.W. (1942) The mechanical properties of glass, J. Appl. Phys., 13:623–634.

Prothero, J. (1995) Bone and fat as a function of body weight in adult mammals, *Comp. Biochem. Physiol.,* 111A:633–639.

Protzen, J.-P. (1985) Inca quarrying and stonecutting, *J. Soc. Arch. Hist.,* 44:161–182.

Pryor, M.G.M. (1940) On the hardening of the oothecac of *Blatta orientalis, Proc. Roy. Soc. Lond.,* 128:378–393.

Pugsley, A.G. (1966) *The Safety of Structures,* Arnold, London.

Puttick, K.E., Rudman, M.R., Smith, K.J., Franks, A., and Lindsey, K. (1989) Single-Point Diamond Machining of Glasses, *Proc. Roy. Soc. Lond.,* A426:19–30.

Puzak, P.P., Eschbacher, E.W., and Pellini, W.S. (1952) Initiation and propagation of brittle fracture in structural steels. *Weld. Journal Res. Suppl.,* 31:561s–562s.

Qi, B., Zhang, Q.X., Bannister, M., and Mai, Y.-W. (2006) Investigation of the mechanical properties of DGEBA-based epoxy resin with nanoclay additives, *Comp. Struct.,* 75:514–519.

Radushkevich, L.V., and Lukyanovich, V.M. (1952) O strukture ugleroda, obrazujucegosja pri termiceskom razlozenii okisi ugleroda na zeleznom kontakte, *Zurn. Fisic. Chim.,* 26:88–95.

Rahman, A. (1964) Correlations in the motion of atoms in liquid argon, *Phys. Rev.,* 136:A405–A411.

Ramberg, W., and Osgood, W.R. (1943) *Description of stress-strain curves by three parameters,* NACA Tech. Note 902.

Rankine, W.J.M. (1843) On the causes of the unexpected breakage of the journals of railway axles; and on the means of preventing such accidents by observing the law of continuity in their construction, *Min. Proc. Inst. Civil Engrs.,* 2:105–107.

Rankine, W.J.M. (1858) *Manual of Applied Mechanics,* London.

Rausing, G. (1967) *The Bow: Some Notes on its Origins and Development,* CWK Glerups Förlag, Lund.

Ravi-Chandar, K. (1998) Dynamic fracture of nominally brittle materials, *Int. J. Fract.,* 90:83–102.

Reddy, T.Y., and Reid, S.R. (1986) Axial splitting of circular metal tubes, *Int. J. Mech. Sci.,* 28:111–131.

Reisner, G.A. (1931) *Mycerinus, The Temples of the Third Pyramid at Giza,* Harvard University Press, Cambridge, Mass.

Rennie, G. (1818) Account of experiments made on the strength of materials, *Proc. Roy. Soc. Lond.*, 108:118–136.

Reuleaux, F. (1900) Über den Taylor Whiteschen Werkzeugstahl, Verein sur Berforderung des Gewerbefleissen in Preussen, *Sitzungsberichete*, 79:179–220.

Rice, J.R. (1965) An examination of the fracture mechanics energy balance from the point of view of continuum mechanics, In *Proc. 1st International Conference on Fracture*, Eds. Yokobori, t., Kanasawa, T., and Swedolw, J.L., Sendai, Japan, Vol 1, pp. 309–340.

Rice, J.R. (1967a) Mechanics of crack tip deformation and extension by fatigue, In *Fatigue Crack Propagation*, ASTM 415, American Society for Testing and Materials, Philadelphia, pp. 247–309.

Rice, J.R. (1967b) Stresses in an infinite strip containing a semi-infinite crack, Discussion to Knauss, W.G. (1966) *J. Appl. Mech.*, 33:356.

Rice, J.R. (1968) A path independent integral and the approximate analysis of strain concentration by notches and cracks, *J. Appl. Mech.*, 35:379–386. Published as a Brown University ARPA SD-86 Report E39 in 1967.

Rice, J.R., Drugan, W.J., and Sham,T. (1980) Elastic-plastic analysis of growing cracks, In *Fracture Mechanic: Twelfth Conference*, ASTM-STP-700, Philadelphia, pp. 189–221.

Rice, J.R., Paris, P.C., and Merkle, J.G. (1973) Some further results of J-integral analysis and estimates, In *Progress in Flaw Growth and Fracture Toughness Testing*, ASTM STP 536, American Society for Testing and Materials, pp. 231–245.

Rice, J.R., and Rosengren, G.F. (1968) Plane strain deformation near a crack tip in a power-law hardening material,, *J. Mech. Phys. Solids*, 16:1–12.

Rice, J.R. and Sih, G.C. (1965) Plane problems of cracks in dissimilar media, *J. Appl. Mech.*, 32:418–423.

Rice, J.R., and Thomson, R. (1974) Ductile versus brittle behaviour of crystals, *Phil. Mag.*, 29:73–97.

Richardson, J.T.H. (1888) *British Patent* 7309.

Ritchie, R.O., Knott, J.F., Rice, J.R. (1973) On the relationship between critical tensile stress and fracture toughness in mild steel, *J. Mech Phys. Solids*, 21:395–410.

Rivlin, R.S., and Thomas, A.G. (1953) Rupture of rubber,. I Characteristic energy for tearing, *J. Polym. Sci.*, 10:291–318.

Roberston, T.S. (1951) Brittle fracture of mild steel, *Engineering*, 172:445–

Roberston, T.S. (1953) Propagation of brittle cracks in steel, *J. Iron & Steel Inst.*, 175:361–374.

Robison, J. (1801) Article on "Strength of materials", In the Supplement to the Encyclopaedia Britannica.

Roelofsen, P.A. (1959) *Encyclopedia of plant anatomy: Part 4. The plant cell wall.* Gebriider Borntraeger, Berlin-Nikolassee, Germany.

Rosakis, A.J., Samudrala, O., and Coker, D. (1999) Cracks faster than the shear wave speed, *Science*, 284:1337–1340.

Rosenfield, A.R. (1976) The crack in the Liberty Bell, *Int. J. Fract.*, 12:791–797.

Rossmanith, H.P. (1995) Translation and introduction to Weighardt's 1907 paper: On splitting and cracking of elastic bodies, *Fatigue Fract. Mater. Strut.*, 18:1367–1404.

Rossmanith, H.P (1997) The struggle for recognition of engineering fracture mechanics, In *Fracture Research in Retrospect: An anniversary volume in Honour of G.R. Irwin's 90th Birthday*, (Ed. Rossmanith, H.P.) A.A. Balkema Rotterdam, pp. 37–94.

Rossmanith, H.P. (1999) Fracture mechanics and materials testing forgotten pioneers of the early 20th century, *Fatigue Fract. Mater. Strut.*, 22:781–797.

Rostoker, W., Bronson, B., and Dvorak, J. (1984) The cast iron bells of China, *Technology and Culture*, 25:750–757.

Rudall, K.M., and Kenchington, W. (1973) The chitin system, *Biol. Rev.*, 48:597–636.

Rühle, M., and Evans, A.G. (1989) High toughness ceramics and ceramic composites, *Prog. Mater. Sci.*, 33:85–167.

Rühle, M., and Evans, A.G.M., McMeeking, R.M., and Charalambides, P.G. (1987) Microcrack toughening in alumina/zirconia, *Acta Metal.*, 35:2701–2710.

Russell S.B. (1898). Experiments with a New Machine for Testing Materials by Impact. *Amer. Soc. Civil Eng.*, 39:.237–250. (reprinted in: *The Pendulum Impact Testing: A Century of Progress*, eds: T.A. Siewert and S. Manahan, 2000:17–45, ASTM STP 1380., Penns.

Saint-Venant, A.B. (1871) Mémoire sur l'établissement des équations différentielles des mouvements intérieurs opérés dans les corps solides ductiles au-delà des limites où l'élasticité pourrait les ramener à leur premier état, *Journal de Mathématiques*, 16:308–316.

Sangoy, L., Meunier, Y., and Pont, G. (1988) Steels for ballistic protection, Israel *J. Technology*, 24:319–326.

Sauer, J.A., Marin, J., and Hsiao, C.C. (1949) Creep and damping properties of polystyrene, *J. Appl. Phys.*, 20:507–517.

Schabtach, C., Fogleman, E.L., Rankin, A.W., Winnie, D.H., Schenecady, N.Y. (1956) Report of the investigation of two generator rotor fractures, *Trans, AMSE*, 78:1567–1584.

Schardin, H., and Struth, W. (1938) Hochfrequenzkinematographische Untersuchung der Bruchvorgänge im Glas. *Glastech. Ber.* 16:219–227.

Scheirs, J., Böhm, L.L., Boot, J.C., and Leevers P.S. (1996) PE100 resins for pipe applications: continuing the development into the 21st century, *Trends Polym. Sci.*, 4:408–415.

Schnadt, H.M. (1944) *Ass. Belge pour l'Étude, l'Essai et l' Emploi des Matériaux*, Brussels.

Schneider, P.J. (1955) *Conduction Heat Transfer*, Addison-Wesley, Reading, MA.

Schrödinger, E. (1926) An undulatory theory of the mechanics of atoms and molecules, *Phys. Rev.* 28:1049–1070.

Schütz, W. (1996) A history of fatigue, *Eng. Fract. Mech.*, 54:263–300.

Semaw, S., Rogers, M.J., Quade, J., Rennie, P.R., Bultler, R.F., Dominguez-Rodrigo, M., Stout, D., Hart, D., Pickering. T., and Simpson, S.W. (2003) 2.6-Million-year-old stone tools and associated bones from OGS-6 and OGS-7, Gona, Afar, Ethiopia, *J. Human Evol.*, 45:169–177.

Seo, M.-K., Lee, J.-R., and Park, S.-J. (2005) Crystallization kinetics and interfacial behaviors of polypropylene composites reinforced with multi-walled carbon nanotubes, *Mater. Sci. Engng.*, A:79–84.

Shadwick R.E., Russell, A.P., and Lauf, RF, (1992) The structure and mechanical design of rhinoceros dermal armour, *Phil. Trans. Roy. Soc. Lond.*, B337:419–428.

Shank, M.E. (1954) A critical survey of brittle fracture in carbon steel plate structures other than ships, *Weld. Res. Council Bull.*, No 17

Sharon, E., and Fineberg, J. (1999) Confirming the continuum theory of dynamic brittle fracture for fast cracks, *Nature*, 397:333–335.

Shaver, W.W. (1964) Recent developments in high-strength glasses and ceramics, *Proc. Roy. Soc. Lond.*, A282:52–56.

Sherby, O.D., and Wadsworth, J. (2001) Ancient blacksmiths, the Iron Age, Damascus steels, and modern metallurgy, *J. Mater. Process. Tech.*, 117:347–353.

Shergold. O.A., and Fleck, N.A. (2004) Mechanisms of deep penetration of soft solids, with application to the injection and wounding of skin, *Proc. Roy. Soc. Lond.*, A460:3037–3058.

Shih, C.F., de Lorenzi, H.G., Andrews, W.R. (1979) Studies on crack initiation and stable crack growth, In *Elastic-Plastic Fracture*, ASTM STP 668, Eds. Landes, J.D., Begley, J.A., and Clarke, G.A., American Society for Testing and Materials, pp. 65–120.

Shih, C.F., O'Dowd, N.P. and Kirk, M.T. (1993) A frame work for quantifying crack tip constraint, In *Constraint Effects in Fracture*, ASTM STP 1171 American Society for Testing and Materials, Philadelphia, , pp. 2–20.

Shipway, J.S. (1989) Tay Rail Bridge Centenary – some notes on its construction 1882–87, *Proc. Instn. Civil Engrs.*, Part 1, 86:1089–1109.

Shipway, J.S. (1990) The Forth Railway Bridge centenary 1890–1990: some notes on its design, *Proc. Instn Civil. Engrs.*, Part 1, 88:1079–1107.

Siewert, T.A., Manahan, M.P., McCowan, C.N., Holt, M., Marsh, F.J., and Ruth, E.A. (2000) The history and importance of impact testing, In *Pendulum Impact Testing: A Century of Testing*, eds. Siewaert, T.A., and Manahan, M.P., ASTM STP 1380, pp. 3–16.

Smedley, G.P. (1981) Material defects and service performance, *Phil. Trans. Roy. Soc. Lond.*, A299:7–17.

Smekal, A. (1950) Verfahren zur Messung von Bruchfortpflanzungs-Geschwindigkeiten an Bruchflächen. *Glastech. Ber.* 23:57–67.

Smekal, A. (1953) Zum Bruchvorgang bei sprödem Stoffverhalten unter ein- und mehrachsiger Beanspruchung, *Österr Ingenieurarch*, 7:49–70.

Smith, H.L., Kies, J.A., and Irwin, G.R. (1951) Rupture of plastic sheets as a function of size and shape, *Phys. Review*, 83:872.

Smith, R.A. (1990) The Versailles railway accident on 1842 and the first research into metal fatigue, In *Fatigue 90* (eds. H. Kitagawa, and T. Tanaka) vol. IV, pp. 2033–2041.

Smith, R.A., and Hillmansen, S. (2004) A brief historical overview of the fatigue of railway axles, *Proc. Inst. Mech. Engrs. Part F*, 218:267–277.

Sneddon, I.N. (1946) The distribution of stress in the neighbourhood of a crack in an elastic solid, *Proc. Roy. Soc. Lond.*, A187:220–260.

Sneddon, I.N. and Elliot, H.A. (1946) The opening of a Griffith crack under internal pressure, *Q. Appl. Mech.*, 4:262–267.

Snodgrass, A.M. (1980) Iron and early metallurgy in the Mediterranean, In: *The Coming of the Iron Age*, Edited by Wertime, TA, and Muhly, JD., Yale University Press, New Haven.

Soderberg, C.R. (1939) Factor of safety and working stress, *Trans. Amer. Soc. Mech. Engngs.* 52:13–28.

Sommer, E. (1969) Formation of lances in glass, *Eng. Fract. Mech.*, 1:539–546.

Spanoudakis, J., and Young, R.J. (1984) Crack propagation in a glass particle-filled epoxy resin. Part 1 Effect of particle volume fraction and size, *J. Mater. Sci.*, 19:473–486.

Special ASTM committee (1960) Fracture testing of high-strength sheet materials (first report of a special QASTM committee), *ASTM Bulletin*, No 243, pp. 29–40, and No 244, pp. 18–28.

Spence, J., and Nash, D.H. (2004) Milestones in pressure vessel technology, *Int. J. Press. and Piping*, 81:89–118.

Spencer, A.J.M. (1965) On the energy of the Griffith crack, *Int. J. Engng. Sci.*, 3:441–449.

Spencer (1928) *Wanderings in Wild Australia*, Macmillan, London.

Stanton, T.E., and Batson, R.G.C. (1920) On the characteristics of notched-bar impact tests, *Min. Proc. Inst. Civil Engrs.*, 99:67–100.

Stevenson, a., and Abmalek, K. (1994) On the puncture mechanics of rubber, *Rubber Chem. Technol.*, 67:743–760.

Stocks, D. (1999) Stone sarcophagus manufacture in ancient Egypt, *Antiquity*, 73:918–922.

Stocks, D. (2001) Testing ancient Egyptian granite working methods in Aswan, Upper Egypt, *Antiquity*, 75:89–94.

Street, W.A.A. (1962) *Metals in the Service of Man*, Penguin Books Ltd., Harmondsworth.

Stroh, A.N. (1957) A theory of the fracture of metals, *Advances in Physics*, 6:418–465.

Sturdie, J. (1693) Extracts from some letters from Mr John Sturdie of Lancashire concerning iron ore; and more particularly of haematites, wrought into iron at Milthrop-Forge in that county, *Trans Phil. Soc.*, 17:695–699.

Sultan, J., and McGarry, A.J. (1973) Effect of rubber particle size on deformation mechanisms in glassy epoxy, *Poly. Engng. Sci.*, 13:29–34.

Sumpter, J.D.G. and Kent, J.S. (2004) Prediction of ship brittle fracture casualty rates by a probabilistic method, *Marine Structures*, 17:575–589.

Suo, Z., and Hutchinson, J.W. (1989) Steady-state cracking in brittle substrates beneath adherent films, *Int. J. Solids Struct.*, 25:1337–1353.

Suresh, S. (1998) *Fatigue of Metals*, 2nd Ed., Cambridge University Press, Cambridge.

Suresh, S., and Ritchie, R.O. (1984) Propagation of short fatigue cracks, *Int. Metals Rev.*, 29:445–476.

Swanson, P.L., Fairbanks, C.J., Lawn, B.R., Mai, Y.W., and Hockey, B.J. (1987) Crack-interface grain bridging as a fracture resistance mechanism in ceramics: I Experimental study on alumina, *J. Am. Ceram. Soc.*, 70:279–289.

Swenson, D.V., and Ingraffea, A.R. The collapse of the Schoharie Creek Bridge: a case study in concrete fracture mechanics, *Eng. Fract. Mech.*, 51:73–92.

Tada, H., Paris, P.C., Irwin, G.R. (1985) *The Stress Analysis of Cracks Handbook*, 2nd edition, The Del Research Corporation, St Louis, Missouri.

Talbot, C.J. (1999) Can field data constrain rock viscosities? *J. Struct. Geol.*, 21:949–957.

Tang, C.A. (1997) Numerical simulation of progressive rock failure and associated seismicity, *Int. J. Rock Mech. Min. Sci.*, 34:249–261.

Tang, Z., Kotov, N.A., Magonov, S., and Ozturk, B. (2003) Nanostructured artificial nacre, *Nature Mater.*, 2:413–418.

Taylor, F.S. (1947) An early satirical poem on the Royal Society, *Notes Rec. Roy. Soc. Lond.*, 5:37–46.

Taylor, G.I. (1948) Formation and enlargement of a circular hole in a thin plastic sheet, *Quart. J .Mech. Appl. Math.*, 1:103–124.

Taylor, G.I., and Quinney, H. (1934) The latent energy remaining in a metal after cold working, *Proc. Roy. Soc. Lond.* A143:307–326.

Terenzio, A. (1934) La restauration du Panthéon de Rome, *La conservation des monuments d'art & d'historie*, Paris, pp. 280–285.

Thio, Y.S., Argon, A.S., Cohen, R.E., and Weinberg, M. (2002) Toughening of isotactic polypropylene with $CaCO_3$ particles, *Polymer*, 43(11):3661–3674.

Thomason, P.F. (1985) A three-dimensional model for ductile fracture by the growth and coalescence of microvoids, *Acta Metall.*, 33:1079–1085.

Thomason, P.F. (1990) *Ductile Fracture of Metals*, Pergamon Press, Oxford.

Thouless, M.D. (1990) Crack spacing in brittle films on elastic substrates, *J. Am. Ceram. Soc.*, 73:2144–2146.

Thouless, M.D. (1993) Combined buckling and cracking of films, *J. Am. Ceram. Soc.*, 76:2936–2938.

Thum, A., and Ude, H. (1930) Mechanische Kennzeichen des Gusseisens, *Zeitschrift des Vereines deutscher Ingenieure*, 74:257–264.

Time, I. (1870) *Resistance of Metals and Wood to Cutting* [in Russian], Dermacow Press House, St. Petersbourg.

Timoshenko, S.P. (1940) *Strength of Materials: Part I Elementary Theory and Problems*, Van Nostrand, New York.

Timoshenko, S. P. (1953) *History of Strength of Materials*, McGraw-Hill Book Co., New York.

Ting, S.K.M. (2003) *The study of crack growth in polyethylene using traction-separation laws*, PhD dissertation, Imperial College, London.

Tipper, C.F. (1948) The fracture of mild steel plate, *Admiralty Ship Welding Committee Report*, R3, HMSO, London.

Tipper, C.F. (1962) *The Brittle Fracture Story*, Cambridge University Press, Cambridge.

Tjong, S.C. (2006) Structural and mechanical properties of polymer nanocomposites, *Mater. Sci. Eng.*, R53:73–197.

Todhunter, I. (1886) *A History of the Theory of Elasticity and of the Strength of Materials, from Galilei to Lord Kelvin*, edited and completed by K. Pearson, Cambridge University Press, Cambridge.

Tredgold, T. (1824) *A Practical Essay on the Strength of Cast Iron*, J. Taylor, London

Tresca, H.E. (1869) *Mémoires sur l'écoulement des Corps Solides, Mémoires présentés par divers savants*, Vol 20.

Tvergaard, V. (1981) Influence of voids on shear band instabilities under plane strain, *Int. J. Fract.*, 17:389:407.

Tvergaard, V., and Hutchinson, J.W. (1992) The relation between crack growth resistance and fracture process parameters in elastic-plastic solids, *J. Mech. Phys. Solids*, 40:1377–1397.

Van der, Giessen, E., and Needleman, A. (1995) Discrete dislocation plasticity: a simple planar model, *Model. Simul. Mater. Sci. Eng.*, 3:689–735.

Vaz, M., Owen, D.R.J., Kalhori,V., Lundblad, M., and Lindgren, L.-E. (2007) Modelling and simulation of machining processes, *Arch. Comput. Methods Eng.*, 14:173–204.

Veronda, D.R., and Westmann, R.A. (1970) Mechanical characterization of skin – finite deformations, *J. Biomech.*, 3:11–124.

Vincent, J.F.V. (1982) The mechanical design of grass, *J. Mater. Sci.*, 17:856–860.

Vincent, J.F.V. (2003) Biomimetic modelling, *Phil. Trans.: Bio. Sci.*, 358:1597–1603.

Vincent, J.F.V., Bogatyreva, O.A., Bogatyrev, N.R., Bowyer, A., and Pahl, A.-K. (2006) Biomimetics: its practice and theory, *J. R. Soc. Interface*, 3:471–482.

Vincent, J.F.V., and Hillerton, J.E. (1979) The tanning of insect cuticle- a critical review and revised mechanism, *J. Insect Physiol.*, 25:653–658.

Vincent, J.F.V., Jeronimidis, G., Khan, A.A., and Luyten, H. (1991) The wedge fracture test a new method for measurement of food texture, *J. Texture Stud.*, 22:45–57.

Vincent, J.F.V., and Wegst, U.G.K. (2004) Design and mechanical properties of insect cuticle, *Arthropod, Struct. Devel.*, 33:187–199.

Virta, J., Koponen, S., and Absetza, I. (2006) Measurement of swelling stresses in spruce (*Picea abies*) samples, *Build. Environ.*, 41:1014–1018.

Volinsky, A.A., Moody, N.R., and Gerberich, W.W. (2002) Interfacial toughness measurements for thin films on substrates, *Acta Mater.*, 50:441–466.

von Kármán, T. (1911) Festigkeitsversuche unter all seitigem, *Druck. Z. Vereins Deut. Ingr.*, 55:1749–1757.

Wainwright, S.A., Biggs, W.D., Currey, J.D., and Gosline, J.M. (1982) *Mechanical Design in Organisms*, Edward Arnold, London.

Wallner, H. (1939) Linienstrukturen an Bruchflächen, *Zeitschrift für Physik*, 114:368–378.

Wang, K., Chen, L., Wu, J., Toh, M.L., He, C., and Yee, A.F. (2005) Epoxy nanocomposites with highly exfoliated clay: mechanical properties and fracture mechanisms, *Macromol.*, 38:788–800.

Webster, J.J., Fairbairn, W., Styffe, M.K., and Sandberg, C.P. (1880) Iron and steel at low temperatures, *Min. Proc. Inst. Civil Engrs.*, 60:161–187.

Wei, Y., and Hutchinson, J.W. (1997) Steady-state crack growth and work of fracture for solids characterised by strain gradient plasticity, *J. Mech. Phys. Solids*, 45:1253–1273.

Weibull, W (1939) *A statistical theory of the strength of materials*, Ingeniörs Vetenskaps Akademiens, Handlingar Nr 151.

Weisberger, G. (2006) The mineral wealth of ancient Arabia and its use. I: Copper mining and smelting at Feinan and Timna – comparison of techniques, production, and strategies, *Arabian Arch. Epigraphy*, 17:1–30.

Wells, A.A. (1952) The mechanics of notch brittle fracture, *Weld. Res.*, 7:34r–56r.

Wells, A.A. (1955) The conditions of fast fracture in aluminium alloys with particular reference to Comet Failures, *British Welding Research Association*, Report NRB 129.

Wells, A.A. (1956a) The 600 ton test rig for brittle fracture research, *Brit. Weld. J.*, 3:25.

Wells, A.A. (1956b) The brittle fracture strengths of welded steel plates, *Trans. Inst. Naval Arch.*, 98:296.

Wells, A.A. (1961a) Unstable crack propagation in metals, *Proceedings of the Crack Propagation Symposium*, Cranfield, pp. 210–230.

Wells, A.A. (1961b) The brittle fracture strength of welded steel plates — tests on five further steels, *Brit. Weld. J.*, 8:259–277.

Wells, A.A. (1963) Application of fracture mechanics at and beyond general yielding, *Brit. Weld. J.*, 10:563–570.

Wells, A.A. (1973) Fracture control: past, present and future, *Exp. Mech.* 13:401–410.

Wells, A.A. (2000) George Rankin Irwin, *Biog. Mem. Fellows Roy. Soc.*, 46:271–283.

Wells, A.A., and Burdekin, F.M. (1963) Effects of thermal stress relief and stress relieving conditions on the fracture of notched and welded steel plates, *Brit. Weld. J.*, 10:270–278.

Westohofen, W. (1890) The Forth Bridge, *Reprinted from Engineering*, Third Edition, pp. 1–72.

Weygand, D., Friedman1, L.H., Giessen, E., Van der, and Needleman, A. (2002) Aspects of boundary-value problem solutions with three-dimensional dislocation dynamics, *Model. Simul. Mater. Sci. Eng.*, 10:437–468.

Wicker, FDP. (1990) *Egypt and the Mountains of the Moon*, Merlin Books, Braunton, Devon.

Wicker, F.D.P. (1998) The road to Punt, *Geog. J.*, 164:155–167.

White, S.R., Sottos, N.R., Geubelle, P.H., Moore, J.S., Kessler, M.R., Sriram, S.R., Brown, E.N., and Viswanathan, S. (2001) Autonomic healing of polymer composites, *Nature*, 704:409:794–797.

Whitney, W. (1963) Observation of deformation bands in amorphous polymers, *J. Appl. Phys.*, 34:3633–3634.

Whittaker, B.N., Singh, R.N., and Sun, G. (1992) *Rock Fracture Mechanics: Principles, Design and Applications*, Elsevier, Amsterdam.

Whittaker, J.C. (2001) 'The oldest British industry': Continuity and obsolescence in a flintknapper's sample set, *Antiquity*, 75:382–390.

Wieghardt, K. (1907) Über das Spalten und Zerreißen elastischer Körper, *Zeitschrift für Mathematik und Physik*, 55:60–103 (English translation by H.P. Rossmanith 1995 On splitting and cracking of elastic bodies, *Fatigue Fract. Eng. Mater. Struct.*, 12:1371–1405).

Wierzbicki, T., and Thomas, P. (1993) Closed-form solution for wedge cutting force through thin sheets. *Int. J. Mech. Sci.*, 35:209–29.

Wierzbicki, T., Trauth, K.A., and Atkins, A.G. (1998) On diverging concertina tearing, *J. Appl. Mech., Trans. ASME.*, 65:990–997.

Williams, A.R., and Maxwell-Hyslop, K.R. (1976) Ancient steel from Egypt, *J. Arch. Sci.*, 3:283–305.

Williams, J.G. (1984) *Fracture Mechanics of Polymers*, Ellis Horwood, Chichester.

Williams, J.G. (1998) Friction and plasticity effects in wedge splitting and cutting fracture tests, *J. Mater. Sci.*, 33:5351–5357.

Williams, J.G., Patel, Y., and Blackman B.R.K. (2009) A fracture mechanics analysis of cutting and machining, *Eng. Fract. Mech.*, In press.

Williams, J.G., and Rink, M. (2007) The standardisation of the EWF test, *Eng. Fract. Mech.*, 74:1009–1017.

Williams, M.L. (1951) Stress singularities resulting from various boundary conditions in angular corners of plates in extension, *J. Appl. Mech.*, 19:526–528.

Williams, M.L. (1957) On the stress distribution at the base of a stationary crack, *Trans. AMSE, J. Appl. Mech.*, 24:109–114.

Williams, M.L. (1959) The stresses around a fault or crack in dissimilar media, *Bull. Seismol. Soc.*, 49:199–204.

Williams, M.L. and Ellinger, G.A. (1953) Investigation of structural failures of welded ships, *Weld. J. Res. Suppl.*, 18:498–527.

Willis, A., and Vincent, J.F.V. (1995) Monitoring cutting forces with an instrumented histological microtome, *J. Microscopy*, 178:56–65.

Wilson, W.M., Hechtman, R.A., and Bruckner, W.H. (1948) Cleavage fracture of ship plate influenced by size effect, *Weld. J. Res .Suppl.*, 27:200.

Winnie, D.H.J., and Wundt, B.H. (1958) Application of the Griffith-Irwin theory of crack propagation to the bursting behaviour of disks including analytical and experimental studies, *Trans. ASME*, 80:1643–1658.

Withey, P.A. (1997) Fatigue failure of the de Havilland Comet 1, *Eng. Fail. Anal.*, 4:147 154.

Wright, S.C., Huang, Y., and Fleck, N.A. (1992) Deep penetration of polycarbonate by a cylindrical punch, *Mech. Mater.*, 13:277–284.

Wu, D.Y., Meure, W.S., and Solomon, D. (2008) Self-healing polymeric materials: A review of recent developments, *Prog. Polym. Sci.*, 33:479–522.

Wu, S. (1985) Phase structure and adhesion in polymer blends: A criterion for rubber toughening, *Polymer*, 26:1855–1863.

Wu, S.X., Mai, Y.W., and Cotterell, B. (1992) A model of fatigue crack growth based on Dugdale model and damage accumulation, *Int. J. Fract.*, 57:253–267.

Wu, S.X., Mai, Y.W., and Cotterell, B. (1995) Prediction of the initiation of ductile fracture, *J. Mech. Phys. Solids*, 1995:43:793–810.

Xia, L., and Shih, C.F. (1995a) Ductile crack growth — I. A numerical study using computational cells with microstructurally-based length scales, *J. Mech. Phys. Solids*, 43:233–259.

Xia, L., and Shih, C.F. (1995b) Ductile crack growth — II. Void nucleation and geometry effects on macroscopic fracture behaviour, *J. Mech. Phys. Solids*, 43:1953–1981.

Xia, L., and Shih, C.F. (1996) Ductile crack growth — III. Transition to cleavage fracture incorporating statistics, *J. Mech. Phys. Solids*, 44:603–639.

Xia, L., Shih, C.F., Hutchinson, J.W. (1995) A computational approach to ductile crack growth under large scale yielding conditions. *J. Mech. Phys. Solids*, 43:389–413.

Xie, X.-L., Mai, Y.-W., Zhou, X.-P. (2005) Dispersion and alignment of carbon nanotubes in polymer matrix: A review, *Mater. Sci. Engng. Rpts.*, 49:89–112.

Yang, K., Yang, Q., Li, G., Sun, y., and Feng, D. (2006) Mechanical properties and morphologies of polypropylene with different sizes of calcium carbonate particles, *Poly. Comp.*, 27:443–450.

Yuan, S.C., and Harrison, J.P. (2006) A review of the state of the art in modelling progressive mechanical breakdown and associated fluid flow in intact heterogeneous rocks, *Int. J. Rock Mech. Min. Sci.*, 43:1001–1022.

Ye, Y., Chen, H., Wu, J., and Ye, L. (2007) High impact strength epoxy nanocomposites with natural nanotubes, *Polymer*, 48:6426–6433.

Yee, A.F., and Pearson, R.A. (1986) Toughening mechanisms in elastomer-modified epoxies: Part 1 Mechanical studies, *J. Mater. Sci.*, 21:2462–2474.

Yoffe, E.H. (1951) The moving Griffith crack, *Phil. Mag.*, 42:739–750.

Yoshiki, M., and Kanazawa, T. (1958) A new testing method to obtain critical stress and limiting temperature for the propagation of brittle crack, 1st *Japan Congress on Testing and Materials – Metallic Materials*, pp. 66–68.

Yoshimasa, C. (1992) Recent archaeological excavations at the Tōdai-ji, *Japanese Journal of Religious Studies*, !9:245–254.

Young, T. (1845) *A Course of Lectures on Natural Philosophy and the Mechanical Arts*, 2 Vols. New Edition, Taylor and Walton, London.

Yu, H.-H., and Hutchinson, J.W. (2002) Influence of substrate compliance on buckling delamination of thin films, *Int. J. Fract.*, 113:39–55.

Zagorskii, F.N. (1982) *An Outline of the History of Metal Cutting Machines to the Middle of the 19th Century*, Amerind Publishing Co. Pvt. Ltd., New Delhi.

Zhang, H., and Zhang, Z. (2007) Impact behaviour of polypropylene filled with multi-walled carbon nanotubes, *Eur. Poly. J.*, 43:3197–3207.

Zhang, H., Zhang, Z., Friedrich, K., and Eger, C. (2006) Property improvements of in situ epoxy nanocomposites with reduced interparticle distance at high nanosilica content, *Acta Mater.*, 54:1833–1842.

Zhang, Q.-X., Yu, Z.-Z., Xie, X.-L., Mai, Y.-W. (2004) Crystallization and impact energy of polypropylene/CaCO$_3$ nanocomposites with non-ionic modifier. *Polymer*, 45:5985–94.

Zhang, T.-Y., and Cao, C.-F. (2004) Fracture behaviors of piezoelectric materials, *Theor. Appl. Fract. Mech.*, 41:339–379.

Zheng, Z.M., and Wierzbicki, T. (1996) A theoretical study of steady-state wedge cutting through metal plates, *Int. J. Fract.*, 78:45–66.

Zuiderduin, W.C.J., Westzaan, C., Huétink, J., and Gaymans, R.J. (2003) Toughening of polypropylene with calcium carbonate particles, *Polymer*, 44:261–275.

Zvorykin, K.A. (1893) *Rabota i Usilie Neobkhodimyya diya Orelenia Metallicheskikh Struzhek*, Moscow.

Name Index

(Only names mentioned in the text are listed)

Abraham, F.F., 380–382
Adams, F.D., 34
Adrian, L.A., 362
Agassiz, L., 46
Airbus A380, 290
Akhenaten, 121
Albert, W.A.J., 174, 205, 226
Aldrovani, U., 87
Aloha Airways Boeing, 737, 21, 282
Altshuller, G.S., 401
Amenhotep III, 121
American Advanced Combat Helmet, 363
American PASGT helmet, 364
Andersson, H., 264
Antikythera instrument, 348
Aquinas, St. T., 37
Archimedes, 160
Argon, A.S., 254
Aristotle, 36
Aristotle, School of, 141, 147
Armstrong, W.G., 364
Aspdin, J., 289
Asscher, J., 243
Astakov, V.P., 351
Aswan, unfinished Obelisk, 128, 129
Atkins, A.G., 16, 20, 24, 268, 269–271,
    315, 337, 338, 340, 341, 345–347, 351,
    353, 358, 362
Auerbach, F., 188
Avicenna, I.S., 36

B-15 iceberg, 49
Bache, A., 174
Backman, M.E., 359
Bacon, J., 354

Baekeland, L.H., 194, 320, 326
Baker, B., 185
Baker, J.F., 208
Barenblatt, G.I., 241, 242, 273
Barlow, P., 165, 166
Barnaby, N., 184
Barthlott, W., 402
Bascom, W.D., 322, 323
Basilica di Santa Maria del Fiore, Florence,
    136, 152
Basquin, O.H., 180, 278
Bastow, F.E., 245
Batson, R.G.C., 199, 200
Battle of Jutland, 365
Bauschinger, J., 189
Bažant, Z.P., 307, 309–311, 404
Benbow, J.J., 243, 244
Bentham, S., 354
Bessemer, H., 181, 183, 184
Bilby, B.A., 243
Binnig, G., 368
Birchall, J.D., 311
Blyth, P.H., 362
Boeing 787, Dreamliner, 290
Böker, R., 43
Boodberg, A., 209
Boscovich, R.G., 155
Boston Molasses Tank, 201
Bowie, O.L., 235
Boyle, R., 149, 151
Boyle, R.W., 241
Braconnott, H., 65
Braithwaite, F., 177
Bramante, D., 152
Brandon, Suffolk, 88

Subject Index

thin films and multilayers cont.,
    delamination and cracking under tension,
        371, 372
    delamination under compression, 372–374
        telephone cord buckles, 372, 373
    interfacial toughness, 369, 370
    mixed mode fracture, 370, 374
    mode-mixity angle, 370, 374

vaults, 117, 120, 132
    stability, 120

walls, 128, 130
welds and welding,
    crack-arresters, 205
    electric, 203
    flaws, 203
    heat affected zones, 203
    residual stresses, 202, 203, 255, 257, 260
    post weld heat treatment, 260

wiggly cuts, 24–25
wire,
    copper, 145
    drawing, 142
    iron, 142, 159
wrought iron,
    armour, 365
    best, 168
    cinder iron, 171
    cold short, 171
    faggoted, 168
    fracture behaviour, 171, 184
    phosphorus, 170, 171, 183, 187
    production, 163
    transition temperature, 171
    riveted, 169, 170, 178
    strength, 169, 170, 196